Elements of Homology Theory

Elements of Homology Theory

V. V. Prasolov

Graduate Studies in Mathematics

Volume 81

American Mathematical Society
Providence, Rhode Island

В. В. Прасолов

ЭЛЕМЕНТЫ ТЕОРИИ ГОМОЛОГИЙ

МЦНМО, Москва, 2005

This work was originally published in Russian by МЦНМО under the title "Элементы теории гомологий" © 2005. The present translation was created under license for the American Mathematical Society and is published by permission.

Translated from the Russian by Olga Sipacheva

2000 *Mathematics Subject Classification.* Primary 55–01.

For additional information and updates on this book, visit
www.ams.org/bookpages/gsm-81

Library of Congress Cataloging-in-Publication Data

Prasolov, V. V. (Viktor Vasil'evich)
[Elementy teorii gomologii. English]
Elements of homology theory / V. V. Prasolov.
p. cm. — (Graduate studies in mathematics ; v. 81)
Includes bibliographical references and index.
ISBN-13: 978-0-8218-3812-9 (alk. paper)
ISBN-10: 0-8218-3812-1 (alk. paper)
1. Homology theory. I. Title.

QA612.3.P73 2007
514′.23—dc22 2006047074

Printed in the United States of America.

∞ The paper used in this book is acid-free and falls within the guidelines
established to ensure permanence and durability.
Visit the AMS home page at `http://www.ams.org/`

10 9 8 7 6 5 4 3 2 1 12 11 10 09 08 07

Contents

Preface vii

Notation ix

Chapter 1. Simplicial Homology 1

§1. Definition and Some Properties 1

§2. Invariance of Homology 6

§3. Relative Homology 12

§4. Cohomology and Universal Coefficient Theorem 21

§5. Calculations 35

§6. The Euler Characteristic and the Lefschetz Theorem 51

Chapter 2. Cohomology Rings 59

§1. Multiplication in Cohomology 59

§2. Homology and Cohomology of Manifolds 69

§3. The Künneth Theorem 95

Chapter 3. Applications of Simplicial Homology 111

§1. Homology and Homotopy 111

§2. Characteristic Classes 131

§3. Group Actions 173

§4. Steenrod Squares 184

Chapter 4. Singular Homology 195

§1. Basic Definitions and Properties 195

§2. The Poincaré and Lefschetz Isomorphisms for Topological Manifolds 227
§3. Characteristic Classes: Continuation 252

Chapter 5. Čech Cohomology and de Rham Cohomology 263
§1. Sheaf Cohomology 263
§2. De Rham Cohomology 275
§3. The de Rham Theorem 289

Chapter 6. Miscellany 301
§1. The Alexander Polynomial 301
§2. The Arf Invariant 317
§3. Embeddings and Immersions 325
§4. Complex Manifolds 339
§5. Lie Groups and $\boldsymbol{H}$-Spaces 344

Hints and Solutions 365

Bibliography 403

Index 411

Preface

This book is a natural continuation of the author's earlier book *Elements of Combinatorial and Differential Topology* (American Mathematical Society, Providence, RI, 2006), which we refer to as Part I here. (A corrected Russian version of Part I is available at `http://www.mccme.ru/prasolov`.)

In Chapter 1, we define simplicial homology and cohomology and give many examples of their calculations and applications. At this point, the book diverges from standard modern courses in algebraic topology, which usually begin with introducing singular homology. Simplicial homology has a simpler and more natural definition. Moreover, it is simplicial homology that is usually involved in calculations. For this reason, we introduce singular homology near the end of the book and use it only when it is indeed necessary, mainly in studying topological manifolds.

Homology and cohomology groups with arbitrary coefficients are expressed in terms of integral homology by means of the functors Tor and Ext. The properties of these functors are very important for homology theory, so we discuss them in detail.

We first prove the Poincaré duality theorem for simplicial (co)homology. This proof applies only to smooth (to be more precise, triangulable) manifolds. There is no triangularization theorem for topological manifolds, and the proof of the Poincaré duality theorem for them uses, of necessity, singular (co)homology. This proof is given in Chapter 4; it is very cumbersome.

Chapter 2 considers an important algebraic structure on cohomology, the cup product of Kolmogorov and Alexander. It is particularly useful in the case of manifolds. Multiplication in cohomology is related to many topological invariants of manifolds, such as the intersection form and signature.

One possible approach to constructing multiplication in cohomology is based on a theorem of Künneth, which expresses the (co)homology of $X \times Y$ in terms of those of X and Y and is of independent interest.

Chapter 3 is devoted to various applications of (simplicial) homology and cohomology. Many of them are related to obstruction theory. One of such applications is the construction of the characteristic classes of vector bundles. Other approaches to constructing characteristic classes (namely, the universal bundle and axiomatic approaches) are also discussed. Then, we consider the (co)homological properties of spaces with actions of groups; we construct transfers and Smith's exact sequences. We conclude the chapter with constructing Steenrod squares, which generalize multiplication in cohomology.

In Chapter 4, we define singular (co)homology and describe some of its applications; in particular, we obtain certain properties of characteristic classes. (Technically, it is more convenient to prove them by using singular cohomology, although the assertions themselves can be stated for simplicial cohomology.)

Chapter 5 considers yet another approach to constructing cohomology theory, namely, Čech cohomology and de Rham cohomology, which are closely related to each other. For the de Rham cohomology, we prove the Poincaré duality theorem. Then, we carry over the construction of de Rham, which was originally introduced for smooth manifolds, to arbitrary simplicial complexes.

The final Chapter 6 is devoted to various applications of homology theory, largely to the topology of manifolds. We begin with a detailed account of the Alexander polynomials, which we construct by using the homology of cyclic coverings; the Arf invariant is also considered. Then, we prove the strong Whitney embedding theorem. We also give a formula for calculating the Chern classes of complete intersections and discuss some homological properties of Lie groups and H-spaces.

The book contains many problems (with solutions) and exercises. The problems are based on the materials of topology seminars for second-year students held by the author at the Independent University of Moscow in 2003.

The basic notation, as well as theorems and other assertions, of Part I are mostly used without explanations; in some cases, we give references to the corresponding places in Part I.

This work was financially supported by the Russian Foundation for Basic Research (project nos. 05-01-01012a, 05-01-02805-NTsNIL_a, and 05-01-02806-NTsNIL_a).

Notation

$H_k(X; G)$, the k-dimensional homology group of X with coefficients in G;
$H^k(X; G)$, the k-dimensional cohomology group of X with coefficients in G;
$\mathrm{Hom}(A, B)$, the group of homomorphisms $A \to B$;
$A \otimes B$, the tensor product of the Abelian groups A and B;
$\mathrm{Tor}(A, B)$, see p. 29;
$\mathrm{Ext}(A, B)$, see p. 29;
$\mathrm{Coker}\,\alpha$, the cokernel of the homomorphism α (see p. 15);
$[M^n]$, the fundamental class of the manifold M^n;
$\chi(X)$, the Euler characteristic of X;
$\Lambda(f)$, the Lefschetz number of the map f;
$\sigma(M^{4n})$, the signature of the manifold M^{4n};
ε^k, the k-dimensional trivial vector bundle;
$w_k(\xi)$, the kth Stiefel–Whitney class of the bundle ξ;
$c_k(\xi)$, the kth Chern class of the bundle ξ;
$p_k(\xi)$, the kth Pontryagin class of the bundle ξ;
Sq^i, the Steenrod square.

Chapter 1

Simplicial Homology

1. Definition and Some Properties

The homology groups of a topological space X can be defined in several ways; different definitions are equivalent only for sufficiently good[1] spaces. The simplest definition is that of simplicial homology. Unfortunately, it has the essential drawback of not being invariant; to be more precise, the proof of its invariance requires some effort. (By invariance here we mean that the homology groups of homeomorphic spaces are isomorphic.) However, the main ideas of homology theory are most transparent at the level of simplicial homology; for this reason, we begin with a detailed discussion of simplicial homology.

1.1. Definition of Homology Groups. Let K be a *simplicial complex* (see Part I, p. 99). We assume that all of its simplices are oriented, i.e., the vertices of each simplex are ordered (the orientations of different simplices are not assumed to be compatible). We denote a simplex with vertices $a_0, a_1, \ldots, a_n$ (in this order) by $[a_0, a_1, \ldots, a_n]$. Two simplices $[a_0, a_1, \ldots, a_n]$ and $[a_{\sigma(0)}, a_{\sigma(1)}, \ldots, a_{\sigma(n)}]$ are said to have the same orientation if $\operatorname{sign}\sigma = 1$; if $\operatorname{sign}\sigma = -1$, then these simplices have opposite orientations.

We define the *boundary* of a simplex as

$$\partial[0, 1, \ldots, n] = \sum (-1)^i [0, \ldots, \hat{i}, \ldots, n],$$

where $[0, \ldots, \hat{i}, \ldots, n] = [0, \ldots, i-1, i+1, \ldots, n]$. At this point, we begin to consider formal sums of simplices. To be more precise, we declare simplices with the same vertices and orientations to be equivalent and consider the

[1]Most of the spaces that arise in topology from geometry are sufficiently good, as opposed to those arising from functional analysis.

set of equivalence classes of simplices. Moreover, we assume that the sum of two simplices with the same vertices and opposite orientations is 0. Thus, we can denote simplices with opposite orientations by Δ and $-\Delta$.

It is convenient to denote the simplex $[a, b]$ by an arrow from its starting vertex a to the end vertex b. In this notation, the boundary of the simplex $[0, 1, 2]$ looks as shown in Figure 1; this fully agrees with the intuitive idea of a boundary.

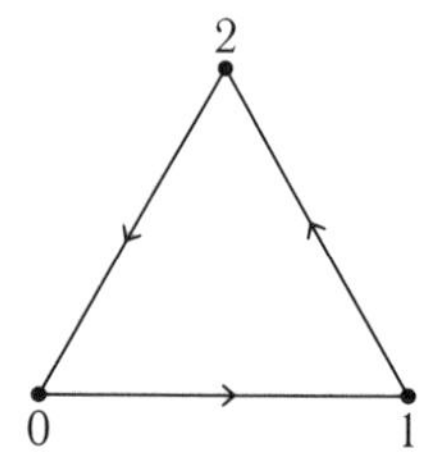

Figure 1. The boundary of the simplex $[0, 1, 2]$

The following assertion is very important for homology theory.

Theorem 1.1. *For any simplex* Δ, $\partial\partial\Delta = 0$.

Proof. Let $i > j$. The simplex $[\dots, \hat{j}, \dots, \hat{i}, \dots]$ occurs in the expression for $\partial\partial[0, \dots, n]$ two times: in $(-1)^i\partial[\dots, \hat{i}, \dots]$ with the sign $(-1)^{i+j}$ and in $(-1)^j\partial[\dots, \hat{j}, \dots]$ with the opposite sign $(-1)^{i+j-1}$. □

We could consider not only linear combinations of simplices with integer coefficients but also finite[2] sums of the form $\sum a_i\Delta_i^k$, where each a_i is an element of some Abelian group G and each Δ_i^k is a simplex of dimension k. The expression $\sum a_i\Delta_i^k$ is called an *n-chain* (with coefficients in the group G). Chains can be added to each other; therefore, they form an Abelian group. The group of k-chains is denoted by $C_k(K; G)$. For brevity, we often denote this group by $C_k(K)$ or simply C_k.

We have defined the map ∂ for simplices. Extending it by linearity, we obtain a map $\partial_k \colon C_k \to C_{k-1}$, which we call the *boundary homomorphism.* For the 0-simplex Δ^0, we set $\partial_0\Delta^0 = 0$.

A chain $c \in C_k$ is called a *cycle* if $\partial_k c = 0$, i.e., $c \in \operatorname{Ker}\partial_k$. The group of k-dimensional cycles is denoted by Z_k. A chain $c \in C_k$ is said to be a *boundary* if $c = \partial_{k+1}c'$ for some chain $c' \in C_{k+1}$, i.e., $c \in \operatorname{Im}\partial_{k+1}$. The group of k-dimensional boundaries is denoted by B_k. It follows from $\partial\partial = 0$ that $B_k \subset Z_k$; therefore, we can consider the quotient group $H_k(K) = Z_k/B_k$. Its elements are equivalence classes of cycles: cycles are equivalent if their

[2]For infinite sums, we could not define the homomorphism ∂ in the case where one simplex is a face for infinitely many simplices.

difference is a boundary; such cycles are said to be *homologous*. The group $H_k(K)$ is called the k-dimensional *simplicial homology group* of the complex K. To indicate the group of coefficients, we use the notation $H_k(K; G)$.

Remark. For the first time, "homological numbers" were mentioned in the works of Riemann (1857) and Betti (1871). The correct definition was given by Poincaré. He realized that, in defining chains, not only the geometric shapes of simplices but also their multiplicities should be taken into account. Initially, homology groups (with coefficients in $\mathbb{Z}$) were described in terms of their ranks (*Betti numbers*) and torsion numbers. In the 1926 short paper [**97**], Emmy Noether made the important observation that the equivalence classes of cycles can be regarded as groups.

It can be seen from the definition that if a simplicial complex K contains no k-simplices, then $H_k(K) = 0$.

Let us calculate homology groups for some simplicial complexes.

Example 1. If K consists of n isolated points, then

$$H_0(K; G) = \underbrace{G \oplus \cdots \oplus G}_{n} \quad \text{and} \quad H_k(K; G) = 0 \quad \text{for } k \geq 1.$$

Proof. Clearly, in the situation under consideration, we have $C_0 = G \oplus \cdots \oplus G$, $C_k = 0$ for $k \geq 1$, $Z_0 = C_0$, and $B_0 = 0$. □

Example 2. Let S^1 be the simplicial complex that is the boundary of the 2-simplex $[0, 1, 2]$. Then $H_0(S^1; G) = H_1(S^1; G) = G$.

Proof. Clearly, $B_1 = 0$, whence $H_1 = Z_1$. Consider an arbitrary 1-chain

$$c = a_0[1, 2] + a_1[2, 0] + a_2[0, 1] \in C_1,$$

where $a_0, a_1, a_2 \in G$. We have

$$\partial c = (a_1 - a_2)[0] + (a_2 - a_0)[1] + (a_0 - a_1)[2].$$

Therefore, Z_1 consists of chains of the form $a[1, 2] + a[2, 0] + a[0, 1]$, where $a \in G$.

The equalities $\partial(a[1, 2] + a[2, 0]) = a[0] - a[1]$ and $\partial(a[2, 1] + a[1, 0]) = a[0] - a[2]$ show that any 0-chain has the form $a[0]$ (up to a boundary). On the other hand, if

$$a[0] = \partial c = (a_1 - a_2)[0] + (a_2 - a_0)[1] + (a_0 - a_1)[2],$$

then $a_2 = a_0$ and $a_0 = a_1$; hence $a = a_1 - a_2 = 0$. Thus, the group $H_0(S^1; G) = C_0/B_0$ is isomorphic to G. □

The argument which we used to calculate $H_0(S^1)$ applies to any connected simplicial complex.

Theorem 1.2. *If K is a connected simplicial complex, then $H_0(K;G) = G$.*

Proof. Arbitrary vertices $[m]$ and $[n]$ can be joined by 1-simplices $[m, i_1]$, $[i_1, i_2]$, ..., $[i_k, n]$; therefore,

$$a[n] - a[m] = \partial(a[m, i_1] + a[i_1, i_2] + \cdots + a[i_k, n]).$$

This means that, up to a boundary, any 0-chain has the form $a[m]$, where $[m]$ is a fixed vertex. It remains to verify that if the chain $a[m]$ is a boundary, then $a = 0$.

Suppose that

$$a[m] = \partial\Big(\sum_\alpha a_\alpha[i_\alpha, j_\alpha]\Big) = \sum_\alpha a_\alpha[j_\alpha] - \sum_\alpha a_\alpha[i_\alpha].$$

The sum of coefficients on the right-hand side vanishes. Therefore, $a = 0$. □

Exercise. Suppose that S^2 is represented as the boundary of the simplex $[0, 1, 2, 3]$. Prove that $\operatorname{Ker}\partial_2$ consists of chains of the form $a\partial[0, 1, 2, 3]$ and $\operatorname{Ker}\partial_1 = \operatorname{Im}\partial_2$.

1.2. Chain Complexes. A *chain complex* is defined as a family of Abelian groups C_k and homomorphisms $\partial_k\colon C_k \to C_{k-1}$ satisfying the relations $\partial_k\partial_{k+1} = 0$. If $C_k = 0$ for $k < 0$, then the chain complex is said to be *nonnegative*.

An Abelian group F is *free* if it contains a set of elements $\{f_\alpha\}$ such that any $f \in F$ has a unique representation in the form $f = n_{\alpha_1} f_{\alpha_1} + \cdots + n_{\alpha_k} f_{\alpha_k}$, where $n_{\alpha_i} \in \mathbb{Z}$ and all $f_{\alpha_1}, \ldots, f_{\alpha_k}$ are distinct. The set $\{f_\alpha\}$ is called a *basis* of the group F. A free Abelian group can be equivalently defined as a (finite or infinite) direct sum of copies of $\mathbb{Z}$. If all of the groups C_k are free, then the chain complex is said to be *free*.

For any chain complex C_*, we can consider the *homology groups* $H_k(C_*) = \operatorname{Ker}\partial_k / \operatorname{Im}\partial_{k+1}$.

A *chain map* of chain complexes C_* and C'_* is a family of homomorphisms $\varphi_k\colon C_k \to C'_k$ satisfying the condition $\partial'_k\varphi_k = \varphi_{k-1}\partial_k$ for each k. Each chain map $\varphi_k\colon C_k \to C'_k$ induces a family of homomorphisms $\varphi_*\colon H_k(C_*) \to H_k(C'_*)$ for which $(\varphi\psi)_* = \varphi_*\psi_*$.

Every simplicial map $f\colon K \to L$ induces the map $f_k\colon C_k(K) \to C_k(L)$ defined by

$$f_k([a_0, \ldots, a_k]) = \begin{cases} [f(a_0), \ldots, f(a_k)] & \text{if } f(a_i) \neq f(a_j) \text{ for } i \neq j, \\ 0 & \text{if } f(a_i) = f(a_j) \text{ for some } i \neq j. \end{cases}$$

If the dimension of the simplex with vertices $f(a_0), \ldots, f(a_k)$ is equal to k or less than $k - 1$, then, obviously, $\partial'_k f_k = f_{k-1}\partial_k$, i.e., the map is chain (in the latter case, both sides of this equality vanish). Suppose that $f(a_0) = f(a_1)$ and the points $f(a_1), \dots, f(a_k)$ are different. In this case, we have $f_k([a_0, \dots, a_k]) = 0$, and

$$f_{k-1}(\partial[a_0, \dots, a_k]) = [f(a_1), f(a_2), \dots, f(a_k)] - [f(a_0), f(a_2), \dots, f(a_k)] = 0,$$

because $f(a_0) = f(a_1)$.

Thus, any simplicial map $f\colon K \to L$ induces a homomorphism $f_*\colon H_k(K)[0] \to H_k(L)$; the identity map induces the identity homomorphism, and $(fg)_* = f_*g_*$.

If $f, g\colon K \to L$ are simplicial maps and there exists a family of homomorphisms $D_k\colon C_k(K) \to C_{k+1}(L)$ satisfying the conditions

$$\partial_{k+1}D_k + D_{k-1}\partial_k = g_k - f_k,$$

then such a family D is called a *chain homotopy* between f and g.

The notion of chain homotopy has a geometric origin. Namely, suppose that f and g are maps joined by a homotopy H, and these maps (including the homotopy) are simplicial. Then the homotopy H determines the chain homotopy that assigns the $(k+1)$-chain $H(\Delta^k)$ to each k-simplex Δ^k. This chain is the curvilinear prism swept out by the image of the simplex during the homotopy. The boundary of the prism consists of the bases $f(\Delta^k)$ and $g(\Delta^k)$ and the lateral surface $H(\partial\Delta^k)$. Taking orientations into account, we obtain precisely the expression for the chain homotopy:

$$\partial H(\Delta^k) - H(\partial\Delta^k) = g(\Delta^k) - f(\Delta^k).$$

Theorem 1.3 (on chain homotopy). *If simplicial maps $f, g\colon K \to L$ are chain homotopic, then $f_* = g_*$.*

Proof. Suppose that $z_k \in C_k(K)$ and $\partial_k z_k = 0$. Then

$$g_k(z_k) - f_k(z_k) = \partial_{k+1}D_k z_k + D_{k-1}\partial_k z_k = \partial_{k+1}(D_k z_k),$$

i.e., the cycles $g_k(z_k)$ and $f_k(z_k)$ are homologous. □

1.3. Homology of Simplices and of Their Boundaries.

Theorem 1.4. *If $k \geq 1$, then $H_k(\Delta^n) = 0$.*

Proof. Note that $\Delta^n = [b, a_1, \dots, a_n]$ is a cone (over $\Delta^{n-1} = [a_1, \dots, a_n]$). To each simplex $[a_{i_1}, \dots, a_{i_k}]$ we assign the simplex $[b, a_{i_1}, \dots, a_{i_k}]$. Extending this map by linearity, we get a homomorphism $C_{k-1}(\Delta^{n-1}) \to C_k(\Delta^n)$; we denote the image of c_{k-1} by $[b, c_{k-1}]$. It is easy to verify that $\partial[b, c_k] = c_k - [b, \partial c_k]$ for $k \geq 1$ and $\partial[b, c_0] = c_0 - \varepsilon(c_0)b$, where $\varepsilon\left(\sum n_p a_{i_p}\right) = \sum n_p$.

Any chain $z_k \in C_k(\Delta^n)$ has a unique representation in the form $z_k = c_k + [b, d_{k-1}]$, where $c_k \in C_k(\Delta^{n-1})$ and $d_{k-1} \in C_{k-1}(\Delta^{n-1})$. Suppose that

$\partial z_k = 0$. Then $\partial c_k + d_{k-1} - [b, \partial d_{k-1}] = 0$ for $k > 1$ and $\partial c_1 + d_0 - \varepsilon(d_0)b = 0$ for $k = 1$. In both cases, $\partial c_k + d_{k-1} = 0$; therefore,

$$\partial[b, c_k] = c_k - [b, \partial c_k] = c_k + [b, d_{k-1}] = z_k,$$

i.e., any cycle $z_k \in C_k(\Delta^n)$ with $k \geq 1$ is a boundary. □

Corollary. *Let $\partial\Delta^n$ be the simplicial complex consisting of all simplices in Δ^n except Δ^n itself. Then $H_{n-1}(\partial\Delta^n) = G$ (if $n \geq 2$) and $H_k(\partial\Delta^n) = 0$ for $0 < k < n - 1$.*

Proof. Up to the dimension $n - 1$, the homology groups of $\partial\Delta^n$ coincide with those of Δ^n. The complex $\partial\Delta^n$ contains no n-simplices; therefore, $\operatorname{Im} \partial_n = 0$, which means that $H_{n-1}(\partial\Delta^n) = \operatorname{Ker} \partial_{n-1}$. But $H_n(\Delta^n) = 0$; hence $\operatorname{Ker} \partial_{n-1} = \operatorname{Im} \partial'_n$, where ∂'_n is the differential in the chain complex for Δ^n. If $\Delta^n = [0, 1, \ldots, n]$, then $\operatorname{Im} \partial'_n$ consists of elements of the form $a\Big(\sum_{i=0}^{n}(-1)^i[0, \ldots, \hat{i}, \ldots, n]\Big)$, where $a \in G$. □

2. Invariance of Homology

First, we prove a theorem about acyclic supports, which allows us to prove the chain homotopy of two chain maps in many cases. Then, we twice apply this theorem in different situations to prove the topological invariance of homology. Finally, using the same theorem, we prove the homotopy invariance of homology.

We can define homology with coefficients in any Abelian group G. But the case most important for applications is the one in which G is the additive group of some commutative ring with identity (e.g., $G = \mathbb{Z}$, $\mathbb{Q}$, or $\mathbb{Z}_p$); on the other hand, some important facts about homology (and especially cohomology) groups can be proved only for such groups of coefficients. For this reason, in what follows, we usually assume the group of coefficients to be the additive group of some commutative ring with identity.

2.1. Acyclic Supports. A simplicial complex is said to be *acyclic* if its homology coincides with that of the singleton. For example, the cone over any simplicial complex is acyclic; this is proved in precisely the same way as the acyclicity of simplices (see Theorem 1.4 on p. 5; the proof of this theorem uses only the representation of a simplex as a cone).

Suppose we have a chain $c_k = \sum a_k\Delta_i^k \in C_k(L)$. We refer to any subcomplex $L' \subset L$ containing all the simplices Δ_i^k as a *support* of c_k.

We say that a chain map φ is *augmentation-preserving*[3] if $\varphi_0\big(\sum a_i\Delta_i^0\big) = \sum b_j\Delta_j^0$, where $\sum a_i = \sum b_j$.

[3]The origin of this term is explained on p. 17.

Theorem 1.5 (on acyclic supports). *Let $\varphi_k, \psi_k \colon C_k(K) \to C_k(L)$ be augmentation-preserving chain maps. Suppose that to any simplex $\Delta \subset K$ there corresponds a complex $L(\Delta) \subset L$ so that the following conditions are satisfied:*

(1) *if $\Delta' \subset \Delta$, then $L(\Delta') \subset L(\Delta)$;*

(2) *the complex $L(\Delta)$ is acyclic;*

(3) *the complex $L(\Delta^k)$ is a support for both chains $\varphi_k(\Delta^k)$ and $\psi_k(\Delta^k)$.*

Suppose also that the coefficient group G is the additive group of a ring with identity. Then the maps φ_k and ψ_k are chain homotopic; in particular, $\varphi_* = \psi_*$.

Proof. We construct a chain homotopy D_k by induction on k. First, suppose that $k = 0$. Let Δ^0 be a vertex of K. The complex $L(\Delta^0)$ is a support for the chains $\varphi_0(\Delta^0)$ and $\psi_0(\Delta^0)$. Since the maps φ_0 and ψ_0 are augmentation-preserving, it follows that $(\varphi_0 - \psi_0)(a\Delta^0) = \sum b_i \Delta_i^0$, where $\sum b_i = 0$. In the acyclic complex $L(\Delta^0)$, the chain $\sum b_i \Delta_i^0$, where $\sum b_i = 0$, is the boundary of some chain; therefore, $(\varphi_0 - \psi_0)(a\Delta^0) = \partial_1 D_0(a\Delta^0)$, where $D_0(a\Delta^0)$ is a 1-chain for which $L(\Delta^0)$ is a support. The equality $D_0(a\Delta^0) + D_0(b\Delta^0) = D_0((a+b)\Delta^0)$ may be false. To make it true, we choose an identity element 1 in the ring G, fix a 1-chain $D_0(1 \cdot \Delta^0)$, and set $D_0(a\Delta^0) = aD_0(1 \cdot \Delta^0)$. In what follows, we denote chains of the form $1 \cdot \Delta$ simply by Δ (note that this notation makes no sense if G is an arbitrary group of coefficients).

Suppose that the required homomorphisms $D_0, \dots, D_{k-1}$ are already constructed and each $L(\Delta^i)$ is a support for the chain $D_i(\Delta^i)$. The only requirement to the homomorphism D_k is that $\partial_{k+1} D_k(\Delta) = c_k$ for any k-simplex Δ in K, where $c_k = \varphi_k(\Delta) - \psi_k(\Delta) - D_{k-1}\partial_k(\Delta)$. All simplices $\partial_k(\Delta)$ are contained in Δ; hence $L(\Delta)$ is a support for the chain $\partial_k(\Delta)$, and therefore $L(\Delta)$ is a support for the chain $D_{k-1}\partial_k(\Delta)$. Thus, $L(\Delta)$ is a support for c_k, and

$$\begin{aligned}\partial_k c_k &= (\partial_k \psi_k - \partial_k \varphi_k - \partial_k D_{k-1} \partial_k)(\Delta) \\ &= (\partial_k \psi_k - \partial_k \varphi_k - (\psi_{k-1}\partial_k - \varphi_{k-1}\partial_k - D_{k-2}\partial_{k-1}\partial_k))(\Delta) = 0.\end{aligned}$$

We have assumed that $k \geq 1$. The acyclicity of the complex $L(\Delta)$ implies $H_k(L(\Delta)) = 0$. Hence there exists a chain $D_k(\Delta)$ for which $L(\Delta)$ is a support and $\partial_{k+1} D_k(\Delta) = c_k$. □

2.2. Topological Invariance of Homology. In this section, we prove that any continuous map $f \colon |K| \to |L|$ of simplicial complexes induces a homomorphism $f_* \colon H_k(K) \to H_k(L)$. Moreover, $(fg)_* = f_* g_*$, which implies the isomorphism of the homology groups of homeomorphic simplicial complexes. To define f_*, we use the simplicial approximation theorem (see Part I, p. 105).

Theorem 1.6. *Suppose that the coefficient group G is the additive group of a ring with identity. Let K' be the barycentric subdivision of a simplicial complex K. Then the homology groups $H_k(K')$ and $H_k(K)$ are isomorphic.*

Proof. First, consider the simplicial map $j\colon K' \to K$ defined as follows. If $K = \Delta^k = [0, 1, \ldots, k]$, then any simplicial map $K' \to K$ can be defined by assigning the vertices (labels) $0, 1, \ldots, k$ to the vertices of the complex K'. To each vertex of K' which is the barycenter of some simplex $[i_0, \ldots, i_p]$ we assign one of the vertices $i_0, \ldots, i_p$. An example of such an assignment is given in Figure 2. It is easy to show that precisely one k-simplex of K' has complete set of labels, and the orientation of this simplex coincides with that of K. Indeed, take the barycenter of K and consider the face whose vertices have labels different from those of the barycenter; take the barycenter of this face, and so on.

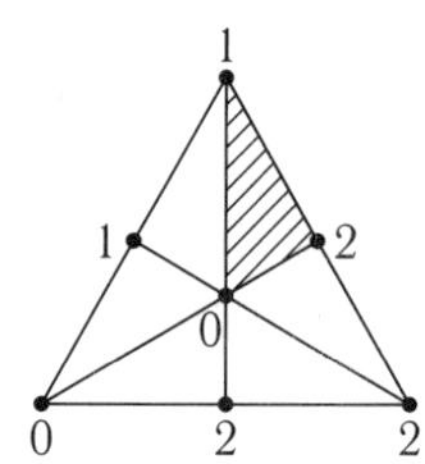

Figure 2. A set of labels

For general simplicial complexes, the map $j\colon K' \to K$ is defined similarly. Note that it is a simplicial approximation of the identity map. The map j induces a chain map $j_k\colon C_k(K') \to C_k(K)$.

Now, let us define a chain map $i_k\colon C_k(K) \to C_k(K')$; it is not induced by a simplicial map. We set $i_0(v) = v$ and $i_1[v_0, v_1] = [b, v_1] - [b, v_0]$, where b is the barycenter of the simplex $[v_0, v_1]$. Formally, the definition of i_1 can be written as $i_1(\Delta^1) = [b, i_0\partial\Delta^1]$. For $k > 1$, we set $i_k(\Delta^k) = [b, i_{k-1}\partial\Delta^k]$, where b is the barycenter of the complex Δ^k. It remains to verify that i_k is a chain map, i.e., $\partial_k i_k = i_{k-1}\partial_k$. Clearly,

$$\partial_k i_k(\Delta^k) = \partial_k[b, i_{k-1}\partial_k\Delta^k] = i_{k-1}\partial_k(\Delta^k) - [b, \partial_{k-1}i_{k-1}\partial_k\Delta^k].$$

Therefore, if $\partial_{k-1}i_{k-1} = i_{k-2}\partial_{k-1}$, then $\partial_{k-1}i_{k-1}\partial_k = i_{k-2}\partial_{k-1}\partial_k = 0$ and $\partial_k i_k = i_{k-1}\partial_k$.

The map j takes exactly one k-simplex from the barycentric subdivision of Δ^k to Δ^k and preserves its orientation; the remaining k-simplices are mapped to simplices of dimension less than k. This implies that $j_k i_k$ is the identity map, and for a support of the chain $i_k j_k(\Delta')$, where Δ' is any simplex from K', we can take the barycentric subdivision of the simplex Δ from K that contains Δ'. The barycentric subdivision of Δ is the cone over $\partial\Delta$; therefore, the barycentric subdivision of a simplex is an acyclic simplicial complex. Thus, the maps $i_k j_k$ and $\mathrm{id}_{C_k(K')}$ have a common acyclic support. Clearly, the map $i_0 j_0$ takes vertices to vertices; hence, it is augmentation-preserving. Therefore, $i_k j_k$ is chain homotopic to the identity.

Thus, the induced homomorphisms $j_* i_*$ and $i_* j_*$ are the identity maps, and hence the groups $H_k(K')$ and $H_k(K)$ are isomorphic. □

Remark 1. The map $i_*\colon H_k(K) \to H_k(K')$ has a simple geometric meaning at the level of chains; namely, to every simplex it assigns the sum of simplices into which it decomposes under the barycentric subdivision. The inverse map $j_*\colon H_k(K') \to H_k(K)$ is canonical[4] only at the level of homology.

Remark 2. It is easy to prove Theorem 1.6 without appealing to acyclic supports. Indeed, it suffices to verify that $i_* j_*$ is the identity map. But all simplices of the barycentric subdivision of Δ must have the same coefficients in any cycle; therefore, the restriction of $i_k j_k$ to Z_k is the identity.

Let $f\colon |K| \to |L|$ be a continuous map of simplicial complexes. We define a homomorphism $f_*\colon H_k(K) \to H_k(L)$ as follows. Take the nth barycentric subdivision $K^{(n)}$ and consider a simplicial approximation $\varphi\colon K^{(n)} \to L$ of f. We set $f_* = \varphi_* i_*^{(n)}$, where $i_*^{(n)}\colon H_k(K) \to H_k(K^{(n)})$ is the canonical isomorphism. We must verify that this map is well defined, i.e., if $\psi\colon K^{(m)} \to L$ is another simplicial approximation of f, then $\varphi_* i_*^{(n)} = \psi_* i_*^{(m)}$. First, suppose that $m = n$.

Theorem 1.7. *If $\varphi, \psi\colon K \to L$ are simplicial approximations of a map $f\colon |K| \to |L|$, then $\varphi_* = \psi_*$.*

Proof. We use only the following property of the simplicial maps φ and ψ. *Let $a_0, \dots, a_k$ be the vertices of some simplex in K. Then the points*

---

[4] We say that a map is canonical if it is uniquely determined by construction or for some other reasons. For example, any group isomorphism $\mathbb{Z}_2 \to \mathbb{Z}_2$ is canonical. Any ring isomorphism $\mathbb{Z} \to \mathbb{Z}$ is canonical also, but a group isomorphism $\mathbb{Z} \to \mathbb{Z}$ is canonical only if it is assumed to take 1 to 1 rather than to -1.

In linear algebra, the most important example of a noncanonical isomorphism is the isomorphism between a linear space V and its dual space V^*. This isomorphism depends on the choice of a basis in V.

In topology, noncanonical isomorphisms often arise between bundles because the fibers of bundles are homeomorphic but the homeomorphisms are not canonical; the canonicity of the homeomorphisms between fibers means the triviality of the bundle. Therefore, the homotopy and homology groups of different fibers are isomorphic, but the isomorphisms are not always canonical because the products of the base and the fibers may be twisted.

$\varphi(a_0), \ldots, \varphi(a_k)$, $\psi(a_0), \ldots, \psi(a_k)$ are the vertices of a simplex[5] *in L.* To prove this assertion, take a point $x \in \operatorname{int}[a_0, \ldots, a_k]$. The point $f(x)$ is interior for a unique simplex $[b_0, \ldots, b_l]$ in L. According to the definition of a simplicial approximation, we have $\varphi(x), \psi(x) \in [b_0, \ldots, b_l]$. Therefore, since the maps φ and ψ are simplicial, it follows that the simplices $\varphi([a_0, \ldots, a_k])$ and $\psi([a_0, \ldots, a_k])$ belong to $[b_0, \ldots, b_l]$. The maps $\varphi_0, \psi_0 \colon C_0(K) \to C_0(L)$ are augmentation-preserving. According to the acyclic support theorem, the maps φ_k and ψ_k are chain homotopic. □

Now, let us prove the equality $\varphi_* i_*^{(n)} = \psi_* i_*^{(m)}$ for $n \neq m$. For definiteness, suppose that $n > m$. We can take $K^{(m)}$ (instead of K) for the initial complex and consider its pth barycentric subdivision for $p = n - m$. Then the required equality takes the form $\varphi_* i_*^{(p)} = \psi_*$.

In the proof of Theorem 1.6, we constructed a map $j \colon K^{(1)} \to K$, which is a simplicial approximation of the identity map. Recall that a composition of simplicial approximations is a simplicial approximation of a composition (see Part I, the corollary of Theorem 3.17 on p. 104). Therefore, the identity map has a simplicial approximation $j^{(p)} \colon K^{(p)} \to K$; the maps $j_*^{(p)}$ and $i_*^{(p)}$ are mutually inverse. Since $\psi j^{(p)}$ is a simplicial approximation of f, it follows that $\varphi_* = \psi_* j^{(p)}$, i.e., $\varphi_* i^{(p)} = \psi_*$.

Theorem 1.8. *If $f \colon |K| \to |L|$ and $g \colon |L| \to |M|$ are continuous maps, then $(gf)_* = g_* f_*$.*

Proof. First, we construct a simplicial approximation $\psi \colon L' \to M$ of g, and then, a simplicial approximation $\varphi \colon K' \to L'$ of f. Let $j^K \colon K' \to K$ and $j^L \colon L' \to L$ be simplicial approximations of the identity maps. These maps form the commutative diagram

$$\begin{array}{ccccc} |K| & \xrightarrow{f} & |L| & \xrightarrow{g} & |M| \\ \uparrow{\scriptstyle j^K} & & \uparrow{\scriptstyle j^L} & & \\ |K'| & \xrightarrow{\varphi} & |L'| & \xrightarrow{\psi} & |M|. \end{array}$$

Since $\psi\varphi$ is a simplicial approximation of gf, we have

$$(gf)_* = (\psi\varphi)_* (j_*^K)^{-1} = \psi_* \varphi_* (j_*^K)^{-1}.$$

The maps $j^L \varphi$ and ψ are simplicial approximations of f and g, respectively; hence $f_* = j_*^L \varphi_* (j_*^K)^{-1}$ and $g_* = \psi_* (j_*^L)^{-1}$. The equality $(gf)_* = g_* f_*$ is easy to verify. □

Corollary. *If simplicial complexes K and L are homeomorphic, then their homology groups are isomorphic.*

[5] The dimension of this simplex may equal any of the numbers $0, 1, \ldots, 2k + 1$.

Proof. If $f\colon |K| \to |L|$ and $g\colon |L| \to |M|$ are mutually inverse continuous maps, then f_* and g_* are mutually inverse homomorphisms of homology groups. □

As an application of the topological invariance of homology, we give yet another proof of the nonexistence of a retraction of the ball onto its boundary.

Theorem 1.9. *There exists no continuous map $r\colon D^n \to S^{n-1} = \partial D^n$ such that $r(x) = x$ for all $x \in S^{n-1}$.*

Proof. Suppose that $r\colon D^n \to S^{n-1}$ is a continuous map for which $r|_{S^{n-1}} = \mathrm{id}_{S^{n-1}}$. Consider the inclusion $i\colon S^{n-1} \hookrightarrow D^n$. The composition ri is the identity map; therefore, $r_*i_*\colon H_{n-1}(S^{n-1};\mathbb{Z}) \to H_{n-1}(S^{n-1};\mathbb{Z})$ is the identity map $\mathbb{Z} \to \mathbb{Z}$ (or $\mathbb{Z}\oplus\mathbb{Z} \to \mathbb{Z}\oplus\mathbb{Z}$, if $n = 1$). On the other hand, the image of $i_*\colon H_{n-1}(S^{n-1};\mathbb{Z}) \to H_{n-1}(D^n;\mathbb{Z})$ is zero (or contained in $\mathbb{Z}$, if $n = 1$) because $H_{n-1}(D^n;\mathbb{Z}) = 0$ for $n \geq 2$ and $H_0(D^1;\mathbb{Z}) = \mathbb{Z}$. □

2.3. Homotopy Invariance of Homology.

Theorem 1.10. *Homotopic maps of simplicial complexes induce the same map on homology.*

Proof. Let $f_0, f_1\colon |K| \to |L|$ be homotopic simplicial maps of simplicial complexes K and L. This means that there is a continuous map $F\colon |K| \times I \to |L|$ for which $F|_{|K|\times\{0\}} = f_0$ and $F|_{|K|\times\{1\}} = f_1$. The Cartesian product $|K| \times I$ can be turned into a simplicial complex; moreover, we can assume that $|K| \times \{0\}$ and $|K| \times \{1\}$ are endowed with the same simplicial structure and for any simplex Δ in K, the set $\Delta \times I$ is a subcomplex in $|K| \times I$ (see Part I, p. 130). Consider the inclusions $i_0\colon K \to |K| \times \{0\} \hookrightarrow |K| \times I$ and $i_1\colon K \to |K| \times \{1\} \hookrightarrow |K| \times I$. Clearly, $f_0 = Fi_0$ and $f_1 = Fi_1$. Therefore, it is sufficient to verify that $i_{0*} = i_{1*}$ (we use the equalities $f_{0*} = F_*i_{0*}$ and $f_{1*} = F_*i_{1*}$).

Let Δ^k be a simplex in K. The chains $i_0(\Delta^k)$ and $i_1(\Delta^k)$ have the common support $\Delta^k \times I$. This support is acyclic because it is homeomorphic to $D^{k+1} \approx \Delta^{k+1}$. □

Corollary. *If simplicial complexes K and L are homotopy equivalent, then their homology groups are isomorphic.*

As an application of the homotopy invariance of homology, we give yet another proof of dimension invariance (one proof was given in Part I on pp. 60–61).

Theorem 1.11. *If $n \neq m$, then the spaces $\mathbb{R}^n$ and $\mathbb{R}^m$ are not homeomorphic.*

Proof. For definiteness, suppose that $n > m \geq 1$; in particular, $n \geq 2$. If the spaces $\mathbb{R}^n$ and $\mathbb{R}^m$ are homeomorphic, then the spaces $\mathbb{R}^n \setminus \{0\} \sim S^{n-1}$ and $\mathbb{R}^m \setminus \{0\} \sim S^{m-1}$ are homeomorphic as well; hence they are homotopy equivalent. Thus, $\mathbb{Z} = H_{n-1}(S^{n-1}; \mathbb{Z}) = H_{n-1}(S^{m-1}; \mathbb{Z})$ and $n - 1 > 0$; therefore, $m = n$. This contradiction completes the proof. □

3. Relative Homology

Let K be a simplicial complex and $L \subset K$ a subcomplex. Then $C_k(L) \subset C_k(K)$, and we can consider the quotient group $C_k(K, L) = C_k(K)/C_k(L)$; the elements of this group are called *relative k-chains*. The homomorphism $\partial_k \colon C_k(K) \to C_{k-1}(K)$ induces a homomorphism $\partial_k \colon C_k(K, L) \to C_{k-1}(K, L)$ with the same property $\partial\partial = 0$. Therefore, we can again take the groups $Z_k(K, L) = \operatorname{Ker} \partial_k$ and $B_k(K, L) = \operatorname{Im} \partial_{k+1}$ and consider the quotient group $H_k(K, L) = Z_k/B_k$, which is called the k-dimensional *relative homology* group.

Example 3. If each path-connected component of a complex K contains at least one point of a subcomplex L, then $H_0(K, L) = 0$.

Proof. Let us join a vertex $x \in K$ with some vertex $y \in L$ by a polygonal line α formed by edges. Then $\partial_1 \alpha = [x] - [y] \sim [x]$; therefore, $B_0 \supset Z_0$. □

Exercise. Prove that $H_0(K, L; G) = \underbrace{G \oplus \dots \oplus G}_{n}$, where n is the number of path-connected components of K containing no points of L.

Problem 1 (excision isomorphism). Suppose that $K \cup M$ is a simplicial complex, K and M are its subcomplexes, and L is a subcomplex in K. Prove that $H_k(K, L) \cong H_k(K \cup M, L \cup M)$ for all k.

A map of simplicial pairs $f \colon (K, L) \to (K', L')$ induces the following homomorphism $f_* \colon H_*(K, L) \to H_*(K', L')$ of relative homology groups. First, we take the map $C_k(K)/C_k(L) \to C_k(K')/fC_k(L)$, and then, using the inclusion $fC_k(L) \subset C_k(L')$, apply the canonical projection

$$C_k(K')/fC_k(L) \to (C_k(K')/fC_k(L))/(C_k(L')/fC_k(L)) \cong C_k(K')/C_k(L').$$

As a result, we obtain a chain map $f_\# \colon C_*(K, L) \to C_*(K', L')$. It induces a homomorphism between the homology groups.

3.1. Exact Homology Sequences of Pairs. The inclusion $i \colon L \to K$ induces a homomorphism $i_* \colon H_k(L) \to H_k(K)$. Moreover, any absolute cycle can be regarded as a relative cycle; thus, we have a homomorphism $p_* \colon H_k(K) \to H_k(K, L)$. Let us construct a *connecting* homomorphism $\partial_* \colon H_k(K, L) \to H_{k-1}(L)$. Take a relative cycle $z_k \in C_k(K, L)$ and its absolute representative $\tilde{z}_k$; this means that $\tilde{z}_k \in C_k(K)$ and $z_k = \tilde{z}_k + C_k(L)$.

By assumption, $\partial_k z_k = 0$, i.e., $\partial_k \tilde{z}_k \in C_{k-1}(L)$. The map $\partial_* \colon H_k(K, L) \to H_{k-1}(L)$ acting as $z_k \mapsto \partial_k \tilde{z}_k$ is well defined because if $y_k \in C_k(L)$, then $\partial_k(\tilde{z}_k + y_k) = \partial_k \tilde{z}_k + \partial_k y_k \sim \partial_k \tilde{z}_k$.

Theorem 1.12. *The sequence of homomorphisms*

$$\cdots \longrightarrow H_k(L) \xrightarrow{i_*} H_k(K) \xrightarrow{p_*} H_k(K, L) \xrightarrow{\partial_*} H_{k-1}(L) \longrightarrow \cdots$$

is exact.

Proof. (1) $\operatorname{Im} i_* \subset \operatorname{Ker} p_*$. Any absolute cycle $\tilde{z}_k \in C_k(L)$ corresponds to the zero relative cycle.

(2) $\operatorname{Ker} p_* \subset \operatorname{Im} i_*$. Suppose that an absolute cycle $\tilde{z}_k \in C_k(K)$ corresponds to a relative cycle homologous to zero. Then $\tilde{z}_k = \tilde{z}'_k + \tilde{z}''_k$, where $\tilde{z}'_k \in C_k(L)$ and $\tilde{z}''_k = \partial \tilde{z}_{k+1}$ for $\tilde{z}_{k+1} \in C_{k+1}(K)$. Therefore, the cycle $\tilde{z}_k$ is homologous to the cycle $\tilde{z}'_k \in C_k(L)$.

(3) $\operatorname{Im} p_* \subset \operatorname{Ker} \partial_*$. Suppose that $\tilde{z}_k \in Z_k(K)$ and $z_k = \tilde{z}_k + C_k(L)$. Then $z_k \mapsto \partial_k \tilde{z}_k = 0$.

(4) $\operatorname{Ker} \partial_* \subset \operatorname{Im} p_*$. Suppose that $z_k = \tilde{z}_k + C_k(L)$ and $\partial_k \tilde{z}_k = 0$. Then for a representative of the relative cycle z_k we can take the absolute cycle $\tilde{z}_k$.

(5) $\operatorname{Im} \partial_* \subset \operatorname{Ker} i_*$. If $\tilde{z}_{k-1} = \partial_k \tilde{z}_k \in C_{k-1}(L)$, then the cycle $\tilde{z}_{k-1}$ is homologous to zero in K.

(6) $\operatorname{Ker} i_* \subset \operatorname{Im} \partial_*$. If $\tilde{z}_{k-1} \in C_{k-1}(L)$ and $\tilde{z}_{k-1} = \partial_k \tilde{z}_k$, where $\tilde{z}_k \in C_k(K)$, then $\tilde{z}_{k-1} = \partial_*(\tilde{z}_k + C_k(L))$. □

Relative homology groups can be expressed in terms of absolute groups; namely, for $k \geq 2$, we have

$$H_k(K, L) \cong H_k(K \cup CL, CL) \cong H_k(K \cup CL),$$

where CL is the cone over L. The first isomorphism is obvious even at the level of relative chains. The second follows from the exact sequence

$$H_k(CL) \xrightarrow{i_*} H_k(K \cup CL) \xrightarrow{p_*} H_k(K \cup CL, CL) \xrightarrow{\partial_*} H_{k-1}(CL)$$

of a pair. Since the cone CL is contractible, it follows that $H_k(CL) = H_{k-1}(CL) = 0$; therefore, the middle map is an isomorphism. Note that the complex $K \cup CL$ is homotopy equivalent to K/L because the complex CL is contractible and, hence, $K \cup L \sim (K \cup CL)/CL \approx K/L$.

Example 4. If $n \geq 1$, then

$$H_k(D^n, S^{n-1}; G) = \begin{cases} G & \text{for } k = n, \\ 0 & \text{for } k \neq n. \end{cases}$$

Proof. The exact sequence

$$H_k(D^n) \longrightarrow H_k(D^n, S^{n-1}) \longrightarrow H_{k-1}(S^{n-1}) \longrightarrow H_{k-1}(D^n)$$

shows that $H_k(D^n, S^{n-1}) \cong H_{k-1}(S^{n-1})$ for $k \geq 2$. For $k = 0$, the required assertion follows from Example 3. For $k = 1$, we have one of the exact sequences

$$0 \longrightarrow H_1(D^n, S^{n-1}) \longrightarrow G \longrightarrow G \longrightarrow 0 \quad (n \geq 2)$$

and

$$0 \longrightarrow H_1(D^1, S^0) \longrightarrow G \oplus G \longrightarrow G \longrightarrow 0.$$

Here the map $G \to G$ is an isomorphism and $G \oplus G \to G$ has the form $(a, b) \mapsto a + b$. The former has zero kernel; the kernel of the latter consists of elements of the form $(a, -a)$ and, therefore, is isomorphic to G. □

Remark. The difficulties involved in the consideration of the case $k = 1$ can be avoided by turning to reduced homology, which is defined on p. 17.

A generalization of exact homology sequences is the following algebraic construction. Suppose we have a short exact sequence of chain maps

$$0 \longrightarrow C'_* \xrightarrow{i_*} C_* \xrightarrow{p_*} C''_* \longrightarrow 0$$

(for example, $C'_k = C_k(L)$, $C_k = C_k(K)$, and $C''_k = C_k(K)/C_k(L)$). Then we can define a *connecting* homomorphism $\partial_* \colon H_k(C''_*) \to H_{k-1}(C'_*)$. Namely, take $z''_k \in C''_k$ for which $\partial''_k z''_k = 0$. We have $z''_k = p_k c_k$ for some $c_k \in C_k$, and $0 = \partial''_k z''_k = p_k(\partial_k c_k)$. Therefore, $\partial_k c_k = i_{k-1} c'_{k-1}$ for some $c'_{k-1} \in C'_{k-1}$. We set $\partial_* z''_k = c'_{k-1}$. It can be proved in precisely the same way as for exact sequences of pairs that this map is well defined at the homology level and the sequence

$$\cdots \xrightarrow{\partial_*} H_k(C'_*) \xrightarrow{i_*} H_k(C_*) \xrightarrow{p_*} H_k(C''_*) \xrightarrow{\partial_*} H_{k-1}(C'_*) \xrightarrow{i_*} \cdots$$

is exact.

Example 5. Let K be a simplicial complex, and let $0 \to G' \to G \to G'' \to 0$ be a short exact sequence of Abelian groups. Then we have a short exact sequence of chain complexes

$$0 \longrightarrow C_k(K; G') \longrightarrow C_k(K; G) \longrightarrow C_k(K; G'') \longrightarrow 0.$$

The connecting homomorphism $\beta_* \colon H_k(K; G'') \to H_{k-1}(K; G)$ is called the *Bockstein homomorphism.*

The most interesting examples of Bockstein homomorphisms can be obtained from the exact sequences

$$0 \longrightarrow \mathbb{Z} \xrightarrow{\times m} \mathbb{Z} \longrightarrow \mathbb{Z}_m \longrightarrow 0 \quad \text{and} \quad 0 \longrightarrow \mathbb{Z}_m \longrightarrow \mathbb{Z}_{m^2} \longrightarrow \mathbb{Z}_m \longrightarrow 0,$$

which determine homomorphisms $H_k(K;\mathbb{Z}_m) \to H_{k-1}(K;\mathbb{Z})$ and $H_k(K;\mathbb{Z}_m) \to H_{k-1}(K;\mathbb{Z}_m)$.

The *cokernel* of a homomorphism $\alpha\colon A \to A'$ is defined as $\operatorname{Coker}\alpha = A'/\operatorname{Im}\alpha$.

Problem 2. Given a commutative diagram of Abelian groups with exact rows

$$\begin{array}{ccccccccc} 0 & \longrightarrow & A & \longrightarrow & B & \longrightarrow & C & \longrightarrow & 0 \\ & & \downarrow{\scriptstyle\alpha} & & \downarrow{\scriptstyle\beta} & & \downarrow{\scriptstyle\gamma} & & \\ 0 & \longrightarrow & A' & \longrightarrow & B' & \longrightarrow & C' & \longrightarrow & 0, \end{array}$$

prove that there is an exact sequence

$$0 \to \operatorname{Ker}\alpha \to \operatorname{Ker}\beta \to \operatorname{Ker}\gamma \to \operatorname{Coker}\alpha \to \operatorname{Coker}\beta \to \operatorname{Coker}\gamma \to 0.$$

For a triple of simplicial complexes $L_1 \subset L_2 \subset K$, we can construct an exact *homology sequence of the triple*

$$\longrightarrow H_k(L_2, L_1) \xrightarrow{i_*} H_k(K, L_1) \xrightarrow{p_*} H_k(K, L_2) \xrightarrow{\partial_*} H_{k-1}(L_2, L_1) \longrightarrow$$

as follows. The isomorphism

$$C_*(K)/C_*(L_2) \cong (C_*(K)/C_*(L_1))/(C_*(L_2)/C_*(L_1))$$

determines a short exact sequence

$$0 \longrightarrow C_*(L_2)/C_*(L_1) \xrightarrow{i} C_*(K)/C_*(L_1) \xrightarrow{p} C_*(K)/C_*(L_2) \longrightarrow 0.$$

This short exact sequence of chain complexes induces the required exact sequence of homology groups.

In dealing with exact sequences, the following assertion is often useful.

Theorem 1.13 (Steenrod's five lemma). *Suppose that*

$$\begin{array}{ccccccccc} A_1 & \xrightarrow{\alpha_1} & A_2 & \xrightarrow{\alpha_2} & A_3 & \xrightarrow{\alpha_3} & A_4 & \xrightarrow{\alpha_4} & A_5 \\ \downarrow{\scriptstyle\varphi_1} & & \downarrow{\scriptstyle\varphi_2} & & \downarrow{\scriptstyle\varphi_3} & & \downarrow{\scriptstyle\varphi_4} & & \downarrow{\scriptstyle\varphi_5} \\ B_1 & \xrightarrow{\beta_1} & B_2 & \xrightarrow{\beta_2} & B_3 & \xrightarrow{\beta_3} & B_4 & \xrightarrow{\beta_4} & B_5 \end{array}$$

is a commutative diagram of Abelian groups with exact rows. If φ_1, φ_2, φ_4, and φ_5 are isomorphisms, then φ_3 is an isomorphism.

Proof. Consider the groups $A_2' = A_2/\operatorname{Im}\alpha_1$, $A_4' = \operatorname{Ker}\alpha_4$, $B_2' = B_2/\operatorname{Im}\beta_1$, and $B_4' = \operatorname{Ker}\beta_4$. It is sufficient to prove the required assertion for the simpler diagram

$$\begin{array}{ccccccccc}
0 & \longrightarrow & A_2' & \xrightarrow{\alpha_2'} & A_3 & \xrightarrow{\alpha_3'} & A_4' & \longrightarrow & 0 \\
 & & \downarrow{\scriptstyle \varphi_2'} & & \downarrow{\scriptstyle \varphi_3} & & \downarrow{\scriptstyle \varphi_4'} & & \\
0 & \longrightarrow & B_2' & \xrightarrow{\beta_2'} & B_3 & \xrightarrow{\beta_3'} & B_4' & \longrightarrow & 0,
\end{array}$$

in which φ_2' and φ_4' are isomorphisms. Clearly,

$$\operatorname{Ker}\varphi_3 \subset \operatorname{Ker}(\beta_3'\varphi_3) = \operatorname{Ker}(\varphi_4'\alpha_3') = \operatorname{Ker}(\alpha_3') = \operatorname{Im}(\alpha_2');$$

therefore, $\operatorname{Ker}\varphi_3 \cong \operatorname{Ker}(\varphi_3\alpha_2') = \operatorname{Ker}(\beta_2'\varphi_2') = 0$, i.e., φ_3 is a monomorphism. Moreover, $\operatorname{Im}(\beta_3'\varphi_3) = \operatorname{Im}(\varphi_4'\alpha_3')$, where α_3' is an epimorphism and φ_4' is an isomorphism. Hence $\beta_3'\varphi_3$ is an epimorphism, and $B_3 = \operatorname{Im}\varphi_3 + \operatorname{Ker}\beta_3'$. Since

$$\operatorname{Ker}\beta_3' = \operatorname{Im}\beta_2' = \operatorname{Im}(\beta_2'\varphi_2') = \operatorname{Im}(\varphi_3\alpha_2') \subset \operatorname{Im}\varphi_3,$$

it follows that $B_3 = \operatorname{Im}\varphi_3$, i.e., φ_3 is an epimorphism. □

Remark. The five lemma remains valid if the diagram is commutative up to sign, i.e., $\varphi_2\alpha_1 = \pm\beta_1\varphi_1$, etc. Indeed, the proof involves only the groups Ker and Im, which do not change under the replacement of φ by $-\varphi$.

Using the five lemma, we can easily prove the following theorem.

Theorem 1.14. *Suppose that $f\colon (K, L) \to (K', L')$ is a map of pairs for which the induced maps $K \to K'$ and $L \to L'$ are homotopy equivalences. Then $f_*\colon H_k(K, L) \to H_k(K', L')$ is an isomorphism.*

Proof. Consider the following commutative diagram with exact rows:

$$\begin{array}{ccccccccc}
H_k(L) & \xrightarrow{i_*} & H_k(K) & \xrightarrow{p_*} & H_k(K,L) & \xrightarrow{\partial_*} & H_{k-1}(L) & \xrightarrow{i_*} & H_{k-1}(K) \\
{\scriptstyle f_*}\downarrow{\scriptstyle \text{iso}} & & {\scriptstyle f_*}\downarrow{\scriptstyle \text{iso}} & & {\scriptstyle f_*}\downarrow{\scriptstyle ?} & & {\scriptstyle f_*}\downarrow{\scriptstyle \text{iso}} & & {\scriptstyle f_*}\downarrow{\scriptstyle \text{iso}} \\
H_k(L') & \xrightarrow{i_*} & H_k(K') & \xrightarrow{p_*} & H_k(K',L') & \xrightarrow{\partial_*} & H_{k-1}(L') & \xrightarrow{i_*} & H_{k-1}(K').
\end{array}$$

The five lemma implies that the middle vertical arrow is an isomorphism. □

Problem 3. Suppose that

$$\begin{array}{ccccccccc}
0 & \longrightarrow & A_2 & \xrightarrow{\alpha_2} & A_3 & \xrightarrow{\alpha_3} & A_4 & \longrightarrow & 0 \\
 & & \downarrow{\scriptstyle \varphi_2} & & \downarrow{\scriptstyle \varphi_3} & & \downarrow{\scriptstyle \varphi_4} & & \\
0 & \longrightarrow & B_2 & \xrightarrow{\beta_2} & B_3 & \xrightarrow{\beta_3} & B_4 & \longrightarrow & 0
\end{array}$$

is a commutative diagram of Abelian groups with exact rows.

(a) Prove that if φ_2 and φ_4 are monomorphisms, then φ_3 is a monomorphism.

(b) Prove that if φ_2 and φ_4 are epimorphisms, then φ_3 is an epimorphism.

Problem 4. Given a commutative diagram

$$\begin{array}{ccccc} A_1 & \xrightarrow{\alpha_1} & A_2 & \xrightarrow{\alpha_2} & A_3 \\ \downarrow{\scriptstyle\varphi_1} & & \downarrow{\scriptstyle\varphi_2} & & \downarrow{\scriptstyle\varphi_3} \\ B_1 & \xrightarrow{\beta_1} & B_2 & \xrightarrow{\beta_2} & B_3 \end{array}$$

with exact rows, prove that φ_2 induces an isomorphism

$$\Phi\colon \operatorname{Ker}(\varphi_3\alpha_2)/(\operatorname{Ker}\alpha_2 + \operatorname{Ker}\varphi_2) \to (\operatorname{Im}\varphi_2 \cap \operatorname{Im}\beta_1)/\operatorname{Im}(\varphi_2\alpha_1).$$

Problem 5 ([**83**]). (a) Given two commutative diagrams satisfying the assumptions of Steenrod's five lemma in which the respective homomorphisms, except for the isomorphisms $\varphi_3' = \eta_1$ and $\varphi_3'' = \eta_2$, coincide, prove that

$$\eta_1^{-1}\eta_2(x) + \eta_2^{-1}\eta_1(x) = 2x$$

for any $x \in A_3$.

(b) Suppose that the diagrams in (a) are diagrams of rings and their homomorphisms. Prove that the isomorphisms η_1 and η_2 in these diagrams can be different if and only if there exists a nontrivial additive homomorphism $\delta\colon A_3 \to A_3$ such that $\delta\alpha_2 = 0$, $\alpha_3\delta = 0$, and $\delta(xy) = (\delta x)y + x\delta y + (\delta x)(\delta y)$ for all $x, y \in A_3$.

(c) Give an example of two diagrams satisfying the assumptions of (a) for which the homomorphisms η_1 and η_2 are different.

3.2. Reduced Homology. The statements and proofs of many theorems can be simplified by considering the reduced homology groups $H_k(K)$ instead of the homology groups $H_k(K)$ themselves; below, we give two equivalent definitions of reduced homology groups.

Definition 1. The *reduced homology group* $\tilde{H}_k(K)$ is the kernel of the homomorphism $p_*\colon H_k(K) \to H_k(*)$, where $p\colon K \to *$ is the map from the complex K to the singleton.

Definition 2. Let us replace the map $\partial_0\colon C_0(K) \to 0$ by $\varepsilon\colon C_0(K) \to \mathbb{Z}$, where $\varepsilon(v) = 1$ for each vertex v. If $c_1 \in C_1(K)$, then $\varepsilon\partial_1 c_1 = 0$; therefore, for the new chain complex, we can also define the reduced homology group $\tilde{H}_k(K)$.

The map ε is called an *augmentation*. This term is used because we augment the chain complex

$$\cdots \longrightarrow C_1(K) \xrightarrow{\partial_0} C_0(K) \longrightarrow 0$$

to the chain complex

$$\cdots \longrightarrow C_1(K) \xrightarrow{\partial_0} C_0(K) \xrightarrow{\varepsilon} C_{-1}(K) \longrightarrow 0,$$

where $C_{-1}(K) = \mathbb{Z}$.

Definitions 1 and 2 are equivalent because $\tilde{H}_k(K) = H_k(K)$ for $k \geq 1$ and the map $p_0\colon C_0(K) \to C_0(*) = \mathbb{Z}$ coincides with ε.

For an arbitrary map $i\colon * \to X$, we have $pi = \mathrm{id}_*$; therefore,

$$H_0(K) = \operatorname{Im} p_* \oplus \operatorname{Ker} i_* = \tilde{H}_0(K) \oplus \mathbb{Z}.$$

Thus, $\tilde{H}_0(K) \cong H_0(K, *)$.

For reduced homology, exact sequences of pairs are also defined.

Example 4 shows that $H_k(D^n, S^{n-1}) \cong \tilde{H}_k(S^n)$ for all k.

We have already shown that $H_k(K, L) \cong H_k(K \cup CL)$ for $k \geq 2$. Using reduced homology, we can also show that $H_k(K, L) \cong \tilde{H}_k(K \cup CL)$ for $k = 0$ and 1.

Exercise. Given a simplicial (possibly, disconnected) complex K embedded in $\mathbb{R}^n$, prove that $H_i(\mathbb{R}^n, K) \cong H_{i-1}(K)$ for $i \geq 2$ and $H_1(\mathbb{R}^n, K) \cong \tilde{H}_0(K)$. Moreover, a path between points of K represents the zero element of the group $H_1(\mathbb{R}^n, K)$ if and only if these points belong to the same connected component of K.

3.3. The Mayer–Vietoris Sequence. The exact sequence of Mayer–Vietoris relates the homology groups of the union of two simplicial complexes K_0 and K_1, of their intersection, and of these complexes themselves. In the simplest case, where K_0 and K_1 consist of finitely many points, we can use the well-known inclusion-exclusion formula $N = N_0 + N_1 - N_{01}$, where N is the number of points in $K_0 \cup K_1$, N_0 and N_1 are the numbers of points in K_0 and K_1, respectively, and N_{01} is the number of points in $K_0 \cap K_1$. In the language of exact sequences, this formula can be written as

$$0 \longrightarrow H_0(K_0 \cap K_1) \longrightarrow H_0(K_0) \oplus H_0(K_1) \longrightarrow H_0(K_0 \cup K_1) \longrightarrow 0.$$

Another simple example of an exact sequence of this form arises in the homology of the wedge $K = K_0 \vee K_1$ of two simplicial complexes. In this case, we have the isomorphism $C_k(K) \cong C_k(K_0) \oplus C_k(K_1)$ for $k \geq 1$ already at the level of chains. Moreover, the kernel of the boundary homomorphism on $C_1(K)$ is the direct sum of the kernels of the boundary homomorphisms on $C_1(K_0)$ and $C_1(K_1)$. Therefore, for any $k \geq 1$, we have the canonical isomorphism $H_k(K) \cong H_k(K_0) \oplus H_k(K_1)$.

These examples suggest the existence of an exact sequence

$$0 \longrightarrow H_k(K_0 \cap K_1) \longrightarrow H_k(K_0) \oplus H_k(K_1) \longrightarrow H_k(K_0 \cup K_1) \longrightarrow 0.$$

However, representing the circle S^1 as the union of two arcs K_0 and K_1 with two-point intersection, we see that no such exact sequence exists for $k = 1$. In reality, the k-dimensional homology of a complex $K_0 \cup K_1$ depends not only on the k-dimensional homology of the complexes $K_0 \cap K_1$, K_0, and K_1 but also on their $(k-1)$-dimensional homology; the correct exact sequence is as follows.

Theorem 1.15 (Mayer–Vietoris [**84**, **142**]). *Suppose that K is a simplicial complex, K_0 and K_1 are subcomplexes of K such $K = K_0 \cup K_1$, and $L = K_0 \cap K_1$. Then there is an exact sequence*

$$\cdots \longrightarrow H_k(L) \longrightarrow H_k(K_0) \oplus H_k(K_1) \longrightarrow H_k(K) \longrightarrow H_{k-1}(L) \longrightarrow \cdots.$$

Proof. The Mayer–Vietoris sequence arises from the exact sequence

$$(1) \qquad 0 \longrightarrow C_*(L) \xrightarrow{(j_0,-j_1)} C_*(K_0) \oplus C_*(K_1) \xrightarrow{(i_0,i_1)} C_*(K) \longrightarrow 0,$$

which we describe below. The complex $C_*(K_0) \oplus C_*(K_1)$ consists of the groups $C_k(K_0) \oplus C_k(K_1)$, and the boundary homomorphism in it is the direct sum of the boundary homomorphisms; i.e., $\partial(c^0, c^1) = (\partial c^0, \partial c^1)$. The maps $j_\alpha \colon L \to K_\alpha$ and $i_\alpha \colon K_\alpha \to K$ are the natural embeddings.

Let us show that (1) is exact. The map $(j_0, -j_1)$ is, obviously, a monomorphism. It follows from $K = K_0 \cup K_1$ that (i_0, i_1) is an epimorphism. Indeed, take $c = \sum a_i \Delta_i^k \in C_k(K)$ and let c^0 be the sum of all terms for which $\Delta_i^k \subset K_0$. Then K_1 is a support of the chain $c - c^0$, and $(i_0, i_1)(c^0, c - c^0) = c^0 + (c - c^0) = c$.

It remains to verify exactness in the middle term. The image of the map $(j_0, -j_1)$ consists of chains of the form $(c, -c)$, where c is a chain in L. The kernel of (i_0, i_1) consists of chains of the form $(c, -c)$, where c is a chain in $K_0 \cap K_1 = L$.

Clearly, the homology of the chain complex $C_*(K_0) \oplus C_*(K_1)$ is isomorphic to $H_*(K_0) \oplus H_*(K_1)$. □

Mayer–Vietoris exact sequences exist also for reduced homology. Indeed, setting $(j_0, -j_1)(n) = (n, -n)$ and $(i_0, i_1)(m, n) = m + n$, we obtain the commutative diagram

$$\begin{array}{ccccccccc}
0 & \longrightarrow & C_0(L) & \xrightarrow{(j_0,-j_1)} & C_0(K_0) \oplus C_0(K_1) & \xrightarrow{(i_0,i_1)} & C_0(K) & \longrightarrow & 0 \\
 & & \downarrow{\scriptstyle \varepsilon_L} & & \downarrow{\scriptstyle \varepsilon_0 \oplus \varepsilon_1} & & \downarrow{\scriptstyle \varepsilon_K} & & \\
0 & \longrightarrow & \mathbb{Z} & \xrightarrow{(j_0,-j_1)} & \mathbb{Z} \oplus \mathbb{Z} & \xrightarrow{(i_0,i_1)} & \mathbb{Z} & \longrightarrow & 0
\end{array}$$

with exact rows.

Using the Mayer–Vietoris exact sequence for reduced homology, we can easily calculate the homology groups of a suspension (see Part I, p. 130).

Theorem 1.16 (the suspension isomorphism). *Let K be a simplicial complex, and let ΣK be the suspension over K. Then, for any $k \geq 1$, there exists a canonical isomorphism $H_k(\Sigma K) \cong \tilde{H}_{k-1}(K)$.*

Proof. We can represent ΣK in the form $K_0 \cup K_1$, where K_0 and K_1 are cones over K and $K_0 \cap K_1 = K$. Let us write the Mayer–Vietoris sequence for reduced homology:

$$\tilde{H}_k(K_0) \oplus \tilde{H}_k(K_1) \to \tilde{H}_k(\Sigma K) \to \tilde{H}_{k-1}(K) \to \tilde{H}_{k-1}(K_0) \oplus \tilde{H}_{k-1}(K_1).$$

The spaces K_0 and K_1 are contractible; therefore, the first and last terms in this sequence are trivial, and the middle map is an isomorphism. □

Remark. We use reduced homology because for ordinary homology, the case $k = 1$ must be considered separately.

Problem 6. Calculate the homology of the torus T^2.

Problem 7. Prove that $H_k(S^p \times S^q, S^p \vee S^q) \cong H_k(S^{p+q})$ for all $k \geq 1$.

Problem 8. Calculate the homology of the space $S^p \times S^q$.

Problem 9. (a) Calculate the homology of the complement of a knot in S^3 (the answer does not depend on the knot).

(b) Consider a link with n connected components in S^3. Calculate the homology of the complement of this link (the answer depends only on n).

Problem 10. Let M^n be a smooth manifold. Prove that $H_k(M^n \setminus \operatorname{int} D^n) \cong H_k(M^n)$ for $1 \leq k \leq n-2$. (Here D^n is a ball contained in some chart of M^n.)

Problem 11. Suppose that K is a simplicial complex and $\mathcal{L} = \{L_1, \ldots, L_n\}$ is a family of its subcomplexes such that they cover K and all their intersections are acyclic. Prove that the homology groups of the complex K are isomorphic to those of the nerve[6] of the cover $\mathcal{L}$.

The Mayer–Vietoris theorem has the following generalization (concerning the *relative Mayer–Vietoris sequence*).

Theorem 1.17. *For any subcomplexes $L_0 \subset K_0$ and $L_1 \subset K_1$ of a simplicial complex K, there is an exact sequence*

$$\begin{aligned} \to H_k(K_0 \cap K_1, L_0 \cap L_1) &\to H_k(K_0, L_0) \oplus H_k(K_1, L_1) \\ &\to H_k(K_0 \cup K_1, L_0 \cup L_1) \to H_{k-1}(K_0 \cap K_1, L_0 \cap L_1) \to . \end{aligned}$$

[6]The definition of the nerve of a cover is given in Part I on p. 108.

Proof. The relative Mayer–Vietoris sequence arises from the short exact sequence
(2)

$$0 \longrightarrow \frac{C_*(K_0 \cap K_1)}{C_*(L_0 \cap L_1)} \xrightarrow{(j_0,-j_1)} \frac{C_*(K_0)}{C_*(L_0)} \oplus \frac{C_*(K_1)}{C_*(L_1)} \xrightarrow{(i_0,i_1)} \frac{C_*(K_0 \cup K_1)}{C_*(L_0 \cup L_1)} \longrightarrow 0$$

defined as follows. For each quadruple of Abelian groups $H_1 \subset H_2$, $G_1 \subset G_2$, where $H_1 \subset G_1$ and $H_2 \subset G_2$, the canonical map $G_1/H_1 \to G_2/H_2$ of their quotients is defined. The maps j_0, j_1 and i_0, i_1 are such canonical maps.

If $c \in C_*(L_0) \cap C_*(L_1)$, then $c \in C_*(L_0 \cap L_1)$. Therefore, $(j_0, -j_1)$ is a monomorphism.

We show that (i_0, i_1) is an epimorphism. Take $c = \sum a_i \Delta_i^k \in C_*(K_0 \cup K_1)$ and let c^0 be the sum of all terms for which $\Delta_i^k \subset K_0$. Then the chain $c - c^0$ is contained in $C_*(K_1)$. Therefore, the pair $(c^0, c - c^0)$ can be associated with an element of the middle group in (2). The image of this element under the map (i_0, i_1) coincides with $c \pmod{C_*(L_0 \cup L_1)}$.

It remains to verify the exactness in the middle term. The image of the map $(j_0, -j_1)$ consists of relative chains $(c \,(\mathrm{mod}\, C_*(L_0)), -c \,(\mathrm{mod}\, C_*(L_1)))$, where c is a chain in $K_0 \cap K_0$. The kernel of (i_0, i_1) consists of relative chains $(c^0 \pmod{C_*(L_0)}, c^1 \pmod{C_*(L_1)})$, where $c^0 + c^1 \in C_*(L_0 \cup L_1)$. Clearly, the image is contained in the kernel. Consider an element of the kernel. Let $\tilde{c}^0$ be the sum of all terms in the chain $c^0 + c^1$ that are contained in $C_*(L_0)$, and let $\tilde{c}^1 = -(c^0 + \tilde{c}^0 + c^1)$. Then $\tilde{c}^1 \in C_*(L_1)$; therefore, for c we can take $c^0 + \tilde{c}^0 = -(c^1 + \tilde{c}^1)$. □

4. Cohomology and Universal Coefficient Theorem

4.1. Cohomology. Suppose that K is a simplicial complex, G an Abelian group, and $C_k(K) = C_k(K; \mathbb{Z})$. A homomorphism $c^k \colon C_k(K) \to G$ is called a k-dimensional *cochain* with coefficients (or values) in G. The group of k-dimensional cochains is denoted by $C^k(K; G) = \mathrm{Hom}(C_k(K), G)$.

The text in small print below refers to infinite simplicial complexes. It can be skipped at the first reading.

For any family of Abelian groups $\{G_\alpha\}$, the *direct sum* $\bigoplus_\alpha G_\alpha$ and the *Cartesian product* $\prod_\alpha G_\alpha$ are defined. Both groups consist of indexed sets (g_α) under componentwise addition. The difference between them is that the group $\bigoplus_\alpha G_\alpha$ consists of the sets (g_α) in which $g_\alpha \neq 0$ only for finitely many indices α, whereas the group $\prod_\alpha G_\alpha$ consists of all such sets. For finite families of groups, Cartesian products coincide with direct sums.

Suppose that the k-simplices in a complex K are indexed by α. According to the definition of groups of chains, we have $C_k(K; G) = \bigoplus_\alpha G_\alpha$, where $G_\alpha \cong G$. But $C^k(K; G) = \prod_\alpha G_\alpha$ for $G_\alpha \cong G$. Indeed, a cochain is a function on a set of simplices with values in G. To each simplex Δ^k_α any value can be assigned; infinitely many values may be nonzero.

For the group of chains a direct sum rather than a Cartesian product is taken because boundary homomorphisms cannot be defined for Cartesian products. The reason for this is that the direct sums of free Abelian groups are free, while the Cartesian products are not.

Suppose that G is the additive group of a ring with identity. Then to each simplex Δ^k_α we can assign the k-cochain $(\Delta^k_\alpha)^*$ which takes the value 1 at Δ^k_α and vanishes at all other k-simplices. Therefore, any k-cochain can be written as a formal sum $\sum g_\alpha(\Delta^k_\alpha)^*$, where $g_\alpha \in G$. This sum can have infinitely many terms.

Take $c_k \in C_k(K)$ and $c^k \in C^k(K;G)$. We denote the value of the homomorphism c^k at the element c_k by $\langle c^k, c_k\rangle$. Such a pairing allows us to associate with the boundary operator $\partial\colon C_{k+1}(K) \to C_k(K)$ the dual operator $\delta\colon C^k(K;G) \to C^{k+1}(K;G)$ defined by

$$\langle \delta c^k, c_{k+1}\rangle = \langle c^k, \partial c_{k+1}\rangle.$$

Note that the operator δ increases dimension, whereas ∂ decreases it.

Exercise. Prove that the value of a cochain δc^k at the simplex $[v_0, \ldots, v_{k+1}]$ equals

$$\sum_{i=0}^{k+1}\langle c^k, [v_0, \ldots, \hat{v}_i, \ldots, v_{k+1}]\rangle.$$

Remark. Sometimes, the operator $\delta\colon C^k(K;G) \to C^{k+1}(K;G)$ is defined by

$$\langle \delta c^k, c_{k+1}\rangle = -(-1)^k\langle c^k, \partial c_{k+1}\rangle.$$

This choice of sign is based on the following convention suggested by MacLane: when the n- and m-dimensional symbols are transposed, the right-hand side must be multiplied by $(-1)^{mn}$; the maps ∂ and δ are then assigned the dimensions -1 and $+1$. We do not adopt this convention for two reasons. First, we use the duality between maps ∂ and δ very often, and the introduction of an additional sign would make it somewhat awkward. Second, to determine how the sign changes under transpositions of symbols, nontrivial considerations are needed every time.

The equality $\partial\partial = 0$ implies $\delta\delta = 0$. Therefore, the *cohomology groups*

$$H^k(K;G) = Z^k(K;G)/B^k(K;G) \quad \text{and} \quad H^0(K;G) = Z^0(K;G)$$

are defined; here Z^k and B^{k+1} are the kernel and the image of the homomorphism $\delta\colon C^k \to C^{k+1}$, respectively. The elements of the groups Z^k and B^k are called *cocycles* and *coboundaries*.

Example 6. If K is a connected simplicial complex, then $H^0(K;G) = G$.

Proof. Let $c_1 = [v_0, v_1]$. Then $\partial c_1 = [v_1] - [v_0]$. Hence $\langle \delta c^0, c_1 \rangle = \langle c^0, \partial c_1 \rangle = \langle c^0, [v_1] \rangle - \langle c^0, [v_0] \rangle$. Thus, the equality $\delta c^0 = 0$ means that the cochain c^0 takes the same value at any two vertices joined by an edge. For a connected simplicial complex K, this means that the cochain c^0 takes the same value at all vertices. Therefore, $Z^0(K; G) = G$. □

Theorem 1.18. *If G is the additive group of a field F, then $H^k(K; G)$ is the dual space of $H_k(K; G)$.*

Proof. The groups $C^k(K; G)$ and $C_k(K; G)$ are linear spaces over the field F, and C^k is the dual space of C_k. The map ∂ is linear, and the map δ is its dual. Therefore, we must prove that if $A\colon U \to V$ and $B\colon V \to W$ are linear maps for which $BA = 0$, then $\operatorname{Ker} A^* / \operatorname{Im} B^*$ is the dual space of $\operatorname{Ker} B / \operatorname{Im} A$.

The linear functions on $\operatorname{Ker} B$ can be interpreted as linear functions on V considered up to functions of the form $g(Bx)$; the functions $g(Bx)$ constitute the space $\operatorname{Im} B^*$. The linear functions on $\operatorname{Ker} B / \operatorname{Im} A$ can be interpreted as the linear functions on $\operatorname{Ker} B$ satisfying the condition $f(Ax) = 0$ for all x; the functions for which $f(Ax) = 0$ form the space $\operatorname{Ker} A^*$. □

Corollary. *If G is the additive group of a field F and one of the linear spaces $H^k(K; G)$ and $H_k(K; G)$ over F is finite-dimensional, then $H^k(K; G) \cong H_k(K; G)$; this isomorphism is not canonical.*

Any simplicial map $\varphi\colon K \to L$ induces a homomorphism $\varphi^*\colon H^k(L; G) \to H^k(K; G)$ acting in the opposite direction. Indeed, each cochain $c^k \in C^k(L; G)$ corresponds to the cochain $\varphi^k c^k \in C^k(K; G)$ defined by $\varphi^k c^k(\Delta) = c^k(\varphi(\Delta))$.

The *reduced cohomology groups* are defined by replacing the augmentation $\varepsilon\colon C_0(K) \to \mathbb{Z}$ by the dual map $\tilde{\varepsilon}\colon G \to C^0(K; G)$, where $\tilde{\varepsilon}(c_0) = g\varepsilon(c_0)$.

The groups of *relative cochains* are defined as

$$C^k(K, L; G) = \operatorname{Hom}(C_k(K, L); G).$$

Each group of relative cochains is a subgroup in the group of absolute cochains. It consists of the cochains vanishing on $C_k(L)$. Recall that the relative chain group is a quotient of the absolute chain group. Under duality, quotient groups correspond to subgroups.

In the relative case, the coboundary and boundary homomorphisms are dual also. The *relative cohomology groups* are defined in a natural way.

Exercise. Prove that if each connected component of $|K|$ contains at least one connected component of $|L|$, then $H^0(K, L; G) = 0$.

The short exact sequence

$$0 \longrightarrow C_k(L) \xrightarrow{i_k} C_k(K) \xrightarrow{p_k} C_k(K,L) \longrightarrow 0 \tag{3}$$

induces the dual sequence

$$0 \longleftarrow C^k(L;G) \xleftarrow{i^k} C^k(K;G) \xleftarrow{p^k} C^k(K,L;G) \longleftarrow 0. \tag{4}$$

This sequence is exact as well, but proving this requires some effort. Moreover, the proof of the exactness of the sequence (4) uses a special feature of the sequence (3), namely, its split property. An exact sequence $0 \to A \xrightarrow{\varphi} B \xrightarrow{\psi} C \to 0$ is said to be *split* if it satisfies any of the three equivalent conditions in Lemma 1.1.

Lemma 1.1. *Let $0 \to A \xrightarrow{\varphi} B \xrightarrow{\psi} C \to 0$ be an exact sequence. Then the following conditions are equivalent:*

(a) *this sequence has the form*[7]

$$0 \longrightarrow A \xrightarrow{i} A \oplus C \xrightarrow{p} C \longrightarrow 0, \tag{5}$$

where i is the natural embedding and p is the natural projection;

(b) *the homomorphism φ has a left inverse, i.e., there exists a homomorphism $\Phi\colon B \to A$ for which $\Phi\varphi = \mathrm{id}_A$;*

(c) *the homomorphism ψ has a right inverse, i.e., there exists a homomorphism $\Psi\colon C \to B$ for which $\psi\Psi = \mathrm{id}_C$.*

Proof. Clearly, conditions (b) and (c) hold for any sequence of the form (5); to see this, it suffices to set $\Phi(a,c) = a$ and $\Psi(c) = (0,c)$. It remains to verify that if (b) or (c) holds, then the sequence has the form (5).

Suppose that $\Phi\varphi = \mathrm{id}_A$. Let us show that $B = \operatorname{Im}\varphi \oplus \operatorname{Ker}\Phi$. Any element $b \in B$ can be represented as $b = \varphi\Phi(b) + (b - \varphi\Phi(b))$, where $\varphi\Phi(b) \in \operatorname{Im}\varphi$ and $b - \varphi\Phi(b) \in \operatorname{Ker}\Phi$. Moreover, if $b \in \operatorname{Im}\varphi \cap \operatorname{Ker}\Phi$, then $b = \varphi(a)$ and $0 = \Phi(b) = \Phi\varphi(a) = a$; hence $b = 0$.

Suppose that $\psi\Psi = \mathrm{id}_C$. Let us show that $B = \operatorname{Ker}\psi \oplus \operatorname{Im}\Psi$. Any element $b \in B$ can be represented as $b = (b - \Psi\psi(b)) + \Psi\psi(b)$, where $b - \Psi\psi(b) \in \operatorname{Ker}\psi$ and $\Psi\psi(b) \in \operatorname{Im}\Psi$. Moreover, if $b \in \operatorname{Ker}\psi \cap \operatorname{Im}\Psi$, then $b = \Psi(c)$ and $0 = \psi(b) = \psi\Psi(c) = c$; hence $b = 0$. □

[7] Formally, this means that there exists an isomorphism $f\colon B \to A \oplus C$ for which the following diagram is commutative:

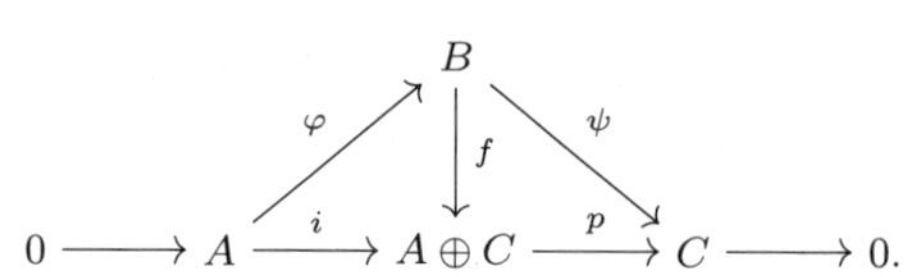

Problem 12. Prove Lemma 1.1 using the five lemma.

Problem 13. Let $p\colon E \to B$ be a fibration[8] with fiber F.

(a) Prove that if there exists a section $s\colon B \to E$ (this means that $p \circ s = \mathrm{id}_B$ is the identity map on B), then $\pi_n(E) \cong \pi_n(B) \oplus \pi_n(F)$.

(b) Prove that if there exists a retraction $r\colon E \to F$, then $\pi_n(E) \cong \pi_n(B) \oplus \pi_n(F)$.

(c) Prove that if the fiber F is contractible in the space E, then $\pi_n(B) \cong \pi_n(E) \oplus \pi_{n-1}(F)$.

It is easy to verify that if the group C is free, then the exact sequence $0 \to A \xrightarrow{\varphi} B \xrightarrow{\psi} C \to 0$ is split. Indeed, the map $\Psi\colon C \to B$ can be constructed as follows. Take a basis in C; for each basis element c, we set $\Psi(c) = b$, where b is any element of $\psi^{-1}(c)$.

Since the group $C_k(K, L)$ is free, it follows that the exact sequence (3) is split.

Theorem 1.19. (a) *If a sequence* $A \xrightarrow{\varphi} B \xrightarrow{\psi} C \to 0$ *is exact, then so is the dual sequence*

$$\mathrm{Hom}(A, G) \xleftarrow{\tilde{\varphi}} \mathrm{Hom}(B, G) \xleftarrow{\tilde{\psi}} \mathrm{Hom}(C, G) \longleftarrow 0.$$

(b) *If a sequence* $0 \to A \xrightarrow{\varphi} B \xrightarrow{\psi} C \to 0$ *is exact and split, then so is the dual sequence*

$$0 \longleftarrow \mathrm{Hom}(A, G) \xleftarrow{\tilde{\varphi}} \mathrm{Hom}(B, G) \xleftarrow{\tilde{\psi}} \mathrm{Hom}(C, G) \longleftarrow 0.$$

Proof. (a) First, let us show that $\mathrm{Ker}\, \tilde{\psi} = 0$. Suppose that $\tilde{c} \in \mathrm{Ker}\, \tilde{\psi}$, i.e., $0 = \tilde{\psi}(\tilde{c}) = \tilde{c} \circ \psi$. This means that $\tilde{c}(\psi(b)) = 0$ for all $b \in B$. By assumption, ψ is an epimorphism; therefore, $\tilde{c} = 0$.

Now, let us show that $\mathrm{Im}\, \tilde{\psi} = \mathrm{Ker}\, \tilde{\varphi}$. The equality $\psi \circ \varphi = 0$ implies $\tilde{\varphi} \circ \tilde{\psi} = 0$, whence $\mathrm{Im}\, \tilde{\psi} \subset \mathrm{Ker}\, \tilde{\varphi}$. Suppose that $\tilde{b} \in \mathrm{Ker}\, \tilde{\varphi}$, i.e., $0 = \tilde{\varphi}(\tilde{b}) = \tilde{b} \circ \varphi$. This means that $\tilde{b}(\mathrm{Im}\, \varphi) = 0$; therefore, $\tilde{b}$ induces a homomorphism $\tilde{b}'\colon B/\mathrm{Im}\, \varphi \to G$. The homomorphism ψ induces an isomorphism $\psi'\colon B/\mathrm{Im}\, \varphi \to C$. Consider the homomorphism $\tilde{b}'(\psi')^{-1} \in \mathrm{Hom}(C, G)$. Clearly,

$$\tilde{\psi}(\tilde{b}'(\psi')^{-1}) = \tilde{b}'(\psi')^{-1}\psi = \tilde{b},$$

[8]The definition of a fibration is given in Part I on p. 162.

because the diagram

$$\begin{array}{ccccc} & & B & & \\ & \tilde{b} \swarrow & \downarrow & \searrow \psi & \\ G & \xleftarrow{\tilde{b}'} & B/\operatorname{Im}\varphi & \xrightarrow{\psi'} & C \end{array}$$

is commutative.

(b) Let $\Phi\colon B \to A$ be a homomorphism for which $\Phi\varphi = \mathrm{id}_A$. Then the composition

$$\operatorname{Hom}(A, G) \xrightarrow{\tilde{\Phi}} \operatorname{Hom}(B, G) \xrightarrow{\tilde{\varphi}} \operatorname{Hom}(A, G)$$

is the identity. Therefore, $\tilde{\varphi}$ is an epimorphism, i.e., the dual sequence is exact. Clearly, the homomorphism $\tilde{\Phi}$ splits this sequence. □

Exercise. Prove that for the exact sequence $0 \to \mathbb{Z} \xrightarrow{\times 2} \mathbb{Z}$, the dual sequence $0 \leftarrow \operatorname{Hom}(\mathbb{Z}, \mathbb{Z}) \leftarrow \operatorname{Hom}(\mathbb{Z}, \mathbb{Z})$ is not exact.

In conclusion, we calculate the groups $\operatorname{Hom}(A, B)$ in some simple cases.

(1) $\operatorname{Hom}(\mathbb{Z}, G) \cong G$ because any homomorphism $\varphi\colon \mathbb{Z} \to G$ is completely determined by the element $\varphi(1) \in G$, and this element can be arbitrary.

(2) $\operatorname{Hom}(\mathbb{Z}_n, \mathbb{Z}) = 0$ because if $\varphi\colon \mathbb{Z}_n \to \mathbb{Z}$ is a homomorphism, then $n\varphi(1) = 0$ and, therefore, $\varphi(1) = 0$.

(3) $\operatorname{Hom}(\mathbb{Z}_n, \mathbb{Z}_m) = \mathbb{Z}_d$, where $d = \mathrm{GCD}(n, m)$. Indeed, any homomorphism $\varphi\colon \mathbb{Z}_n \to \mathbb{Z}_m$ is completely determined by the element $\varphi(1)$, which must satisfy the relation $n\varphi(1) \equiv 0 \pmod{m}$, i.e., $\varphi(1) \equiv 0 \pmod{m/d}$. Thus,

$$\varphi(1) \in \left\{0, \frac{m}{d}, \frac{2m}{d}, \dots, \frac{(d-1)m}{d}\right\}.$$

We also have $\operatorname{Hom}(\mathbb{Z}_n, G) \cong \operatorname{Ker}(G \xrightarrow{\times n} G)$; the proof is similar to that of (3).

Exercise. Write the Mayer–Vietoris exact sequence for cohomology using Theorem 1.19.

Problem 14. Let A be an Abelian group and $mA = \{ma \mid a \in A\}$. Prove that $\operatorname{Hom}(A, \mathbb{Z}) \cong \operatorname{Hom}(mA, \mathbb{Z})$.

Problem 15. Prove that the group $\operatorname{Hom}(\mathbb{Q}, \mathbb{Q}/\mathbb{Z})$ is uncountable. (Here $\mathbb{Q}$ is the group of rational numbers under addition.)

4.2. Tensor Product and Homology with Arbitrary Coefficients. We have defined the cohomology groups $H^k(K; G)$ with arbitrary coefficients using chain groups with coefficients in $\mathbb{Z}$. Similarly, the homology groups $H_k(K; G)$ can be defined using chain groups with coefficients in $\mathbb{Z}$. The groups $\operatorname{Hom}(C_k(K), G)$ are replaced by $C_k \otimes G$, where $\otimes$ is the *tensor product* of Abelian groups, which is defined as follows. Suppose that A and B are Abelian groups, $F(A, B)$ is the free Abelian group with basis $A \times B$, and $R(A, B)$ is the subgroup in $F(A, B)$ generated by the elements of the form $(a + a', b) - (a, b) - (a', b)$ and $(a, b + b') - (a, b) - (a, b')$. Then $A \otimes B = F(A, B)/R(A, B)$. The coset containing an element (a, b) is denoted by $a \otimes b$. Less formally, $A \otimes B$ is the group with generators $a \otimes b$ and defining relations $(a + a') \otimes b = a \otimes b + a' \otimes b$ and $a \otimes (b + b') = a \otimes b + a \otimes b'$.

Remark. The group $A \otimes B$ may contain an element $a_1 \otimes b_1 + \cdots + a_k \otimes b_k$ that cannot be represented in the form $a \otimes b$.

It is easy to show that $\mathbb{Z} \otimes G \cong G$; the isomorphism is defined by $n \otimes g \mapsto ng$. Indeed, $n \otimes g = 1 \otimes g + \cdots + 1 \otimes g = 1 \otimes (ng)$. Therefore, any element of the group $\mathbb{Z} \otimes G \cong G$ has the form $1 \otimes g$; there are no relations between these elements.

A little more involved argument proves that $\mathbb{Z}_n \otimes G \cong G/nG$. Indeed, any element of the group $\mathbb{Z}_n \otimes G$ has the form $1 \otimes g$, and $1 \otimes ng = 0$. In particular, $\mathbb{Z}_n \otimes \mathbb{Z}_m \cong \mathbb{Z}_d$, where $d = \operatorname{GCD}(n, m)$. Moreover, $\mathbb{Z}_n \otimes \mathbb{Q} = 0$.

The very definition of tensor product readily implies the following two assertions.

1. If a map $\varphi \colon A \times B \to C$ is *bilinear*, i.e., $\varphi(a + a', b) = \varphi(a, b) + \varphi(a', b)$ and $\varphi(a, b + b') = \varphi(a, b) + \varphi(a, b')$, then there exists a homomorphism $\tilde{\varphi} \colon A \otimes B \to C$ for which $\tilde{\varphi}(a \otimes b) = \varphi(a, b)$.

2. For any two homomorphisms $\varphi \colon A \to A'$ and $\psi \colon B \to B'$, there exists a homomorphism $\varphi \otimes \psi \colon A \otimes B \to A' \otimes B'$ for which $(\varphi \otimes \psi)(a \otimes b) = \varphi(a) \otimes \psi(b)$. Indeed,

$$\begin{aligned}(\varphi \otimes \psi)&[(a + a') \otimes b - a \otimes b - a' \otimes b] \\ &= [\varphi(a) + \varphi(a')] \otimes \psi(b) - \varphi(a) \otimes \psi(b) - \varphi(a') \otimes \psi(b) = 0.\end{aligned}$$

For the tensor product, the following theorem is valid; it is dual to Theorem 1.19.

Theorem 1.20. (a) *If a sequence $A \xrightarrow{\varphi} B \xrightarrow{\psi} C \to 0$ is exact, then so is the sequence*

$$A \otimes G \xrightarrow{\varphi \otimes 1} B \otimes G \xrightarrow{\psi \otimes 1} C \otimes G \longrightarrow 0,$$

where $1 = \mathrm{id}_G$.

(b) *If a sequence $0 \to A \xrightarrow{\varphi} B \xrightarrow{\psi} C \to 0$ is exact and split, then so is the sequence*

$$0 \longrightarrow A \otimes G \xrightarrow{\varphi \otimes 1} B \otimes G \xrightarrow{\psi \otimes 1} C \otimes G \longrightarrow 0.$$

Proof. (a) First, let us show that $\psi \otimes 1$ is an epimorphism and its kernel coincides with $\operatorname{Ker}\psi \otimes G$. The homomorphism $\psi \otimes 1$ induces a homomorphism

$$\widetilde{\psi \otimes 1}\colon B \otimes G/(\operatorname{Ker}\psi \otimes G) \to C \otimes G.$$

It is sufficient to verify that $\widetilde{\psi \otimes 1}$ is an isomorphism. Consider the map $\Psi\colon C \times G \to B \otimes G/(\operatorname{Ker}\psi \otimes G)$ defined by $\Psi(c, g) = b \otimes g + \operatorname{Ker}\psi \otimes G$, where $\psi(b) = c$. This map is well defined because if $\psi(b') = c$, then $b \otimes g - b' \otimes g = (b - b') \otimes g$, where $b - b' \in \operatorname{Ker}\psi$. The map Ψ is bilinear; therefore, it induces a homomorphism $\tilde{\Psi}\colon C \otimes G \to B \otimes G/(\operatorname{Ker}\psi \otimes G)$. The maps $\widetilde{\psi \otimes 1}$ and $\tilde{\Psi}$ are mutually inverse.

Thus, $\operatorname{Ker}(\psi \otimes 1) = \operatorname{Ker}\psi \otimes G = \operatorname{Im}\varphi \otimes G = \operatorname{Im}(\varphi \otimes 1)$; the last equality holds because the group $\operatorname{Im}(\varphi \otimes 1)$ is generated by all elements of the form $\varphi(a) \otimes g$.

(b) Let $\Phi\colon B \to A$ be a homomorphism for which $\Phi\varphi = \mathrm{id}_A$. Then

$$(\Phi \otimes 1) \circ (\varphi \otimes 1) = \Phi\varphi \otimes 1 = \mathrm{id}_A \otimes \mathrm{id}_G = \mathrm{id}_{A \otimes G};$$

therefore, $\varphi \otimes 1$ is a monomorphism, and the map $\Phi \otimes 1$ splits the exact sequence of tensor products. □

Exercise. Prove that the sequence $0 \to \mathbb{Z} \otimes \mathbb{Z}_2 \to \mathbb{Q} \otimes \mathbb{Z}_2$ induced by the exact sequence $0 \to \mathbb{Z} \to \mathbb{Q}$ is not exact.

4.3. The Groups Tor and Ext. The homology and cohomology groups with arbitrary coefficients can be expressed in terms of integral homology groups. The expressions include operations Tor and Ext, which assign Abelian groups $\operatorname{Tor}(A, B)$ and $\operatorname{Ext}(A, B)$ to pairs of Abelian groups A and B. In cmputations, it is sufficient to know $\operatorname{Tor}(A, B)$ and $\operatorname{Ext}(A, B)$ for $A, B = \mathbb{Z}$, $\mathbb{Z}_m$, $\mathbb{Q}$, and $\mathbb{R}$.

The groups $\operatorname{Tor}(A, B)$ and $\operatorname{Ext}(A, B)$ are defined as follows. The Abelian group A can be specified by generators and defining relations. This means that there exists an exact sequence

$$0 \to R \xrightarrow{i} F \xrightarrow{p} A \to 0, \tag{6}$$

where F and R are free Abelian groups (the group F is defined by generators and R, by relations; the group R is free because it is a subgroup of a free group). The exact sequence (6) is called a *free resolution* of the Abelian group A. The exact sequence (6) induces exact sequences

$$R \otimes B \xrightarrow{i \otimes 1} F \otimes B \xrightarrow{p \otimes 1} A \otimes B \longrightarrow 0$$

and

$$\operatorname{Hom}(R, B) \xleftarrow{\tilde{\imath}} \operatorname{Hom}(F, B) \xleftarrow{\tilde{p}} \operatorname{Hom}(A, B) \longleftarrow 0.$$

We set $\operatorname{Tor}(A, B) = \operatorname{Ker}(i \otimes 1)$ and $\operatorname{Ext}(A, B) = \operatorname{Coker} \tilde{\imath} = \operatorname{Hom}(R, B)/\operatorname{Im} \tilde{\imath}$ (note that both definitions use only the map $i\colon R \to F$). Thus, the groups $\operatorname{Tor}(A, B)$ and $\operatorname{Ext}(A, B)$ complete the exact sequences under consideration to the exact sequences

$$0 \longrightarrow \operatorname{Tor}(A, B) \longrightarrow R \otimes B \xrightarrow{i\otimes 1} F \otimes B \xrightarrow{p\otimes 1} A \otimes B \longrightarrow 0$$

and

$$0 \longleftarrow \operatorname{Ext}(A, B) \longleftarrow \operatorname{Hom}(R, B) \xleftarrow{\tilde{\imath}} \operatorname{Hom}(F, B) \xleftarrow{\tilde{p}} \operatorname{Hom}(A, B) \longleftarrow 0.$$

First, we must check that the groups $\operatorname{Tor}(A, B)$ and $\operatorname{Ext}(A, B)$ thus defined do not depend on the choice of a free resolution.

Lemma 1.2. (a) *For an arbitrary homomorphism $\varphi\colon A \to A'$ of Abelian groups A and A', any free resolutions of A and A' can be completed to a commutative diagram*

$$\begin{array}{ccccccccc}
0 & \longrightarrow & R & \xrightarrow{i} & F & \xrightarrow{p} & A & \longrightarrow & 0 \\
 & & \downarrow{\scriptstyle \varphi_1} & & \downarrow{\scriptstyle \varphi_0} & & \downarrow{\scriptstyle \varphi} & & \\
0 & \longrightarrow & R' & \xrightarrow{i'} & F' & \xrightarrow{p'} & A' & \longrightarrow & 0.
\end{array}$$

(b) *If ψ_0 and ψ_1 are other homomorphisms completing the free resolutions to a commutative diagram, then there exists a homomorphism $D_0\colon F \to R'$ for which $i'D_0 = \varphi_0 - \psi_0$ and $D_0 i = \varphi_1 - \psi_1$.*

Proof. (a) Take bases $\{r_\alpha\}$ and $\{f_\beta\}$ in the free Abelian groups R and F. The map p' is an epimorphism; therefore, F' contains elements f'_β such that $p'(f'_\beta) = \varphi p(f_\beta)$. For the basis elements, we set $\varphi_0(f_\beta) = f'_\beta$; then, we extend φ_0 to a group homomorphism $F \to F'$. We have $p'\varphi_0 = \varphi p$.

By assumption, $pi = 0$; hence $p'\varphi_0 i(r_\alpha) = \varphi pi(r_\alpha) = 0$. It follows from $\operatorname{Ker} p' = \operatorname{Im} i'$ that R' has an element r'_α for which $i'(r'_\alpha) = \varphi_0 i(r_\alpha)$. We set $\varphi_1(r_\alpha) = r'_\alpha$. Then $i'\varphi_1 = \varphi_0 i$.

(b) Choose a basis $\{f_\alpha\}$ in F. By assumption, $p'(\varphi_0 - \psi_0)(f_\alpha) = \varphi p(f_\alpha) - \varphi p(f_\alpha) = 0$. Therefore, R' has an element $\{r'_\alpha\}$ for which $i'(r'_\alpha) = (\varphi_0 - \psi_0)(f_\alpha)$. We set $D_0(f_\alpha) = r'_\alpha$. Then $i'D_0 = \varphi_0 - \psi_0$, whence $i'D_0 i = (\varphi_0 - \psi_0)i = i'(\varphi_1 - \psi_1)$. The map i is a monomorphism; therefore, $D_0 i = \varphi_1 - \psi_1$. □

The awkward formulation of Lemma 1.2 has a very natural interpretation in the language of homological algebra. Namely, consider the sequence of homomorphisms

$$\cdots \longrightarrow 0 \longrightarrow 0 \longrightarrow R \xrightarrow{i} F$$

as a chain complex C_* (here $C_0 = F$, $C_1 = R$, and $C_k = 0$ for $k > 1$). Lemma 1.2 asserts that any homomorphism $\varphi\colon A \to A'$ induces a chain map $C_* \to C'_*$, which is unique up to chain homotopy. Thus, the induced homology map $H_*(C_*) \to H_*(C'_*)$ is uniquely determined. Moreover, the corresponding homology group homomorphisms $H_*(C_* \otimes G) \to H_*(C'_* \otimes G)$ and $H_*(\mathrm{Hom}(C'_*, G)) \to H_*(\mathrm{Hom}(C_*, G))$ are uniquely determined as well. It is these homomorphisms that we are interested in because $H_1(C_* \otimes G) = \mathrm{Tor}(A, G)$ and $H_1(\mathrm{Hom}(C_*, G)) = \mathrm{Ext}(A, G)$. Note also that $H_0(C_* \otimes G) = F \otimes G/R \otimes G \cong A \otimes G$ and $H_0(\mathrm{Hom}(C_*, G)) \cong \mathrm{Hom}(A, G)$.

Now, the relations $(\mathrm{id})_* = \mathrm{id}$ and $(\varphi\psi)_* = \varphi_*\psi_*$ imply that the group $\mathrm{Tor}(A, B)$ is well defined. For $\mathrm{Ext}(A, B)$, the proof is similar; the only difference is that $(\varphi\psi)_* = \psi_*\varphi_*$ for this group.

Let us calculate the groups $\mathrm{Tor}(A, B)$ and $\mathrm{Ext}(A, B)$ in simple cases. Note that if the group A is free, then $\mathrm{Tor}(A, B) = 0$ and $\mathrm{Ext}(A, B) = 0$. In particular, $\mathrm{Tor}(\mathbb{Z}, B) = 0$ and $\mathrm{Ext}(\mathbb{Z}, B) = 0$.

For the group $A = \mathbb{Z}_n$, consider the free resolution

$$0 \longrightarrow \mathbb{Z} \xrightarrow{\times n} \mathbb{Z} \longrightarrow \mathbb{Z}_n \longrightarrow 0.$$

First, we calculate $\mathrm{Tor}(\mathbb{Z}_n, B)$. Any element of the group $\mathbb{Z} \otimes B$ can be represented in the form $1 \otimes b$. We must calculate the kernel of the map $1 \otimes b \mapsto n \otimes b = 1 \otimes nb$. Clearly, it consists of the elements $1 \otimes b$ for which $nb = 0$. Thus, the group $\mathrm{Tor}(\mathbb{Z}_n, B)$ is isomorphic to $\mathrm{Ker}(B \xrightarrow{\times n} B)$. In particular, $\mathrm{Tor}(\mathbb{Z}_n, \mathbb{Z}) = 0$ and $\mathrm{Tor}(\mathbb{Z}_n, \mathbb{Z}_m) = \mathbb{Z}_{(n,m)}$.

Now, we calculate $\mathrm{Ext}(\mathbb{Z}_n, B)$. Clearly, $\mathrm{Hom}(\mathbb{Z}, B) \cong B$, and to multiplication by n in $\mathbb{Z}$ corresponds the homomorphism $B \to B$ that takes each b to nb. Therefore, $\mathrm{Ext}(\mathbb{Z}_n, B) = B/nB$. In particular, $\mathrm{Ext}(\mathbb{Z}_n, \mathbb{Z}) = \mathbb{Z}_n$ and $\mathrm{Ext}(\mathbb{Z}_n, \mathbb{Z}_m) = \mathbb{Z}_{(n,m)}$.

Problem 16. Let $A = \mathbb{Z}^k \oplus T$, where T is a finite Abelian group. Prove that $\mathrm{Ext}(A, \mathbb{Z}) \cong T$.

Theorem 1.21. (a) *If A is an Abelian group such that $\mathrm{Ker}(A \xrightarrow{\times n} A) = 0$ for any $n \in \mathbb{N}$, then $\mathrm{Tor}(A, B) = 0$ for any group B.*

(b) *If B is an Abelian group such that $nB = B$ for any $n \in \mathbb{N}$, then $\mathrm{Ext}(A, B) = 0$ for any group A.*

Proof. (a) Let A' be an arbitrary finitely generated subgroup in A. Then A' is a free Abelian group; therefore, for any free resolution

$$0 \longrightarrow R \xrightarrow{i} F \xrightarrow{p} B \longrightarrow 0,$$

we have the commutative diagram

$$\begin{array}{ccc} A' \otimes R & \xrightarrow{1' \otimes i} & A' \otimes F \\ \downarrow & & \downarrow \\ A \otimes R & \xrightarrow{1 \otimes i} & A \otimes F \end{array} , \tag{7}$$

in which the map $1' \otimes i$ is a monomorphism. The vertical arrows in this diagram are monomorphisms too because R and F are free groups. Let us prove that the map $1 \otimes i$ is a monomorphism. Take any element $\omega = a_1 \otimes r_1 + \cdots + a_k \otimes r_k$ in $A \otimes R$. The group A' can be chosen so as to contain the elements $a_1, \ldots, a_k$. Since the diagram (7) is commutative and, moreover, $1' \otimes i$ and the right vertical arrow are monomorphisms, it follows that $(1 \otimes i)(\omega) \neq 0$.

(b) Any free resolution

$$0 \longrightarrow R \xrightarrow{i} F \xrightarrow{p} A \longrightarrow 0$$

determines a homomorphism $\operatorname{Hom}(F, B) \to \operatorname{Hom}(R, B)$; we have to prove that this is an epimorphism. Let $\varphi\colon R \to B$ be a homomorphism. Take $y \in F \setminus R$ and extend φ to the group generated by R and y as follows:

(1) if $my \notin R$ for all $m \in \mathbb{N}$, then we set $\tilde{\varphi}(y) = 0$;

(2) if $my \in R$ for some $m \in \mathbb{N}$, then we choose the least number $n \in \mathbb{N}$ with this property and put $\tilde{\varphi}(y) = b$, where b is the element of B such that $nb = \varphi(ny)$.

An extension of φ over the entire group F can be constructed by induction (if the group F/R is not finitely generated, then the induction is transfinite). □

Corollary. *If* $G = \mathbb{Q}$, $\mathbb{R}$, *or* $\mathbb{C}$ (*we regard these fields as groups under addition*), *then* $\operatorname{Tor}(G, B) = 0$ *and* $\operatorname{Ext}(A, G) = 0$ *for any Abelian groups* A *and* B.

In calculating homology and cohomology groups with coefficients in G by using the universal coefficient theorem, the groups $\operatorname{Tor}(A, G)$ and $\operatorname{Ext}(A, G)$ arise. If $G = \mathbb{Q}$, $\mathbb{R}$, or $\mathbb{C}$, then we have $\operatorname{Tor}(A, G) = 0$ because $\operatorname{Tor}(A, B) \cong \operatorname{Tor}(B, A)$ for any Abelian groups A and B. Note that Ext does not have this property; for example, $\operatorname{Ext}(\mathbb{Z}, \mathbb{Z}_n) = 0$ and $\operatorname{Ext}(\mathbb{Z}_n, \mathbb{Z}) = \mathbb{Z}_n$.

Problem 17. Prove that $\operatorname{Tor}(A, B) \cong \operatorname{Tor}(B, A)$, and the isomorphism is canonical.

An Abelian group T is said to be *periodic* if for any $t \in T$, there exists a positive integer n for which $nt = 0$.

Problem 18. Prove that if $\operatorname{Ext}(T, \mathbb{Z}) = 0$ and T is a periodic group, then $T = 0$.

Problem 19. Prove that if a sequence $0 \to A \xrightarrow{\varphi} B \xrightarrow{\psi} C$ is exact, then so is the sequence

$$0 \longrightarrow \operatorname{Hom}(G, A) \xrightarrow{\tilde{\varphi}} \operatorname{Hom}(G, B) \xrightarrow{\tilde{\psi}} \operatorname{Hom}(G, C)$$

for any Abelian group G.

An Abelian group G is said to be *divisible* if for any positive integer n, the map $G \to G$ defined by $g \mapsto ng$ is an epimorphism.

Problem 20. Let $0 \to A \xrightarrow{\varphi} B \xrightarrow{\psi} C \to 0$ be an exact sequence.

(a) Prove that, for any free Abelian group F, the sequence

$$0 \longrightarrow \operatorname{Hom}(F, A) \xrightarrow{\tilde{\varphi}} \operatorname{Hom}(F, B) \xrightarrow{\tilde{\psi}} \operatorname{Hom}(F, C) \longrightarrow 0$$

is exact.

(b) Prove that, for any divisible group G, the sequence

$$0 \longleftarrow \operatorname{Hom}(A, G) \xleftarrow{\tilde{\varphi}} \operatorname{Hom}(B, G) \xleftarrow{\tilde{\psi}} \operatorname{Hom}(C, G) \longleftarrow 0$$

is exact.

A free resolution of an Abelian group A is often called a *projective resolution* because there is the dual notion of injective resolution; an *injective resolution* of an Abelian group A is an exact sequence $0 \to A \to G \to H \to 0$ in which G and H are divisible Abelian groups.

Any Abelian group A has an injective resolution. Indeed, any quotient of a divisible group is divisible. Therefore, it is sufficient to embed A in a divisible group G. Suppose that $A = F/R$, where F is a free group. The group F is a direct sum of a number of copies of $\mathbb{Z}$; hence it is a subgroup of the group G defined as the direct sum of the same number of copies of $\mathbb{Q}$. The group G has a subgroup corresponding to R. The quotient group G/R contains A as a subgroup.

The group $\operatorname{Ext}(A, B)$ can be defined not only by using a projective resolution of A but also by using an injective resolution of B. Namely, let $0 \to B \to G \to H \to 0$ be an injective resolution of the group B. According to Problem 19, the induced sequence $0 \to \operatorname{Hom}(A, B) \to \operatorname{Hom}(A, G) \to \operatorname{Hom}(A, H)$ is exact. Therefore, it can be completed to an exact sequence of the form

$$0 \longrightarrow \operatorname{Hom}(A, B) \longrightarrow \operatorname{Hom}(A, G) \longrightarrow \operatorname{Hom}(A, H) \longrightarrow \widetilde{\operatorname{Ext}}(A, B) \longrightarrow 0.$$

Problem 21. Prove that $\widetilde{\operatorname{Ext}}(A, B) \cong \operatorname{Ext}(A, B)$.

Problem 22. Given an exact sequence of Abelian groups $0 \to A \to B \to C \to 0$, prove that, for any Abelian group X, there are exact sequences

$$\begin{aligned} 0 \longrightarrow \mathrm{Hom}(X, A) &\longrightarrow \mathrm{Hom}(X, B) \longrightarrow \mathrm{Hom}(X, C) \\ &\longrightarrow \mathrm{Ext}(X, A) \longrightarrow \mathrm{Ext}(X, B) \longrightarrow \mathrm{Ext}(X, C) \longrightarrow 0 \end{aligned}$$

and

$$\begin{aligned} 0 \longleftarrow \mathrm{Ext}(A, X) &\longleftarrow \mathrm{Ext}(B, X) \longleftarrow \mathrm{Ext}(C, X) \\ &\longleftarrow \mathrm{Hom}(A, X) \longleftarrow \mathrm{Hom}(B, X) \longleftarrow \mathrm{Hom}(C, X) \longleftarrow 0. \end{aligned}$$

4.4. Universal Coefficient Theorem. Homology and cohomology with arbitrary coefficients can be expressed in terms of integral homology. Namely, the following theorem is valid.

Theorem 1.22 (universal coefficient formulas). *For any Abelian group G, there are exact sequences*

$$0 \longrightarrow H_k(K) \otimes G \longrightarrow H_k(K; G) \longrightarrow \mathrm{Tor}(H_{k-1}(K), G) \longrightarrow 0$$

and

$$0 \longleftarrow \mathrm{Hom}(H_k(K), G) \longleftarrow H^k(K; G) \longleftarrow \mathrm{Ext}(H_{k-1}(K), G) \longleftarrow 0,$$

where $H_k(K) = H_k(K; \mathbb{Z})$. These exact sequences split; hence

$$H_k(K; G) \cong (H_k(K) \otimes G) \oplus \mathrm{Tor}(H_{k-1}(K), G)$$

and

$$H^k(K; G) \cong \mathrm{Hom}(H_k(K), G) \oplus \mathrm{Ext}(H_{k-1}(K), G)$$

(the isomorphisms are not canonical).

Proof. Consider the exact sequence of chain complexes

$$\begin{array}{ccccccccc} & & \downarrow & & \downarrow & & \downarrow & & \\ 0 & \longrightarrow & Z_k & \xrightarrow{i} & C_k & \xrightarrow{\partial} & B_{k-1} & \longrightarrow & 0 \\ & & \downarrow{\scriptstyle \partial=0} & & \downarrow{\scriptstyle \partial} & & \downarrow{\scriptstyle \partial=0} & & \\ 0 & \longrightarrow & Z_{k-1} & \xrightarrow{i} & C_{k-1} & \xrightarrow{\partial} & B_{k-2} & \longrightarrow & 0 \\ & & \downarrow & & \downarrow & & \downarrow & & \end{array}$$

It is assumed that $C_k = C_k(K; \mathbb{Z})$, etc.

The group B_{k-1} is free; therefore, tensoring with G, we obtain the exact sequence

$$0 \longrightarrow Z_k \otimes G \xrightarrow{i \otimes 1} C_k \otimes G \xrightarrow{\partial \otimes 1} B_{k-1} \otimes G \longrightarrow 0.$$

Any short exact sequence of chain complexes induces an exact sequence of homology groups. Taking into account the fact that the homology groups of a chain complex with zero boundary homomorphisms coincide with the chain groups, we obtain the exact sequence

$$\longrightarrow B_k \otimes G \longrightarrow Z_k \otimes G \longrightarrow H_k(C_* \otimes G) \longrightarrow B_{k-1} \otimes G \longrightarrow Z_{k-1} \otimes G \longrightarrow .$$

It follows directly from the definition of connecting homomorphisms in exact sequences of homology groups that the map $B_k \otimes G \to Z_k \otimes G$ in this sequence is induced by the embedding $j\colon B_k \to Z_k$. Thus, we obtain the exact sequence

$$0 \to Z_k \otimes G / B_k \otimes G \longrightarrow H_k(C_* \otimes G) \longrightarrow \operatorname{Ker}(B_{k-1} \otimes G \xrightarrow{j\otimes 1} Z_{k-1} \otimes G) \to 0.$$

Here $Z_k \otimes G/B_k \otimes G = H_k \otimes G$ and $H_k(C_* \otimes G) = H_k(K; G)$. Moreover, $\operatorname{Ker}(B_{k-1} \otimes G \xrightarrow{j\otimes 1} Z_{k-1} \otimes G) = \operatorname{Tor}(H_{k-1}(K), G)$ because any exact sequence

$$0 \longrightarrow B_{k-1} \longrightarrow Z_{k-1} \longrightarrow H_{k-1} \longrightarrow 0$$

is a free resolution for the group $H_{k-1} = H_{k-1}(K; \mathbb{Z})$.

Similarly, for cohomology, we have the exact sequence

$$0 \longleftarrow \operatorname{Hom}(H_k(K), G) \longleftarrow H^k(K; G) \longleftarrow \operatorname{Ext}(H_{k-1}(K), G) \longleftarrow 0.$$

It remains to show that these exact sequences split. In the exact sequence of homology groups, at the level of cycles, the homomorphism $\varphi\colon H_k(K) \otimes G \to H_k(K; G)$ has the structure $\varphi((\sum n_i \Delta_i^k) \otimes g) = \sum n_i g \Delta_i^k$. It is required to construct a homomorphism $\Phi\colon H_k(K; G) \to H_k(K) \otimes G$ for which $\Phi\varphi = \mathrm{id}$. The exact sequence

$$0 \longrightarrow Z_k \xrightarrow{i} C_k \xrightarrow{\partial} B_{k-1} \longrightarrow 0$$

splits because the group B_{k-1} is free. Suppose that $I: C_k \to Z_k$ is a splitting map, i.e., $Ii = \mathrm{id}_{Z_k}$. Restricting the composition

$$C_k \otimes G \xrightarrow{I\otimes 1} Z_k \otimes G \longrightarrow H_k \otimes G$$

to the cycles of the complex $C_* \otimes G$, we obtain a map that induces a homomorphism $\Phi\colon H_k(K; G) \to H_k(K) \otimes G$ since $I \otimes 1$ takes boundaries to boundaries. Clearly, if $\sum n_i \Delta_i^k$ is a cycle in C_k, then $\Phi\varphi((\sum n_i \Delta_i^k) \otimes g) = \Phi(\sum n_i g \Delta_i^k) = (\sum n_i \Delta_i^k) \otimes g$.

For cohomology, the fact that the exact sequence is split follows from Theorem 1.19, (b). □

Corollary. *If $G = \mathbb{Q}$, $\mathbb{R}$, or $\mathbb{C}$, then $H_k(K; G) \cong H_k(K) \otimes G$ and $H^k(K; G) \cong \operatorname{Hom}(H_k(K), G)$.*

In calculating the groups Tor and Ext, the following fairly obvious isomorphisms are useful:

$$\operatorname{Ext}(A_1 \oplus A_2, B) \cong \operatorname{Ext}(A_1, B) \oplus \operatorname{Ext}(A_2, B),$$
$$\operatorname{Tor}(A_1 \oplus A_2, B) \cong \operatorname{Tor}(A_1, B) \oplus \operatorname{Tor}(A_2, B).$$

Remark. For infinite families of Abelian groups, we have

$$\operatorname{Ext}\Big(\bigoplus_\alpha A_\alpha, B\Big) = \prod_\alpha \operatorname{Ext}(A_\alpha, B), \qquad \operatorname{Ext}\Big(A, \prod_\alpha B\Big) = \prod_\alpha \operatorname{Ext}(A, B_\alpha),$$
$$\operatorname{Tor}\Big(\bigoplus_\alpha A_\alpha, B\Big) = \bigoplus_\alpha \operatorname{Tor}(A_\alpha, B), \quad \operatorname{Tor}\Big(A, \bigoplus_\alpha B\Big) = \bigoplus_\alpha \operatorname{Tor}(A, B_\alpha).$$

Problem 23. Prove that if $H_1(K;\mathbb{Z}) \cong \mathbb{Z}^r \oplus T_1$, where T_1 is a finite group, then $H^1(K;\mathbb{Z}) \cong \mathbb{Z}^r$.

Problem 24. Suppose that $H_i(K;\mathbb{Z}) = \mathbb{Z}^{n_i} \oplus T_i$ and $H^i(K;\mathbb{Z}) = \mathbb{Z}^{m_i} \oplus T^i$, where T_i and T^i are finite groups. Prove that $m_i = n_i$ and $T^i \cong T_{i-1}$.

Problem 25. Derive the assertion of Problem 24 directly from the definition of homology and cohomology groups, without using the universal coefficient theorem.

Problem 26. Let X and Y be finite simplicial complexes, and let $f\colon X \to Y$ be a continuous map. Prove that if the map $f_*\colon H_1(X) \to H_1(Y)$ is zero, then so is $f^*\colon H^1(Y) \to H^1(X)$.

5. Calculations

5.1. Fundamental Classes. Many homological properties of manifolds are consequences of the fact that for an orientable closed manifold M^n, the group $H_n(M^n;\mathbb{Z})$ is nontrivial and generated by one homology class; a similar assertion is true for the $\mathbb{Z}_2$-homology of an arbitrary closed manifold. Pseudomanifolds[9] have the same homological properties.

Theorem 1.23. *Let M^n be a pseudomanifold, and let G be any Abelian group.*

(a) *If M^n has nonempty boundary, then $H_n(M^n;G) = 0$.*

(b) *If M^n is closed, then*

$$H_n(M^n;G) \cong \begin{cases} G & \text{if } M^n \text{ is orientable,} \\ \operatorname{Ker}(G \xrightarrow{\times 2} G) & \text{if } M^n \text{ is nonorientable.} \end{cases}$$

[9]The definition of a pseudomanifold was given in Part I on p. 109. Recall that any smooth manifold can be triangulated, and a triangulated manifold is a pseudomanifold. Therefore, all theorems on pseudomanifolds presented below are also valid for smooth manifolds.

Proof. The group $H_n(M^n; G)$ consists of chains $c_n = \sum a_i \Delta_i^n$ with $a_i \in G$ for which $\partial c_n = 0$. Suppose that simplices Δ_i^n and Δ_j^n have a common face Δ_s^{n-1}. Since any pseudomanifold must be unramified, it follows that Δ_s^{n-1} is a face for only these two simplices; therefore, it is contained in ∂c_n with coefficient $\pm a_i \pm a_j$. The coefficient is $\pm(a_i - a_j)$ if the orientations of the simplices Δ_i^n and Δ_j^n are compatible, and $\pm(a_i + a_j)$ otherwise. The coefficient $\pm a_i \pm a_j$ can vanish only if $a_i = \pm a_j$. Therefore, according to the strong connectedness condition, if $\partial c_n = 0$, then $c_n = \sum \pm a \Delta_i^n$, where the summation is over all n-simplices.

If a simplex Δ^{n-1} belongs to the boundary of M^n, then it is contained in ∂c_n with coefficient $\pm a$. Therefore, $\partial c_n = 0$ implies $a = 0$; i.e., $H_n(M^n; G) = 0$ for any coefficient group G.

Changing (if necessary) the orientations of some simplices, we can assume that $c_n = \sum a \Delta_i^n$. If the pseudomanifold M^n is closed and $2a \neq 0$, then $\partial c_n = 0$ if and only if the orientations of all simplices Δ_i^n are compatible; if $2a = 0$, then $\partial c_n = 0$. Therefore, if M^n is orientable, then $H_n(M^n; G) \cong G$, and if M^n is nonorientable, then $H_n(M^n; G) \cong \operatorname{Ker}(G \xrightarrow{\times 2} G)$. □

For any closed pseudomanifold M^n, the homology class of the cycle $\sum \Delta_i^n$ in $H_n(M^n; \mathbb{Z}_2)$ is called the *fundamental class* of M^n (more formally, this cycle should be written as $\sum 1 \cdot \Delta_i^n$, where $1 \in \mathbb{Z}_2$). The term *fundamental class* is also used for the homology class of the cycle $\sum \Delta_i^n$ in $H_n(M^n; \mathbb{Z})$ provided that the closed pseudomanifold M^n is oriented (the orientations of all simplices Δ_i^n are compatible with that of M^n). The fundamental class of a pseudomanifold M^n is denoted by $[M^n]$; for the coefficient group $\mathbb{Z}_2$, the notation $[M^n]_2$ is sometimes used.

Problem 27. Prove that if M^n is a closed orientable manifold, then the assertion of Problem 10 on p. 20 remains valid for $k = n - 1$. Is it essential that the manifold M^n is closed? orientable?

For pseudomanifolds with boundary, the same argument proves that

$$H_n(M^n, \partial M^n; G) \cong \begin{cases} G & \text{if } M^n \text{ is orientable,} \\ \operatorname{Ker}(G \xrightarrow{\times 2} G) & \text{if } M^n \text{ is nonorientable.} \end{cases}$$

Indeed, if a simplex Δ^{n-1} belongs to ∂M^n, then its contribution to the relative chain is zero. The definition of fundamental classes in relative homology is the same as in absolute homology.

Let M^n and N^n be closed oriented pseudomanifolds. A map $f \colon M^n \to N^n$ takes the fundamental class $[M^n]$ to an element of the group $H_n(N^n; \mathbb{Z})$, and any element of this group has the form $k[N^n]$, where $k \in \mathbb{Z}$. The integer $\deg f$ for which $(\deg f)[N^n] = f_*([M^n])$ is called the *degree* of the map f. If the pseudomanifolds M^n and N^n are closed (but not necessarily orientable), then the *degree* of f *modulo* 2 can be defined.

Remark. This definition is equivalent to that given in Part I on p. 111, where the degree of a map was defined as the number of preimages of points counted with signs.

If M^n and N^n are pseudomanifolds with boundary, then we can also define the degree (or degree modulo 2) of a map $f\colon (M^n, \partial M^n) \to (N^n, \partial N^n)$ by considering the image of the fundamental class $[M^n]$ in the group $H_n(N^n, \partial N^n; \mathbb{Z})$ (or in $H_n(N^n, \partial N^n; \mathbb{Z}_2)$).

Theorem 1.24. *Let* $i_*\colon H_n(\partial W^{n+1}) \to H_n(W^{n+1})$ *be the homomorphism induced by the natural embedding. If the pseudomanifold* W^{n+1} *is orientable and its boundary* ∂W^{n+1} *is connected, then* $i_* = 0$ *for the coefficient groups* $\mathbb{Z}$ *and* $\mathbb{Z}_2$*, and if it is nonorientable, then* $i_* = 0$ *for the coefficient group* $\mathbb{Z}_2$.

Proof. The group $H_n(\partial W^{n+1})$ is cyclic; therefore, it suffices to show that the fundamental cycle $[\partial W^{n+1}]$ is the boundary of some cycle in W^{n+1}. Clearly, it is the boundary of the fundamental cycle $[W^{n+1}]$. If the pseudomanifold W^{n+1} is nonorientable, then its fundamental cycle is defined only for the coefficient group $\mathbb{Z}_2$. □

Corollary. *If* $f\colon \partial W^{n+1} \to M^n$ *is the restriction of some continuous map* $F\colon W^{n+1} \to M^n$*, the pseudomanifold* W^{n+1} *is orientable, and its boundary* ∂W^{n+1} *is connected, then* $\deg f = 0$.

Proof. The homomorphism $f_*\colon H_n(\partial W^{n+1}) \to H_n(M^n)$ can be represented as the composition $f_* = F_* i_*$; hence $f_* = 0$. □

Problem 28. Let $f\colon S^n \to S^n$ be a map, and let $\Sigma f\colon \Sigma S^n \to \Sigma S^n$ be the map whose restriction to $S^n \times \{t\}$ coincides with f for each t. Prove that $\deg f = \deg \Sigma f$ by using the suspension isomorphism.

5.2. Cellular Homology. If a simplicial complex K is endowed with the structure of a CW-complex (see Part I, p. 118) X and each skeleton X^k is a subcomplex in K, then the integral homology of K can be calculated by using the chain complex C_* in which every C_k is the free group whose generators are in one-to-one correspondence with the k-cells of the complex X. This substantially facilitates calculating the homology because the number of cells in X may be many times smaller than the number of simplices in K. For example, a minimal triangulation of the sphere S^n contains $2^{n+1}-1$ simplices (of all dimensions), while a minimal cell decomposition consists of only two cells (n-dimensional and zero-dimensional).

Lemma. *For $i \neq k$, $H_i(X^k, X^{k-1}) = 0$, and $H_k(X^k, X^{k-1})$ is the free Abelian group whose generators are in one-to-one correspondence with the k-cells of X.*

Proof. Any chain $c_i \in C_i(X^k, X^{k-1})$ has a unique representation in the form of the (finite) sum of chains $c_{i\alpha}$, where the $\{e^k_\alpha\}$ are the k-cells of X. Moreover, $\partial c_i = \sum \partial c_{i\alpha}$. Any group $C_i(e^k_\alpha, \partial e^k_\alpha)$ has the form $C_i(D^k, S^{k-1})$ for some triangulation of the disk D^k. Therefore, $H_i(X^k, X^{k-1})$ is the direct sum of the groups $H_i(D^k, S^{k-1})$. □

The exact sequence

$$H_{i+1}(X^{k+1}, X^k) \longrightarrow H_i(X^k) \longrightarrow H_i(X^{k+1}) \longrightarrow H_i(X^{k+1}, X^k)$$

for the pair (X^{k+1}, X^k) shows that $H_i(X^k) \cong H_i(X^{k+1})$ for $i \neq k, k+1$. Hence $H_{k-1}(X^k) \cong H_{k-1}(X^{k+1}) \cong H_{k-1}(X^{k+2}) \cong \dots \cong H_{k-1}(X)$.

Let $C_k = H_k(X^k, X^{k-1})$. Consider the map $\partial_k \colon C_k \to C_{k-1}$ defined as follows. We arrange the exact sequences for the pairs (X^k, X^{k-1}) and (X^{k-1}, X^{k-2}) along horizontal and vertical lines:

$$\begin{array}{ccccccccc}
 & & & & 0 & & & & \\
 & & & & \downarrow & & & & \\
0 \longrightarrow & H_k(X^k) & \longrightarrow & C_k & \longrightarrow H_{k-1}(X^{k-1}) & \longrightarrow & H_{k-1}(X^k) & \longrightarrow & 0 \\
 & & & & \downarrow & & & & \\
 & & & & C_{k-1} & & & & \\
 & & & & \downarrow & & & & \\
 & & & & H_{k-2}(X^{k-2}). & & & &
\end{array}$$

The map $\partial_k \colon C_k \to C_{k-1}$ is the composition of the horizontal and vertical arrows. The group $H_k(X^k)$ can be regarded as a subgroup in C_k. Moreover, we can identify $H_k(X^k)$ with $\operatorname{Ker} \partial_k$ because if $\varphi = h\varphi'$ for a monomorphism h, then $\operatorname{Ker} \varphi = \operatorname{Ker} \varphi'$. Identifying $H_{k-1}(X^{k-1})$ with $\operatorname{Ker} \partial_{k-1}$ in a similar way, we obtain $H_{k-1}(X^k) \cong \operatorname{Ker} \partial_{k-1} / \operatorname{Im} \partial_k$.

The map $\partial_k \colon C_k \to C_{k-1}$ acts as follows. Suppose that to a generator $c^k_\alpha \in C_k$ corresponds the cell e^k_α. Consider an absolute chain $[e^k_\alpha]$ representing the relative fundamental class $(e^k_\alpha, \partial e^k_\alpha)$. At the level of $H_{k-1}(X^{k-1})$, we have $\partial [e^k_\alpha] = \sum n_{\alpha\beta} [e^{k-1}_\beta]$. Thus, $\partial c^k_\alpha = \sum n_{\alpha\beta} c^{k-1}_\beta$. The number $n_{\alpha\beta}$ is equal to the degree of the map $S^{k-1} \to e^{k-1}_\beta / \partial e^{k-1}_\beta = S^{k-1}$ induced by the characteristic map of the cell e^k_α.

The equality $\partial\partial = 0$ follows from $\partial\partial[e^k_\alpha] = 0$.

The homology with coefficients in $\mathbb{Z}_2$ is calculated similarly; in this case, the degree modulo 2 should be considered.

As an example, we calculate the homology of closed two-dimensional surfaces. We have already calculated the groups H_0 and H_2; let us calculate H_1.

Example 7. $H_1(nT^2) = \mathbb{Z}^{2n}$ and $H_1(nT^2; \mathbb{Z}_2) = \mathbb{Z}_2^{2n}$ (nT^2 is a special case of the surface $S^2 \# pT^2 \# qK \# rP^2$ defined in Part I on p. 142).

Proof. Consider the standard representation of nT^2 by using a $4n$-gon. For this representation, the complex for calculating the cellular homology has the form

$$\mathbb{Z} \xrightarrow{\partial_2} \mathbb{Z}^{2n} \xrightarrow{\partial_1} \mathbb{Z} \longrightarrow 0.$$

Here $\partial_1 = 0$ because there is one zero-dimensional cell, and $\partial_2 = 0$ because each 1-cell occurs two times with opposite orientations in the boundary of the 2-cell.

For the coefficient group $\mathbb{Z}_2$, the calculation is similar. □

For a sphere with n handles, the $2n$ cycles generating the one-dimensional homology group can be chosen as shown in Figure 3. To prove this, we must

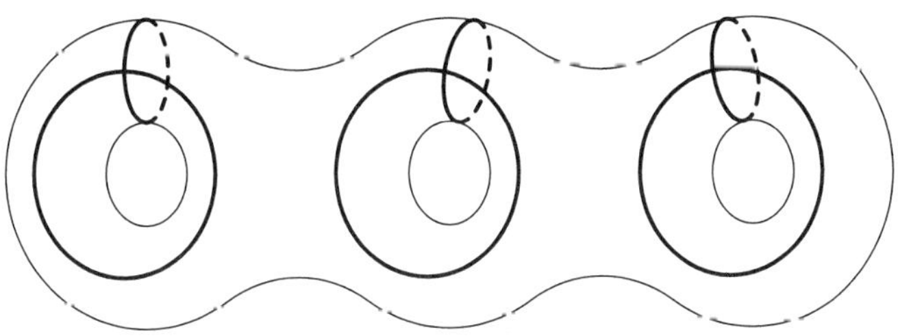

Figure 3. Basis cycles

verify that these cycles are homologous to the sides of the $4n$-gon from which the sphere with handles is obtained. First, consider a handle from which a disk is removed (see Figure 4). Clearly, to the cycles in Figure 3 correspond those in Figure 5; the latter cycles are homologous to the sides of the polygon.

Example 8. $H_1(mP^2) = \mathbb{Z}^{m-1} \oplus \mathbb{Z}_2$ and $H_1(mP^2; \mathbb{Z}_2) = \mathbb{Z}_2^m$.

Proof. The complex for calculating the cellular homology of mP^2 has the form

$$\mathbb{Z} \xrightarrow{\partial_2} \mathbb{Z}^{m} \xrightarrow{\partial_1} \mathbb{Z} \longrightarrow 0.$$

We again have $\partial_1 = 0$, but this time, $\partial_2(1) = (2, \dots, 2)$ because each 1-cell occurs two times with the same orientation in the boundary of the 2-cell. We

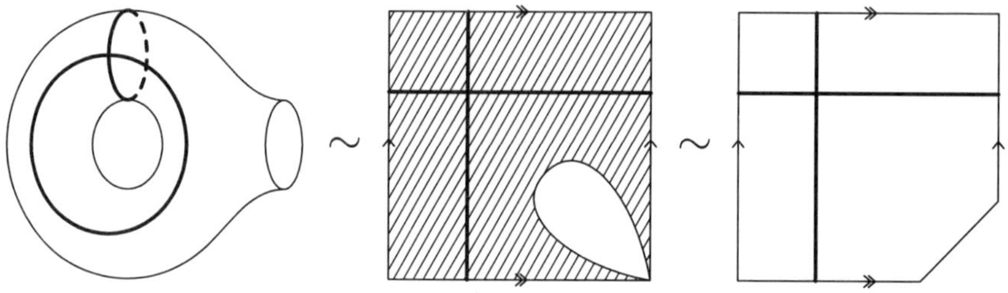

Figure 4. Basis cycles on a handle

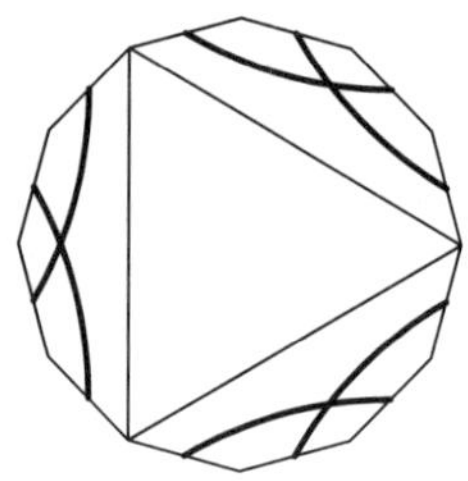

Figure 5. Basis cycles on the polygon

must take the quotient group of $\mathbb{Z}^m$ modulo the subgroup generated by the element $(2, \ldots, 2)$. Consider $e_1 = (1, 0, \ldots, 0), \ldots, e_{m-1} = (0, \ldots, 0, 1, 0)$, and $e_m = (1, \ldots, 1)$. We have

$$(a_1, \ldots, a_m) = (a_1 - a_m)e_1 + \cdots + (a_{m-1} - a_m)e_m + a_m e_m.$$

The condition $(a_1, \ldots, a_m) \in \operatorname{Im} \partial_2$ is equivalent to $a_1 - a_m = 0, \ldots, a_{m-1} - a_m = 0$ and $a_m \equiv 0 \pmod 2$. Therefore, the quotient group $\mathbb{Z}^m / \operatorname{Im} \partial_2$ is isomorphic to $\mathbb{Z}^{m-1} \oplus \mathbb{Z}_2$; the group $\mathbb{Z}^{m-1}$ is generated by $e_1, \ldots, e_{m-1}$, and the group $\mathbb{Z}_2$ is generated by e_m. As before, for the coefficient group $\mathbb{Z}_2$, we have $\partial_2 = 0$. □

Cycles α and β generating the integral homology group for the Klein bottle are shown in Figure 6. At the level of homology, we have $2\beta = 0$ because the boundary of the Möbius band hatched in Figure 6a is homologous to 2β.

Problem 29. Calculate the homology groups of closed two-dimensional surfaces with coefficients in $\mathbb{Z}_p$ for $p \neq 2$.

Problem 30. Calculate the cohomology groups of closed two-dimensional surfaces with coefficients in $\mathbb{Z}$.

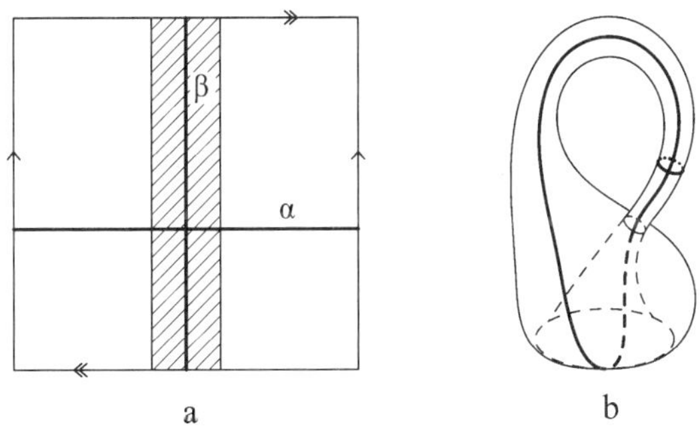

Figure 6. Basis cycles on the Klein bottle

Problem 31. Let H be a handlebody of genus g, that is, a three-dimensional body bounded by a sphere with g handles standardly embedded in $\mathbb{R}^3$. Suppose that H is smoothly embedded in S^3 and $X = \overline{S^3 \setminus H}$. Prove that $H_1(X) \cong \mathbb{Z}^g$; describe geometrically the generators of this group.

Example 9. The following equality holds:

$$H_k(\mathbb{C}P^n) = \begin{cases} \mathbb{Z} & \text{if } k = 0, 2, 4, \ldots, 2n, \\ 0 & \text{otherwise.} \end{cases}$$

Proof. The space $\mathbb{C}P^n$ has the structure of a CW-complex with cells $\mathbb{C}P^k \setminus \mathbb{C}P^{k-1}$ (see Part I, pp. 118–119). We can construct a triangulation of the manifold $\mathbb{C}P^n$ for which every skeleton $X^{2k+1} = X^{2k} = \mathbb{C}P^k$ is a simplicial subcomplex in the same way as the triangulation of a manifold with boundary was constructed in Part I.

The chain complex for calculating the cellular homology of the space $\mathbb{C}P^n$ has the form

$$\mathbb{Z} \longrightarrow 0 \longrightarrow \mathbb{Z} \longrightarrow 0 \longrightarrow \cdots \longrightarrow 0 \longrightarrow \mathbb{Z} \longrightarrow 0 \longrightarrow \mathbb{Z} \longrightarrow 0.$$

The homology groups of this chain complex coincide with the chain groups. □

Example 10. (a) For $k > 0$, $H_{2k}(\mathbb{R}P^n) = 0$, and for $0 \le k < \frac{n-1}{2}$, $H_{2k+1}(\mathbb{R}P^n) = \mathbb{Z}_2$.

(b) For $0 \le k \le n$, $H_k(\mathbb{R}P^n; \mathbb{Z}_2) = \mathbb{Z}_2$.

Proof. Suppose that $\mathbb{R}P^n$ is endowed with the structure of the CW-complex with cells $\mathbb{R}P^k \setminus \mathbb{R}P^{k-1}$ (see Part I, pp. 118–119). Then $C_k = \mathbb{Z}$ for $0 \le k \le n$. The central symmetry of the sphere S^{k-1} preserves orientation for even k and reverses it for odd k. Therefore, the boundary homomorphism $\partial_k \colon C_k \to C_{k-1}$ takes 1 to $1 + 1 = 2$ for even k and to $1 - 1 = 0$

for odd k. The chain complex for calculating the cellular homology of the complex $\mathbb{R}P^n$ for even n has the form

$$\mathbb{Z} \xrightarrow{\times 2} \mathbb{Z} \xrightarrow{0} \mathbb{Z} \xrightarrow{\times 2} \mathbb{Z} \longrightarrow \cdots \longrightarrow \mathbb{Z} \xrightarrow{\times 2} \mathbb{Z} \xrightarrow{0} \mathbb{Z} \longrightarrow 0;$$

for odd n, it has the form

$$\mathbb{Z} \xrightarrow{0} \mathbb{Z} \xrightarrow{\times 2} \mathbb{Z} \longrightarrow \cdots \longrightarrow \mathbb{Z} \xrightarrow{\times 2} \mathbb{Z} \xrightarrow{0} \mathbb{Z} \longrightarrow 0.$$

Therefore,

$$H_k(\mathbb{R}P^n) = \begin{cases} \mathbb{Z} & \text{if } k = 0 \text{ or } k = n \text{ and } n \text{ is odd;} \\ \mathbb{Z}_2 & \text{is } k \text{ is odd and } 0 < k < n; \\ 0 & \text{otherwise.} \end{cases}$$

For coefficients in $\mathbb{Z}_2$, all boundary homomorphisms are zero. □

Problem 32. Calculate the integral homology groups of the n-torus T^n.

Problem 33. Calculate the integral cohomology groups of the complex $\mathbb{R}P^n$.

Problem 34. Give an example of a noncontractible acyclic two-dimensional CW-complex with 1-skeleton $S^1 \vee S^1$.

5.3. The Intersection Number and the Poincaré Duality Isomorphism. Suppose that a closed oriented manifold M^n is decomposed into cells in two different ways so that there is a one-to-one correspondence between the open k-cells of one decomposition and the open $(n-k)$-cells of the other decomposition. We denote the corresponding open cells by σ_i and σ_i^*, respectively. Suppose also that $\sigma_i \cap \sigma_j^* = \varnothing$ for $i \neq j$ and the cells σ_i and σ_i^* transversally intersect each other in one point for all i. We assume that the cells σ_i and σ_i^* are oriented so that if $e_1, \dots, e_k$ and $\varepsilon_1, \dots, \varepsilon_{n-k}$ are positively oriented bases[10] for σ_i and σ_i^*, then $e_1, \dots, e_k, \varepsilon_1, \dots, \varepsilon_{n-k}$ is a positively oriented basis for M^n. Finally, suppose that the coefficient group is the additive group of an associative commutative ring R with identity. In this case, we can consider the *intersection number* for the chains $\sum a_i\sigma_i$ and $\sum b_i\sigma_i^*$; it is defined as

$$\left\langle\left\langle \sum a_i\sigma_i, \sum b_i\sigma_i^* \right\rangle\right\rangle = \sum \delta_{ij}a_ib_j = \sum a_ib_i.$$

The required pair of cell decompositions K and K^* can be constructed, e.g., as follows. For K we take an arbitrary triangulation of the manifold M^n. Let K' be the barycentric subdivision of this triangulation. For each simplex Δ^k in the triangulation of K, take all simplices from K' that intersect Δ^k

[10] For a simplex $[v_0, \dots, v_k]$, the basis $v_1 - v_0, \dots, v_k - v_0$ is oriented positively.

precisely in its barycenter. Their union is a closed $(n-k)$-cell; we denote it by $(\Delta^k)^*$. These cells form the decomposition K^*.

Using the intersection number, we can construct a bilinear map

$$\varphi\colon H_k(M^n;R)\times H_{n-k}(M^n;R)\to R.$$

Namely, we take the cell decompositions K and K^*, choose the cycles $\sum a_i\sigma_i$ and $\sum b_i\sigma_i^*$ representing homology classes $\alpha\in H_k(M^n;R)$ and $\beta\in H_{n-k}(M^n;R)$, and define $\varphi(\alpha,\beta)$ to be the intersection number of these chains (cycles). We must verify that this intersection number does not depend on the cycles representing α and β.

Lemma. *If Δ_i and Δ_j are simplices of dimensions k and $k-1$, respectively, then*

$$\langle\langle\partial\Delta_i,\Delta_j^*\rangle\rangle=(-1)^k\langle\langle\Delta_i,\partial(\Delta_j^*)\rangle\rangle.$$

Proof. First, note that the expressions on both sides of this equality are nonzero if and only if Δ_j is a face of the simplex Δ_i. Moreover, $\langle\langle\partial\Delta_i,\Delta_j^*\rangle\rangle=\langle\langle\pm\Delta_j,\Delta_j^*\rangle\rangle=\pm1$ and $\langle\langle\Delta_i,\partial(\Delta_j^*)\rangle\rangle=\langle\langle\Delta_i,\pm\Delta_i^*\rangle\rangle=\pm1$. It remains to determine the signs of Δ_j in $\partial\Delta_i$ and of Δ_i^* in $\partial(\Delta_j^*)$.

Suppose that v_0 is the barycenter of the simplex Δ_i and v_0' is the barycenter of Δ_j. Then Δ_i and $\partial(\Delta_j^*)$ intersect in the point v_0, and $\partial\Delta_i$ and Δ_j^* intersect in v_0'. Let $v_1,\dots,v_k$ be the vertices of Δ_j. We can assume that the orientations of the simplices $[v_0,v_1,\dots,v_k]$ and Δ_i are compatible and, moreover, $\partial\Delta_i=\Delta_j+\cdots$, i.e., $\Delta_j=[v_1,\dots,v_k]$. Suppose that $[v_0',v_0,v_{k+1},\dots,v_n]$ is one of the simplices that form the cell Δ_j^*, and its orientation is compatible with that of this cell. Then $\partial(\Delta_j^*)=[v_0,v_{k+1},\dots,v_n]+\cdots$, where the dots denote the cells disjoint from Δ_i.

Augmenting a positively oriented basis of the simplex Δ_i by a positively oriented basis of the cell $\partial(\Delta_j^*)$, we obtain a positively oriented basis of the simplex $[v_0,v_1,\dots,v_n]$. We now find out what is the result of augmenting a positively oriented basis of $\Delta_j=[v_1,\dots,v_k]$ by a positively oriented basis of the simplex $[v_0',v_0,v_{k+1},\dots,v_n]$. First, note that the simplex $[v_0',v_2,\dots,v_k]$ is contained in the simplex $[v_1,\dots,v_k]$, and the orientations of these simplices are compatible. Thus, we obtain a positively oriented basis of the simplex

$$[v_0',v_2,\dots,v_k,v_{k+1},\dots,v_n]=(-1)^k[v_0,v_0',v_2,\dots,v_n].$$

It remains to note that $[v_0,v_0',v_2,\dots,v_n]$ is contained in $[v_0,v_1,v_2,\dots,v_n]$, and the orientations of these simplices are compatible. □

It follows from this lemma that the map $\varphi(\alpha,\beta)$ is well defined because if $c_1=\partial c'$ and $\partial c_2=0$, then $\langle\langle c_1,c_2\rangle\rangle=\langle\langle\partial c',c_2\rangle\rangle=\pm\langle\langle c',\partial c_2\rangle\rangle=0$.

Using the cell decompositions K and K^*, we can obtain a correspondence between the chains $\sum a_i\sigma_i$ and $\sum a_i\sigma_i^*$. However, this correspondence is not interesting from the point of view of homology because the chains $(\partial\sigma)^*$ and $\partial(\sigma^*)$ belong to different spaces (if $\dim\sigma = k$, then $\dim(\partial\sigma)^* = n-k+1$, whereas $\dim\partial(\sigma^*) = n-k-1$). It is more natural to associate each chain $c_k = \sum a_i\sigma_i$ with a cochain c^{n-k} for which $\langle c^{n-k}, \sigma_i^*\rangle = a_i$, i.e.,

$$\langle c^{n-k}, \sum b_i\sigma_i^*\rangle = \sum a_ib_i = \langle\langle c_k, \sum b_i\sigma_i^*\rangle\rangle.$$

(In other words, the value of the cochain c^{n-k} at the given chain is equal to the intersection number of c^{n-k} with c_k; such a cochain exists and is unique.) We have

$$\begin{aligned}\langle\delta c^{n-k}, d_{n-k+1}\rangle &= \langle c^{n-k}, \partial d_{n-k+1}\rangle \\ &= \langle\langle c_k, \partial d_{n-k+1}\rangle\rangle = (-1)^k\langle\langle\partial c_k, d_{n-k+1}\rangle\rangle;\end{aligned}$$

i.e., associated with each chain of the form ∂c_k is the cochain $\pm\delta c^{n-k}$. This means, in particular, that to cycles and boundaries correspond cocycles and coboundaries, respectively. Therefore, this isomorphism between the space of k-chains in K and the space of $(n-k)$-cochains in K^* induces an isomorphism between the groups $H_k(M^n; R)$ and $H^{n-k}(M^n; R)$.

For the coefficient group $\mathbb{Z}_2$, we do not have to worry about orientations; therefore, for closed but not necessarily orientable manifolds, we can consider intersection numbers modulo 2 and obtain isomorphism $H_k(M^n; \mathbb{Z}_2) \cong H^{n-k}(M^n; \mathbb{Z}_2)$.

We have proved the following assertion.

Theorem 1.25 (the Poincaré duality isomorphism). (a) *If M^n is a closed orientable manifold and R is the additive group of a unital associative commutative ring, then $H_k(M^n; R) \cong H^{n-k}(M^n; R)$.*

(b) *If M^n is a closed manifold, then $H_k(M^n; \mathbb{Z}_2) \cong H^{n-k}(M^n; \mathbb{Z}_2)$.*

Corollary. (a) *If M^n is a closed orientable manifold and F is the additive group of a field, then $H_k(M^n; F) \cong H_{n-k}(M^n; F)$.*

(b) *If M^n is a closed manifold, then $H_k(M^n; \mathbb{Z}_2) \cong H_{n-k}(M^n; \mathbb{Z}_2)$.*

The subgroup formed by the finite-order elements in a group is called the *torsion subgroup* of this group.

Problem 35 (Poincaré duality). Suppose that M^n is a closed orientable manifold, $H_k(M^n) \cong \mathbb{Z}^{a_k} \oplus T_k$, and $H^k(M^n) \cong \mathbb{Z}^{b_k} \oplus T^k$, where T_k and T^k are the torsion subgroups. Prove that $a_k = a_{n-k}$, $b_k = b_{n-k}$, $T_k \cong T_{n-k-1}$, $T^k \cong T^{n-k+1}$, and $T^1 = 0$.

The Poincaré duality isomorphism and the universal coefficient theorem imply the following properties of the homology and cohomology groups of manifolds.

Theorem 1.26. (a) *If M^n is a closed (connected) manifold and R is the additive group of a unital associative commutative ring, then*

$$H^n(M^n; R) \cong \begin{cases} R & \text{if } M^n \text{ is orientable,} \\ R/2R & \text{if } M^n \text{ is nonorientable.} \end{cases}$$

(b) *If M^n is a closed orientable manifold, then the group $H_{n-1}(M^n; \mathbb{Z})$ is torsion-free, and if M^n is a closed connected (nonorientable) manifold, then the torsion subgroup of $H_{n-1}(M^n; \mathbb{Z})$ is isomorphic to $\mathbb{Z}_2$.*

Proof. The group $Z^n(M^n; R)$ is generated by cochains dual to n-simplices in M^n with the same orientation. Moreover, two cochains dual to n-simplices with the same orientation sharing an $(n-1)$-face differ by a coboundary. Therefore, we have $H^n(M^n; R) \cong R$ if the manifold is orientable and $H^n(M^n; R) \cong R/2R$ otherwise; indeed, in the nonorientable case, every cochain c^n dual to a simplex satisfies the relation $2c^n = 0$.

For orientable manifolds, the Poincaré duality isomorphism implies $T_{n-1} \cong T^1 = 0$, and for nonorientable manifolds, Problem 24 implies $T_{n-1} \cong T^n = \mathbb{Z}_2$. □

Exercise. Let M^4 be a closed orientable manifold. Prove that if the group $H_1(M^4)$ is torsion-free, then so are all other groups $H_k(M^4)$.

Problem 36. Suppose that M^n is a closed orientable manifold and its suspension ΣM^n is homeomorphic to a closed orientable manifold. Prove that M^n is a *homology sphere*, i.e., $H_k(M^n) \cong H_k(S^n)$ for all k.

Remark. A homology sphere may or may not be homeomorphic to the sphere. The best-known homology sphere not homeomorphic to the sphere is the Poincaré 3-sphere; its various descriptions can be found, e.g., in [**108**]. There are infinitely many pairwise nonhomeomorphic homology 3-spheres. Interestingly, the double suspension over any homology 3-sphere is homeomorphic to the sphere S^5 (this was proved by Edwards).

The bilinear map $\varphi\colon H_k(M^n) \times H_{n-k}(M^n) \to \mathbb{Z}$ constructed by using the intersection number has the following nondegeneracy property.

Theorem 1.27. *If M^n is a closed oriented manifold, then any homomorphism $h\colon H_k(M^n) \to \mathbb{Z}$ can be represented as $h(\alpha) = \varphi(\alpha, \beta_h)$, where $\beta_h \in H_{n-k}(M^n)$. The element β_h is determined by the homomorphism h uniquely up to a finite-order element.*

Proof. The exact sequence

$$0 \longrightarrow \operatorname{Ext}(H_{k-1}(M^n), \mathbb{Z}) \longrightarrow H^k(M^n; \mathbb{Z}) \longrightarrow \operatorname{Hom}(H_k(M^n), \mathbb{Z}) \longrightarrow 0$$

shows that the homomorphism h corresponds to an element of the group $H^k(M^n;\mathbb{Z})$ determined up to an element of the group $\operatorname{Ext}(H_{k-1}(M^n),\mathbb{Z})$; this group is finite. Clearly, $h(\alpha) = \varphi(\alpha,\beta)$ for all α if and only if the cochain corresponding to a representative of the homology class β belongs to the cohomology class mapped to h.

Note that if $m\beta = 0$, then $m\varphi(\alpha,\beta) = \varphi(\alpha, m\beta) = 0$; therefore, $\varphi(\alpha,\beta) = 0$. □

If M^n is a closed orientable manifold and the coefficient group F is the additive group of some field (or $F = \mathbb{Z}_2$ if M^n nonorientable), then the bilinear map $\varphi\colon H_k(M^n;F)\times H_{n-k}(M^n;F)\to F$ is nondegenerate in the usual sense.

Theorem 1.28. *Any linear map $h\colon H_k(M^n;F)\to F$ can be represented as $h(\alpha) = \varphi(\alpha,\beta_h)$, where $\beta_h \in H_{n-k}(M^n;F)$ is determined uniquely.*

Proof. The linear map h is an element of the space dual to $H_k(M^n;F)$, that is, of $H^k(M^n;F)$, which is isomorphic to $H_{n-k}(M^n;F)$. The isomorphism takes h to β_h. □

Corollary. *In the spaces $H_k(M^n;F)$ and $H_{n-k}(M^n;F)$, there exist bases $\alpha_1,\dots,\alpha_m$ and $\beta_1,\dots,\beta_m$ such that $\langle\langle\alpha_i,\beta_j\rangle\rangle = \delta_{ij}$.*

Proof. We choose an arbitrary basis $\alpha_1,\dots,\alpha_m$ and take the element corresponding to the linear map $h_j\left(\sum x_i\alpha_i\right) = x_j$ for β_j. For these bases, we have $\langle\langle\alpha_i,\beta_j\rangle\rangle = \varphi(\alpha_i,\beta_j) = h_j(\alpha_i) = \delta_{ij}$. □

Similarly, if M^n is a closed orientable manifold, then the free groups $H_k(M^n)/T_k$ and $H_{n-k}(M^n)/T_{n-k}$, where T_k and T_{n-k} are the torsion subgroups, have bases $\alpha_1,\dots,\alpha_m$ and $\beta_1,\dots,\beta_m$ for which $\langle\langle\alpha_i,\beta_j\rangle\rangle = \delta_{ij}$.

Linking Number. Let c_1 and c_2 be closed oriented curves in S^3. They determine one-dimensional cycles (with coefficients in $\mathbb{Z}$), which we also denote by c_1 and c_2. Let d_1 be a two-dimensional chain such that $\partial d_1 = c_1$. The intersection number $\langle\langle d_1, c_2\rangle\rangle$ is called the *linking number* of the curves c_1 and c_2 and denoted by $\operatorname{lk}(c_1,c_2)$. It is easy to verify that the intersection number $\langle\langle d_1,c_2\rangle\rangle$ does not depend on the choice of d_1. Indeed, if d_1' is another chain for which $\partial d_1' = \partial d_1$, then $d_1' - d_1$ is a cycle and, hence, a boundary (because the two-dimensional homology group of the sphere S^3 is trivial). Let e be a three-dimensional chain for which $d_1' - d_1 = \partial e$. Then $\langle\langle d_1' - d_1, c_2\rangle\rangle = \langle\langle\partial e, c_2\rangle\rangle = \langle\langle e,\partial c_2\rangle\rangle = 0$.

It is easy to show that this definition of linking number is equivalent to that given in PSknots, say. In [**108**], the linking number of closed oriented curves J and K was defined as follows. Consider the diagram of the oriented link formed by these curves. Look at the crosses where J passes under K. There are two types of such crosses (see Figure 7). For each cross under consideration, we set $\varepsilon_i = 1$ or -1, depending on its type. The sum of all numbers ε_i thus obtained is called the linking number of the closed oriented curves J and K.

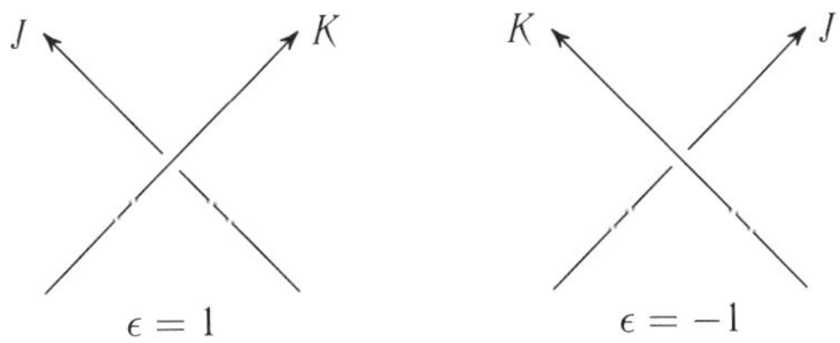

Figure 7. Two types of crosses

To show that the two definitions are equivalent, we arrange the curves c_1 and c_2 so that they leave the plane of the diagram only in small neighborhoods of the crosses. We can also assume that the surface d_1 is located above the plane of the diagram, except in small neighborhoods of the crosses (see Figure 8). We are interested only in the crosses where the curve c_2 passes

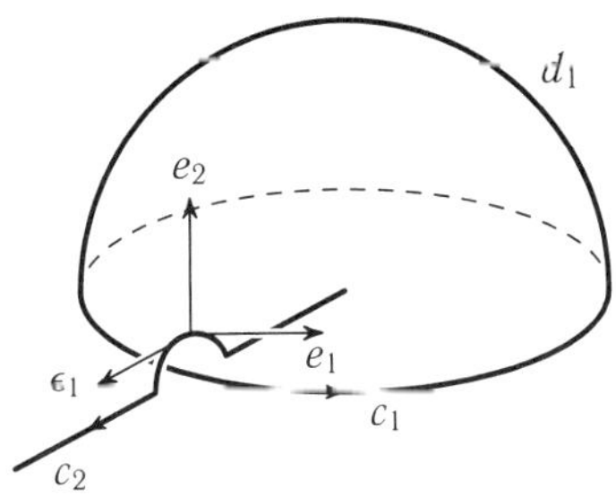

Figure 8. The surface d_1

over c_1. These are precisely the places where c_2 intersects the surface d_1. The signs do not cause any problem either. For example, in Figure 8, the basis e_1, e_2, ε_1 is oriented positively (we use the same notation as in the definition of the intersection number); this corresponds to a positive cross.

There are many other definitions of the linking number of curves J and K. For example, consider the map $f\colon J \times K \to S^2$ defined by

$$f(x, y) = \frac{K(y) - J(x)}{\|K(y) - J(x)\|}$$

(J and K are treated as circles and, simultaneously, as maps $J, K\colon S^1 \to \mathbb{R}^3$).

Problem 37. Prove that $\deg f = \pm \operatorname{lk}(J, K)$. Define orientations on the torus $J \times K$ and on the sphere S^2 so that $\deg f = \operatorname{lk}(J, K)$.

Cohomology with Compact Support and Homology with Closed Support. The Poincaré isomorphism theorem as stated above does not apply to noncompact manifolds. It is false even for the manifold $\mathbb{R}^n$. Nevertheless, for noncompact manifolds, we can construct dual decompositions and try to establish a correspondence between chains c_k and cochains c^{n-k}. Of course, such an effort cannot succeed, but we can see where it fails and we can try to remedy the situation.

Recall that, by definition, any chain c_k is a finite sum of simplices; therefore, when we assign a cochain c^{n-k} to a chain c_k in the case of a noncompact manifold, we obtain only cochains that vanish on simplices outside some compact set, rather than all cochains. Such cochains are called *cochains with compact supports.* Clearly, applying the operator δ to a cochain with compact support, we obtain a cochain with compact support. Hence we can define *cohomology groups* $H_c^k(K)$ *with compact supports* for compactly supported cochains. If K and K^* are dual decompositions of an orientable manifold, then we obtain an isomorphism $H_k(K; R) \cong H_c^{n-k}(K^*; R)$; this isomorphism holds even at the level of (co)chains.

Now, let us try to assign a chain c_k to a cochain c^{n-k}. When we assigned cochains to chains, we obtained something less than the set of all cochains. Now, on the contrary, we obtain something more than the set of all chains. Namely, instead of finite sums of simplices $\sum a_i \Delta_i^k$, we obtain arbitrary sums of k-simplices, in which each simplex is contained with a certain coefficient but the number of simplices may be infinite. Such "chains" are called *chains with closed supports* because the union of all simplices with nonzero coefficients in such a chain is a closed set (possibly noncompact).

Note that in defining chains with closed supports for any simplicial complexes, we must additionally require that each point belong to only finitely many simplices in the sum $\sum a_i \Delta_i^k$. Otherwise, some simplex may be a face for infinitely many simplices in the sum $\sum a_i \Delta_i^k$, and the chain $\partial(\sum a_i \Delta_i^k)$ is then not defined. But for triangulated manifolds, each simplex is a face for only finitely many simplices, and the two conditions are equivalent.

Using chains with closed supports, we can construct *homology groups* $H_k^{\mathrm{cl}}(K)$ *with closed supports.* If K and K^* are dual decompositions of an orientable manifold, we obtain an isomorphism $H_k^{\mathrm{cl}}(K; R) \cong H^{n-k}(K^*; R)$; this isomorphism holds even at the level of (co)chains.

Remark. Note that the groups $H_c^*(K)$ and $H_*^{\mathrm{cl}}(K)$, in contrast to the ordinary cohomology and homology groups, are not homotopy invariants of the space K. For example, $H_c^n(\mathbb{R}^n; \mathbb{Z}) \cong \mathbb{Z}$ and $H_n^{\mathrm{cl}}(\mathbb{R}^n; \mathbb{Z}) \cong \mathbb{Z}$.

5.4. Realization of Homology Classes of Surfaces. Any smooth curve $f\colon S^1 \to M^2$ induces a homomorphism $f_*\colon H_1(S^1) \to H_1(M^2)$. We say that the curve f *realizes* the homology class $f_*[S^1] \in H_1(M^2)$. It is fairly obvious that allowing self-intersecting curves, we can realize any homology class of a closed two-dimensional manifold M^2. For this reason, we study realization of homology classes by only self-avoiding curves. We consider only orientable manifolds.

We say that a homology class α over $\mathbb{Z}$ is *primitive* if $\alpha \neq m\beta$, where $m \in \mathbb{N}$, $m > 1$, and β is a homology class over $\mathbb{Z}$. Note that if $m\alpha = 0$, then $\alpha = (m+1)\alpha$; thus, an element of finite order cannot be primitive.

Theorem 1.29. *If a closed self-avoiding curve f realizes a homology class $\alpha \in H_1(M^2)$, where M^2 is a closed orientable manifold, then either $\alpha = 0$ or α is a primitive homology class.*

Proof (Samelson). If the curve f separates M^2, then $\alpha = 0$. Indeed, let us represent one of the parts into which f divides M^2 in the form of a two dimensional chain. The boundary of this chain is a representative of the homology class α or $-\alpha$.

If the curve f does not divide M^2, then there exists a closed self-avoiding curve g transversally intersecting f at precisely one point. To construct such a curve, it suffices to take a small interval transversally intersecting f and join its ends by a curve in $M^2 \setminus f$ (see Figure 9). Let γ be the homology class realized by g. Then $\langle\langle\alpha,\gamma\rangle\rangle = \pm 1$. On the other hand, if $\alpha = m\beta$, then $\langle\langle\alpha,\gamma\rangle\rangle = m\langle\langle\beta,\gamma\rangle\rangle$. □

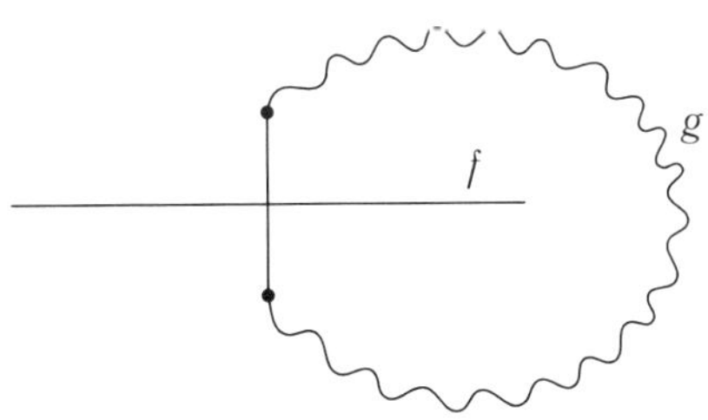

Figure 9. The construction of the curve g

Remark. The orientability of the manifold was used to define the integer $\langle\langle\alpha,\gamma\rangle\rangle$. The generator of the group $H_1(\mathbb{R}P^2) = \mathbb{Z}_2$ is not primitive, but it is realized by a closed self-avoiding curve.

Theorem 1.30 ([85]). *Any primitive homology class of a closed orientable manifold M^2 can be realized by a closed self-avoiding curve.*

Proof. To simplify notation, we denote curves and the homology classes they represent by the same symbols. Take curves $\alpha_1, \dots, \alpha_g$, $\beta_1, \dots, \beta_g$, and $\gamma_1, \dots, \gamma_{g-1}$ on the sphere with g handles as shown in Figure 10 (for $g = 4$). The homology classes $\alpha_1, \dots, \alpha_g, \beta_1, \dots, \beta_g$ generate the one-dimensional homology group. We denote the homology class $a_1\alpha_1 + \dots + a_g\alpha_g + b_g\beta_1 + \dots + b_g\beta_g$ by $(a_1, b_1, \dots, a_g, b_g)$. This class is primitive if and only if $\mathrm{GCD}(a_1, b_1, \dots, a_g, b_g) = 1$.

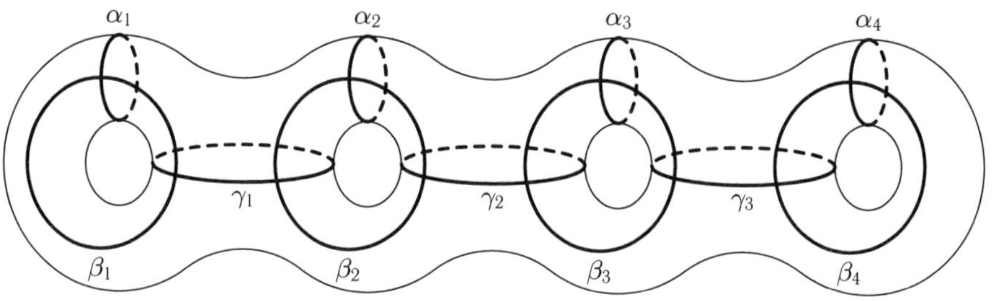

Figure 10. Curves on a sphere with handles

For each closed self-avoiding curve γ on a two-dimensional orientable surface M^2, a homeomorphism $M^2 \to M^2$, called the *Dehn twist* along γ, is defined. Namely, consider a small neighborhood of γ and choose one of the two parts into which it divides γ (in Figure 11, left, the chosen part is hatched). Let us make a cut along the curve γ and deform the chosen part of the neighborhood so that the points of the outer boundary remain fixed and the points of γ slide on γ. After a rotation through 360°, the points of the cut return to their initial positions; as a result, we obtain a homeomorphism. The Dehn twists are discussed in detail in [**105**, **108**].

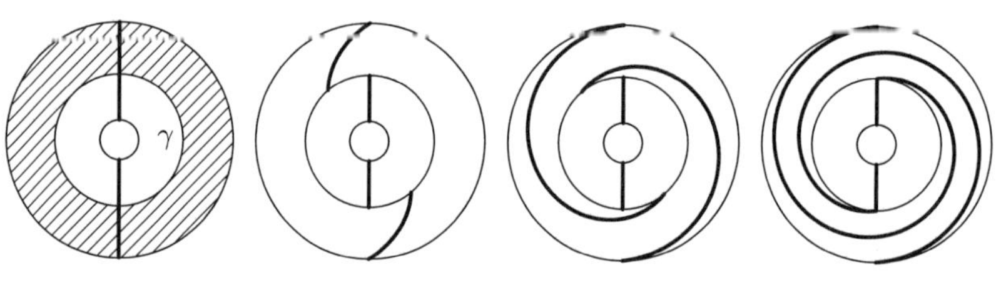

Figure 11. A Dehn twist

If a curve α transversally intersects γ in one point, then any Dehn twist along γ takes the homology class of α to $\alpha \pm \gamma$ (the sign is determined by the direction of rotation in the Dehn twist). It is easy to verify that the homology class of γ_k is $\pm(\alpha_k - \alpha_{k+1})$ (see Figure 12). Thus, the Dehn twists

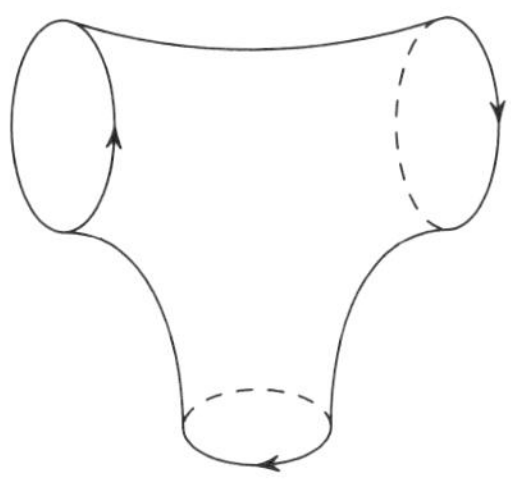

Figure 12. Homologous cycles

act on homology classes as follows:

$$\alpha_k \xrightarrow{\beta_k} \alpha_k \pm \beta_k, \quad \beta_k \xrightarrow{\alpha_k} \beta_k \pm \alpha_k;$$
$$\beta_k \xrightarrow{\gamma_k} \beta_k \pm (\alpha_k - \alpha_{k+1}), \quad \beta_{k+1} \xrightarrow{\gamma_k} \beta_{k+1} \pm (\alpha_k - \alpha_{k+1}).$$

Let $(a_1, b_1, \dots, a_g, b_g)$ be an element of the group $H_1(M^2)$. By using Euclid's algorithm and employing twists along α_1 and β_1, we can transform this element into $(d_1, 0, a_2, b_2, \dots, a_g, b_g)$, where $d_1 = \pm\,\mathrm{GCD}(a_1, b_1)$. The element thus obtained can be transformed into $(d_1, 0, d_2, 0, \dots, d_g, 0)$, where $d_k = \pm\,\mathrm{GCD}(a_k, b_k)$, by similar procedures.

Using the operations $(0, d_k) \xrightarrow{\beta_k} (d_k, d_k) \xrightarrow{\alpha_k} (d_k, 0)$ and $(d_k, 0, 0, d_{k+1}) \xrightarrow{\gamma_k} (d_k, \pm d_k, 0, d_{k+1} \pm d_k)$, we can apply Euclid's algorithm to the numbers d_k and d_{k+1}. As a result, we successively obtain the elements

$$(0, 0, \pm\,\mathrm{GCD}(d_1, d_2), 0, d_3, \dots), \dots, (0, 0, \dots, 0, \pm\,\mathrm{GCD}(d_1, \dots, d_g), 0).$$

If the initial element $(a_1, b_1, \dots, a_g, b_g)$ is primitive, then

$$\mathrm{GCD}(d_1, \dots, d_g) = \pm 1;$$

hence the resulting element is represented by one of the curves $\pm\alpha_g$. This means that there exists a homeomorphism $h\colon M^2 \to M^2$ such that $h_*\colon H_1(M^2) \to H_1(M^2)$ takes each element $(a_1, \dots, b_g)$ to an element represented by a self avoiding curve. The image of this curve under the homeomorphism h^{-1} represents the element $(a_1, \dots, b_g)$. □

6. The Euler Characteristic and the Lefschetz Theorem

6.1. The Euler Characteristic. Let K be a finite simplicial complex of dimension n. Its *Euler characteristic* is defined as

$$\chi(K) = a_0 - a_1 + a_2 - \dots + (-1)^n a_n,$$

where a_i is the number of i-simplices in K.

Theorem 1.31. *If $b_k = \dim H_k(K;F)$, where F is the additive group of a field, then*

$$\chi(K) = b_0 - b_1 + b_2 - \cdots + (-1)^n b_n.$$

Proof. The exactness of the sequence

$$0 \longrightarrow Z_k(K;F) \xrightarrow{\ i\ } C_k(K;F) \xrightarrow{\ \partial\ } B_{k-1}(K;F) \longrightarrow 0$$

implies $\dim Z_k + \dim B_{k-1} = \dim C_k = a_k$. Therefore,

$$\begin{aligned}\chi(K) &= \sum(-1)^k \dim Z_k + \sum(-1)^k \dim B_{k-1}\\ &= \sum(-1)^k(\dim Z_k - \dim B_k)\\ &= \sum(-1)^k \dim H_k(K;F)\end{aligned}$$

because $H_k(K;F) = Z_k/B_k$. □

Corollary. *Homotopy equivalent simplicial complexes have equal Euler characteristics.*

Example 11. $\chi(D^n) = 1$.

Example 12. $\chi(S^n) = 1 + (-1)^n = \begin{cases} 0 & \text{for odd } n, \\ 2 & \text{for even } n. \end{cases}$

Example 13. $\chi(\mathbb{R}P^n) = \begin{cases} 0 & \text{for odd } n, \\ 1 & \text{for even } n. \end{cases}$

Example 14. $\chi(\mathbb{C}P^n) = n + 1$.

Exercise. (a) Suppose that a finite simplicial complex K is covered by two subcomplexes K_1 and K_2. Prove that

$$\chi(K) = \chi(K_1) + \chi(K_2) - \chi(K_1 \cap K_2)$$

in two ways, by directly counting simplices in these complexes and by using the Mayer–Vietoris sequence.

(b) Suppose that a finite simplicial complex K is covered by subcomplexes $K_1, K_2, \ldots, K_m$. Prove that

$$\chi(K) = \sum \chi(K_i) - \sum_{i<j} \chi(K_i \cap K_j) + \sum_{i<j<k} \chi(K_i \cap K_j \cap K_k) - \cdots .$$

Problem 38. Given finite simplicial complexes K and L and an n-fold covering $p\colon K \to L$, prove that $\chi(K) = n\chi(L)$.

Problem 39. Given finite simplicial complexes K and L, prove that $\chi(K \times L) = \chi(K)\chi(L)$.

Problem 40. Suppose that K is a finite simplicial complex and $f\colon K \to K$ is a continuous map for which $f^n = \underbrace{f \circ \cdots \circ f}_{n}$ is the identity. Prove that if $\chi(K) \not\equiv 0 \pmod{n}$, then f has a fixed point.

The Euler characteristic of a closed manifold M^n was defined in Part I as the sum of the indices of singular points of any smooth vector field (with isolated singular points) on M^n. On the other hand, we can triangulate M^n and calculate the Euler characteristic of the simplicial complex thus obtained. The equivalence of these two definitions is a corollary of the following theorem.

Theorem 1.32. *Suppose that M^n is a triangulated closed manifold. Then there exists a smooth vector field on M^n whose singular points coincide with the barycenters of the simplices in the triangulation; moreover, the barycenter of each k-simplex has index $(-1)^k$.*

Proof. Let us construct a vector field on the triangulated manifold M^n by induction on the dimension of skeletons so that, under the passage from the $(k-1)$-skeleton to the k-skeleton, the additional integral trajectories be incoming for the barycenters of k-simplices ("new" singular points) and outgoing for the barycenters of simplices of dimension less than k ("old" singular points). The construction of this vector field for simplices of dimensions 1, 2, and 3 is shown in Figure 13.

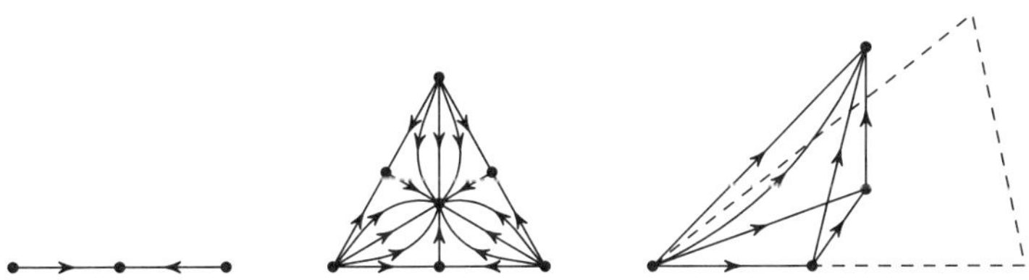

Figure 13. Construction of a vector field

In a neighborhood of the barycenter of a k-simplex, the constructed vector field has the form $v(x) = Ax$, where

$$A = \operatorname{diag}(\underbrace{-1, \dots, -1}_{k}, 1, \dots, 1).$$

The index of such a singular point equals $(-1)^k$. □

Theorem 1.33. *If M^{2n+1} is a closed manifold, then $\chi(M^{2n+1}) = 0$.*

Proof. One proof of this theorem has already been given in Part I. Another proof can be obtained by using the isomorphism

$$H_k(M^{2n+1}; \mathbb{Z}_2) \cong H_{2n+1-k}(M^{2n+1}; \mathbb{Z}_2).$$

Indeed, let $b_k = \dim H_k(M^{2n+1};\mathbb{Z}_2)$. Then $b_k = b_{2n+1-k}$, whence

$$\chi(M^{2n+1}) = \sum_{k=0}^{2n+1}(-1)^k b_k = \sum_{k=0}^{n}((-1)^k b_k + (-1)^{2n+1-k} b_{2n+1-k})$$
$$= \sum_{k=0}^{n}(-1)^k(b_k - b_{2n+1-k}) = 0. \qquad \square$$

Theorem 1.34. *If M^{2n} is a closed manifold, then we have $\chi(M^{2n}) \equiv \dim H_n(M^{2n};\mathbb{Z}_2) \pmod 2$.*

Proof. Let $b_k = \dim H_k(M^{2n};\mathbb{Z}_2)$. Then $b_k = b_{2n-k}$; therefore,

$$\chi(M^{2n}) = \sum_{k=0}^{2n}(-1)^k b_k = b_n + \sum_{k=0}^{n-1}((-1)^k b_k + (-1)^{2n-k} b_{2n-k})$$
$$= b_n + \sum_{k=0}^{n-1}(-1)^k(b_k + b_{2n-k}) \equiv b_n \pmod 2. \qquad \square$$

Theorem 1.35. *If a closed manifold M^m is the boundary of a compact manifold W^{m+1}, then $\chi(M^m)$ is even.*

Proof. It is sufficient to consider the case $m = 2n$ because otherwise $\chi(M^m) = 0$. Let us triangulate the manifold W^{2n+1} and consider the closed manifold $\tilde{W}^{2n+1}$ obtained from two copies of W^{2n+1} by gluing together their boundaries M^{2n}. The manifold $\tilde{W}^{2n+1}$ also has the structure of a simplicial complex with a subcomplex M^{2n}, and all simplices not contained in M^{2n} come in pairs. Therefore, $\chi(\tilde{W}^{2n+1}) \equiv \chi(M^{2n}) \pmod 2$. But $\chi(\tilde{W}^{2n+1}) = 0$. $\square$

Corollary. *Neither the manifold $\mathbb{R}P^{2n}$ nor $\mathbb{C}P^{2n}$ is the boundary of a compact manifold.*

On the other hand, let us show that the manifolds $\mathbb{R}P^{2n+1}$ and $\mathbb{C}P^{2n+1}$ are the boundaries of compact manifolds. For this purpose, we can use, e.g., the following assertion.

Theorem 1.36. *Suppose that M^n is a closed manifold and $\sigma\colon M^n \to M^n$ is a fixed-point-free smooth involution, i.e., $\sigma(\sigma(x)) = x$ and $\sigma(x) \neq x$ for all $x \in M^n$. Then M^n is the boundary of a compact manifold.*

Proof. Consider the coset space

$$W^{n+1} = (M^n \times [0,1])/(x,0) \sim (\sigma(x),0).$$

The involution σ has no fixed points; therefore, any point $(x,0)$ in W^{n+1} has a neighborhood homeomorphic to the ball obtained by gluing together two half-disks D_+^{n+1} and σD_+^{n+1}. Hence W^{n+1} is a manifold with boundary $M^n \times \{1\}$. $\square$

Fixed-point-free involutions on $\mathbb{R}P^{2n+1}$ and $\mathbb{C}P^{2n+1}$ can be defined as

$$\sigma(x_0, x_1, \dots, x_{2n}, x_{2n+1}) = (-x_1, x_0, \dots, -x_{2n+1}, x_{2n}),$$
$$\sigma(z_0, z_1, \dots, z_{2n}, z_{2n+1}) = (-\bar{z}_1, \bar{z}_0, \dots, -\bar{z}_{2n+1}, \bar{z}_{2n}).$$

Exercise. Prove that the orientation covering manifold (Part I, p. 207) for a closed manifold is the boundary of a compact manifold.

Theorem 1.37. *If a closed manifold M^{2n} is the boundary of a compact manifold, then the dimension of the space $H_n(M^{2n}; \mathbb{Z}_2)$ is even.*

Proof. According to Theorem 1.34, we have $\chi(M^{2n}) \equiv \dim H_n(M^{2n}; \mathbb{Z}_2)$ (mod 2), and Theorem 1.35 implies that $\chi(M^{2n})$ is even. □

Problem 41. Given a finite exact sequence of vector spaces

$$\cdots \longrightarrow V_{i+1} \xrightarrow{f_{i+1}} V_i \xrightarrow{f_i} V_{i-1} \longrightarrow \cdots,$$

prove that $\sum (-1)^i \dim V_i = 0$.

Problem 42. Prove that the Euler characteristic of the complement of any link in S^3 is zero.

If A is a finite simplicial complex and B is its subcomplex, then we can define the *Euler characteristic of the pair* (A, B) by

$$\chi(A, B) = \sum (-1)^i \dim H_i(A, B),$$

where the $H_i(A, B)$ are the homology groups with coefficients in some field.

Problem 43. Prove that $\chi(A, B) = \chi(A) - \chi(B)$.

Problem 44*([143]). Let K be a Euclidean CW-complex for which $|K| \approx M^n$, where M^n is a closed manifold. Consider the alternating sum

$$\Psi(K) = \sum_{i-0}^{n} \sum_{j-0}^{n} (-1)^{i+j} \alpha_{ij},$$

where each α_{ij} is the number of ordered pairs of disjoint (closed) faces of dimensions i and j in the complex K. (Note that $\alpha_{ij} = \alpha_{ji}$ and α_{ii} is twice the number of unordered pairs of disjoint i-faces.) Prove that $\Psi(K) = \chi^2(M^n) - \chi(M^n)$, where $\chi(M^n)$ is the Euler characteristic of the manifold M^n.

Remark. The number $\Psi(K)$ defined in Problem 44* can be interpreted as the Euler characteristic of the deleted product of the manifold M^n. (The definition of deleted product is given in Part I on p. 161.)

Geometric conditions often impose various constraints on the Euler characteristic. As an example, consider transnormal embeddings (see [**110**]). An embedding of a manifold M^k into $\mathbb{R}^n$ is said to be *transnormal* if for any points $x, y \in M^k$ such that x belongs to the normal subspace at y, the normal subspaces at the points x and y coincide. Examples of transnormal embeddings are the standard embeddings of $\mathbb{R}^k$ and S^k into $\mathbb{R}^n$.

Problem 45. Prove that the number of intersection points of transnormally embedded (connected) manifolds with a normal subspace does not depend on the normal subspace. (If this number is equal to r, then the embedding is said to be r*-transnormal.*)

Problem 46. Prove that if a closed manifold admits an r-transnormal embedding, then the number r is even.

Problem 47. Prove that if a closed manifold M^k admits an r-transnormal embedding, then $\chi(M^k) = 0$ or r.

Problem 48. Let M^k be a closed manifold admitting a transnormal embedding. Prove that if $\chi(M^k) \neq 0$, then $H_i(M^k; G) = 0$ for all odd i.

6.2. The Lefschetz Fixed Point Theorem. Let K be a finite simplicial complex, and let $f\colon |K| \to |K|$ be a continuous map. The induced map $(f_*)_k\colon H_k(|K|; \mathbb{R}) \to H_k(|K|; \mathbb{R})$ is a linear operator on a finite-dimensional space; therefore, we can consider its trace $\operatorname{tr}(f_*)_k$. The *Lefschetz number* of the map f is defined by $\Lambda(f) = \sum_{k \geq 0} (-1)^k \operatorname{tr}(f_*)_k$.

If the complex K is path-connected, then $(f_*)_0\colon \mathbb{R} \to \mathbb{R}$ is the identity map, and hence $\operatorname{tr}(f_*)_0 = 1$.

Example 15. If the complex K is acyclic with respect to the coefficient group $\mathbb{R}$, then $\Lambda(f) = 1$ for any continuous map $f\colon |K| \to |K|$.

Example 16. The equality $\Lambda(\mathrm{id}_K) = \chi(K)$ holds.

Theorem 1.38 (Lefschetz fixed point theorem, [**77**])**.** *Let K be a finite simplicial complex. If $\Lambda(f) \neq 0$, then the map $f\colon K \to K$ has a fixed point.*

Proof. Suppose that the map f has no fixed points. Since the complex K is compact, one of its subdivisions K' of K has the following property: for any point $x \in |K|$, the distance between x and $f(x)$ is much larger than the maximum diameter of a simplex from K'. In turn, this property implies that if K'' is a subdivision of K' and $\varphi\colon K'' \to K'$ is a simplicial approximation of f, then, for any simplex $\Delta'' \subset K''$, the sets Δ'' and $\varphi(\Delta'')$ are disjoint. Thus, to prove the Lefschetz theorem, it suffices to interpret the number $\Lambda(\varphi)$ so that if $\Delta'' \cap \varphi(\Delta'') = \varnothing$ for all simplices $\Delta'' \subset K''$, then $\Lambda(\varphi) = 0$.

Lemma. (a) (Hopf [**57**]) *If K is a finite simplicial complex and $\varphi_k\colon C_k(K,\mathbb{R}) \to C_k(K,\mathbb{R})$ is a chain map, then*

$$\sum_{k\geq 0}(-1)^k \operatorname{tr}\varphi_k = \sum_{k\geq 0}(-1)^k \operatorname{tr}(\varphi_*)_k.$$

(b) ([**120**]) *If K is a finite simplicial complex, K' is its finite subdivision, and $\varphi\colon K' \to K$ is a simplicial map, then*

$$\Lambda(\varphi) = \sum_{k\geq 0}(-1)^k\Phi_k,$$

where Φ_k is the number of k-simplices $\Delta^k \subset K'$ for which $\Delta^k \subset \varphi(\Delta^k)$ with signs taken into account.[11]

Proof. (a) It suffices to show that

$$\operatorname{tr}\varphi_k = \operatorname{tr}(\varphi_k|_{B_k}) + \operatorname{tr}(\varphi_k)_* + \operatorname{tr}(\varphi_{k-1}|_{B_{k-1}}).$$

Let $C_k = Z_k \oplus \tilde{C}_k$. The linear operator φ_k maps Z_k to Z_k; hence it induces an operator $\tilde{\varphi}_k\colon \tilde{C}_k \to \tilde{C}_k$, and $\operatorname{tr}\varphi_k = \operatorname{tr}(\varphi_k|_{Z_k}) + \operatorname{tr}\tilde{\varphi}_k$.

Let $Z_k = B_k \oplus \tilde{Z}_k$. Then $\tilde{Z}_k \cong H_k(K;\mathbb{R})$, and the operator induced by φ_k on $H_k(K;\mathbb{R})$ coincides with $(\varphi_*)_k$. Therefore, $\operatorname{tr}(\varphi_k|_{Z_k}) = \operatorname{tr}(\varphi_k|_{B_k}) + \operatorname{tr}(\varphi_*)_k$.

The map $\partial\colon C_k \to B_{k-1}$ induces an isomorphism $\tilde{C}_k \to B_{k-1}$. The equality $\partial\varphi_k = \varphi_{k-1}\partial$ shows that the operator corresponding to $\tilde{\varphi}_k$ under this isomorphism is φ_{k-1}.

(b) The simplicial map $\varphi\colon K' \to K$ induces a chain map $\varphi_k\colon C_k(K';\mathbb{R}) \to C_k(K;\mathbb{R})$. Consider also the chain map $i_k\colon C_k(K;\mathbb{R}) \to C_k(K';\mathbb{R})$ that takes each simplex $\Delta \subset K$ cut into simplices $\Delta_1,\dots,\Delta_s$ to the sum $\Delta_1 + \dots + \Delta_s$ (the orientations of these simplices are compatible with that of the simplex Δ; a formal definition of the map i_k for the barycentric subdivision was given on p. 8). The composition $\varphi_k i_k\colon C_k(K;\mathbb{R}) \to C_k(K;\mathbb{R})$ takes each simplex Δ to the sum of simplices $\varphi(\Delta_1) + \dots + \varphi(\Delta_s)$. Therefore, $\Phi_k = \operatorname{tr}(\varphi_k i_k)$. According to (a), we have

$$\sum_{k\geq 0}(-1)^k \operatorname{tr}(\varphi_k i_k) = \sum_{k\geq 0}(-1)^k \operatorname{tr}(\varphi_*)_k;$$

this follows from the fact that, under the identification of $H_k(K';\mathbb{R})$ with $H_k(K;\mathbb{R})$ by means of the isomorphism i_*, the map in homology induced by the chain map $\varphi_k i_k$ is identified with $(\varphi_*)_k$. □

The Lefschetz theorem follows from part (b) of the lemma because if $\Delta' \cap \varphi(\Delta') = \varnothing$ for all simplices $\Delta' \subset K'$, then $\Phi_k = 0$ for all k. □

[11] If the simplices Δ^k and $\varphi(\Delta^k)$ have the same orientation, then the plus sign is taken, and if their orientations are opposite, then they are taken with the minus sign.

Example 17. For a map $f\colon S^n \to S^n$, we have $\Lambda(f) = 1 + (-1)^n \deg f$; hence if $\deg f \neq (-1)^{n+1}$, then the map f has a fixed point.

Corollary. *On the sphere S^{2n}, there is no continuous vector field without singular points.*

Proof. Suppose that v is a vector field on S^{2n}, $v(x)$ is the vector of this field at a point $x \in S^{2n}$, and $f(x)$ is the intersection point of S^{2n} with the ray from the center of the sphere to the endpoint of $v(x)$ (see Figure 14). Clearly, $f \sim \mathrm{id}_{S^{2n}}$, whence $\deg f = 1 \neq -1 = (-1)^{2n+1}$. Thus, the map f has a fixed point x_0, and $v(x_0) = 0$. □

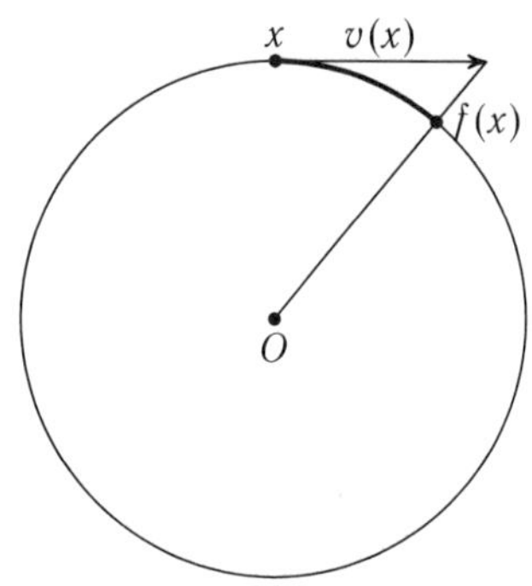

Figure 14. The construction of the map f

Remark. Recall that, in Part I (pp. 203–204), the nonexistence of a continuous vector field without singular points on the sphere S^{2n} was proved by more elementary methods.

Example 18. For the coefficient group $\mathbb{R}$, the space $\mathbb{R}P^{2n}$ is acyclic; therefore, any map $f\colon \mathbb{R}P^{2n} \to \mathbb{R}P^{2n}$ has a fixed point.

Problem 49 (**[100]**). Given a finite simplicial complex K, a continuous map $f\colon K \to K$, and a prime p, prove that $\Lambda(f^p) \equiv \Lambda(f) \pmod p$ (here $f^p = \underbrace{f \circ \cdots \circ f}_{p}$).

Chapter 2

Cohomology Rings

1. Multiplication in Cohomology

The set $H^*(K) = \bigoplus H^k(K)$ has not only a group structure but also a ring structure; it is this structure that is largely used in the applications of cohomology. The ring structure is determined by the *Kolmogorov–Alexander multiplication*, which is now known as the *cup product*. This operation was first introduced by Kolmogorov [**72**, **73**] and Alexander [**6**]. Later, the definition was improved by Čech [**21**] and Whitney [**151**].

Before Kolmogorov and Alexander, there was a multiplication operation on the homology of manifolds (intersection of cycles); the Poincaré duality isomorphism transforms it into Kolmogorov–Alexander multiplication. Certainly, for many applications of multiplication in cohomology, this long-known multiplication is sufficient, but there are also applications not related to manifolds.

To define the cup product for a simplicial complex, we must order the vertices of this complex. Such a definition is not invariant at the level of cochains, but it is invariant at the cohomology level. The proof of its invariance uses an auxiliary object, a total chain complex.

1.1. Homology of Total Chain Complexes. In some cases, it is convenient to assume that the vertices of simplices are ordered. The most important instances of such situations are those of defining multiplication in cohomology and proving the isomorphism between singular and simplicial homology. Let us try to define the homology of a simplicial complex by considering, instead of oriented simplices, ordered sets $(v_0, \ldots, v_k)$, where $v_0, \ldots, v_k$ are pairwise distinct vertices of some simplex. A simple example shows that this attempt will be unsuccessful.Indeed, for $K = [v_0, v_1]$,

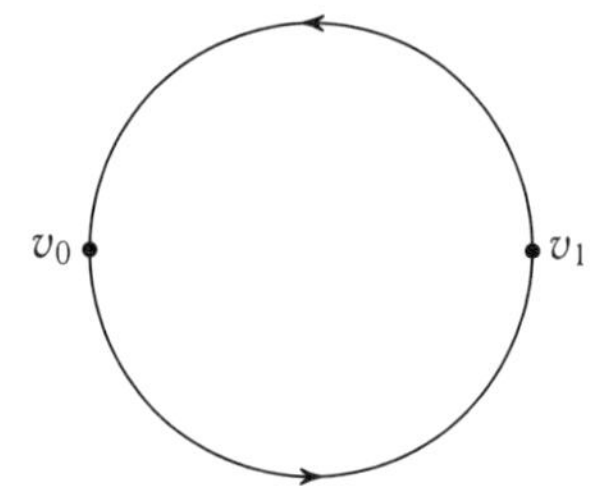

Figure 1. A CW-complex homeomorphic to the circle

the group $\operatorname{Ker}\partial_1$ is generated by the element $(v_0, v_1) + (v_1, v_0)$, whereas $\operatorname{Im}\partial_2 = 0$.

The reason for the failure is fairly clear: in fact, we try to calculate the homology of a CW-complex homeomorphic to the circle (see Figure 1). Surprisingly, this situation is easy to remedy; it suffices to dispense with the seemingly natural condition that the vertices $v_0, \dots, v_k$ are pairwise distinct. This is indeed an unexpected phenomenon, because now the calculation of the homology of the complex $K = [v_0, v_1]$ involves an infinite chain complex with 2^{k+1} generators of dimension k for each $k = 0, 1, 2, \dots$. But in return, at least the one-dimensional homology group is as required. Indeed, the group $\operatorname{Ker}\partial_1$ is generated by the elements $(v_0, v_1) + (v_1, v_0)$, (v_0, v_0), and (v_1, v_1); all of them are contained in $\operatorname{Im}\partial_2$: $(v_0, v_0) = \partial(v_0, v_0, v_0)$, $(v_1, v_1) = \partial(v_1, v_1, v_1)$, and $(v_0, v_1) + (v_1, v_0) = \partial(v_0, v_1, v_0) + \partial(v_0, v_0, v_0)$.

The chain complex $\hat{C}_*(K)$ generated by the ordered sets $(v_0, \dots, v_k)$, where $v_0, \dots, v_k$ are vertices of the same simplex in K (some of them may coincide), is called a *total*, or *ordered*, chain complex. The boundary homomorphism for $\hat{C}_*(K)$ is defined in precisely the same way as for $C_*(K)$.

Theorem 2.1. *Let K be a simplicial complex with ordered vertices. Then the homology groups of the chain complexes $\hat{C}_*(K)$ and $C_*(K)$ are canonically isomorphic.*

Proof. Consider the homomorphism $\varphi\colon \hat{C}_*(K) \to C_*(K)$ taking $(v_0, \dots, v_k)$ to $[v_0, \dots, v_k]$ if the points $v_0, \dots, v_k$ are pairwise distinct, and to 0 if some of these points coincide. This homomorphism is a chain map. The proof is similar to the argument showing that the simplicial map $K \to L$ induces a chain map $C_*(K) \to C_*(L)$; the case where there are precisely two coinciding points among $v_0, \dots, v_k$ is again to be handled separately. We also consider the homomorphism $\Phi\colon C_*(K) \to \hat{C}_*(K)$ induced by the ordering of vertices. Namely, we assume that the orientation of each simplex is determined by a set of vertices arranged in increasing order, and to each oriented simplex we assign the set of its vertices ordered accordingly. The homomorphism Φ is a chain map as well.

The map $\varphi\Phi\colon C_*(K) \to C_*(K)$ is the identity. The map $\Phi\varphi$ is not the identity, but it is chain homotopic to the identity. To prove this, we use the same argument as in the proof of the acyclic support theorem (Theorem 1.5 on p. 7) with the only difference that instead of assigning the acyclic subcomplex $L(\Delta)$ to each simplex $\Delta \subset K$, we associate with the ordered set of vertices $v = (v_0, \dots, v_k)$ of each simplex $\Delta_v \subset K$ a chain subcomplex $\hat{C}_*(\Delta_v) \subset \hat{C}_*(K)$ so that the following conditions hold:

(1) if $\Delta' \subset \Delta$, then $\hat{C}_*(\Delta') \subset \hat{C}_*(\Delta)$;

(2) the chain complex $\hat{C}_*(\Delta_v)$ is a support for the chains $\Phi\varphi(v)$ and $v = \mathrm{id}(v)$.

If v_0 is a vertex, then $\Phi\varphi(v_0) = v_0 = \mathrm{id}(v_0)$. Thus, it remains to verify that if z is a cycle in $\hat{C}_k(\Delta_v)$ for $k \geq 1$, then $z = \partial d$, where $d \in \hat{C}_{k+1}(\Delta_v)$. We associate with every chain $c \in \hat{C}_k(\Delta_v)$ the chain $(v_0, c) \in \hat{C}_{k+1}(\Delta_v)$. For $k \geq 1$, we have $\partial(v_0, c) = c - (v_0, \partial c)$. If z is a cycle, then $z = \partial(v_0, z)$ because $\partial z = 0$. □

Any simplicial map $f\colon K \to L$ induces the homomorphism $\hat{f}_k\colon \hat{C}_k(K) \to \hat{C}_k(L)$ defined by

$$\hat{f}_k((v_0, \dots, v_k)) = (f(v_0), \dots, f(v_k)).$$

It is easy to verify that $\hat{f}_k$ is a chain map (it does not matter whether the points $f(v_i)$ and $f(v_j)$ are different).

It is also easy to check that the diagram

$$\begin{array}{ccc} H_k(\hat{C}_*(K)) & \xrightarrow{\hat{f}_*} & H_k(\hat{C}_*(L)) \\ \big\downarrow{\scriptstyle \varphi_*} & & \big\downarrow{\scriptstyle \varphi_*} \\ H_k(K) & \xrightarrow{f_*} & H_k(L) \end{array}$$

is commutative. Indeed, even at the level of chains, we have $f_k\varphi = \varphi\hat{f}_k$. The equality $f_*\varphi_* = \varphi_*\hat{f}_*$ implies $\Phi_* f_* = \hat{f}_*\Phi_*$ because $\Phi_* = \varphi_*^{-1}$ (but only at the level of homology, not at the level of chains).

1.2. Definition of Multiplication in Cohomology. Any total chain complex naturally determines a cochain complex. The cohomology groups of a total chain complex are canonically isomorphic to the simplicial cohomology groups because chain homotopic chain maps induce not only the same maps of homology groups but also the same maps of cohomology groups. Let us prove this. First, note that any chain map $\varphi_k\colon C_k \to C'_k$ induces a cochain map $\varphi^k\colon \mathrm{Hom}(C'_k, G) \to \mathrm{Hom}(C_k, G)$. Suppose that φ_k and ψ_k are chain homotopic chain maps and let $\alpha = \varphi^k - \psi^k$. If c_2 is a cycle, then

$$\langle \alpha c^1, c_2 \rangle = \langle c^1, (\partial D + D\partial)c_2 \rangle = \langle c^1, \partial D c_2 \rangle = \langle \delta c^1, D c_2 \rangle.$$

Clearly, if c^1 is a cocycle, then the last quantity vanishes.

Suppose that the coefficient group R is the additive group of a commutative associative ring with identity. Then the *cup product* $c^p \smile c^q \in \hat{C}^{p+q}(K; R)$ of cochains $c^p \in \hat{C}^p(K; R)$ and $c^q \in \hat{C}^q(K; R)$ can be defined as

$$\langle c^p \smile c^q, (v_0, \dots, v_{p+q}) \rangle = \langle c^p, (v_0, \dots, v_p) \rangle \cdot \langle c^q, (v_p, v_{p+1}, \dots, v_{p+q}) \rangle;$$

here $\langle c^p, (v_0, \dots, v_p) \rangle$ and $\langle c^q, (v_p, v_{p+1}, \dots, v_{p+q}) \rangle$ are multiplied as ring elements.

Under the multiplication of cochains thus defined, the identity element is the cochain c^0 taking the value $1 \in R$ at each vertex v. This is a cocycle. Indeed, if $c_1 = \sum a_i [v_{i1}, v_{i2}]$, then $\partial c_1 = \sum a_i v_{i1} - \sum a_i v_{i2}$; therefore,

$$\langle \delta c^0, c_1 \rangle = \langle c^0, \partial c_1 \rangle = \sum a_i - \sum a_i = 0.$$

Remark. The cup product can also be defined in a somewhat more general situation. Namely, the product of $c^p \in \hat{C}^p(K; G)$ and $c^q \in \hat{C}^q(K; G')$ is the cochain $c^p \smile c^q \in \hat{C}^{p+q}(K; G \otimes G')$ defined by

$$\langle c^p \smile c^q, (v_0, \dots, v_{p+q}) \rangle = \langle c^p, (v_0, \dots, v_p) \rangle \otimes \langle c^q, (v_p, v_{p+1}, \dots, v_{p+q}) \rangle.$$

If $G = G' = R$ is the additive group of a unital commutative ring, then ring multiplication is a map $R \otimes R \to R$, and we arrive at the preceding definition.

Clearly, the cup product is bilinear and associative. It turns out that it can be carried over to cohomology due to the following property.

Lemma. $\delta(c^p \smile c^q) = (\delta c^p) \smile c^q + (-1)^p c^p \smile (\delta c^q)$.

Proof. The values of the cochains $(\delta c^p) \smile c^q$ and $(-1)^p c^p \smile (\delta c^q)$ at $(v_0, \dots, v_{p+q+1})$ are equal to

$$\sum_{0 \le i \le p+1} (-1)^i c^p(v_0, \dots, \hat{v}_i, \dots, v_{p+1}) c^q(v_{p+1}, \dots, v_{p+q+1})$$

and

$$(-1)^p \sum_{p \le i \le p+q+1} (-1)^{i-p} c^p(v_0, \dots, v_p) c^q(v_p, \dots, \hat{v}_i, \dots, v_{p+q+1}),$$

respectively. The last term in the first sum is canceled by the first term in the second sum, and the sum of the remaining terms in these sums equals the value of the cochain $\delta(c^p \smile c^q)$ at $(v_0, \dots, v_{p+q+1})$. □

The expression for $\delta(c^p \smile c^q)$ shows that if z^p and z^q are cocycles, then so is $\delta(z^p \smile z^q)$. Moreover, the cohomology class of the cup product of two cocycles depends only on the cohomology classes of these cocycles because

$$(z^p + \delta c^{p-1}) \smile z^q = z^p \smile z^q + \delta(c^{p-1} \smile z^q)$$

and

$$z^p \smile (z^q + \delta c^{q-1}) = z^p \smile z^q + (-1)^p \delta(z^p \smile c^{q-1}).$$

Thus, cup product is defined in the entire cohomology.

It follows directly from the definitions that if $f\colon K \to L$ is a simplicial map, then $\hat{f}^*(c^p \smile c^q) = (\hat{f}^* c^p) \smile (\hat{f}^* c^q)$, where $\hat{f}^*$ is the induced map of the cohomology groups of the total chain complex.

The isomorphism $\varphi^*\colon H^*(K) \to H^*(\hat{C}_*(K))$ allows us to transfer cup product to simplicial cohomology. Namely, for $\alpha, \beta \in H^*(K)$, we define $\alpha \smile \beta$ to be the element of $H^*(K)$ for which $\varphi^*(\alpha \smile \beta) = (\varphi^*\alpha) \smile (\varphi^*\beta)$. At the level of cocycles, the cup product in simplicial cohomology can be described as follows: the cocycle $z^p \smile z^q$ belongs to the same cohomology class as the cocycle z determined by

$$\langle z, [v_0, \ldots, v_{p+q}]\rangle = \langle z^p, [v_0, \ldots, v_p]\rangle \cdot \langle z^q, [v_p, v_{p+1}, \ldots, v_{p+q}]\rangle. \tag{8}$$

It is assumed that the vertices of the simplicial complex K are ordered and the vertices of the oriented simplex are written in increasing order; otherwise, formula (8) does not define any cocycle z because the right-hand side changes unpredictably under permutations of the vertices $v_0, \ldots, v_{p+q}$.

The total chain complex is needed only for purely technical purposes. In simplicial cohomology, we can define the cup product by numbering the vertices and setting

$$\langle c^p \smile c^q, [v_{i_0}, \ldots, v_{i_{p+q}}]\rangle = \langle c^p, [v_{i_0}, \ldots, v_{i_p}]\rangle \cdot \langle c^q, [v_{i_p}, v_{i_{p+1}}, \ldots, v_{i_{p+q}}]\rangle,$$

where $i_0 < i_1 < \cdots < i_{p+q}$. Such a multiplication of cochains determines a well-defined multiplication of cohomology classes, which does not depend on the numbering. The total chain complex is only needed to prove this fact and the naturality of the cup product in the sense that if $f\colon |K| \to |L|$ is a continuous map, then $f^*(\alpha \smile \beta) = f^*(\alpha) \smile f^*(\beta)$.

Multiplication in cohomology has the following *anticommutativity* (or *skew-commutativity*) property.

Theorem 2.2. *If $\alpha \in H^p(K)$ and $\beta \in H^q(K)$, then $\alpha \smile \beta = (-1)^{pq}\beta \smile \alpha$.*

Proof. We take cocycles z^p and z^q representing α and β, and order the vertices of K arbitrarily. We calculate $z^p \smile z^q$ using this ordering, and

$z^q \smile z^p$ using the reverse ordering. We obtain

$$\langle z^p \smile z^q, [v_0, \ldots, v_{p+q}] \rangle = \langle z^p, [v_0, \ldots, v_p] \rangle \cdot \langle z^q, [v_p, \ldots, v_{p+q}] \rangle$$

and

$$\langle z^q \smile z^p, [v_{p+q}, \ldots, v_0] \rangle = \langle z^q, [v_{p+q}, \ldots, v_p] \rangle \cdot \langle z^p, [v_p, \ldots, v_0] \rangle.$$

It is clear that $[v_r, \ldots, v_0] = (-1)^{r(r+1)/2}[v_0, \ldots, v_r]$ and $(p+q)(p+q+1) - p(p+1) - q(q+1) = 2pq$. □

The definition of cup product in relative cohomology is precisely the same as in absolute cohomology. If $\alpha \in H^p(K, L_1)$ and $\beta \in H^q(K, L_2)$, then $\alpha \smile \beta \in H^{p+q}(K, L_1 \cup L_2)$. Indeed, take cocycles z^p and z^q representing α and β and consider the simplex $[v_0, \ldots, v_{p+q}]$ in $L_1 \cup L_2$. This simplex, as well as any other simplex in $L_1 \cup L_2$, is contained entirely either in L_1 or in L_2. In the former case, $[v_0, \ldots, v_p]$ is contained in L_1, and in the latter, $[v_p, \ldots, v_{p+q}]$ is contained in L_2. Therefore,

$$\langle z^p \smile z^q, [v_0, \ldots, v_{p+q}] \rangle = \langle z^p, [v_0, \ldots, v_p] \rangle \cdot \langle z^q, [v_p, \ldots, v_{p+q}] \rangle = 0.$$

Thus, the cocycle $z^p \smile z^q$ is contained in $C^{p+q}(K, L_1 \cup L_2)$.

Problem 50. Given $\alpha, \beta \in H^*(K, L)$, prove that $(\alpha|_K) \smile \beta = \alpha \smile \beta$, where $\alpha|_K$ is the image of α under the natural homomorphism $H^*(K, L) \to H^*(K)$.

Problem 51. (a) Given $K = \bigcup_{i=1}^n L_i$, where the L_i are contractible subcomplexes, prove that the product $\alpha_1 \smile \cdots \smile \alpha_n$ vanishes for any n elements $\alpha_i \in H^{p_i}(K)$, where $p_i > 0$.

(b) Prove that cohomology multiplication in ΣK is trivial, i.e., the cup product of any two cohomology classes of positive dimension vanishes.

Problem 52. Given a manifold M^k embedded in S^n, prove that the product of any two classes of positive dimension in the ring $H^*(S^n, M^k)$ vanishes.

Problem 53. Let a subcomplex $A \subset X$ be a retract.

(a) Prove that $H_*(X) = \operatorname{Im} i_* \oplus \operatorname{Ker} r_*$ and $H^*(X) = \operatorname{Ker} i^* \oplus \operatorname{Im} r^*$, where $i\colon A \to X$ is the natural embedding and $r\colon X \to A$ is a retraction.

(b) Prove that for cohomology, $\operatorname{Ker} i^*$ is an ideal and $\operatorname{Im} r^*$ is a subring.

Problem 54. Given multiplications in the rings $H^*(X)$ and $H^*(Y)$, describe the multiplication in the ring $H^*(X \vee Y)$.

Problem 55. Given $p, q \geq 1$, prove that $S^p \vee S^q = (S^p \times \{x_0\}) \cup (\{y_0\} \times S^q)$ is not a retract of $S^p \times S^q$.

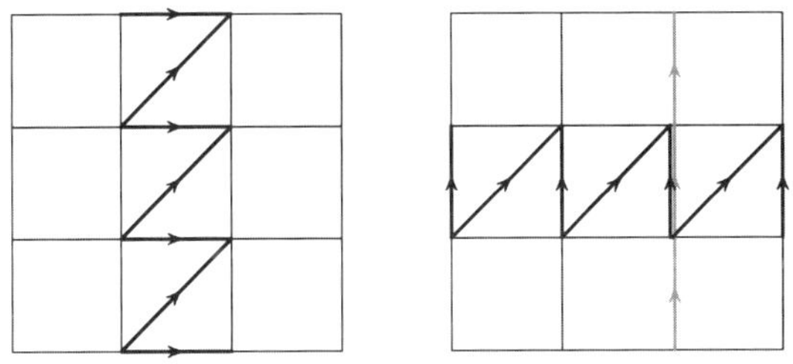

a b

Figure 2. Basis cocycles on the torus

Problem 56. (a) Prove that for $m < n$, the standardly embedded space $\mathbb{R}P^m \subset \mathbb{R}P^n$ is not a retract.

(b) Prove that for $m < n$, the standardly embedded space $\mathbb{C}P^m \subset \mathbb{C}P^n$ is not a retract.

1.3. Cohomology Rings of Two-Dimensional Surfaces. In this section, we calculate the cohomology rings of closed two-dimensional surfaces directly from the definition of cup multiplication. Note that the required result is easy to obtain by using the Poincaré isomorphism and interpreting the multiplication of cocycles in the cohomology of a two-dimensional surface as the intersection of the dual cycles (see Section 2.2).

For two-dimensional surfaces, we must calculate only the cup products of elements of H^1, because the product of an element of H^1 and an element of H^2 is contained in $H^3 = 0$. Before proceeding further, let us discuss the geometric meaning of cocycles and coboundaries.

Each (oriented) k-simplex in a simplicial complex determines the dual k-cochain, which takes the value 1 at this simplex and vanishes at all of the other k-simplices. We consider only cochains that are sums of cochains dual to various simplices. We identify such cochains with sets of oriented simplices.

The formula $\langle \delta c^1, c_2 \rangle = \langle c^1, \partial c_2 \rangle$ shows that a one-dimensional cochain c^1 of this form is a cocycle if and only if, for any 2-simplex, either none of the three 1-simplices constituting its boundary is contained in c^1 or c^1 contains precisely two of the three boundary simplices; moreover, in the latter case, if the coefficient group is $\mathbb{Z}$, then the orientations of the two boundary cycles in c^1 must be opposite to that of the boundary of the 2-simplex. An example of a cocycle on the torus is shown in Figure 2a; another cocycle, together with a cycle dual to it at the homology level, is shown in Figure 2b. The cycles dual to these two cocycles form a basis in the homology of the torus;

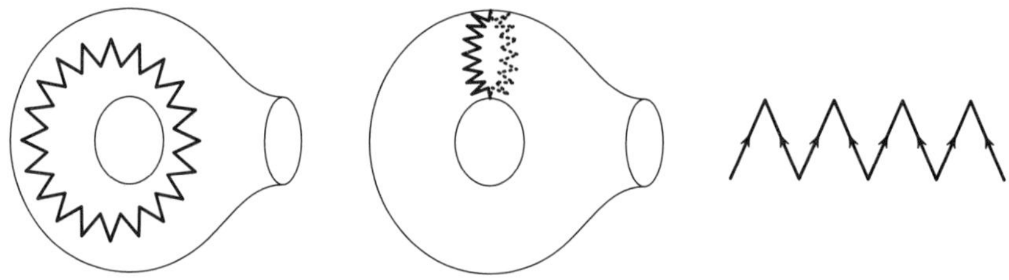

Figure 3. Basis cocycles on a handle

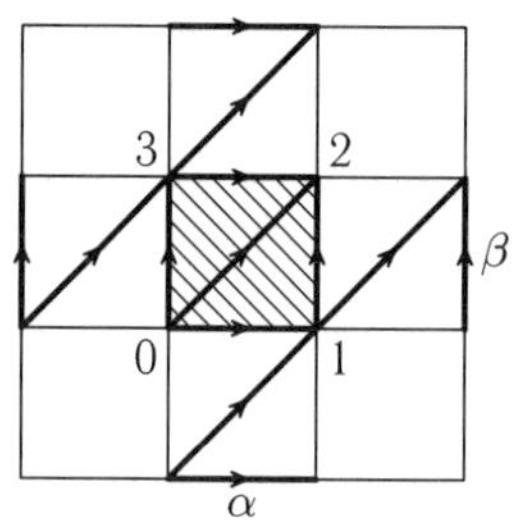

Figure 4. Calculation of $\alpha \smile \beta$ for the torus

therefore, they form a basis in its cohomology (for both coefficient groups $\mathbb{Z}$ and $\mathbb{Z}_2$).

A basis for the cohomology of a sphere with any number of handles is constructed similarly. In Figure 3, basis cocycles for a separate handle are presented; on the right, the orientations of the links of the polygonal line are shown.

Any cochain dual to a 2-simplex generates the two-dimensional cohomology group of a sphere with handles. If simplices $\Delta_1^2, \dots, \Delta_p^2$ have the same orientation, then the cochain $\sum a_i(\Delta_i^2)^*$ is a generator if and only if $\sum a_i = \pm 1$, because cochains dual to 2-simplices with the same orientation are cohomologous. For example, the difference of two cochains dual to 2-simplices with the same orientation sharing a 1-simplex Δ^1 is the coboundary of the cochain dual to Δ^1.

The formula $\langle \alpha \smile \beta, [ijk] \rangle = \langle \alpha, [ij] \rangle \cdot \langle \beta, [jk] \rangle$ shows that in calculating $\alpha \smile \beta$, it suffices to consider only the 2-simplices adjacent to both α and β. For the chosen basis cocycles in the one-dimensional cohomology group of the torus, there are two such 2-simplices (they are hatched in Figure 4). We have

$$\langle \alpha, [01] \rangle \cdot \langle \beta, [12] \rangle = 1 \quad \text{and} \quad \langle \alpha, [02] \rangle \cdot \langle \beta, [23] \rangle = 0$$

because $\langle \beta, [23] \rangle = 0$. Thus, $\alpha \smile \beta$ generates the two-dimensional cohomology group of the torus.

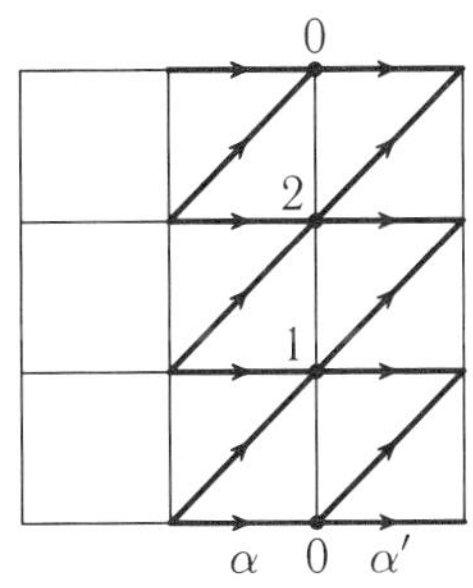

Figure 5. Cohomologous cocycles on the torus

For the coefficient group $\mathbb{Z}$, the antisymmetry of the cup product implies $\alpha \smile \alpha = 0$ for any cohomology class of odd dimension. But for the coefficient group $\mathbb{Z}_2$, there are one-dimensional cohomology classes for which $\gamma \smile \gamma \neq 0$. For this reason, below we give a proof of the equalities $\alpha \smile \alpha = 0$ and $\beta \smile \beta = 0$ for the chosen basis cocycles α and β in the one-dimensional cohomology group of the torus that applies to both coefficient groups $\mathbb{Z}$ and $\mathbb{Z}_2$.

In Figure 5, the cocycle α is cohomologous to the cocycle α'. To prove this, we investigate the coboundary of a cochain that is dual to a point. The relations

$$\langle \delta c^0, [v_0, v_1] \rangle = \langle c^0, \partial[v_0, v_1] \rangle = \langle c^0, v_1 \rangle - \langle c^0, v_0 \rangle$$

show that the coboundary of a cochain dual to a point w_0 is a sum of the cochains dual to 1-simplices of the form $[v, w_0]$, i.e., intervals directed to w_0. Therefore, subtracting the coboundary of the sum of cochains dual to 0, 1, and 2 from α, we obtain the cocycle α'.

At the level of cohomology, we have $\alpha \smile \alpha = \alpha \smile \alpha'$. However, none of the 2-simplices is adjacent simultaneously to α and α'. Therefore, $\alpha \smile \alpha' = 0$.

Applying a similar argument to separate handles and using the fact that any cochain dual to a 2-simplex is a generator of the two-dimensional cohomology group, we obtain the following theorem.

Theorem 2.3. *The one-dimensional cohomology group of a sphere with n handles (with coefficients in $\mathbb{Z}$ or $\mathbb{Z}_2$) has a basis $\alpha_1, \dots, \alpha_n, \beta_1, \dots, \beta_n$ such that $\alpha_i \smile \alpha_j = 0$ and $\beta_i \smile \beta_j = 0$ for all i and j, $\alpha_i \smile \beta_j = 0$ for $i \neq j$, and $\alpha_i \smile \beta_i = \Lambda$, where Λ is a generator of the two-dimensional cohomology group.*

The cup product of cocycles pertaining to different handles vanishes already at the level of cochains for obvious reasons.

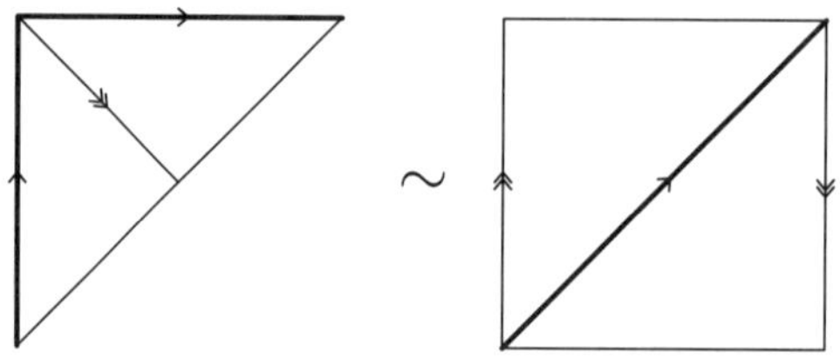

Figure 6. A basis cycle on the Möbius band

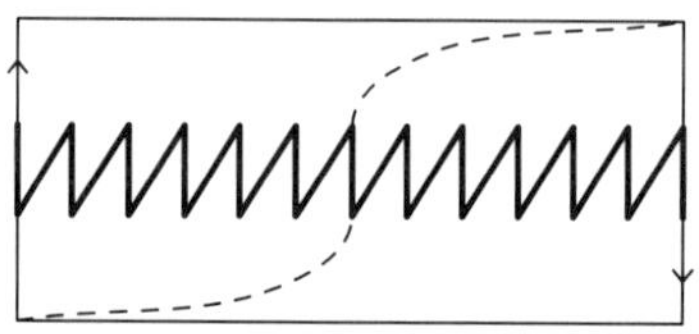

Figure 7. The basis cocycle on the Möbius band

Now, consider the cohomology of nonorientable two-dimensional surfaces. Unlike the homology groups of a closed nonorientable surface with coefficients in $\mathbb{Z}$, the two-dimensional cohomology group with coefficients in $\mathbb{Z}_2$ is nontrivial. Its generator is as follows. On the polygon from which the nonorientable two-dimensional surface is built by gluing, all simplices $\Delta_1^2, \ldots, \Delta_p^2$ can be endowed with compatible orientations. The cochains $(\Delta_i^2)^*$ and $(\Delta_j^2)^*$ are then cohomologous, as in the orientable case. But in the nonorientable case, there exist two simplices, say Δ_i^2 and Δ_j^2, which have opposite orientations on the surface itself. Hence, at the cohomology level, we have $(\Delta_i^2)^* = (\Delta_j^2)^*$ and $(\Delta_i^2)^* = -(\Delta_j^2)^*$; therefore, $2(\Delta_i^2)^* = 0$. Thus, a cochain $\sum a_i(\Delta_i^2)^*$ generates the cohomology group if and only if the number $\sum a_i$ is odd; if $\sum a_i$ is even, then this cochain represents the zero cohomology class.

Let us calculate the cohomology ring of the surface mP^2 with coefficients in $\mathbb{Z}_2$. Each Möbius band attached to S^2 corresponds to one direct summand $\mathbb{Z}_2$ in the one-dimensional homology and cohomology groups. For a Möbius band attached to the sphere, a basis cycle is represented by the diagonal of a rectangle (see Figure 6). In Figure 7, the basis cycle is shown by a dashed line and the dual cocycle α, by a polygonal line.

The product $\alpha \smile \alpha$ can be calculated in the same way as in the case of a handle, i.e., by replacing α with a cocycle α' obtained from α by adding some coboundaries of cochains dual to points. However, in the nonorientable case, it is impossible to completely separate the cocycles, because A and B in Figure 8a represent the same point. Supplementing α with the coboundaries of the cochains dual to the points $A_1, A_2, \ldots, A_k$, we obtain the cocycle α'

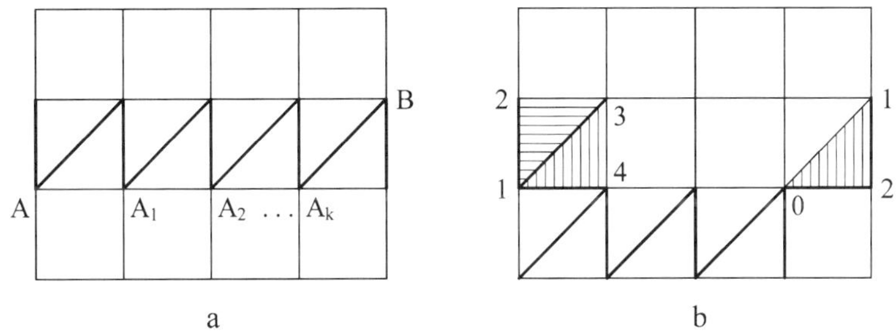

Figure 8. The basis cocycle α' on the Möbius band

shown in Figure 8b. The cocycles α and α' share their boundaries with three 2-simplices (they are hatched in Figure 8b). Let us number the vertices of these simplices. For the numbering shown in Figure 8b, we obtain

$$\begin{aligned}&\langle \alpha \smile \alpha', [012]\rangle = 1;\\ &\langle \alpha \smile \alpha', [123]\rangle = 0, \text{ because } \langle \alpha', [23]\rangle = 0;\\ &\langle \alpha \smile \alpha', [134]\rangle = 0, \text{ because } \langle \alpha', [34]\rangle = 0.\end{aligned}$$

Therefore, the cocycle $\alpha \smile \alpha'$ generates the two-dimensional cohomology group. We have proved the following theorem.

Theorem 2.4. *The one-dimensional $\mathbb{Z}_2$-cohomology group of the sphere with m Möbius bands attached has generators $\alpha_1, \ldots, \alpha_m$ such that $\alpha_i \smile \alpha_j = 0$ for $i \neq j$ and $\alpha_i \smile \alpha_i = \Lambda$, where Λ is the generator of the two-dimensional cohomology group.*

Problem 57. Let M_1^2 and M_2^2 be closed orientable two-dimensional surfaces different from the sphere S^2.

(a) Let $f\colon M_1^2 \to M_2^2$ be a continuous map. Prove that the homomorphism $f_2^*\colon H^2(M_2^2) \to H^2(M_1^2)$ is completely determined by the homomorphism $f_1^*\colon H^1(M_2^2) \to H^1(M_1^2)$.

(b) Prove that there exists a homomorphism $h\colon H^1(M_2^2) \to H^1(M_1^2)$ which is not induced by any continuous map $f\colon M_1^2 \to M_2^2$.

2. Homology and Cohomology of Manifolds

The most important role in studying the homology and cohomology of manifolds is played by the Poincaré duality isomorphism, which can be expressed in terms of cap product. Cap product is defined for any simplicial complexes, but its applications are related mainly to manifolds. The identification of cohomology and homology by means of the Poincaré isomorphism transforms the cap product of manifolds into the map $H_k(M^n) \times H_l(M^n) \to H_{k+l-n}(M^n)$ that sends each pair of cycles to their intersection. Historically, cap product arose as a generalization of intersections of cycles in manifolds to arbitrary complexes.

2.1. Cap Product. The cap product is a bilinear map

$$\frown : H^p(K;R) \times H_{p+q}(K;R) \to H_q(K;R),$$

where R is the additive group of an associative commutative ring with identity. We could define cap product for total chain complexes and then carry the definition over to simplicial homology and cohomology, but we shall not repeat this procedure; instead of checking that we have the well-defined notions (at the level of homology and cohomology), we order the vertices of our complex and consider simplicial chains and cochains.

We assume that the simplicial complex K is path-connected. For $c^p \in C^p(K;R)$, we set

$$c^p \frown [v_0, \dots, v_{p+q}] = \langle c^p, [v_q, \dots, v_{p+q}]\rangle [v_0, \dots, v_q];$$

it is assumed that the vertices $v_0, \dots, v_{p+q}$ are numbered in increasing order. For $q = 0$, we set

$$c^p \frown [v_0, \dots, v_p] = \langle c^p, [v_0, \dots, v_p]\rangle \in R \cong C_0(K;R).$$

Extending this map by linearity, we obtain a bilinear map

$$\frown : C^p(K;R) \times C_{p+q}(K;R) \to C_q(K;R).$$

It induces a map of (co)homology groups due to the following property.

Lemma. $\partial(d^p \frown c_{p+q}) = (-1)^q(\delta d^p \frown c_{p+q}) + d^p \frown \partial c_{p+q}$.

Proof. It is sufficient to check the required equality for $c_{p+q} = [v_0, \dots, v_{p+q}]$. By definition, we have

$$\begin{aligned} d^p &\frown \partial[v_0, \dots, v_{p+q}] \\ &= \sum_{0 \le i \le q-1} (-1)^i \langle d^p, [v_q, \dots, v_{p+1}]\rangle [v_0, \dots, \hat{v}_i, \dots, v_q] \\ &\quad + \sum_{q \le i \le p+q} (-1)^i \langle d^p, [v_{q-1}, \dots, \hat{v}_i, \dots, v_{p+q}]\rangle [v_0, \dots, v_{q-1}] \end{aligned}$$

and

$$\begin{aligned} (-1)^q &\delta d^p \frown [v_0, \dots, v_{p+q}] \\ &= (-1)^q \langle \delta d^p, [v_{q-1}, \dots, v_{p+q}]\rangle [v_0, \dots, v_{q-1}] \\ &= \sum_{q-1 \le i \le p+q} (-1)^{i+1} \langle d^p, [v_{q-1}, \dots, \hat{v}_i, \dots, v_{p+q}]\rangle [v_0, \dots, v_q]. \end{aligned}$$

When we add the second sum in the first expression to the second expression, all terms except for the one with $i = q$ cancel out. It is also clear that adding the term with $i = q$ to the first sum (in the first expression), we obtain $\partial(d^p \frown [v_0, \dots, v_{p+q}])$. $\square$

The formula for $\partial(d^p \frown c_{p+q})$ shows that the cap product is well defined for (co)homology.

Theorem 2.5. *Cap product is natural in the sense that if $f\colon |K| \to |L|$ is a continuous map, then $f_*(f^*\alpha \frown \beta) = \alpha \frown f_*(\beta)$.*

Proof. The required equality holds even at the level of (co)chains:

$$\begin{aligned} f_*(f^*c^p \frown [v_0, \dots, v_{p+q}]) &= f_*(\langle f^*c^p, [v_q, \dots, v_{p+q}]\rangle [v_0, \dots, v_q]) \\ &= \langle f^*c^p, [v_q, \dots, v_{p+q}]\rangle f_*[v_0, \dots, v_q] \\ &= \langle c^p, f_*[v_q, \dots, v_{p+q}]\rangle f_*[v_0, \dots, v_q] \\ &= c^p \frown f_*[v_q, \dots, v_{p+q}]. \end{aligned}$$ $\square$

Theorem 2.6. *The cap product is dual to the cup product in the sense that*

$$\langle \alpha^q, \beta^p \frown \gamma_{p+q}\rangle = \langle \alpha^q \smile \beta^p, \gamma_{p+q}\rangle.$$

Proof. The required equality holds even at the level of (co)chains:

$$\begin{aligned} \langle c^q, c^p \frown [v_0, \dots, v_{p+q}]\rangle &= \langle c^q, \langle c^p, [v_q, \dots, v_{p+q}]\rangle [v_0, \dots, v_q]\rangle \\ &= \langle c^q, [v_0, \dots, v_q]\rangle \langle c^p, [v_q, \dots, v_{p+q}]\rangle \\ &= \langle c^q \smile c^p, [v_0, \dots, v_{p+q}]\rangle. \end{aligned}$$ $\square$

Theorem 2.7. *The cap and cup products are related by*

$$\alpha^p \frown (\beta^q \frown \gamma_{p+q+r}) = (\alpha^p \smile \beta^q) \frown \gamma_{p+q+r}.$$

Proof. The required equality holds even at the level of (co)chains. If $\gamma_{p+q+r} = [v_0, \dots, v_{p+q+r}]$, then both expressions are equal to

$$\langle \alpha^p, [v_r, \dots, v_{r+p}]\rangle \langle \beta^q, [v_{r+p}, \dots, v_{p+q+r}]\rangle [v_0, \dots, v_r].$$ $\square$

In the relative case, a similar construction of cap product yields a map

$$\frown\colon H^p(K, L_1) \times H_{p+q}(K, L_1 \cup L_2) \to H_q(K, L_2).$$

Indeed, let $[v_0, \dots, v_{p+q}]$ be a simplex in $L_1 \cup L_2$. It is contained entirely either in L_1 or in L_2. If it is contained in L_1, then any cochain from $C^p(K, L_1)$ vanishes at the simplex $[v_q, \dots, v_{p+q}]$, and if it is contained in L_2, then the simplex $[v_0, \dots, v_q]$ corresponds to the zero element of the group $C_q(K, L_2)$.

The important role played by the cap product in the topology of manifolds is related to the following description of the Poincaré isomorphism $H^{n-k}(M^n; R) \to H_k(M^n; R)$ (see p. 44), where M^n is a closed manifold and $R = \mathbb{Z}_2$ if M^n is nonorientable, in terms of the cap product.

Theorem 2.8. *The Poincaré duality isomorphism has the form* $\alpha^{n-k} \mapsto \alpha^{n-k} \frown [M^n]$, *where* $[M^n]$ *is the fundamental class of the manifold* M^n.

Proof. Consider a triangulation K of the manifold M^n and its barycentric subdivision K'. Let us number the vertices of K' so that the least numbers are assigned to the vertices of K, next follow the barycenters of the 1-simplices in K, then the barycenters of the 2-simplices are numbered, etc. For a representative of the homology class $[M^n]$ we take the sum $\sum \pm[v_0, \dots, v_n]$, where the summation is over all simplices K' whose vertices are written in the order of increasing numbers; the signs indicate whether the orientations of the simplex and manifold are compatible.

The vertices of the complex K' are numbered in such a way that a simplex $[v_0, \dots, v_k]$ is contained in one of the k-simplices σ_i in K. Each vertex v_k uniquely determines a k-simplex σ_i, and each $(n-k)$-cell σ_i^* is represented as $\sum \pm[v_k, \dots, v_n]$; here the vertices are again written in the order of increasing numbers.

At the level of (co)chains, the Poincaré isomorphism associates to each chain $c_k = \sum a_i\sigma_i$ a cochain c^{n-k} for which $\langle c^{n-k}, \sigma_i^*\rangle = a_i$. Therefore,

$$\begin{aligned} c^{n-k} \frown \sum \pm[v_0, \dots, v_n] &= \sum \langle c^{n-k}, \pm[v_k, \dots, v_n]\rangle [v_0, \dots, v_k] \\ &= \langle c^{n-k}, \sigma_i^*\rangle \sigma_i = \sum a_i\sigma_i = c_k, \end{aligned}$$

as required. □

When combined together, the two approaches to the Poincaré duality isomorphism have the following important consequences.

(1) The map $\alpha^{n-k} \mapsto \alpha^{n-k} \frown [M^n]$ is an isomorphism between the groups $H^{n-k}(M^n; R)$ and $H_k(M^n; R)$;

(2) The correspondence between the cochains c^{n-k} and the chains c_k, which is not invariant at the level of (co)chains, is invariant at the (co)homology level.

The invariant definition of the Poincaré isomorphism allows us to give an invariant definition of the intersection number of the cycles representing the homology classes of cycles $\alpha_k \in H_k(M^n; R)$ and $\beta_{n-k} \in H_{n-k}(M^n; R)$. Namely, for $\alpha_k = \alpha^{n-k} \frown [M^n]$ and $\beta_{n-k} = \beta^k \frown [M^n]$, we have

$$(9) \qquad \langle\langle \alpha_k, \beta_{n-k}\rangle\rangle = (\alpha^{n-k} \smile \beta^k) \frown [M^n] = \langle \alpha^{n-k} \smile \beta^k, [M^n]\rangle.$$

Indeed, suppose that $\alpha_k = \sum a_i\sigma_i$ and $\beta_{n-k} = \sum b_i\sigma_i^*$ at the level of chains. Then

$$\begin{aligned} (\alpha^{n-k} \smile \beta^k) \frown [M^n] &= \alpha^{n-k} \frown (\beta^k \frown [M^n]) = \alpha^{n-k} \frown \beta_{n-k} \\ &= \left\langle \alpha^{n-k}, \sum b_i\sigma_i^*\right\rangle = \sum b_i\langle \alpha^{n-k}, \sigma_i^*\rangle = \sum b_i a_i. \end{aligned}$$

Formula (9) can be rewritten so as to express the cup product in terms of the intersection number. Namely, let Λ be an element of $H^n(M^n; R)$ such that $\Lambda \frown [M^n] = 1 \in H_0(M^n; R) \cong R$. Then

$$\alpha^{n-k} \smile \beta^k = \langle\langle \alpha_k, \beta_{n-k} \rangle\rangle \Lambda. \tag{10}$$

Relation (9) and anticommutativity of the cup product imply

$$\langle\langle \alpha_k, \beta_{n-k} \rangle\rangle = (-1)^{k(n-k)} \langle\langle \beta_{n-k}, \alpha_k \rangle\rangle. \tag{11}$$

To see that (11) holds at the level of chains, it suffices to take chains representing the classes α_k and β_{n-k} in the chain complexes $C_k(K)$ and $C_{n-k}(K^*)$. Therefore, equality (11) follows directly from the invariance of the definition of intersection number.

The relation $\dim(V + W) = \dim V + \dim W - \dim(V \cap W)$, well-known in linear algebra, shows that if $\dim V = k$, $\dim W = l$, and the spaces V and W are subspaces of an n-space, then $\dim(V \cap W) \geq k + l - n$. If $\dim(V \cap W) = k + l - n$, then the subspaces V and W are said to be *transversal*.

In agreement with this terminology, we say that a k- and an l-dimensional cycles in an n-manifold are *transversal* if the intersection of any two simplices contained in different cycles is either empty or of dimension $k+l-n$. In particular, cycles of complementary dimensions are transversal if and only if they intersect in finitely many points.

Relation (9) has the following geometric interpretation: if α_k and β_{n-k} are two transversal cycles of complementary dimensions, then the cohomology class dual to their intersection is the cup product of the classes α^{n-k} and β^k dual to α_k and β_{n-k}. This statement is true not only for cycles of complementary dimensions but also for any transversal cycles. Let us prove this. Take cycles α_k and β_l; their dual cocycles are α^{n-k} and β^{n-l}. We have

$$\begin{aligned}\langle \alpha^{n-k} \smile \beta^{n-l}, [v_0, \dots, v_{2n-k-l}] \rangle \\ = \langle \alpha^{n-k}, [v_0, \dots, v_{n-k}] \rangle \cdot \langle \beta^{n-l}, [v_{n-k}, \dots, v_{2n-k-l}] \rangle.\end{aligned}$$

Suppose that $\alpha_k = \sum x_i \sigma_i^k$ and $\beta_l = \sum y_j \sigma_j^l$. Then $\alpha^{n-k} = \sum x_i (\sigma_i^k)^*$ and $\beta^{n-l} = \sum y_j (\sigma_j^l)^*$. For transversal cycles, we have $\alpha_k \cap \beta_l = \gamma_{k+l-n} = \sum \alpha_i \beta_j \sigma_i^k \cap \sigma_j^l$; therefore, $\gamma^{2n-k-l} = \sum \alpha_i \beta_j (\sigma_i^k \cap \sigma_j^l)^*$. Thus, we must show that if $[v_0, \dots, v_{n-k}) = (\sigma^k)^*$ and $[v_{n-k}, \dots, v_{2n-k-l}] = (\sigma^l)^*$, then $[v_0, \dots, v_{2n-k-l}] = (\sigma^k \cap \sigma^l)^*$. In the barycentric coordinates, the cell σ^k is determined by the equations $x_0 = x_1 = \cdots = x_{n-k}$, and the cell σ^l is determined by the equations $x_{n-k} = \cdots = x_{2n-k-l}$. Hence the cell $\sigma^k \cap \sigma^l$ is determined by the equations $x_0 = x_1 = \cdots = x_{2n-k-l}$. Now, it is clear that $(\sigma^k \cap \sigma^l)^*$ coincides with the simplex $[v_0, \dots, v_{2n-k-l}]$.

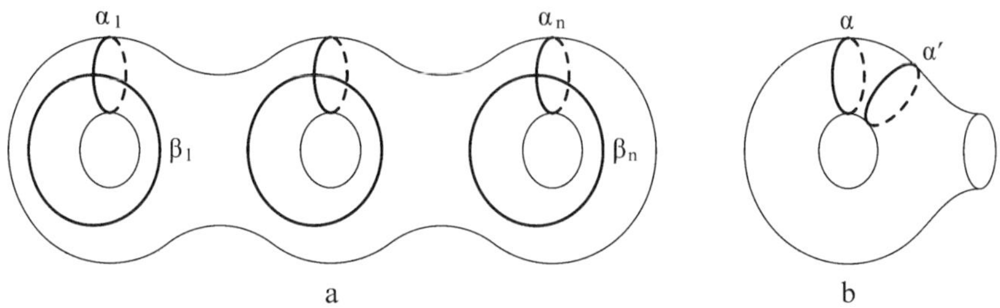

Figure 9. The basis cycles on a sphere with handles

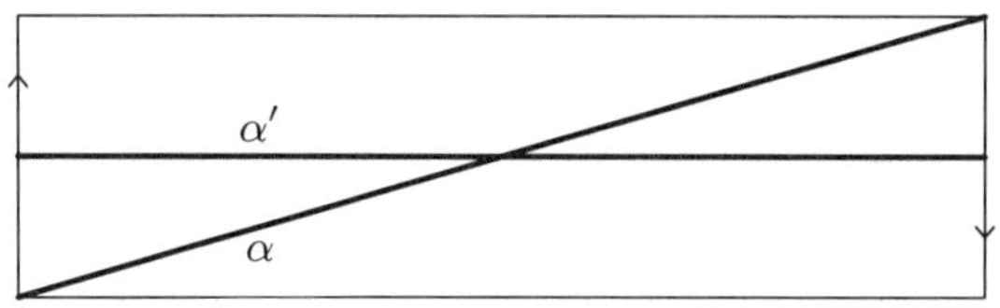

Figure 10. The basis cycle on the Möbius band

Problem 58. (a) Calculate the integral cohomology ring of $S^k \times S^k$.

(b) Find the self-intersection number of the diagonal in $S^k \times S^k$.

Problem 59. Prove that for positive m and n, any map $f\colon S^{n+m} \to S^n \times S^m$ induces the trivial map $f^*\colon H^{n+m}(S^n \times S^m) \to H^{n+m}(S^{n+m})$.

2.2. Cohomology Rings of Manifolds. The Poincaré isomorphism and the relation

$$\alpha^{n-k} \smile \beta^k = \langle\langle \alpha_k, \beta_{n-k} \rangle\rangle \Lambda \tag{12}$$

substantially facilitate calculating cohomology rings for manifolds (here $\Lambda \in H^n(M^n;\mathbb{R})$ is the element for which $\Lambda \frown [M^n] = 1$, i.e., $\langle \Lambda, [M^n] \rangle = 1$). For example, the calculation performed above for closed two-dimensional surfaces becomes almost obvious, because it suffices to determine the arrangement of the dual cycles.

Figure 9a shows basis cycles $\alpha_1, \ldots, \alpha_n$ and $\beta_1, \ldots, \beta_n$ for a sphere with n handles. Two cycles intersect only if they have equal subscripts. However, it is seen from Figure 9b that each cycle α is homologous to a cycle α' disjoint from it.

For the Möbius band attached to the sphere, the basis cycle α is represented by the diagonal (see Figure 10). It is homologous to a cycle α' which transversally intersects α in precisely one point. For the projective plane, α and α' correspond to two projective lines.

An important information about the structure of the cohomology ring of a closed manifold is provided by the following assertion, which we have essentially proved already.

Theorem 2.9. *Suppose that M^n is a closed manifold and F is a field; if M^n is nonorientable, then we assume in addition that $F = \mathbb{Z}_2$. Then the spaces $H^{n-k}(M^n; F)$ and $H^k(M^n; F)$ have bases $\alpha_1, \dots, \alpha_m$ and $\beta_1, \dots, \beta_m$ such that $\alpha_i \smile \beta_j = \delta_{ij}\Lambda$, where Λ is a basis element of the one-dimensional space $H^n(M^n; F)$.*

Proof. The Poincaré isomorphism and relation (12) show that the required assertion can be stated as follows: *The spaces $H_k(M^n; F)$ and $H_{n-k}(M^n; F)$ have bases $\alpha_1^*, \dots, \alpha_m^*$ and $\beta_1^*, \dots, \beta_m^*$ such that $\langle\langle \alpha_i^*, \beta_j^* \rangle\rangle = \delta_{ij}$.* This assertion coincides with the corollary of Theorem 1.28 on p. 46. □

A similar argument, together with Theorem 1.27, proves the following assertion.

Theorem 2.10. *Let M^n be a closed orientable manifold. Then the free Abelian groups $H^{n-k}(M^n; \mathbb{Z})/T^{n-k}$ and $H^k(M^n; \mathbb{Z})/T^k$, where T^{n-k} and T^k are the torsion subgroups, have bases $\alpha_1, \dots, \alpha_m$ and $\beta_1, \dots, \beta_m$ such that $\alpha_i \smile \beta_j = \delta_{ij}\Lambda$, where Λ is a generator of the group $H^n(M^n; \mathbb{Z})$.*

Using Theorems 2.9 and 2.10, we can calculate the cohomology rings $H^*(\mathbb{R}P^n; \mathbb{Z}_2)$ and $H^*(\mathbb{C}P^n; \mathbb{Z})$.

Theorem 2.11. *For $0 \le k \le n$, the group $H^k(\mathbb{R}P^n; \mathbb{Z}_2)$ contains only one nonzero element α^k, and*

$$\alpha^k \smile \alpha^l = \begin{cases} \alpha^{k+l} & \text{if } k+l \le n, \\ 0 & \text{if } k+l > n. \end{cases}$$

Proof. We already know that $H_k(\mathbb{R}P^n; \mathbb{Z}_2) \cong \mathbb{Z}_2$ for $0 \le k \le n$; therefore $H^k(\mathbb{R}P^n; \mathbb{Z}_2) \cong \mathbb{Z}_2$ for $0 \le k \le n$. It remains to verify that $\alpha^k \smile \alpha^l \ne 0$ for $k + l \le n$. Theorem 2.9 implies $\alpha^k \smile \alpha^l \ne 0$ for $k+l = n$, because α^k and α^l are the only nonzero elements of the groups $H^k(\mathbb{R}P^n; \mathbb{Z}_2)$ and $H^l(\mathbb{R}P^n; \mathbb{Z}_2)$. Suppose that $k + l = m < n$. The homomorphism $H^*(\mathbb{R}P^n) \to H^*(\mathbb{R}P^m)$ induced by the natural embedding is a ring isomorphism in dimensions at most m. Since the product $\alpha^k \smile \alpha^l$ is nonzero in the ring $H^*(\mathbb{R}P^m)$, it follows that it is nonzero in $H^*(\mathbb{R}P^n)$. □

Theorem 2.12. *For $k = 0, 2, 4, \dots, 2n$, $H^k(\mathbb{C}P^n; \mathbb{Z}) = \mathbb{Z}$, and for all other k, the cohomology group is trivial. Moreover, there exist generators $\alpha^2, \alpha^4, \dots, \alpha^{2n}$ such that $\alpha^{2k} \smile \alpha^{2l} = \alpha^{2(k+l)}$ for $k + l \le n$.*

Proof. Cohomology groups are easy to obtain from homology groups by using the universal coefficient theorem. The cohomology groups of $\mathbb{C}P^n$ are torsion-free; therefore, Theorem 2.10 applies to the cohomology groups themselves. Let $\beta^2, \beta^4, \ldots, \beta^{2n}$ be their generators. According to Theorem 2.10, we have $\beta^{2k} \smile \beta^{2(n-k)} = \pm\beta^{2n}$. For $k + l = m < n$, we have $\beta^{2k} \smile \beta^{2l} = \pm\beta^{2m}$; the proof is similar to that in the case of $\mathbb{R}P^n$. Now, for generators with correct signs we can take $\alpha^2 = \beta^2, \alpha^4 = \beta^2 \smile \beta^2 = \pm\beta^4, \alpha^6 = \beta^2 \smile \beta^2 \smile \beta^2 = \pm\beta^6, \ldots$. □

Problem 60. (a) Prove that if $m > n$, then any continuous map $f\colon \mathbb{R}P^m \to \mathbb{R}P^n$ induces the zero map $f^*\colon H^k(\mathbb{R}P^n; \mathbb{Z}_2) \to H^k(\mathbb{R}P^m; \mathbb{Z}_2)$ for each $k \geq 1$.

(b) Prove that if $m > n$, then any continuous map $f\colon \mathbb{R}P^m \to \mathbb{R}P^n$ induces the zero map $f_*\colon H_k(\mathbb{R}P^m; \mathbb{Z}_2) \to H_k(\mathbb{R}P^n; \mathbb{Z}_2)$ for each $k \geq 1$.

Problem 61. Prove that if $m > n \geq 1$, then there exists no continuous map $g\colon S^m \to S^n$ such that $g(-x) = -g(x)$ for all $x \in S^n$ (this fact is known as the *Borsuk–Ulam theorem*).

Problem 62. (a) Prove that $\mathbb{R}P^n$ and $\mathbb{C}P^n$ cannot be represented as the union of n contractible subcomplexes.

(b) Represent $\mathbb{R}P^n$ and $\mathbb{C}P^n$ as the union of $n+1$ contractible subcomplexes.

Problem 63. Given a continuous map $f\colon \mathbb{C}P^n \to \mathbb{C}P^n$, prove that $\Lambda(f) = 1 + \lambda + \lambda^2 + \cdots + \lambda^n$, where $\lambda \in \mathbb{Z}$.

Problem 64. Prove that if n is even, then any continuous map $f\colon \mathbb{C}P^n \to \mathbb{C}P^n$ has a fixed point.

Problem 65. Prove that the degree of any map $f\colon \mathbb{C}P^n \to \mathbb{C}P^n$ has the form λ^n, where $\lambda \in \mathbb{Z}$.

Problem 66. (a) Prove that for even n, there exists no orientation-reversing diffeomorphism $f\colon \mathbb{C}P^n \to \mathbb{C}P^n$.

(b) Construct an orientation-reversing diffeomorphism $\mathbb{C}P^{2n+1} \to \mathbb{C}P^{2n+1}$.

Problem 67. Prove that the spaces $\mathbb{C}P^2$ and $S^2 \vee S^4$ are not homotopy equivalent; using this fact, show that the Hopf fibration $S^3 \to S^2$ is not homotopic to a constant map.

2.3. Two Examples. In this section, we give two examples showing that cohomology *rings* are not determined by data that completely determine homology *groups*. First, *CW*-complexes with isomorphic cellular chain complexes may have nonisomorphic ring cohomologies. Thus, to define multiplication in cohomology, it is necessary to consider the simplicial structure; the cell structure alone[1] is not sufficient for this purpose. Second, the cohomology ring with coefficients in $\mathbb{Z}$ does not determine cohomology rings with other coefficients; spaces with isomorphic integral cohomology rings may have nonisomorphic $\mathbb{Z}_2$-cohomology rings.

We use the fact that for any coefficient ring and any $i > 0$, we have $H^i(X \vee Y) = H^i(X) \oplus H^i(Y)$, and, moreover, $(x + y) \smile (x' + y') = x \smile x' + y \smile y'$ for cohomology classes of positive dimension (see Problem 54).

Example 19. The cellular chain complexes for the torus T^2 and the space $S^1 \vee S^1 \vee S^2$ are isomorphic, but the cohomology rings of these spaces are not.

Proof. The cellular chain complexes for both spaces have the form

$$\mathbb{Z} \longrightarrow \mathbb{Z} \oplus \mathbb{Z} \longrightarrow \mathbb{Z} \longrightarrow 0;$$

all boundary homomorphisms are zero.

For the space $S^1 \vee S^1 \vee S^2$, the one-dimensional cohomology group is isomorphic to $H^1(S^1 \vee S^1)$ because $H^1(S^2) = 0$. Therefore, the product of any one-dimensional classes of cocycles vanishes. On the other hand, for the torus, the product of the generators of $H^1(T^2)$ does not vanish. □

Problem 68. Prove that the spaces $S^m \times S^n$ and $S^m \vee S^n \vee S^{n+m}$ have equal cohomology groups but different cohomology rings.

Example 20. The cohomology rings of the spaces $\mathbb{R}P^3$ and $\mathbb{R}P^2 \vee S^3$ with coefficients in $\mathbb{Z}$ are isomorphic, whereas their cohomology rings with coefficients in $\mathbb{Z}_2$ are not.

Proof. Both spaces have trivial one-dimensional integral cohomology group; therefore, multiplication in cohomology is trivial. The $\mathbb{Z}_2$-cohomology ring of space $\mathbb{R}P^3$ has an element α for which $\alpha \smile \alpha \smile \alpha \neq 0$, whereas the $\mathbb{Z}_2$-cohomology ring of $P^2 \vee S^3$ has no such element. Indeed, the only nonzero elements of this ring are $\alpha' \in H^1(\mathbb{R}P^2; \mathbb{Z}_2)$, $\alpha' \smile \alpha' \in H^2(\mathbb{R}P^2; \mathbb{Z}_2)$, and $\beta \in H^3(S^3; \mathbb{Z}_2)$. □

2.4. The Lefschetz Isomorphism. The Lefschetz isomorphism generalizes the Poincaré duality isomorphism to compact orientable manifolds with boundary. This is an isomorphism between the groups $H^{n-k}(M^n)$ and $H_k(M^n, \partial M^n)$ and between the groups $H^{n-k}(M^n, \partial M^n)$ and $H_k(M^n)$. Both isomorphisms are given by the same formula $x \mapsto x \frown [M^n]$, where

[1] By the cell structure we mean the structure determined by the boundary homomorphism, i.e., that of the cellular chain complex. Certainly, if not only the boundary homomorphism but also the characteristic maps of cells are given, then the CW-complex is determined up to homeomorphism, and thereby its cohomology ring is determined uniquely.

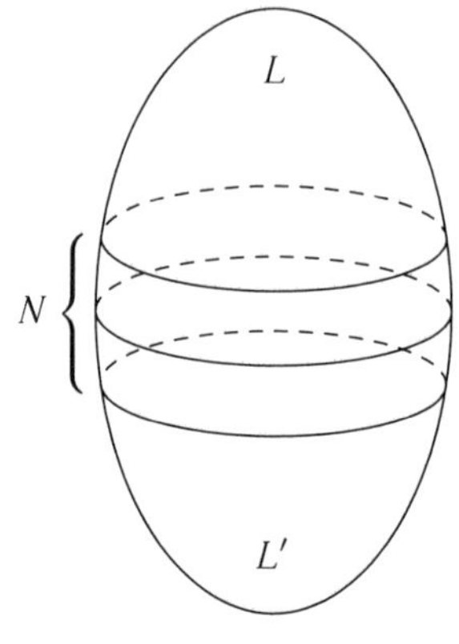

Figure 11. The manifold $\tilde{M}$

$[M^n] \in H_n(M^n, \partial M^n)$ is the fundamental class. The point is that the cap multiplication

$$\frown : H^{n-k}(K, L_1) \times H_n(K, L_1 \cup L_2) \to H_k(K, L_2),$$

where $K = M^n$ and $L_1 \cup L_2 = \partial M^n$, can be defined in two different ways, by setting $L_1 = \varnothing$ and $L_2 = \partial M^n$ or $L_1 = \partial M^n$ and $L_2 = \varnothing$.

To prove a theorem about the Lefschetz isomorphism, we need the following assertion, which is often useful in dealing with manifolds with boundary.

Theorem 2.13 (on collar). *If M^n is a compact manifold with boundary ∂M^n, then there exists a smooth embedding $F\colon \partial M^n \times [0, 1) \to M^n$ such that $F(x, 0) = x$ for $x \in \partial M^n$.*

Proof. Choose finitely many coordinate neighborhoods $U_1, \ldots, U_m$ in M^n covering ∂M^n. Let $\{\lambda_i\}$ be a smooth partition of unity subordinate to the cover $\{U_i \cap \partial M^n\}$. Each set U_i can be identified with a subset of $\mathbb{R}^n$ such that ∂M^n is determined by the equation $x_1 = 0$ and the vector $e_1 = (1, 0, \ldots, 0)$ is directed inside M^n. Consider the vector fields $w_i(x_1, \ldots, x_n) = \lambda_i(x_2, \ldots, x_n)e_i$ on the U_i. On the set $U = \bigcup_{i=1}^m U_i$, the vector field $\sum_{i=1}^m w_i$ is defined; it vanishes nowhere and is directed inside M^n on ∂M^n. Let $H_t(x)$ be the displacement of the point x during time t along the trajectory of this vector field. Since the manifold ∂M^n is compact, we can choose $\varepsilon > 0$ so that $H_t(x)$ is defined for all $x \in \partial M^n$ and all $t \in [0, \varepsilon]$. The map $F(x, t) = H_t(x)$ is the required smooth embedding $\partial M^n \times [0, \varepsilon) \to M^n$. $\square$

Let M be a compact manifold with boundary ∂M. Consider the closed manifold $\tilde{M}$ obtained from two copies of M by identifying the respective points of their boundaries; we denote these copies by M and M'. The collar theorem implies the existence of a submanifold $N \approx \partial M^n \times [-1,1]$ in $\tilde{M}$ (∂M corresponds to $\partial M \times \{0\}$) such that the closure of $\tilde{M} \setminus N$ consists of two disjoint subsets L and L' homeomorphic to M (see Figure 11). The manifold $\tilde{M}$ can be triangulated so that its submanifolds N, L, and L' are subcomplexes.

Consider the relative Mayer–Vietoris sequence

$$\cdots \to H_k(\tilde{M}, L \cap L') \longrightarrow H_k(\tilde{M}, L') \oplus H_k(\tilde{M}, L) \longrightarrow H_k(\tilde{M}, L \cup L') \to \cdots .$$

In this sequence, $H_k(\tilde{M}, L \cap L') \cong H_k(\tilde{M})$ because $L \cap L' = \varnothing$. To transform the remaining homology groups, we use the isomorphism $H_*(A, B) \cong H_*(A \cup C, B \cup C)$, where $A \cup C$ is a simplicial complex, A and C are subcomplexes in $A \bigcap C$, and B is a subcomplex in A. This isomorphism is obvious even at the level of relative chains.

Clearly, the pairs $(\tilde{M}, M')$ and $(\tilde{M}, L')$ are homotopy equivalent and $(\tilde{M}, M') = (M \cup M', \partial M \cup M')$; therefore, $H_k(\tilde{M}, L') \cong H_k(M, \partial M)$. Moreover, $(\tilde{M}, L \cup L') = (N \cup (L \cup L'), \partial N \cup (L \cup L'))$, whence $H_k(\tilde{M}, L \cup L') \cong H_k(N, \partial N)$. Let us show that $H_k(N, \partial N) \cong H_{k-1}(\partial M)$ for $k > 1$. Attaching two copies of the cone over ∂M to N, we obtain a simplicial complex homeomorphic to $\Sigma(\partial M)$. Hence $H_k(N, \partial N) \cong H_k(\Sigma(\partial M), C \cup C')$, where $C' \approx C = C(\partial M)$. The exact sequence of the pair $(\Sigma(\partial M), C \cup C')$ shows that $H_k(N, \partial N) \cong H_k(\Sigma(\partial M))$ for $k > 1$. Moreover, $H_k(\Sigma(\partial M)) \cong H_{k-1}(\partial M)$.

In the new exact sequence

$$\begin{aligned} \longrightarrow H_k(\tilde{M}) \longrightarrow H_k(M, \partial M) &\oplus H_k(M', \partial M') \\ &\longrightarrow H_{k-1}(\partial M) \longrightarrow H_{k-1}(\tilde{M}) \longrightarrow, \end{aligned}$$

the homomorphism $H_k(M, \partial M) \to H_{k-1}(\partial M)$ is the composition of the homomorphisms

$$H_k(M, \partial M) \longrightarrow H_k(N, \partial N) \longrightarrow H_k(\Sigma(\partial M), C \cup C') \xrightarrow{\partial_*} H_{k-1}(\partial M);$$

therefore, it coincides with the homomorphism ∂_* sending each relative cycle to its absolute boundary. The remaining two homomorphisms are induced by inclusions.

Consider the absolute Mayer–Vietoris sequence

$$\cdots \longrightarrow H^{n-k}(\tilde{M}) \longrightarrow H^{n-k}(M) \oplus H^{n-k}(M') \longrightarrow H^{n-k}(\partial M) \longrightarrow \cdots .$$

This exact sequence can be mapped to the one considered above as follows:

$$\begin{array}{ccc}
\to H^{n-k}(\tilde{M}) & \longrightarrow & H^{n-k}(M) \oplus H^{n-k}(M') \\
\downarrow{\frown[\tilde{M}]} & & \downarrow{\frown[M]\oplus[M']} \\
\longrightarrow H_k(\tilde{M}) & \longrightarrow & H_k(M, \partial M) \oplus H_k(M', \partial M')
\end{array}$$

$$\begin{array}{ccccc}
\longrightarrow H^{n-k}(\partial M) & \longrightarrow & H^{n-k+1}(\tilde{M}) & \longrightarrow \\
\downarrow{\frown[\partial M]} & & \downarrow{\frown[\tilde{M}]} & \\
\longrightarrow H_{k-1}(\partial M) & \longrightarrow & H_{k-1}(\tilde{M}) & \longrightarrow .
\end{array}$$

In the resulting diagram, the left square is commutative because all operations in it (including the passage from $[\tilde{M}]$ to $[M]$ and $[M']$) are induced by inclusions and restrictions.

The middle square commutes as well because if z^{n-k} is a cocycle, then

$$\partial(z^{n-k} \frown [M]) = (-1)^k(\delta z^{n-k} \frown [M]) + z^{n-k} \frown \partial[M] = z^{n-k} \frown [\partial M].$$

In the right square, the upper map $H^{n-k}(\partial M) \to H^{n-k+1}(\tilde{M})$ acts as follows. Take a cocycle $z^{n-k} \in C^{n-k}(\partial M)$ and consider a cochain $c^{n-k} \in C^{n-k}(M)$ whose restriction to ∂M coincides with z^{n-k}. Let d^{n-k+1} be the cochain in $C^{n-k+1}(\tilde{M})$ which coincides with δc^{n-k} on M and vanishes outside M (it exists because δc^{n-k} vanishes on ∂M). The map in question has the form $[z^{n-k}] \mapsto [d^{n-k+1}]$. The lower map $i_*\colon H_{k-1}(\partial M) \to H_{k-1}(\tilde{M})$ in this square is induced by an inclusion. Clearly, $d^{n-k+1} \frown [\tilde{M}] = \delta c^{n-k} \frown [M]$ and

$$\begin{aligned}
i_*(z^{n-k} \frown [\partial M]) &= i_*(i^*(c^{n-k}) \frown [\partial M]) \\
&= c^{n-k} \frown i_*[\partial M] = c^{n-k} \frown [\partial M].
\end{aligned}$$

Finally, the equality

$$\partial(c^{n-k} \frown [M]) = (-1)^k(\delta c^{n-k} \frown [M]) + c^{n-k} \frown [\partial M]$$

shows that the chains $(-1)^{k+1}\delta c^{n-k} \frown [M]$ and $c^{n-k} \frown [\partial M]$ differ by a boundary, i.e., belong to the same homology class. Thus, the right square is commutative up to sign.

We have constructed a diagram commutative up to sign. The maps $\frown [\tilde{M}]$ and $\frown [\partial M]$ in this diagram are isomorphisms by the Poincaré duality theorem. Therefore, by the five lemma, the map $H^{n-k}(M) \xrightarrow{\frown[M]} H_k(M, \partial M)$ is an isomorphism as well.

Similarly, in the diagram

$$\begin{array}{ccccc} \longrightarrow H^{n-k+1}(\partial M) & \longrightarrow & H^{n-k}(M,\partial M)\oplus H^{n-k}(M',\partial M) & & \\ \Big\downarrow{\scriptstyle\frown[\partial M]} & & \Big\downarrow{\scriptstyle\frown[M]\oplus[M']} & & \\ \longrightarrow H_k(\partial M) & \longrightarrow & H_k(M)\oplus H^k(M') & & \end{array}$$

$$\begin{array}{c} \longrightarrow H^{n-k}(\tilde{M}) \longrightarrow \\ \Big\downarrow{\scriptstyle\frown[\tilde{M}]} \\ \longrightarrow H_k(\tilde{M}) \longrightarrow \;, \end{array}$$

the map $H^{n-k}(M,\partial M)\xrightarrow{\frown[M]}H_k(M)$ is an isomorphism. We have proved the following theorem.

Theorem 2.14 (the Lefschetz isomorphism [**78**]). *Let M^n be a compact orientable manifold with boundary ∂M^n. Then the maps $H^{n-k}(M)\xrightarrow{\frown[M]}H_k(M,\partial M)$ and $H^{n-k}(M,\partial M)\xrightarrow{\frown[M]}H_k(M)$ are isomorphisms.*

The same isomorphisms hold in (co)homology with coefficients in R, where R is the additive group of a ring.

In the case of a nonorientable manifold, such isomorphisms hold for (co)homology with coefficients in $\mathbb{Z}_2$.

Problem 69. Prove that if a manifold M^n is contractible, then its boundary is a homology $(n-1)$-sphere, i.e., $H_k(\partial M^n)\cong H_k(S^{n-1})$ for all k.

Problem 70. Given a compact orientable manifold M^3 with nonempty boundary ∂M^3, prove that for its rational homology, the dimension of the image of the map $\partial\colon H_2(M^3,\partial M^3)\to H_1(\partial M^3)$ equals half the dimension of the space $H_1(\partial M^3)$.

2.5. Alexander Duality. The Lefschetz isomorphism theorem has an important consequence, namely, the following duality theorem.

Theorem 2.15 (Alexander duality [**4**, **5**]). *If $M\subsetneq S^n$ is a closed submanifold, then $\tilde{H}^k(M)\cong\tilde{H}_{n-k-1}(S^n\setminus M)$ and $\tilde{H}_k(M)\cong H^{n-k-1}(S^n\setminus M)$ for $0\le k\le n-1$.*

Proof. We begin by proving that $H^k(S^n,M)\cong H_{n-k}(S^n\setminus M)$. Let us remove a point $x\notin M$ from S^n, identify $S^n\setminus\{x\}$ with $\mathbb{R}^n$, and apply the tubular neighborhood theorem to $M\subset\mathbb{R}^n$. If M_ε is an open tubular neighborhood of M, then the pair (S^n,M) is homotopy equivalent to (S^n,M_ε), and the space $S^n\setminus M$ is homotopy equivalent to $S^n\setminus M_\varepsilon$. According to the Lefschetz isomorphism theorem, we have $H^k(S^n\setminus M_\varepsilon,\partial M_\varepsilon)\cong$

$H_{n-k}(S^n \setminus M_\varepsilon)$. Moreover, $H^k(S^n \setminus M_\varepsilon, \partial M_\varepsilon) \cong H^k(S^n, M_\varepsilon) \cong H^k(S^n, M)$ and $H_{n-k}(S^n \setminus M_\varepsilon) \cong H_{n-k}(S^n \setminus M)$.

To obtain the first of the required isomorphisms, consider the exact sequence of the pair (S^n, M):

$$\tilde{H}^k(S^n) \longrightarrow \tilde{H}^k(M) \xrightarrow{\delta^*} H^{k+1}(S^n, M) \longrightarrow \tilde{H}^{k+1}(S^n).$$

For $k \neq n-1$, we have $\tilde{H}^k(S^n) = \tilde{H}^{k+1}(S^n) = 0$, and δ^* is an isomorphism. Hence $\tilde{H}^k(M) \cong H^{k+1}(S^n, M) \cong H_{n-k-1}(S^n \setminus M)$; here $n-k-1 \neq 0$ and, therefore, $H_{n-k-1} \cong \tilde{H}_{n-k-1}$.

For $k = n-1$, the exact sequence takes the form

$$0 \longrightarrow \tilde{H}^{n-1}(M) \xrightarrow{\delta^*} H^n(S^n, M) \xrightarrow{i^*} H^n(S^n) \cong \mathbb{Z}.$$

Here the homomorphism i^* is induced by the inclusion of pairs $i\colon (S^n, \varnothing) \hookrightarrow (S^n, M)$. Consider also the inclusion $j\colon (S^n \setminus M) \hookrightarrow S^n$ and the homomorphism of the zero-dimensional homology groups which it induces. The homomorphisms i^* and j_* enter the diagram

$$\begin{array}{ccc} H^n(S^n, M) & \xrightarrow{i^*} & H^n(S^n) \\ \downarrow \cong & & \downarrow \cong \\ H_0(S^n \setminus M) & \xrightarrow{j_*} & H_0(S^n), \end{array}$$

in which the vertical arrows are the Lefschetz and Poincaré isomorphisms. This diagram is commutative up to sign; hence $\operatorname{Ker} i^* \cong \operatorname{Ker} j_*$. The homomorphism $p_*\colon H_0(S^n \setminus M) \to H_0(*)$, where $*$ is a point from $S^n \setminus M$, can be represented as the composition $H_0(S^n \setminus M) \xrightarrow{j_*} H_0(S^n) \xrightarrow{\cong} H_0(*)$, whence $\tilde{H}_0(S^n \setminus M) = \operatorname{Ker} p_* = \operatorname{Ker} j_*$. Thus, $\tilde{H}^{n-1}(M) \cong \operatorname{Im} \delta^* = \operatorname{Ker} i^* \cong \operatorname{Ker} j_* = \tilde{H}_0(S^n \setminus M)$.

The second isomorphism is established in a similar way. □

Corollary 1. *If L is the image, under an embedding into S^3, of a disconnected manifold that consists of n circles, then $H_1(S^3 \setminus L) \cong H^1(L) \cong \mathbb{Z}^n$ and $H_2(S^3 \setminus L) \cong \tilde{H}^0(L) \cong \mathbb{Z}^{n-1}$.*

Corollary 2 ([**19**]). *If $M^{n-1} \subset S^n$ is a closed connected submanifold, then M^{n-1} is orientable and $S^n \setminus M^{n-1}$ has two connected components.*

Proof. According to the Alexander duality theorem, we have $H^{n-1}(M^{n-1}) \cong \tilde{H}_0(S^n \setminus M^{n-1})$. If M^{n-1} is a closed connected nonorientable manifold, then $H^{n-1}(M^{n-1}) \cong \mathbb{Z}_2$ (by Theorem 1.26 on p. 45). On the other hand, $\tilde{H}_0(S^n \setminus M^{n-1})$ is a free Abelian group. Therefore, the manifold M^{n-1} is orientable, and $H^{n-1}(M^{n-1}) \cong \mathbb{Z}$. The isomorphism $\tilde{H}_0(S^n \setminus M^{n-1}) \cong \mathbb{Z}$ implies that $S^n \setminus M^{n-1}$ has two path-connected components. □

Remark. The Alexander duality holds not only for submanifolds but also for any subcomplex $K \subsetneq S^n$ (see Theorem 5.7).

Problem 71. Suppose that m pairwise disjoint copies of the sphere S^{n-2} are embedded in S^n, where $n \geq 3$. Let X be the complement of the embedded spheres. Calculate the homology of the space X.

Problem 72. Given the sphere S^p arbitrarily embedded in S^{n+1}, prove that the space $X = S^{n+1} \setminus S^p$ has the same homology as S^{n-p}.

Problem 73. Given the spheres S^p and S^q arbitrarily embedded in S^{n+1}, prove that the space $X = S^{n+1} \setminus (S^p \sqcup S^q)$ has the same homology as $S^{n-p} \vee S^{n-q} \vee S^n$.

The Linking Number and Multiplication in Cohomology. The linking number can be defined not only for circles embedded in S^3 but also for any closed connected oriented manifolds M_1^p and M_2^q embedded in S^n, where $n = p + q + 1$. The definition repeats almost word for word the one given on p. 46. Namely, consider fundamental cycles $[M_1^p]$ and $[M_2^q]$ as cycles in S^n. Since the p-dimensional homology group of S^n is trivial, it follows that there exists a chain W_1^{p+1} for which $\partial W_1^{p+1} = [M_1^p]$. The *linking number* $\operatorname{lk}(M_1^p, M_2^q)$ is defined to be equal to the intersection number $\langle\langle W_1^{p+1}, [M_2^q]\rangle\rangle$. This number does not depend on the choice of W_1^{p+1}; the proof repeats that for circles embedded in S^3. Taking a chain W_2^{q+1} for which $\partial W_2^{q+1} = [M_2^q]$ and considering the intersection number $\langle\langle [M_1^p], W_2^{q+1}\rangle\rangle$, we obtain the same number (up to sign); this follows from the lemma on p. 43.

The Alexander duality makes it possible to interpret linking number in terms of multiplication in cohomology. This can be performed as follows. Let $X = M_1^p \sqcup M_2^q$. The Alexander duality implies $\tilde{H}^k(S^n \setminus X) \cong \tilde{H}_{p+q-k}(X)$ (recall that $n = p + q + 1$). Therefore, there is a class in $H^q(S^n \setminus X)$ that corresponds to the fundamental class $[M_1^p]$, and there is a class in $H^p(S^n \setminus X)$ that corresponds to the fundamental class $[M_2^q]$. The product of these cohomology classes belongs to $H^{p+q}(S^n \setminus X) \cong \tilde{H}_0(X) \cong \mathbb{Z}$, i.e., it is an integer. This integer is equal to the linking number (up to sign). To prove this, note that the Alexander duality isomorphism is the composition of isomorphisms

$$H^{p+q}(S^n \setminus X) \longrightarrow H_1(S^n, X) \xrightarrow{\partial} \tilde{H}_0(X).$$

The product of the cohomology classes under consideration corresponds the intersection of the chains W_1^{p+1} and W_2^{q+1} in $H_1(S^n, M_1^p \sqcup M_2^q)$ (these chains are assumed to be transversal). Indeed, the Alexander duality isomorphism is the composition of isomorphisms

$$H^q(S^n \setminus X) \longrightarrow H_{p+1}(S^n, X) \xrightarrow{\partial} \tilde{H}_p(X);$$

therefore, the fundamental class $[M_1^p]$ corresponds to the homology class in $H_{p+1}(S^n, X)$ that is represented by the chain W_1^{p+1}. Taking the boundary of the intersection $W_1^{p+1} \cap W_2^{q+1}$ (which consists of arcs of curves), we obtain the union of the two intersections $W_1^{p+1} \cap [M_2^q]$ and $[M_1^p] \cap W_2^{q+1}$. However, we not only take the boundary but also pass to the reduced homology; hence only one of these intersections (say, the former) remains. Indeed, the contribution of each arc from $W_1^{p+1} \cap W_2^{q+1}$ whose endpoints belong to the same manifold (M_1^p or M_2^q) is zero. Nonzero contributions are made only by arcs having one endpoint in M_1^p and the other in M_2^q. To calculate such contributions, one of the two intersections is sufficient.

2.6. The Triple Massey Product. Take three cohomology classes $[\alpha^p] \in H^p(X)$, $[\beta^q] \in H^q(X)$, and $[\gamma^r] \in H^r(X)$ represented by three cocycles α^p, β^q, and γ^r. Suppose that $[\alpha^p] \smile [\beta^q] = 0$ and $[\beta^q] \smile [\gamma^r] = 0$. Then $\alpha^p \smile \beta^q = \delta c^{p+q-1}$ and $\beta^q \smile \gamma^r = \delta d^{q+r-1}$ for some cochains c^{p+q-1} and d^{q+r-1}. Consider the cochain

$$z = c^{p+q-1} \smile \gamma^r + (-1)^p \alpha^p \smile d^{q+r-1}.$$

It is easy to verify that this is a cocycle. Indeed, we have $\delta\gamma^r = 0$ and $\delta\alpha^p = 0$, whence

$$\begin{aligned} \delta^z &= \delta c^{p+q-1} \smile \gamma^r + (-1)^{p-1}(-1)^p \alpha^p \smile \delta d^{q+r-1} \\ &= \alpha^p \smile \beta^q \smile \gamma^r - \alpha^p \smile \beta^q \smile \gamma^r = 0. \end{aligned}$$

The cocycle z corresponds to the cohomology class $[z] \in H^{p+q+r-1}(X)$. This class is not uniquely determined because we can add coboundaries to the cocycles α^p, β^q, and γ^r and cocycles to the cochains c^{p+q-1} and d^{q+r-1}. The set of all cohomology classes $[z]$ thus obtained is called the *triple Massey product* of the classes $[\alpha^p]$, $[\beta^q]$, and $[\gamma^r]$ and denoted by $\langle [\alpha^p], [\beta^q], [\gamma^r] \rangle$.

Theorem 2.16. *The triple Massey product of cohomology classes $[\alpha^p]$, $[\beta^q]$, and $[\gamma^r]$ is an element of the quotient group of $H^{p+q+r-1}(X)$ modulo $[\alpha^p] \smile H^{q+r-1}(X) + H^{p+q-1}(X) \smile [\gamma^r]$.*

Proof. Choose cochains c_1^{p+q-1} and c_2^{p+q-1} so that $\alpha^p \smile \beta^q = \delta c_1^{p+q-1} = \delta c_2^{p+q-1}$. Let z_1 and z_2 be the cocycles constructed by using these cochains. Then $z_1 - z_2 = (c_1^{p+q-1} - c_2^{p+q-1}) \smile \gamma^r$. Here $c_1^{p+q-1} - c_2^{p+q-1}$ is a cocycle; therefore, $[z_1] - [z_2] \in H^{p+q-1}(X) \smile [\gamma^r]$. Similarly, for different cochains d_1^{q+r-1} and d_2^{q+r-1}, we have $[z_1] - [z_2] \in [\alpha^p] \smile H^{q+r-1}(X)$.

Now, suppose that $\alpha_2^p = \alpha_1^p + \delta a^{p-1}$. If $\alpha_1^p \smile \beta^q = \delta c_1^{p+q-1}$, then $\alpha_2^p \smile \beta^q = \delta c_1^{p+q-1} + \delta a^{p-1} \smile \beta^q = \delta(c_1^{p+q-1} + a^{p-1} \smile \beta^q)$ because $\delta\beta^q = 0$.

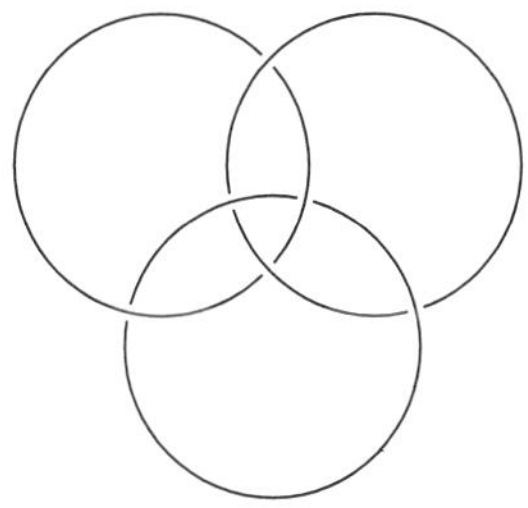

Figure 12. The Borromean rings

Thus, we can set $c_2^{p+q-1} = c_1^{p+q-1} + a^{p-1} \smile \beta^q$ and obtain

$$\begin{aligned} z_2 - z_1 &= a^{p-1} \smile \beta^q \smile \gamma^r + (-1)^p \delta a^{p-1} \smile d^{q+r-1} \\ &= a^{p-1} \smile \delta d^{q+r-1} + (-1)^p \delta a^{p-1} \smile d^{q+r-1} \\ &= (-1)^p \delta (a^{p-1} \smile d^{q+r-1}), \end{aligned}$$

whence $[z_2] = [z_1]$. The case where the cocycle γ^r changes is handled similarly.

Finally, suppose that $\beta_2^q = \beta_1^q + \delta b^{q-1}$. In this case, we have $c_2^{p+q-1} = c_1^{p+q-1} + (-1)^p \alpha^p \smile b^{q-1}$ and $d_2^{q+r-1} = d_1^{q+r-1} + b^{q-1} \smile \gamma^r$. Therefore, $z_2 - z_1 - 2(-1)^p \alpha^p \smile b^{q-1} \smile \gamma^r$. □

The triple Massey product can be used to prove that the *Borromean rings* are linked (see Figure 12). It should be mentioned that for usual Borromean rings, there is a quite elementary proof, which is based on colorings of the link diagram (see [**105**], p. 31). However, for the multidimensional generalization of the Borromean rings, the triple product Massey must be applied.

For calculations, it is convenient to specify the Borromean rings by the equations

$$\begin{aligned} x = 0, &\quad y^2 + \frac{z^2}{4} = 1 \quad \text{(circle } S_1\text{)}; \\ y = 0, &\quad z^2 + \frac{x^2}{4} = 1 \quad \text{(circle } S_2\text{)}; \\ z = 0, &\quad x^2 + \frac{y^2}{4} = 1 \quad \text{(circle } S_3\text{)}. \end{aligned}$$

Figure 13 shows that these equations do indeed determine the Borromean rings.

Instead of the space $\mathbb{R}^3$ in which the given circles are embedded, it is more convenient to consider its one-point compactification S^3. By the Alexander duality theorem, $\tilde{H}^k(S^3 \setminus (S_1 \sqcup S_2 \sqcup S_3)) \cong \tilde{H}_{2-k}(S_1 \sqcup S_2 \sqcup S_3)$.

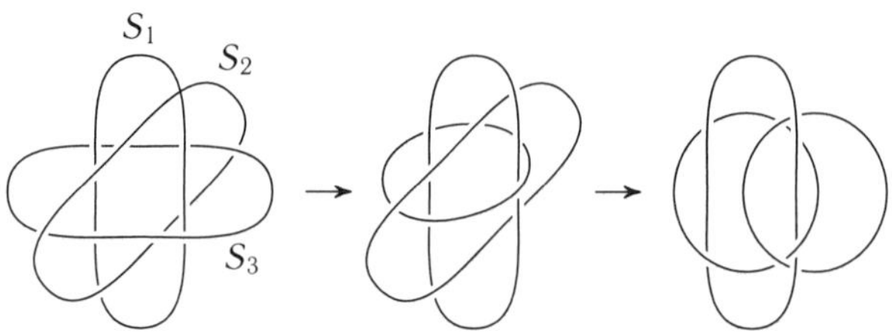

Figure 13. The transformation of the diagram

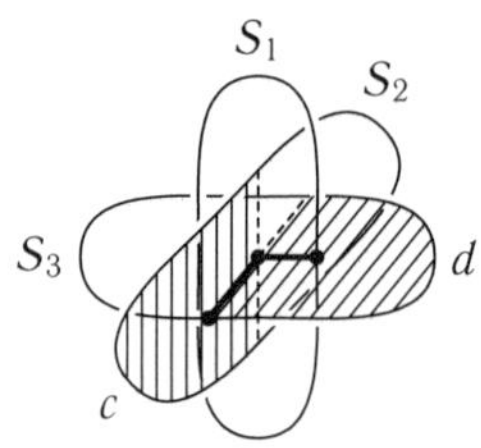

Figure 14. The intersections of chains

The circles S_1, S_2, and S_3 are not pairwise linked; therefore, the pairwise products of the cohomology classes α^1, β^1, and γ^1 corresponding to the fundamental classes of S_1, S_2, and S_3 under the Alexander duality are zero. Thus, the triple product $\langle \alpha^1, \beta^1, \gamma^1 \rangle$ is defined.

Let us show that for the coefficient group $\mathbb{Z}_2$, this triple Massey product is nonzero. First, note that it belongs to $H^2(S^3 \setminus (S_1 \sqcup S_2 \sqcup S_3)) \cong \tilde{H}_0(S_1 \sqcup S_2 \sqcup S_3)$ because the group $\alpha^1 \smile H^1(X) + H^1(X) \smile \gamma^1$, where $X = S^3 \setminus (S_1 \sqcup S_2 \sqcup S_3)$, is trivial. Indeed, not only the pairwise products of the classes α^1, β^1, and γ^1 but also the products $\alpha^1 \smile \alpha^1$, $\beta^1 \smile \beta^1$, and $\gamma^1 \smile \gamma^1$ vanish because the circles S_1, S_2, and S_3 are unknotted and, therefore, S_i can be transformed into S_i' by a small deformation so that the disks D_i and D_i' spanned by S_i and S_i' are disjoint.

Let w_1, w_2, and w_3 be the chains (relative cycles) corresponding to the disks spanned by S_1, S_2, and S_3. Under the isomorphism $H_2(S^3, S_1 \sqcup S_2 \sqcup S_3) \cong H^1(S^3 \setminus (S_1 \sqcup S_2 \sqcup S_3))$, the homology classes of these relative cycles correspond to α^1, β^1, and γ^1. Let c and d be the half-disks such that $w_1 \cap w_2 = \partial c$ and $w_2 \cap w_3 = \partial d$ in the group of relative chains (see Figure 14). Then the triple Massey product $\langle \alpha^1, \beta^1, \gamma^1 \rangle$ corresponds to the relative cycle $c \cap w_3 + w_1 \cap d$ in $H_1(S^3, S_1 \sqcup S_2 \sqcup S_3)$; this cycle is shown by the heavy line in Figure 14. The isomorphism

$$H_1(S^3, S_1 \sqcup S_2 \sqcup S_3) \xrightarrow{\partial} \tilde{H}_0(S_1 \sqcup S_2 \sqcup S_3)$$

takes the homology class of this cycle to a nonzero element because the cycle endpoints belong to different connected components: one of them belongs to S_1 and the other to S_3.

On the other hand, if the link $S_1 \sqcup S_2 \sqcup S_3 \subset S^3$ is trivial (in particular, S_1, S_2, and S_3 are contained in three pairwise disjoint balls), then the disks w_1, w_2, and w_3 can be chosen to be pairwise disjoint, so that the triple Massey product $\langle \alpha^1, \beta^1, \gamma^1 \rangle$ vanishes. Thus, the Borromean rings form a nontrivial link.

A similar argument proves the nontriviality of the following multidimensional generalization of the Borromean rings. Suppose that $n = p + q + r$, $x = (x_1, \dots, x_p)$, $y = (y_1, \dots, y_q)$, and $z = (z_1, \dots, z_r)$. Consider the three spheres in $\mathbb{R}^n$ determined by the equations

$$x = 0, \quad \|y\|^2 + \frac{\|z\|^2}{4} = 1 \quad (\text{sphere } S_1^{q+r-1});$$

$$y = 0, \quad \|z\|^2 + \frac{\|x\|^2}{4} = 1 \quad (\text{sphere } S_2^{p+r-1});$$

$$z = 0, \quad \|x\|^2 + \frac{\|y\|^2}{4} = 1 \quad (\text{sphere } S_3^{p+q-1}).$$

Any two of these spheres are separated by an $(n-1)$-sphere. For example, S_1^{q+r-1} lies outside the sphere

$$\frac{\|x\|^2}{3^2} + \frac{\|y\|^2}{(1/2)^2} + \frac{\|z\|^2}{(3/2)^2} = 1,$$

while S_3^{p+q-1} lies inside it. Thus, the spheres under consideration are not pairwise linked. Therefore, the pairwise products of the cohomology classes α^p, β^q, and γ^r corresponding to the fundamental classes of S_1^{q+r-1}, S_2^{p+r-1}, and S_3^{p+q-1} under the Alexander duality vanish. The products $\alpha^p \smile \alpha^p$, $\beta^q \smile \beta^q$, and $\gamma^r \smile \gamma^r$ vanish as well. Let us show that for the coefficient group $\mathbb{Z}_2$, the triple product $\langle \alpha^p, \beta^q, \gamma^r \rangle \in H^{n-1}(S^n \setminus S_1 \sqcup S_2 \sqcup S_3) \cong \tilde{H}_0(S_1 \sqcup S_2 \sqcup S_3)$ does not vanish. Consider the disks w_1, w_2, and w_3 spanned by S_1, S_2, and S_3, that is,

$$w_1 = \left\{ (x, y, z) \mid x = 0, \|y\|^2 + \frac{\|z\|^2}{4} \leq 1 \right\},$$

$$w_2 = \left\{ (x, y, z) \mid y = 0, \|z\|^2 + \frac{\|x\|^2}{4} \leq 1 \right\},$$

$$w_3 = \left\{ (x, y, z) \mid z = 0, \|x\|^2 + \frac{\|y\|^2}{4} \leq 1 \right\}.$$

Clearly,

$$w_1 \cap w_2 = \{(x, y, z) \mid x = 0, y = 0, \|z\| \leq 1\}$$

and

$$w_2 \cap w_3 = \{(x, y, z) \mid y = 0, z = 0, \|x\| \leq 1\}.$$

We set

$$c = \Big\{(x, y, z) \mid x_1 \geq 0, x_2 = \cdots = x_p = 0, y = 0, \|z\|^2 + \frac{x_1^2}{4} \leq 1\Big\}$$

and

$$d = \Big\{(x, y, z) \mid y_1 \geq 0, y_2 = \cdots = y_q = 0, z = 0, \|x\|^2 + \frac{y_1^2}{4} \leq 1\Big\}.$$

For relative chains, we have $w_1 \cap w_2 = \partial c$ and $w_2 \cap w_3 = \partial d$. Therefore, the triple Massey product $\langle \alpha^p, \beta^q, \gamma^r \rangle$ corresponds to the one-dimensional relative cycle $c \cap w_3 + w_1 \cap d$. It is easy to verify that

$$c \cap w_3 = \{(x, y, z) \mid 0 \leq x_1 \leq 1, x_2 = \cdots = x_p = 0, y = 0, z = 0\}$$

and

$$w_1 \cap d = \{(x, y, z) \mid 0 \leq y_1 \leq 1, y_2 = \cdots = y_q = 0, x = 0, z = 0\}.$$

Hence the relative cycle under consideration is a path joining different connected components: one of its endpoints belongs to S_1^{q+r-1} and the other to S_3^{p+q-1}. Such a cycle is a representative of a nonzero homology class.

2.7. Intersection Forms and the Signature of a Manifold. Let M^{2n} be a closed oriented manifold of even dimension, and let F be any field. On the space $H^n(M^{2n}; F)$, we define the bilinear form

$$f(\alpha^n, \beta^n) = \langle \alpha^n \smile \beta^n, [M^{2n}] \rangle,$$

where $[M^{2n}]$ is the fundamental class of the manifold M^{2n}. For the opposite orientation, the fundamental class changes sign; therefore, the bilinear form f changes sign as well, i.e., $f_- = -f$, where f_- is the form for the manifold with the opposite orientation.

On dual space $H_n(M^{2n}; F)$, there is the dual form

$$f^*(\alpha_n, \beta_n) = \langle\langle \alpha_n, \beta_n \rangle\rangle.$$

Recall that the intersection number $\langle\langle \alpha_n, \beta_n \rangle\rangle$ depends on the orientation of the manifold, and for the opposite orientation, the intersection number is of the opposite sign.

The bilinear forms f and f^* are called the *intersection forms.*

The intersection form can be defined not only over the field F but also over the ring $\mathbb{Z}$. For $\mathbb{Z}$, $H_n(M^{2n})$ must be replaced by the free Abelian group $H_n(M^{2n})/T_n$, where T_n is the torsion subgroup, because the intersection form vanishes at the elements of finite order.

We shall consider only intersection forms over $\mathbb{R}$ and $\mathbb{Z}$. The universal coefficient theorem implies

$$H_n(M^{2n};\mathbb{R}) \cong H_n(M^{2n}) \otimes \mathbb{R} \cong (H_n(M^{2n})/T_n) \otimes \mathbb{R}.$$

The elements of finite order do not affect the intersection number, and it is seen directly from the definition of the intersection number that if $\alpha_n, \beta_n \in C_n(M^{2n})$ and $r\alpha_n, s\beta_n \in C_n(M^{2n};\mathbb{R})$, then

$$\langle\langle r\alpha_n, s\beta_n\rangle\rangle = rs\langle\langle \alpha_n, \beta_n\rangle\rangle.$$

Therefore, the intersection forms over $\mathbb{R}$ and $\mathbb{Z}$ have the same matrix. This implies, in particular, that $H_n(M^{2n};\mathbb{R})$ has a basis in which the intersection form has integer coefficients.

Since the cup product is anticommutative, it follows that the intersection form is symmetric for even n and antisymmetric for odd n.

Let Q be the matrix of the intersection form over $\mathbb{Z}$, and let $m = \dim H_n(M^{2n};\mathbb{R})$. Theorem 2.10 can be reformulated as follows: there exist $m \times m$ matrices A and B over $\mathbb{Z}$ for which $A^TQB = I_m$ is the identity matrix. Therefore, $\det Q = \pm 1$.

The following theorem collects the properties of the intersection form proved above.

Theorem 2.17. (a) *The intersection form is determined by a nonsingular integer matrix.*

(b) *For a $4k$-manifold, the matrix of the intersection form is symmetric, and for a $(4k+2)$-manifold, it is antisymmetric.*

(c) *For the same manifold with the opposite orientation, the intersection form has the opposite sign.*

Remark. The intersection form can also be defined for nonorientable and nonclosed manifolds, but it may be degenerate in this case. Moreover, the definitions of the intersection form in terms of multiplication in cohomology and in terms of intersections of cycles are not equivalent for such manifolds. For example, defining the intersection form for the cylinder $I \times S^1$ and for the Möbius band in terms of multiplication in cohomology, we obtain the zero quadratic form in both cases, while defining it in terms of intersections of cycles, we obtain the zero quadratic form for the cylinder and a nonzero form for the Möbius band. Indeed, on the cylinder, the cycles $\{t_1\} \times S^1$ and $\{t_2\} \times S^1$ with $t_1 \neq t_2$ are disjoint, while on the Möbius band, the cycles α and α' shown in Figure 10 intersect at one point.

The properties of the intersection form specified in Theorem 2.17 have the following consequences.

Theorem 2.18. *Let M^{4k+2} be a closed orientable manifold. Then the dimension* $\dim H_{2k+1}(M^{4k+2};\mathbb{R})$ *is even.*

Proof. The matrix of the intersection form of the manifold M^{4k+2} is antisymmetric; therefore, it has even rank (see [**104**], p. 102). Since this matrix is nondegenerate, it follows that its rank is equal to $\dim H_{2k+1}(M^{4k+2};\mathbb{R})$. □

Corollary. *If M^{4k+2} is a closed orientable manifold, then its Euler characteristic $\chi(M^{4k+2})$ is even.*

Proof. For a closed orientable manifold, we have

$$\chi(M^{2n}) \equiv \dim H_n(M^{2n};\mathbb{R}) \pmod 2.$$

The proof of this relation is similar to that of Theorem 1.34 (for orientable manifolds, the argument applies not only to the coefficient group $\mathbb{Z}_2$ but also to $\mathbb{R}$). □

The *signature* of a manifold M^{4k} is defined to be equal to the signature of its intersection form. Recall that the signature of a symmetric bilinear form over $\mathbb{R}$ is defined as follows. Using symmetry, we choose a basis for which $f(\sum x_ie_i, \sum y_ie_i) = \sum \lambda_i x_i y_i$. The signature is equal to the difference between the numbers of positive and negative coefficients λ_i. The signature of a form does not depend on the choice of a basis; therefore, the signature of the oriented manifold M^{4k}, which we denote by $\sigma(M^{4k})$, is an invariant of this manifold. Moreover, $\sigma(-M^{4k}) = -\sigma(M^{4k})$, where $-M^{4k}$ is the manifold M^{4k} with the opposite orientation. We have proved the following theorem.

Theorem 2.19. *If $\sigma(M^{4k}) \neq 0$, then there is no orientation-reversing homeomorphism $f\colon M^{4k} \to M^{4k}$ (i.e., such that $f_*[M^{4k}] = -[M^{4k}]$).*

Theorem 2.19 implies that there exists no orientation-reversing homeomorphism $\mathbb{C}P^{2n} \to \mathbb{C}P^{2n}$ because $\sigma(\mathbb{C}P^{2n}) = \pm 1$. (Another proof of this assertion is given in the solution of Problem 66.)

There is a distinguished orientation on the manifold $\mathbb{C}P^m$, because $\mathbb{C}P^m \setminus \mathbb{C}P^{m-1} = \mathbb{C}^m$ and complex space does have a distinguished orientation (all complex transformations regarded as transformations over $\mathbb{R}$ have positive determinants). It is usually assumed that $\mathbb{C}P^m$ is endowed with this orientation. In this case, $\sigma(\mathbb{C}P^{2n}) = 1$, because the orientation determined by a pair of transversally intersecting n-dimensional complex subspaces in $\mathbb{C}^{2n}$ is compatible with that of $\mathbb{C}^{2n}$.

Problem 74 (Novikov–Rokhlin). Given closed orientable manifolds M_1^{4n} and M_2^{4n}, prove that $\sigma(M_1^{4n} \# M_2^{4n}) = \sigma(M_1^{4n}) + \sigma(M_2^{4n})$.

Using the Lefschetz isomorphism, we can prove the following property of the signature of a manifold that is the boundary of another manifold.

Theorem 2.20 (Thom [**137**]). *If M^{4k} is a closed orientable manifold which is the boundary of a compact orientable manifold W^{4k+1}, then $\sigma(M^{4k}) = 0$.*

Proof. Consider the cohomology and homology sequences of the pair (W^{4k+1}, M^{4k}) and the Poincaré and Lefschetz isomorphisms from the first sequence to the second:

$$\begin{array}{ccccc} H^{2k}(W^{4k+1}) & \xrightarrow{i^*} & H^{2k}(M^{4k}) & \xrightarrow{\delta^*} & H^{2k+1}(W^{4k+1}, M^{4k}) \\ \downarrow{\frown[W^{4k+1}]} & & \downarrow{\frown[M^{4k}]} & & \downarrow{\frown[W^{4k+1}]} \\ H_{2k+1}(W^{4k+1}, M^{4k}) & \xrightarrow{\partial_*} & H_{2k}(M^{4k}) & \xrightarrow{i_*} & H^{2k}(W^{4k+1}). \end{array}$$

In the proof of the Lefschetz isomorphism theorem, we showed that this diagram is commutative up to sign. Exactness and commutativity imply $\operatorname{Im} i^* \cong \operatorname{Im} \partial_* = \operatorname{Ker} i_*$.

If $\alpha, \beta \in \operatorname{Im} i^*$, then $\langle \alpha \smile \beta, [M^{4k}] \rangle = 0$. Indeed, if $\alpha = i^*(a)$ and $\beta = i^*(b)$, then

$$\langle \alpha \smile \beta, [M^{4k}] \rangle = \langle i^*(a \smile b), \partial_*[W^{4k+1}] \rangle = \langle \delta^* i^*(a \smile b), [W^{4k+1}] \rangle = 0$$

because $\delta^* i^* = 0$.

The next step of the proof is performed most conveniently in the absence of torsion, i.e., for (co)homology with coefficients in $\mathbb{R}$. In this case, $H^{2k}(M^{4k})$ and $H^{2k}(W^{4k+1})$ are linear spaces, and they are dual to $H_{2k}(M^{4k})$ and $H_{2k}(W^{4k+1})$. Moreover, the maps i_* and i^* are dual to each other.

Lemma. *If a map $f^* \colon W^* \to V^*$ is dual to $f \colon V \to W$, then $(\operatorname{Im} f)^* \cong W^* / \operatorname{Ker} f^*$.*

Proof. The linear functions on $\operatorname{Im} f$ coincide with the linear functions on W up to functions vanishing on $\operatorname{Im} f$. The equality $\langle f^*(\eta), v \rangle = \langle \eta, f(v) \rangle$ shows that $\operatorname{Ker} f^*$ is a function on W vanishing on $\operatorname{Im} f$. □

Using this lemma, we obtain $(\operatorname{Im} i^*)^* \cong H_{2k}(M^{4k}; \mathbb{R}) / \operatorname{Ker} i_*$. Since $\operatorname{Im} i^* \cong \operatorname{Ker} i_*$, it follows that $\dim \operatorname{Im} i^* = \dim \operatorname{Ker} i_* = \frac{1}{2} \dim H_{2k}(M^{4k}; \mathbb{R})$.

The intersection form is nondegenerate; therefore, the dimension of the subspace on which it vanishes is at most half the dimension of the entire space, and the equality holds only if the signature vanishes. □

2.8. The Bockstein Homomorphism and Poincaré Isomorphism. The main theorem of this section is applied in the next section to classifying lens spaces up to homotopy equivalence.

Suppose that K is a simplicial complex and $0 \to G' \to G \to G'' \to 0$ is an exact sequence of Abelian groups. It induces an exact sequence of chain complexes

$$0 \longrightarrow C_k(K; G') \longrightarrow C_k(K; G) \longrightarrow C_k(K; G'') \longrightarrow 0.$$

As we have mentioned earlier, the connecting homomorphism $\beta_*\colon H_k(K; G'') \to H_{k-1}(K; G')$ is called the *Bockstein homomorphism*. For cochain complexes, we also have an exact sequence

$$0 \longrightarrow \operatorname{Hom}(C_k, G') \longrightarrow \operatorname{Hom}(C_k, G) \longrightarrow \operatorname{Hom}(C_k, G'') \longrightarrow 0,$$

where $C_k = C_k(K)$. The connecting homomorphism $\beta^*\colon H^k(K; G'') \to H^{k+1}(K; G')$ of cohomology groups is also called the *Bockstein homomorphism*.

Theorem 2.21. *For any closed orientable manifold M^n, the Bockstein homomorphism commutes with the Poincaré isomorphism up to sign; i.e., if $\alpha^k \in H^k(M^n; G'')$, then*

$$(\beta^* \alpha^k) \frown [M^n] = \pm \beta_*(\alpha^k \frown [M^n]),$$

where $[M^n] \in H_n(M^n; G')$ is a fundamental class. (It is assumed that G', G, and G'' are the additive groups of unital commutative rings.)

Proof. At the level of (co)cycles, the homology and cohomology Bockstein homomorphisms can be described as follows. In the case of homology, for a cycle z_k'' with coefficients in $G'' = G/G'$, a chain z_k with coefficients in G which represents this cycle is chosen, and to this chain ∂z_k (which is a cycle with coefficients in $G' \subset G$) is assigned. In the case of cohomology, for a cocycle $(z^k)''$ with values in G'', a cochain z^k with values in G which represents this cycle is chosen and sent to the cocycle δz^k with values in $G' \subset G$.

At the level of (co)chains, the Poincaré isomorphism acts as follows. Let K and K^* be dual decompositions of M^n. The Poincaré isomorphism associates each chain $c_k \in C_k(K)$ to the cochain $c^{n-k} \in C^{n-k}(K^*)$, and associated with the chain ∂c_k is the cochain $\pm \delta c^{n-k}$ (to be more precise, the cochain $(-1)^k \delta c^{n-k}$). Calculating the homology Bockstein homomorphism from the decomposition K and the cohomology Bockstein homomorphism from K^*, we easily obtain the required result. □

Problem 75. Given $a^p \in H^p(K; G'')$ and $b^q \in H^q(K; G'')$, prove that

$$\beta^*(a^p \smile b^q) = \beta^*(a^p) \smile b^q + (-1)^p a^p \smile \beta^*(b^q).$$

2.9. Lens Spaces. Suppose that p and q are coprime positive integers and $p \geq 3$. Consider the fixed-point-free action of the group $\mathbb{Z}_p$ with generator σ on the unit sphere $S^3 \subset \mathbb{C}^2$ defined by

$$\sigma(z, w) = \left(e^{\frac{2\pi i}{p}} z, e^{\frac{2\pi i q}{p}} w\right).$$

The quotient of S^3 by this action is a 3-manifold; it is called a *lens space* and denoted by $L(p, q)$.

In [**108**], it was explained in detail that the lens space $L(p, q)$ can be represented in the form of a CW-complex with one cell in each dimension 0, 1, 2, and 3 as follows. We take a ball D^3, divide its equator into p equal parts, and identify the points of the lower hemisphere with the points of the upper hemisphere by rotating the lower hemisphere through an angle of $2\pi q/p$ and reflecting it in the equator. Under this operation, the p points dividing the equator merge into one point; the p arcs into which the equator is partitioned are identified as well. As a result, we obtain a 0-cell and a 1-cell; for a 2-cell we take, e.g., the lower hemisphere. The boundary of the cell D^3 contains two copies of the 2-cell with opposite orientations. Therefore, the chain complex for calculating the cell homology of $L(p, q)$ has the form

$$C_3 \xrightarrow{0} C_2 \xrightarrow{\times p} C_1 \xrightarrow{0} C_0 \longrightarrow 0,$$

where $C_i \cong \mathbb{Z}$, and the homology groups of $L(p, q)$ with coefficients in $\mathbb{Z}$ are

$$H_3 = \mathbb{Z}, \quad H_2 = 0, \quad H_1 = \mathbb{Z}_p, \quad H_0 = \mathbb{Z}.$$

For $L(p, q)$, the homology and cohomology with coefficients in $\mathbb{Z}_p$ can easily be calculated without applying the universal coefficient theorem. Indeed, the chain complex with coefficients in $\mathbb{Z}_p$ has a very simple form, namely,

$$C_3 \xrightarrow{0} C_2 \xrightarrow{0} C_1 \xrightarrow{0} C_0 \longrightarrow 0,$$

where $C_i \cong \mathbb{Z}_p$. Hence $H_k(L(p, q); \mathbb{Z}_p) \cong H^k(L(p, q); \mathbb{Z}_p) \cong \mathbb{Z}_p$ for $k = 0$, 1, 2, 3.

The homology and cohomology groups are not sufficient to distinguish between the lens spaces $L(p, q_1)$ and $L(p, q_2)$. But we can distinguish between them in some cases by using the Bockstein homomorphism β_* associated to the exact sequence

$$0 \longrightarrow \mathbb{Z}_p \xrightarrow{\times p} \mathbb{Z}_{p^2} \longrightarrow \mathbb{Z}_p \longrightarrow 0.$$

Theorem 2.22. *The group $H_2(L(p, q); \mathbb{Z}_p)$ has an element a such that $q\langle\langle a, \beta_*(a)\rangle\rangle = \pm 1 \in \mathbb{Z}_p$.*

Proof. Let c be the 2-cell of the CW-complex $L(p, q)$ described above, and let d be its 1-cell. For a we take the homology class represented by the cycle $1 \cdot c$ (modulo p), where $1 \in \mathbb{Z}_p$. The element $\beta_*(a) \in H_1(L(p, q); \mathbb{Z}_p)$ is

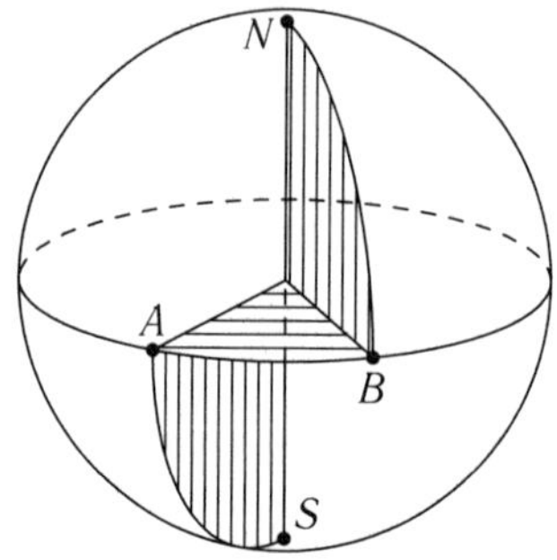

Figure 15. A 2-chain

constructed as follows. Consider the cycle $1 \cdot c$ as a chain with coefficient in $\mathbb{Z}_{p^2}$, i.e., assume that $1 \in \mathbb{Z}_{p^2}$ (1 can be replaced by any element of the form $1+kp$). By assumption, $\partial(1 \cdot c)$ is a cycle with coefficients in $\mathbb{Z}_p \subset \mathbb{Z}_{p^2}$; the subgroup $\mathbb{Z}_p$ of $\mathbb{Z}_{p^2}$ consists of all elements of the form lp for $l \in \mathbb{Z}_p$. Indeed, $\partial(1 \cdot c) = p \cdot d$. The element $\beta_*(a)$ is represented by the cycle $p \cdot d$ treated as a cycle with coefficients in $\mathbb{Z}_p$ ($lp \in \mathbb{Z}_{p^2}$ should be replaced by $l \in \mathbb{Z}_p$). Thus, $\beta_*(a)$ is represented by the cycle $1 \cdot d$.

The intersection number $\langle\langle c, d\rangle\rangle$ cannot be calculated directly, because the cycles c and d are not transversal. For this reason, instead of the cycle d, we consider another cycle, which is homologous to qd and intersects c transversally.

Suppose that N and S are the north and the south poles, A is a point on the equator, and B is the point on the equator obtained by rotating A through the angle $2\pi q/p$ (see Figure 15). The boundary of the 2-chain hatched in Figure 15 consists of the interval NS and the arcs SA, AB, and BN. In $L(p,q)$, the arcs SA and BN are identified, and they have opposite orientations; clearly, the points N and S are identified as well. Therefore, the cycle NS is homologous to AB, which consists of q cycles d, and it transversally intersects c in one point; therefore,

$$q\langle\langle c, d\rangle\rangle = \langle\langle c, qd\rangle\rangle = \langle\langle c, NS\rangle\rangle = \pm 1$$

(all of the equalities are modulo p). □

Corollary. *The group $H^1(L(p,q);\mathbb{Z}_p)$ has an element α such that $q\alpha \smile \beta^*(\alpha) = \pm\Lambda_p$, where $\Lambda_p \in H^3(L(p,q);\mathbb{Z}_p)$ is the element for which $\Lambda_p \frown [L(p,q)]_p = 1$.*

Proof. The manifold $L(p,q)$ is orientable because $H_3(L(p,q)) = \mathbb{Z}$. Therefore, the homology and cohomology Bockstein homomorphisms commute with the Poincaré isomorphism up to sign. □

Theorem 2.22 and its corollary readily imply the following assertion.

Theorem 2.23. *If lens spaces $L(p, q_1)$ and $L(p, q_2)$ are homotopy equivalent, then $q_1 \equiv \pm a^2 q_2 \pmod p$.*

Proof. Choose elements α_1 and α_2 in $H^1(L(p, q_1); \mathbb{Z}_p)$ and $H^1(L(p, q_2); \mathbb{Z}_p)$, respectively, so that

$$q_1\alpha_1 \smile \beta^*(\alpha_1) = \pm\Lambda_{p,1} \quad \text{and} \quad q_2\alpha_2 \smile \beta^*(\alpha_2) = \pm\Lambda_{p,2}.$$

Let $f\colon L(p, q_1) \to L(p, q_2)$ be a homotopy equivalence. Then $f^*(\Lambda_{p,2}) = \pm\Lambda_{p,1}$. Let us write $f^*(\alpha_2)$ in the form $a\alpha_1$, where $a \in \mathbb{Z}_p$. Then

$$q_1\alpha_1 \smile \beta^*(\alpha_1) = \pm\Lambda_{p,1} = \pm f^*(\Lambda_{p,2}) = \pm a^2 q_2\alpha_1 \smile \beta^*(\alpha_1);$$

hence $q_1 \equiv \pm a^2 q_2 \pmod p$. (The equality $f^*\beta^* = \beta^* f^*$, which we use, is easy to derive at the level of cochains.) □

Remark. In Part I (Chapter 5, Section 4.6), we proved the converse of Theorem 2.23: If $q_1 \equiv \pm a^2 q_2 \pmod p$, then the lens spaces $L(p, q_1)$ and $L(p, q_2)$ are homotopy equivalent.

3. The Künneth Theorem

The Künneth theorem allows us to express the homology of $K \times L$ in terms of the homology of K and L. It consists of two parts. One, relatively simple, expresses the chain complex $C_*(K \times L)$ in terms of the chain complexes $C_*(K)$ and $C_*(L)$; the other, somewhat more complicated algebraic part, expresses the homology of the complex $C_*(K \times L)$ in terms of the homology of $C_*(K)$ and $C_*(L)$.

3.1. The Chain Complex $C_*(K \times L)$. Knowing chain complexes $C_*(K)$ and $C_*(L)$, we can construct the chain complex $C_*(K \times L)$ for calculating the cellular homology of $K \times L$. The cellular homology of $K \times L$ is more convenient to calculate than the simplicial homology because the product of any two open cells is an open cell, whereas the product of simplices may not be a simplex.

We assume that K and L are finite CW-complexes because the topology of the product $K \times L$ of infinite CW-complexes may not coincide with that of $K \times L$ as a CW-complex.

The open k-cells in $K \times L$ are the products $\sigma^p \times \tau^q$, where $p + q = k$ and σ^p and τ^q are open cells in K and L. The main difficulty occurs when we try to express the boundary homomorphism in $C_*(K \times L)$ in terms of the boundary homomorphisms in $C_*(K)$ and $C_*(L)$.

It is clear that $\partial(D^p \times D^q) = (D^p \times \partial D^q) \cup (\partial D^p \times D^q)$ and $(D^p \times \partial D^q) \cap (\partial D^p \times D^q) = S^{p-1} \times S^{q-1}$; here ∂ is the geometric boundary. The calculation

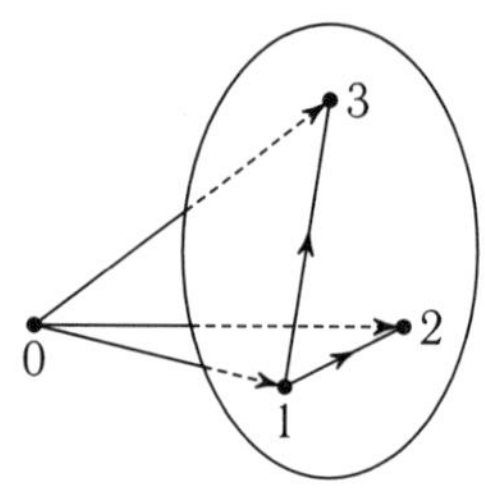

Figure 16. The orientation of the boundary of a cell

of $\partial(\sigma^p \times \tau^q)$ does not involve cells of dimension $(p-1)+(q-1)=k-2$; therefore,

$$\partial(\sigma^p \times \tau^q) = \pm\partial\sigma^p \times \tau^q \pm \sigma^p \times \partial\tau^q.$$

It remains to determine the signs.

First, we need an agreement on how the orientation of a cell or, more generally, of a manifold induces the orientation of its boundary. For a simplex $[v_0,\ldots,v_n]$, the orientation of the boundary is determined by $\partial[v_0,\ldots,v_n] = \sum_{i=0}^n(-1)^i[v_0,\ldots,\hat{v}_i,\ldots,v_n]$. We assume that each simplex $[v_0,\ldots,v_n]$ is oriented so that the basis $e_1 = v_1 - v_0,\ldots,e_n = v_n - v_0$ has positive orientation. Moreover, the simplex $[v_1,\ldots,v_n]$, which is contained in the boundary of $[v_0,\ldots,v_n]$, is oriented so that the basis $e_2 - e_1 = v_2 - v_1,\ldots,e_n - e_1 = v_n - v_1$ has positive orientation (i.e., the basis $e_2,\ldots,e_n$ has positive orientation). This convention on induced orientation is easy to transfer to any oriented manifold M^n with boundary ∂M^n. Namely, suppose that the orientation of a manifold M^n is determined by a basis $e_1,\ldots,e_n$ of the tangent space T_xM^n at some point $x \in \partial M^n$, the vector e_1 is directed inside the manifold, and the vectors $e_2,\ldots,e_n$ belong to the tangent space to the boundary. Then the orientation of the boundary is determined by the basis $e_2,\ldots,e_n$.

Suppose that the positive orientations of cells σ^p and τ^q are determined by bases $e_1,\ldots,e_p$ and $\varepsilon_1,\ldots,\varepsilon_q$. We endow the cell $\sigma^p \times \tau^q$ with the orientation determined by the basis $e_1,\ldots,e_p$, $\varepsilon_1,\ldots,\varepsilon_q$. The orientation of $\partial\sigma^p$ is determined by the basis $e_2 - e_1,\ldots,e_p - e_1$ (see Figure 16); the orientation of $\partial\sigma^p \times \tau^q$ is determined by the basis

$$e_2 - e_1,\ldots,e_p - e_1, \varepsilon_1 - e_1,\ldots,\varepsilon_q - e_1 \tag{13}$$

(this orientation coincides with that of $\partial(\sigma^p \times \tau^q)$); and the orientation of $\sigma^p \times \partial\tau^q$ is determined by the basis

$$e_1 - \varepsilon_1, e_2 - \varepsilon_1,\ldots,e_p - \varepsilon_1, \varepsilon_2 - \varepsilon_1,\ldots,\varepsilon_q - \varepsilon_1. \tag{14}$$

Let us move $\varepsilon_1 - e_1$ in the basis (13) to the first position. The transition matrix from the basis thus obtained to (14) is

$$\begin{pmatrix} -1 & -1 & -1 & \dots & -1 \\ 0 & 1 & 0 & \dots & 0 \\ 0 & 0 & 1 & \dots & 0 \\ \dots & \dots & \dots & \dots & \dots \\ 0 & 0 & 0 & \dots & 1 \end{pmatrix}.$$

Therefore, the sign of the orientation of the basis (14) with respect to the basis (13) is $(-1)^p$.

As a result, we obtain

$$\partial(\sigma^p \times \tau^q) = \partial\sigma^p \times \tau^q + (-1)^p \sigma^p \times \partial\tau^q.$$

Now the problem of calculating the homology of $K \times L$ becomes purely algebraic.

3.2. The Algebraic Künneth Theorem. Let C'_* and C''_* be nonnegative chain complexes with boundary homomorphisms ∂' and ∂''. Their *tensor product* is the chain complex C_*, where $C_k = \bigoplus_{p+q=k} C'_p \otimes C''_q$ and

$$\partial(c'_p \otimes c''_q) = \partial' c'_p \otimes c''_q + (-1)^p c'_p \otimes \partial'' c''_q.$$

The sign $(-1)^p$ in the second term ensures that $\partial\partial = 0$.

We denote the chain complex C_* by $C'_* \otimes C''_*$. A similar notation is used for any graded Abelian groups $A_* = \bigoplus A_p$ and $B_* = \bigoplus B_q$; i.e., it is assumed that $(A_* \otimes B_*)_k = \bigoplus_{p+q=k} A_p \otimes B_q$. Similarly, we define $(\operatorname{Tor}(A_*, B_*))_k = \bigoplus_{p+q=k} \operatorname{Tor}(A_p, B_q)$.

Theorem 2.24. *Let C'_* and C''_* be free nonnegative chain complexes. Then the sequence*

$$0 \longrightarrow (H_*(C'_*) \otimes H_*(C''_*))_k \longrightarrow H_k(C'_* \otimes C''_*) \longrightarrow (\operatorname{Tor}(H_*(C'_*), H_*(C''_*)))_{k-1} \longrightarrow 0$$

is exact. This exact sequence splits, but not canonically.

Proof. Consider the exact sequence

$$0 \longrightarrow Z'_p \xrightarrow{i'} C'_p \xrightarrow{\partial'} B'_{p-1} \longrightarrow 0.$$

This sequence splits because the group B'_{p-1} is free. Therefore, tensor multiplication by C''_q preserves exactness. The direct sum of two exact sequences is an exact sequence; hence the sequence

(15)

$$0 \longrightarrow \bigoplus_{p+q=k} (Z'_p \otimes C''_q) \xrightarrow{i' \otimes 1} \bigoplus_{p+q=k} (C'_p \otimes C''_q) \xrightarrow{\partial' \otimes 1} \bigoplus_{p+q=k-1} (B'_p \otimes C''_q) \longrightarrow 0$$

is exact.

Consider the chain complexes $Z'_* \otimes C''_*$ and $B'_* \otimes C''_*$; in these complexes, the boundary homomorphism has the form $\pm 1 \otimes \partial''$. Indeed, $\partial(z'_p \otimes c''_q) = (-1)^p z'_p \otimes \partial'' c''_q$, because $\partial' z'_p = 0$. The homomorphisms $i' \otimes 1$ and $\partial' \otimes 1$ are chain maps. Thus, from the short exact sequence (15) we obtain the exact homology sequence

$$\text{(16)} \quad \cdots \longrightarrow H_k(B'_* \otimes C''_*) \xrightarrow{\alpha_k} H_k(Z'_* \otimes C''_*) \longrightarrow H_k(C'_* \otimes C''_*) \\ \longrightarrow H_{k-1}(B'_* \otimes C''_*) \xrightarrow{\alpha_{k-1}} H_{k-1}(Z'_* \otimes C''_*) \longrightarrow \cdots .$$

Let us show that $H_k(Z'_* \otimes C''_*) \cong (Z'_* \otimes H_*(C''_*))_k$. The group Z'_p is free; therefore, the exact sequences

$$0 \longrightarrow Z''_q \longrightarrow C''_q \xrightarrow{\partial''_q} B''_{q-1} \longrightarrow 0$$

and

$$0 \longrightarrow B''_{q-1} \longrightarrow Z''_{q-1} \longrightarrow H''_{q-1}(C''_*) \longrightarrow 0$$

yield the exact sequences

$$\text{(17)} \qquad 0 \longrightarrow Z'_p \otimes Z''_q \longrightarrow Z'_p \otimes C''_q \xrightarrow{1 \otimes \partial''_q} Z'_p \otimes B''_{q-1} \longrightarrow 0$$

and

$$\text{(18)} \quad 0 \longrightarrow Z'_p \otimes B''_{q-1} \longrightarrow Z'_p \otimes Z''_{q-1} \longrightarrow Z'_p \otimes H''_{q-1}(C''_*) \longrightarrow 0.$$

The exact sequence (17) implies

$$\operatorname{Ker}(\pm 1 \otimes \partial''_q) = \operatorname{Ker}(1 \otimes \partial''_q) = Z'_p \otimes Z''_q$$

and

$$\operatorname{Im}(\pm 1 \otimes \partial''_q) = \operatorname{Im}(1 \otimes \partial''_q) = Z'_p \otimes B''_{q-1}.$$

Hence the exact sequence (18) has the form

$$0 \longrightarrow \operatorname{Im}(\pm 1 \otimes \partial''_q) \longrightarrow \operatorname{Ker}(\pm 1 \otimes \partial''_{q-1}) \longrightarrow Z'_p \otimes H''_{q-1}(C''_*) \longrightarrow 0.$$

It follows that $\operatorname{Ker}(\pm 1 \otimes \partial''_{q-1}) / \operatorname{Im}(\pm 1 \otimes \partial''_q) \cong Z'_p \otimes H''_{q-1}(C''_*)$ and, consequently, $H_k(Z'_* \otimes C''_*) \cong (Z'_* \otimes H_*(C''_*))_k$.

Similarly, $H_k(B'_* \otimes C''_*) \cong (B'_* \otimes H_*(C''_*))_k$. It follows directly from the definition of the connecting homomorphism α_k that, under the above identifications, this homomorphism takes the form

$$j \otimes 1 \colon (B'_* \otimes H_*(C''_*))_k \to (Z'_* \otimes H_*(C''_*))_k,$$

where $j \colon B'_* \to Z'_*$ is the natural embedding. Therefore, the free resolution

$$0 \longrightarrow B'_* \longrightarrow Z'_* \longrightarrow H(C'_*) \longrightarrow 0$$

determines the exact sequence

$$0 \longrightarrow (\operatorname{Tor}(H_*(C'_*), H_*(C''_*)))_k \longrightarrow (B'_* \otimes H_*(C''_*))_k \\ \xrightarrow{\alpha_k} (Z'_* \otimes H_*(C''_*))_k \longrightarrow (H_*(C'_*) \otimes H_*(C''_*))_k \longrightarrow 0,$$

which implies that $\operatorname{Ker}\alpha_k = (\operatorname{Tor}(H_*(C'_*), H_*(C''_*)))_k$ and $\operatorname{Coker}\alpha_k = (H_*(C'_*)\otimes H_*(C''_*))_k$. Summarizing, we see that the exact homology sequence (16) gives rise to the exact sequence

$$0 \longrightarrow (H_*(C'_*) \otimes H_*(C''_*))_k \longrightarrow H_k(C'_* \otimes C''_*) \\ \longrightarrow (\operatorname{Tor}(H_*(C'_*), H_*(C''_*)))_{k-1} \longrightarrow 0 .$$

It remains to verify that this exact sequence splits. Recall that the chain complex C''_* is free (so far, we have used only the assumption that the chain complex C'_* is free). Let $I'\colon C'_* \to Z'_*$ and $I''\colon C''_* \to Z''_*$ be splitting maps. Then the restriction of $I' \otimes I''$ to $Z'_* \otimes Z''_*$ induces a homomorphism

$$H_*(C'_* \otimes C''_*) \to \frac{Z'_* \otimes Z''_*}{B'_* \otimes Z''_* + Z'_* \otimes B''_*} \cong H_*(C'_*) \otimes H_*(C''_*).$$

This homomorphism is splitting. □

3.3. The Homology of Products. Combining the algebraic Künneth theorem with the calculation of the chain complex $C_*(K \times L)$, we obtain the following assertion.

Theorem 2.25 (Künneth [**74**], [**75**]). *If K and L are finite simplicial complexes, then the sequence*

$$0 \longrightarrow (H_*(K)\otimes H_*(L))_k \longrightarrow H_k(K \times L) \longrightarrow (\operatorname{Tor}(H_*(K), H_*(L)))_{k-1} \longrightarrow 0$$

is exact. This exact sequence splits, but not canonically.

Remark. Both Künneth's papers appeared before Emmy Noether noticed the group structure in homology. For this reason, Künneth considered the Betti and torsion numbers rather than groups.

Example 21. The equality $H_k(T^n) = \mathbb{Z}^{\binom{n}{k}}$ holds.

Proof. We prove the required equality by induction on n. Note that $T^{n+1} = T^n \times S^1$. For $n = 1$, the assertion is obvious. We have $\operatorname{Tor}(H_*(T^n), \mathbb{Z}) = 0$; therefore,

$$\begin{aligned} H_k(T^{n+1}) &\cong (H_k(T^n) \otimes H_0(S^1)) \oplus (H_{k-1}(T^n) \otimes H_1(S^1)) \\ &\cong H_k(T^n) \oplus H_{k-1}(T^n). \end{aligned}$$

It remains to note that $\binom{n+1}{k} = \binom{n}{k} + \binom{n}{k-1}$. □

The proof of the algebraic Künneth theorem remains valid for homology with coefficients in the additive group of the field F; moreover, the argument becomes simpler, because we deal with linear maps of vector spaces. For vector spaces V and W over F, the tensor product $V \otimes_F W$ is defined as the group obtained from $V \otimes W$ by introducing the additional relation

$\lambda v \otimes w = v \otimes \lambda w$ for any $\lambda \in F$; the group $V \otimes_F W$ is a vector space over F. We repeat the proof of the algebraic Künneth theorem with the free chain complexes C'_* and C''_* replaced by $C'_* \otimes F$ and $C''_* \otimes F$ and the operation $\otimes$, by $\otimes_F$. Up to the calculation of the kernel of α_k, no changes in the proof are needed. But now, the exact sequence obtained from

$$0 \longrightarrow B'_p \otimes F \longrightarrow Z'_p \otimes F \longrightarrow H_p(C'_* \otimes F) \longrightarrow 0$$

has the form

$$\begin{aligned} 0 \longrightarrow (B'_p \otimes F) \otimes_F H_q(C''_* \otimes F) \xrightarrow{\alpha_k} (Z'_p \otimes F) \otimes_F H_q(C''_* \otimes F) \\ \longrightarrow H_p(C'_* \otimes F) \otimes_F H_q(C''_* \otimes F) \longrightarrow 0; \end{aligned}$$

i.e., $\operatorname{Ker} \alpha_k = 0$ and the group Tor disappears. As a result, instead of a short exact sequence, we obtain a canonical isomorphism

$$\bigoplus_{p+q=k} H_p(C'_* \otimes F) \otimes_F H_q(C''_* \otimes F) \to H_k(C'_* \otimes C''_* \otimes F).$$

Since $C'_* \otimes C''_* \otimes F \cong (C'_* \otimes F) \otimes_F (C''_* \otimes F)$, it follows that

$$\bigoplus_{p+q=k} H_p(K; F) \otimes_F H_q(L; F) \cong H_k(K \times L; F).$$

Problem 76. Prove that the product of two closed manifolds is orientable if and only if both manifolds are orientable.

Problem 77. Prove that the sphere S^n is not a product of two manifolds of positive dimension.

Problem 78. Given $n > m > 1$, prove that all homotopy groups of the spaces $S^n \times \mathbb{R}P^m$ and $S^m \times \mathbb{R}P^n$ are isomorphic, while the homology groups are different.

Problem 79. Prove that all homotopy groups of the spaces $S^2 \times \mathbb{R}P^\infty$ and $\mathbb{R}P^2$ isomorphic, while the homology groups are different.

Problem 80 ([86]). Given finite connected simplicial complexes A and B, prove that

$$H_{r+1}(A * B) = \bigoplus_{i+j=r} (\tilde{H}_i(A) \otimes \tilde{H}_j(B)) \oplus \bigoplus_{i+j=r-1} \operatorname{Tor}(H_i(A), H_j(B)).$$

3.4. The Künneth Theorem for Cohomology. The cohomology of a product can be calculated using the universal coefficient formulas and the Künneth theorem for homology, but it is also possible to directly obtain an exact sequence in cohomology similar to the Künneth exact sequence in homology. We consider only the coefficient groups $\mathbb{Z}$ and F (the additive group of a field).

Suppose that $C'_* = C_*(K;\mathbb{Z})$ and $C''_* = C_*(L;\mathbb{Z})$, where K and L are finite simplicial complexes. Consider the cochain complexes $C'^* = \operatorname{Hom}(C'_*, \mathbb{Z})$ and $C''^* = \operatorname{Hom}(C''_*, \mathbb{Z})$ and their tensor product $C'^* \otimes C''^*$. Any cochain complex C^* can be regarded formally as the chain complex D_* in which $D_k = C^{-k}$ (this change of numbering is needed for the boundary homomorphism to act in the required direction). The condition $D_k = 0$ for $k \le c$ is equivalent to $C^l = 0$ for $l \ge -c$; it holds for all cochain complexes under consideration, because K and L are finite simplicial complexes.

We apply the algebraic Künneth theorem, formally treating cochain complexes as chain complexes. As a result, we obtain an exact sequence

$$\begin{aligned} 0 \longrightarrow (H_*(C'^*) \otimes H_*(C''^*))_k &\longrightarrow H_k(C'^* \otimes C''^*) \\ &\longrightarrow (\operatorname{Tor}(H_*(C'^*), H_*(C''^*)))_{k+1} \longrightarrow 0. \end{aligned}$$

Note that the number $k-1$ is replaced by $k+1$ because of the change of numbering: if $(-p)+(-q) = (-k)-1$, then $p+q = k+1$.

Unfortunately, this exact sequence is not quite what we want. In this sequence, $H_p(C'^*) = H^p(K;\mathbb{Z})$ and $H_q(C''^*) = H^q(L;\mathbb{Z})$, as required, but the cohomology of the product $K \times L$ is calculated using the cochain complex $\operatorname{Hom}(C'_* \otimes C''_*, \mathbb{Z})$ rather than $C'^* \otimes C''^* = \operatorname{Hom}(C'_*, \mathbb{Z}) \otimes \operatorname{Hom}(C''_*, \mathbb{Z})$.

Consider the homomorphism

$$\theta\colon \operatorname{Hom}(C'_*, \mathbb{Z}) \otimes \operatorname{Hom}(C''_*, \mathbb{Z}) \to \operatorname{Hom}(C'_* \otimes C''_*, \mathbb{Z})$$

defined by

$$\langle \theta(c'^p \otimes c''^q), c'_r \otimes c''_s \rangle = \langle c'^p, c'_r \rangle \langle c''^q, c''_s \rangle;$$

we assume that $\langle c'^p, c'_r \rangle = 0$ if $p \ne r$ and $\langle c''^q, c''_s \rangle = 0$ if $q \ne s$.

Lemma. *The homomorphism θ is a cochain map, i.e., $\delta\theta = \theta\delta$.*

Proof. By definition,

$$\delta(c'^p \otimes c''^q) = \delta' c'^p \otimes c''^q + (-1)^p c'^p \otimes \delta'' c''^q.$$

Therefore,

$$\begin{aligned} \langle \delta\theta(c'^p \otimes c''^q), c'_r \otimes c''_s \rangle &= \langle \theta(c'^p \otimes c''^q), \partial(c'_r \otimes c''_s) \rangle \\ &= \langle c'^p, \partial' c'_r \rangle \langle c''^q, c''_s \rangle + (-1)^r \langle c'^p, c'_r \rangle \langle c''^q, \partial'' c''_s \rangle \end{aligned}$$

and

$$\begin{aligned} \langle \theta\delta(c'^p \otimes c''^q), c'_r \otimes c''_s \rangle &= \langle \delta' c'^p, c'_r \rangle \langle c''^q, c''_s \rangle + (-1)^p \langle c'^p, c'_r \rangle \langle \delta'' c''^q, c''_s \rangle \\ &= \langle c'^p, \partial' c'_r \rangle \langle c''^q, c''_s \rangle + (-1)^p \langle c'^p, c'_r \rangle \langle c''^q, \partial'' c''_s \rangle. \end{aligned}$$

The second terms in these expressions may have different signs, but if $p \ne r$, then $\langle c'^p, c'_r \rangle = 0$. □

We show that θ is an isomorphism. First, note that if $e_1, \dots, e_n$ is a basis of the group $\mathbb{Z}^n$, then the maps ε^i such that $\varepsilon^i(e_j) = \delta_{ij}$ form a basis in $\mathrm{Hom}(\mathbb{Z}^n, \mathbb{Z})$. Therefore, $\mathrm{Hom}(\mathbb{Z}^n, \mathbb{Z}) \cong \mathbb{Z}^n$, and the isomorphism is uniquely determined by the choice of a basis. Choosing bases in C'_p and C''_q, we obtain isomorphisms $\mathrm{Hom}(C'_p, \mathbb{Z}) \cong C'_p$ and $\mathrm{Hom}(C''_q, \mathbb{Z}) \cong C''_q$. Bases in C'_p and C''_q determine bases in $C'_p \otimes C''_q$; hence

$$\mathrm{Hom}(C'_p, \mathbb{Z}) \otimes \mathrm{Hom}(C''_q, \mathbb{Z}) \cong C'_p \otimes C''_q \cong \mathrm{Hom}(C'_p \otimes C''_q, \mathbb{Z}).$$

Clearly, the isomorphism thus defined coincides with θ.

As a result, we obtain the canonical isomorphism

$$H_k(C'^* \otimes C''^*) \to H^k(K \times L; \mathbb{Z})$$

induced by θ.

If the coefficient group is the additive group of a field F rather than $\mathbb{Z}$, then, repeating the argument for linear spaces, we obtain an isomorphism

$$\bigoplus_{p+q=k} H^p(K; F) \otimes_F H^q(L; F) \to H^k(K \times L; F).$$

3.5. Multiplication in Cohomology and the Künneth Theorem. The diagonal map $d\colon |K| \to |K \times K|$ defined by $d(x) = (x, x)$ induces homomorphisms $H_*(K) \xrightarrow{d_*} H_*(K \times K)$ and $H^*(K \times K) \xrightarrow{d^*} H^*(K)$ in homology and cohomology, respectively. For cohomology, we can consider the composition of d^* and the canonical monomorphism $H^*(K) \otimes H^*(K) \to H^*(K \times K)$ from the Künneth theorem. As a result, we obtain a canonical homomorphism $H^*(K) \otimes H^*(K) \to H^*(K)$. The image of an element $\alpha \otimes \beta$ under this homomorphism coincides with $\alpha \smile \beta$, but the proof of this assertion is complicated because the map d is not simplicial. In this section, we give a proof based on acyclic models.

This approach does not apply to homology because the homomorphism d_* acts in the opposite direction:

$$H_*(K) \otimes H_*(K) \longrightarrow H_*(K \times K) \xleftarrow{d_*} H_*(K).$$

For this reason, multiplication in homology, unlike in cohomology, can be defined only in some special cases.

We proved the Künneth theorem not only for integer coefficients but also for coefficients in F, where F is the additive group of a field. The proof used the multiplicative structure of the field (e.g., in the definition of the homomorphism similar to θ for integer coefficients). Therefore, for cohomology with coefficients in F, as well as for integral cohomology, we have a canonically defined multiplication $H^*(K; F) \otimes_F H^*(K; F) \to H^*(K; F)$; it coincides with cup product also.

The Acyclic Model Theorem. The method of acyclic models, which was developed by Eilenberg and MacLane in [**33**], generalizes that of acyclic supports. The acyclic model theorem acquires the most natural formulation when stated in the language of categories and functors.

A *category* $\mathcal{C}$ consists of *objects* and *morphisms* $\hom(X,Y)$ defined for each pair of objects X and Y. For any two morphisms $f \in \hom(X,Y)$ and $g \in \hom(Y,Z)$, a morphism $g \circ f \in \hom(X,Z)$ must be defined; moreover, it is required that

(1) $(h \circ g) \circ f = h \circ (g \circ f)$ and

(2) $\hom(X,X)$ has an element id_X such that $f \circ \mathrm{id}_X = f$ and $\mathrm{id}_X \circ g = g$ for all $f \in \hom(X,Y)$ and $g \in \hom(Y,X)$.

A *covariant functor* T from a category $\mathcal{C}$ to a category $\mathcal{D}$ takes each object X of the category $\mathcal{C}$ to an object $T(X)$ of the category $\mathcal{D}$ and each morphism $f \in \hom(X,Y)$ to a morphism $T(f) \in \hom(T(X),T(Y))$. It must satisfy the conditions $T(\mathrm{id}_X) = \mathrm{id}_{T(X)}$ and $T(f \circ g) = T(f) \circ T(g)$.

A *contravariant functor* differs from a covariant functor in that $T(f) \in \hom(T(Y),T(X))$ and $T(f \circ g) = T(g) \circ T(f)$.

Let T_1 and T_2 be covariant functors from a category $\mathcal{C}$ to a category $\mathcal{D}$. A *natural transformation*$\tau\colon T_1 \to T_2$ assigns a morphism $\tau(X)\colon T_1(X) \to T_2(X)$ to each object X of the category $\mathcal{C}$ in such a way that for any morphism $f\colon X \to Y$ (i.e., $f \in \hom(X,Y)$), the following diagram is commutative:

$$\begin{array}{ccc} T_1(X) & \xrightarrow{T_1(f)} & T_1(Y) \\ \Big\downarrow{\scriptstyle \tau(X)} & & \Big\downarrow{\scriptstyle \tau(Y)} \\ T_2(X) & \xrightarrow{T_2(f)} & T_2(Y). \end{array}$$

A *category with models* is a category in which a class of objects $\mathcal{M} = \{M_\alpha\}$ is distinguished; the distinguished objects are called *models.* A (covariant) functor T from a category with models to the category of free nonnegative chain complexes and chain maps is said to be *free* with respect to the models from $\mathcal{M}$ if for each k, there exists a subset $\mathcal{M}_k \subset \mathcal{M}$ such that every group $C_k(M_\alpha)$, where $M_\alpha \in \mathcal{M}_k$, has an element e_α^k for which the elements $T(f)(e_\alpha^k) \in C_k(X)$ are pairwise distinct and form a basis of the group $C_k(X)$. We assume that f ranges over all elements of the set $\hom(e_\alpha^k, X)$, and α ranges over all indices of the elements of $\mathcal{M}_k$.

A functor T is said to be *acyclic* with respect to $\mathcal{M}$ if $H_k(T(M_\alpha)) = 0$ for $k > 0$.

If a functor is free with respect to the models from $\mathcal{M}$, then it is also free with respect to the models from any $\mathcal{M}' \supset \mathcal{M}$, and if a functor is acyclic with respect to the models from $\mathcal{M}$, then it is also acyclic with respect to the models from any $\mathcal{M}' \subset \mathcal{M}$.

Example 22. Consider the category of simplicial complexes with ordered vertices and simplicial maps preserving the order of vertices. The functor that takes each simplicial complex to its chain complex is free and acyclic with respect to the models from $\mathcal{M} = \{[0, 1, \dots, k]\}$, where $k = 0, 1, \dots$.

Theorem 2.26 (on acyclic models). *If T is a free functor and T' is a functor acyclic with respect to the models from $\mathcal{M}$, then any natural transformation $\varphi\colon H_0(T) \to H_0(T')$ is induced by some natural chain map $\tau\colon T \to T'$. Moreover, any two natural chain maps $\tau, \bar{\tau}\colon T \to T'$ inducing φ are joined by a natural chain homotopy.*

Proof. The group H_0 is a quotient of the zero-dimensional chains; hence there exist epimorphisms $p\colon T_0(M_\alpha) \to H_0(T(M_\alpha))$ and $p'\colon T_0'(M_\alpha) \to H_0(T'(M_\alpha))$. For each element $e_\alpha^0 \in T_0(M_\alpha)$, there exists a $\tau_0(e_\alpha^0) \in T_0'(M_\alpha)$ such that $p'\tau_0(e_\alpha^0) = \varphi p(e_\alpha^0)$.

The elements $T(f)(e_\alpha^0)$, where $f \in \hom(M_\alpha, X)$, form a basis of the group $T_0(X)$. The formula $\tau_0(T(f)(e_\alpha^0)) = T'(f)(\tau_0(e_\alpha^0))$ uniquely determines a homomorphism $\tau_0\colon T_0(X) \to T_0'(X)$. Since the transformation φ is natural and $T(f)$ and $T'(f)$ are chain maps, the homomorphism τ_0 induces the initial map $\varphi\colon H_0(T_0(X)) \to H_0(T_0'(X))$ in the zero-dimensional homology group.

We construct homomorphisms $\tau_k\colon T_k(X) \to T_k'(X)$ for which $\partial\tau_k = \tau_{k-1}\partial$ by induction on k. For each element $e_\alpha^k \in T_k(M_\alpha)$, consider $\tau_{k-1}(\partial e_\alpha^k) \in T_{k-1}'(M_\alpha)$. There exists a chain $c_\alpha^k \in H_k(M_\alpha)$ with $\partial c_\alpha^k = \tau_{k-1}(\partial e_\alpha^k)$. Indeed, for $k = 1$ it exists because τ_0 induces a map of zero-dimensional homology groups, and for $k > 1$, because $\partial\tau_{k-1}(\partial e_\alpha^k) = \tau_{k-2}(\partial\partial e_\alpha^k) = 0$ and, by assumption, $H_{k-1}(T'(M_\alpha)) = 0$. We set $\tau_k(e_\alpha^k) = c_\alpha^k$ and define a homomorphism $\tau_k\colon T_k(X) \to T_k'(X)$ by using the construction described above.

For chain maps $\tau, \bar{\tau}\colon T \to T'$ that induce the same natural transformation $\varphi\colon H_0(T) \to H_0(T')$, the chain homotopy $D_k\colon T_k(X) \to T_{k+1}'(X)$ is constructed in a similar way. For the equalities $\partial D_0 = \tau_0 - \bar{\tau}_0$ and $\partial D_k = \tau_k - \bar{\tau}_k - D_{k-1}\partial$ to hold for $k > 1$, it is sufficient for the chain $\tau_k(e_\alpha^k) - \bar{\tau}_k(e_\alpha^k) - D_{k-1}\partial e_\alpha^k$ (the chain $\tau_0(e_\alpha^0) - \bar{\tau}_0(e_\alpha^0)$ for $k = 0$) to be the boundary for any element e_α^k. This condition holds for $k = 0$ because τ_0 and $\bar{\tau}_0$ induce the same map in homology; for $k > 1$, this follows from the equality

$$\partial\tau_k - \partial\bar{\tau}_k - \partial D_{k-1}\partial = \tau_{k-1}\partial - \bar{\tau}_{k-1}\partial - \partial D_{k-1}\partial = D_{k-2}\partial\partial = 0$$

and the acyclicity of the functor T'. $\square$

The Alexander–Whitney Diagonal Approximation. Let $d\colon |K| \to |K \times K|$ be a diagonal map, and let

$$d_*\colon H_*(K) \to H_*(K \times K) \cong H_*(C_*(K) \otimes C_*(K))$$

be the induced map of homology groups. At the level of chains, d_0 takes each vertex v to the chain $v \otimes v$. We refer to any natural chain map $\tau\colon C_*(K) \to C_*(K) \otimes C_*(K)$ that takes v to $v \otimes v$ as a *diagonal approximation.*

The functor $C_*(K)$ is free with respect to the models from $\mathcal{M} = \{\Delta^k\}$ (see Example 22), and the functor $C_*(K) \otimes C_*(K)$ is acyclic with respect to these models. Therefore, according to the acyclic model theorem, any diagonal approximation induces a map d_* in homology.

In calculations, it is convenient to use the *Alexander–Whitney diagonal approximation*, which is defined by

$$\tau([v_0, v_1, \ldots, v_n]) = \sum_{i=0}^{n} [v_0, \ldots, v_i] \otimes [v_i, \ldots, v_n].$$

We must verify only that this is a chain map, i.e., $\tau\partial = \partial\tau$. Clearly,

$$\begin{aligned}\tau(\partial[0, 1, \ldots, n]) &= \tau\Big(\sum (-1)^i [0, \ldots, \hat{\imath}, \ldots, n]\Big) \\ &= \sum_{i<j} (-1)^i [0, \ldots, \hat{\imath}, \ldots, j] \otimes [j, \ldots, n] \\ &\quad + \sum_{i<j} (-1)^i [0, \ldots, j] \otimes [j, \ldots, \hat{\imath}, \ldots, n]\end{aligned}$$

and

$$\begin{aligned}\partial(\tau[0, 1, \ldots, n]) &= \partial \sum [0, \ldots, j] \otimes [j, \ldots, n] \\ &= \sum_i (-1)^i [0] \otimes [0, \ldots, \hat{\imath}, \ldots, n] \\ &\quad + \sum_{i \le j,\, j \ne 0} (-1)^i [0, \ldots, \hat{\imath}, \ldots, j] \otimes [j, \ldots, n] \\ &\quad + \sum_{i \ge j,\, j \ne n} (-1)^i [0, \ldots, j] \otimes [j, \ldots, \hat{\imath}, \ldots, n] \\ &\quad + \sum_i (-1)^i [0, \ldots, \hat{\imath}, \ldots, n] \otimes [n].\end{aligned}$$

The second expression differs from the first by the additional sum

$$
\begin{array}{ll}
 & +[0]\otimes[1,\dots,n] \\
-[0]\otimes[1,\dots,n] & -[0,1]\otimes[2,\dots,n] \\
+[0,1]\otimes[2,\dots,n] & +[0,1,2]\otimes[3,\dots,n] \\
\dots\dots\dots & \dots\dots\dots \\
+(-1)^{n-1}[0,1,\dots,n-2]\otimes[n-1,n] & +(-1)^{n-1}[0,1,\dots,n-1]\otimes[n] \\
+(-1)^{n}[0,1,\dots,n-1]\otimes[n], &
\end{array}
$$

but this sum vanishes.

The map $C_*(K)\to C_*(K\times K)$ defined as the composition

$$[0,1,\dots,n]\mapsto\sum[0,\dots,i]\otimes[i,\dots,n]\mapsto\sum[0,\dots,i]\times[i,\dots,n]$$

induces the map d_* in homology and d^* in cohomology. It easily follows that the composition

$$H^*(K)\otimes H^*(K)\longrightarrow H^*(K\times K)\xrightarrow{d^*}H^*(K)$$

is a cup product. Indeed, at the level of cochains, the cochain c^{p+q} corresponding to $c^p\otimes c^q$ has the property

$$
\begin{aligned}
\langle c^{p+q},[v_0,v_1,\dots,v_{p+q}]\rangle &= \Big\langle\theta(c^p\otimes c^q),\sum[v_0,\dots,v_i]\otimes[v_i,\dots,v_{p+q}]\Big\rangle \\
&= \sum\langle c^p,[v_0,\dots,v_p]\rangle\langle c^q,[v_p,\dots,v_{p+q}]\rangle.
\end{aligned}
$$

The last equality holds because the remaining terms vanish. Thus, $c^{p+q}=c^p\smile c^q$, as required.

3.6. Cohomology Cross Product. Suppose that K and L are finite simplicial complexes, $C^*(K)$ and $C^*(L)$ are simplicial cochain complexes, and $C^*(K\times L)$ is the cellular cochain complex. Two cochains $c^p\in C^p(K)$ and $c^q\in C^q(L)$ determine the cochain $c^p\times c^q\in C^{p+q}(K\times L)$ by the rule

$$\langle c^p\times c^q,\Delta^{p'}\times\Delta^{q'}\rangle=\delta_{pp'}\delta_{qq'}\langle c^p,\Delta^{p'}\rangle\langle c^q,\Delta^{q'}\rangle.$$

The equality $\partial(\Delta^p\times\Delta^q)=\partial\Delta^p\times\Delta^q+(-1)^p\Delta^p\times\partial\Delta^q$ implies $\delta(c^p\times c^q)=\delta c^p\times c^q+(-1)^pc^p\times\delta c^q$. We obtain a group homomorphism $H^p(K)\times H^q(L)\to H^{p+q}(K\times L)$, which is called the *cohomology cross product*, or *external cup product.*

It is easy to show that cohomology cross product has the following properties:

(1) $(\alpha\times\beta)\times\gamma=\alpha\times(\beta\times\gamma)$;

(2) if $p_K\colon K\times L\to K$ is the natural projection, then $p_K^*(\alpha^p)=\alpha^p\times 1_L$;

(3) if a map $T\colon X\times Y\to Y\times X$ is given by $T(x,y)=(y,x)$, then $T^*(\beta^q\times\alpha^p)=(-1)^{pq}\alpha^p\times\beta^q$.

The cohomology cross product is closely related to the cup product. Namely, $\alpha^p \smile \beta^q = d^*(\alpha^p \times \beta^q)$, where $d^*\colon H^{p+q}(K \times K) \to H^{p+q}(K)$ is induced by the diagonal map d. Indeed, $\langle d^*(\alpha^p \times \beta^q), \Delta^{p+q}\rangle = \langle \alpha^p \times \beta^q, d(\Delta^{p+q})\rangle$. The map d is not cellular; therefore, we must replace $d(\Delta^{p+q})$ by the Alexander–Whitney diagonal approximation. As a result, we obtain

$$\begin{aligned}\langle d^*(\alpha^p \times \beta^q), [v_0, \dots, v_{p+q}]\rangle &= \left\langle \alpha^p \times \beta^q, \sum_{i=0}^{p+q} [v_0, \dots, v_i] \times [v_i, \dots, v_{p+q}]\right\rangle \\ &= \langle \alpha^p \times \beta^q, [v_0, \dots, v_p] \times [v_p, \dots, v_{p+q}]\rangle \\ &= \langle \alpha^p \smile \beta^q, [v_0, \dots, v_{p+q}]\rangle,\end{aligned}$$

as required.

In calculating the cohomology rings of products of simplicial complexes, the following assertion is often useful.

Theorem 2.27. *In the ring $H^*(K \times L)$,*

$$(\alpha_1^{p_1} \times \beta_1^{q_1}) \smile (\alpha_2^{p_2} \times \beta_2^{q_2}) = (-1)^{q_1 p_2} (\alpha_1^{p_1} \smile \alpha_2^{p_2}) \times (\beta_1^{q_1} \smile \beta_2^{q_2}).$$

Proof. We have

$$\begin{aligned}(\alpha_1^{p_1} \times \beta_1^{q_1}) \smile (\alpha_2^{p_2} \times \beta_2^{q_2}) &= d_{K\times L}^*((\alpha_1^{p_1} \times \beta_1^{q_1}) \times (\alpha_2^{p_2} \times \beta_2^{q_2})) \\ &= d_{K\times L}^*(\alpha_1^{p_1} \times \beta_1^{q_1} \times \alpha_2^{p_2} \times \beta_2^{q_2})\end{aligned}$$

and

$$(\alpha_1^{p_1} \smile \alpha_2^{p_2}) \times (\beta_1^{q_1} \smile \beta_2^{q_2}) = d_K^*(\alpha_1^{p_1} \times \alpha_2^{p_2}) \times d_L^*(\beta_1^{q_1} \times \beta_2^{q_2}).$$

Clearly, $d_{K\times L} = (\mathrm{id}_K \times T \times \mathrm{id}_L) \circ (d_K \times d_L)$. Therefore,

$$\begin{aligned}&(\alpha_1^{p_1} \times \beta_1^{q_1}) \smile (\alpha_2^{p_2} \times \beta_2^{q_2}) \\ &\qquad = (d_K \times d_L)^*(\mathrm{id}_K \times T \times \mathrm{id}_L)^*(\alpha_1^{p_1} \times \beta_1^{q_1} \times \alpha_2^{p_2} \times \beta_2^{q_2}) \\ &\qquad = (-1)^{q_1 p_2}(d_K \times d_L)^*(\alpha_1^{p_1} \times \alpha_2^{p_2} \times \beta_1^{q_1} \times \beta_2^{q_2}) \\ &\qquad = (-1)^{q_1 p_2}(d_K \times d_L)^* d_K^*(\alpha_1^{p_1} \times \alpha_2^{p_2}) \times d_L^*(\beta_1^{q_1} \times \beta_2^{q_2}). \qquad \square\end{aligned}$$

Corollary. *If $\alpha \in H^*(K)$ and $\beta \in H^*(L)$, then $p_K^*\alpha \smile p_L^*\beta = \alpha \times \beta$.*

Proof. We have $p_K^*\alpha = \alpha \times 1_L$ and $p_L^*\beta = 1_K \times \beta$. Taking into account the fact that $q_1 = p_2 = 0$ in the case under consideration, we obtain $(\alpha \times 1_L) \smile (1_K \times \beta) = (\alpha \smile 1_L) \times (1_K \smile \beta) = \alpha \times \beta$. □

Problem 81. (a) Calculate the cohomology ring of $\mathbb{R}P^{n_1} \times \dots \times \mathbb{R}P^{n_k}$ with coefficients in $\mathbb{Z}_2$.

(b) Calculate the cohomology ring $\mathbb{C}P^{n_1} \times \dots \times \mathbb{C}P^{n_k}$ with coefficients in $\mathbb{Z}$.

Problem 82. Prove that if at least one of the groups $H^*(K)$ and $H^*(L)$ is torsion-free, then multiplication in the ring $H^*(K \times L)$ is completely determined by multiplications in $H^*(K)$ and $H^*(L)$.

Problem 83. Given $K = \mathbb{R}P^2$, $L_1 = \mathbb{R}P^2 \vee S^3$, and $L_2 = \mathbb{R}P^3$, prove that the rings $H^*(L_1)$ and $H^*(L_2)$ are isomorphic, whereas the rings $H^*(K \times L_1)$ and $H^*(K \times L_2)$ are not.

The Signature of Cartesian Products. Cohomology cross product can also be used to prove the following theorem concerning the signature of the Cartesian product of two manifolds. We assume that $\sigma(M^n) = 0$ if n is not divisible by 4.

Theorem 2.28. *The equality $\sigma(M^p \times N^q) = \sigma(M^p)\sigma(N^q)$ holds.*

Proof. If $p+q$ is not divisible by 4, then one of the numbers p and q is not divisible by 4 either; hence $\sigma(M^p \times N^q) = 0$ and $\sigma(M^p)\sigma(N^q) = 0$. In what follows, we assume that $p + q = 4k$. For cohomology with coefficients in $\mathbb{R}$, the Künneth theorem implies

$$H^{2k}(M^p \times N^q) \cong \bigoplus_{i=0}^{2k} (H^i(M^p) \otimes_{\mathbb{R}} H^{2k-i}(N^q)).$$

If $i + j \neq p$, then the intersection form vanishes at the elements $\alpha_1^i \otimes \beta_1^{2k-i}$ and $\alpha_2^j \otimes \beta_2^{2k-j}$. We decompose the space $H^{2k}(M^p \times N^q)$ into the sum of spaces

$$(H^i(M^p) \otimes_{\mathbb{R}} H^{2k-i}(N^q)) \oplus (H^{p-i}(M^p) \otimes_{\mathbb{R}} H^{2k+i-p}(N^q)), \tag{19}$$

where $i < p$, and the space $H^{p/2}(M^p) \otimes_{\mathbb{R}} H^{q/2}(N^q)$ (if p and q are even). These spaces are mutually orthogonal with respect to the intersection form.

We prove that the intersection form has zero signature on each subspace of the form (19), i.e., the signature of $\sigma(M^p \times N^q)$ equals that of the restriction of the intersection form to the subspace $H^{p/2}(M^p) \otimes_{\mathbb{R}} H^{q/2}(N^q)$. Let $i < p/2$. Choose a basis $\{\alpha_s\}$ in $H^i(M^p)$ and let $\{\alpha_s^*\}$ be the dual basis in the space $H^{p-i}(M^p)$; then $\langle \alpha_s \smile \alpha_r^*, [M^p]\rangle = \delta_{rs}$. Similarly, choose a basis $\{\beta_t\}$ in $H^{2k-i}(N^q)$ and the dual basis $\{\beta_t^*\}$ in $H^{2k+i-p}(N^q)$. For a basis in the space (19) we can take the vectors $\alpha_s \otimes \beta_t$ and $\alpha_r^* \otimes \beta_u^*$. The intersection form vanishes at all pairs of basis vectors except at the pairs $(\alpha_s \otimes \beta_t, \alpha_s^* \otimes \beta_t^*)$. According to Theorem 2.27, at such pairs of basis vectors, the intersection form takes the value $(-1)^{\dim \beta_t \dim \alpha_s^*} = (-1)^{(2k-i)(p-i)}$. Clearly, this form is symmetric. Therefore, in the chosen basis, the restriction of the intersection form to the space (19) is determined by the direct sum of matrices $(-1)^d \left(\begin{smallmatrix} 0 & 1 \\ 1 & 0 \end{smallmatrix}\right)$, where $d = (2k-i)(p-i)$. Each of these matrices has signature zero.

If the numbers p and q are odd, then the space $H^{p/2}(M^p) \otimes_{\mathbb{R}} H^{q/2}(N^q)$ is undefined, and $\sigma(M^p \times N^q) = 0$.

If each of the numbers p and q is congruent to 2 modulo 4, then the intersection forms on the spaces $H^{p/2}(M^p)$ and $H^{q/2}(N^q)$ are antisymmetric. Therefore, these spaces have bases in which the intersection form is determined by the direct sum of matrices $\left(\begin{smallmatrix} 0 & 1 \\ -1 & 0 \end{smallmatrix}\right)$. The restriction of the intersection form to the tensor product of two two-dimensional spaces with such intersection matrices is given by

$$-\begin{pmatrix} 0 & 1 \\ -1 & 0 \end{pmatrix} \otimes \begin{pmatrix} 0 & 1 \\ -1 & 0 \end{pmatrix} = \begin{pmatrix} 0 & 0 & 0 & -1 \\ 0 & 0 & 1 & 0 \\ 0 & 1 & 0 & 0 \\ -1 & 0 & 0 & 0 \end{pmatrix};$$

the minus sign appears because the number $\frac{p}{2} \cdot \frac{q}{2}$ is odd. It is easy to verify that this matrix has signature zero. Therefore, $\sigma(M^p \times N^q) = 0$.

Finally, suppose that both numbers p and q are divisible by 4. Choose bases in the spaces $H^{p/2}(M^p)$ and $H^{q/2}(N^q)$ so that the intersection matrices in these bases are diagonal. Take the tensor product of these bases and consider the matrix A of the restriction of the intersection form to the space $H^{p/2}(M^p) \otimes H^{q/2}(N^q)$. Let r and s be the numbers of positive diagonal elements for the intersection forms on $H^{p/2}(M^p)$ and $H^{q/2}(N^q)$, respectively. The matrix A is diagonal; its diagonal has $rs + \left(\frac{p}{2} - r\right)\left(\frac{q}{2} - s\right)$ positive elements and $r\left(\frac{q}{2} - s\right) + s\left(\frac{p}{2} - r\right)$ negative elements. The signature of this matrix equals $\left(r - \left(\frac{p}{2} - r\right)\right)\left(s - \left(\frac{q}{2} - s\right)\right) = \sigma(M^p)\sigma(N^q)$. □

Nondegenerate Bilinear Maps. A bilinear map $f\colon \mathbb{R}^r \times \mathbb{R}^s \to \mathbb{R}^n$ is said to be *nondegenerate* if $f(x, y) = 0$ implies that $x = 0$ or $y = 0$. To any nondegenerate bilinear map $f\colon \mathbb{R}^r \times \mathbb{R}^s \to \mathbb{R}^n$ we associate a map $F\colon \mathbb{R}P^{r-1} \times \mathbb{R}P^{s-1} \to \mathbb{R}P^{n-1}$ by setting $F(\{\lambda x\}, \{\mu y\}) = \{\alpha f(x, y)\}$. This map is well defined because $f(\lambda x, \mu y) = \lambda\mu f(x, y)$.

According to the solution of Problem 81, the cohomology ring of the space $\mathbb{R}P^n \times \mathbb{R}P^m$ with coefficients in $\mathbb{Z}_2$ is isomorphic to the quotient of the polynomial ring $\mathbb{Z}_2[u, v]$ by the relations $u^{n+1} = 0$ and $v^{m+1} = 0$. This property determines constraints on the numbers r, s, and n that ensure the existence of a nondegenerate bilinear map $\mathbb{R}^r \times \mathbb{R}^s \to \mathbb{R}^n$. These constraints have been obtained simultaneously by Stiefel [**133**] and Hopf [**62**]; they submitted their papers to the journal *Commentarii Mathematici Helvetici* on the same day.

Theorem 2.29 (Stiefel–Hopf). *If there exists a nondegenerate bilinear map* $f\colon \mathbb{R}^r \times \mathbb{R}^s \to \mathbb{R}^n$, *then the number* $\binom{n}{k}$ *is even for all* k *such that* $n - s < k < r$.

Proof. Consider the homology classes a and b in $H_1(\mathbb{R}P^{r-1} \times \mathbb{R}P^{s-1}; \mathbb{Z}_2)$ that are represented by the cycles $\mathbb{R}P^1 \times \{*\}$ and $\{*\} \times \mathbb{R}P^1$. The cohomology classes α and β dual (in the sense of linear algebra) to a and b multiplicatively generate the cohomology ring of $\mathbb{R}P^{r-1} \times \mathbb{R}P^{s-1}$. Let c and γ be the generators of the one-dimensional homology and cohomology groups of $\mathbb{R}P^{n-1}$. It is sufficient to prove that $F^*\gamma = \alpha + \beta$. Indeed, this equality implies $0 = F^*(\gamma^n) = (\alpha + \beta)^n = \sum_{k=0}^{n} \binom{n}{k} \alpha^k \beta^{n-k}$. Therefore, if $k < r$ and $n - k < s$ (i.e., $n - s < k < r$), then the number $\binom{n}{k}$ is even.

First, we show that $F_* a = c$. It suffices to verify that $F_* a \neq 0$. The homology class represented by a loop in $\mathbb{R}P^{n-1}$ is nonzero if and only if the lifting of this loop to S^{n-1} is a nonclosed path. In the situation under consideration, the condition $f(-x, y) = -f(x, y)$ implies that the liftings of loops are nonclosed. Hence $F_* a = c$ and, similarly, $F_* b = c$. Thus, $\langle F^*\gamma, a\rangle = \langle \gamma, F_* a\rangle = \langle \gamma, c\rangle = 1$ and $\langle F^*\gamma, b\rangle = 1$. Therefore, $F^*\gamma = \alpha + \beta$. □

Problem 84. Prove that if there exists a nondegenerate bilinear map $\mathbb{R}^n \times \mathbb{R}^n \to \mathbb{R}^n$, then $n = 2^k$ for some k.

Problem 85. (a) Prove that there exists a nondegenerate bilinear multiplication $\mathbb{R}^r \times \mathbb{R}^s \to \mathbb{R}^{r+s-1}$.

(b) Prove that if r and s are even numbers, then there exists a nondegenerate bilinear multiplication $\mathbb{R}^r \times \mathbb{R}^s \to \mathbb{R}^{r+s-2}$.

(c) Prove that if r and s are numbers divisible by 4, then there exists a nondegenerate bilinear multiplication $\mathbb{R}^r \times \mathbb{R}^s \to \mathbb{R}^{r+s-4}$.

(d) Prove that if r and s are numbers divisible by 8, then there exists a nondegenerate bilinear multiplication $\mathbb{R}^r \times \mathbb{R}^s \to \mathbb{R}^{r+s-8}$.

Chapter 3

Applications of Simplicial Homology

This chapter discusses various applications of simplicial homology. They are based on studying the properties of simplicial homology and cohomology in more detail and introducing additional structures.

1. Homology and Homotopy

In this section, we consider relations between homology and homotopy. First, we prove a theorem of Hurewicz, which asserts that the groups $\pi_n(K)$ and $H_n(K)$ are isomorphic for any $(n-1)$-connected simplicial complex K. Then, we construct the apparatus of obstruction theory, which is applied to interpret cohomology groups as homotopy classes of maps to $K(\pi, n)$ spaces.

1.1. The Hurewicz Theorem. Suppose that K is a simplicial complex and $x_0 \in K$. Consider the map $h\colon \pi_n(K, x_0) \to H_n(K)$ defined as follows. Take a spheroid $\varphi\colon (S^n, s_0) \to (K, x_0)$. It induces the homomorphism $\varphi_*\colon H_n(S^n) \to H_n(K)$ of homology groups. We set $h(\varphi) = \varphi_*([S^n])$, where $[S^n]$ is the fundamental class. Homotopic maps φ_0 and φ_1 induce the same homomorphism in homology; therefore, $h(\varphi_0) = h(\varphi_1)$, i.e., the map h is indeed defined for the homotopy group $\pi_n(K, x_0)$.

Let us show that h is a group homomorphism. We regard spheroids φ and ψ as maps $(D^n, \partial D^n) \to (K, x_0)$. We assume that the sphere S^n is obtained by gluing together two balls D^n with triangulations so fine that the maps φ and ψ are homotopic to simplicial maps; we also assume that the union of the triangulations of the balls is a triangulation of the sphere. In this case, the homology classes $h(\varphi)$ and $h(\psi)$ are represented by the cycles

$\sum_\alpha \varphi(\Delta^n_{1\alpha})$ and $\sum_\beta \varphi(\Delta^n_{2\beta})$, where the $\Delta^n_{1\alpha}$ and $\Delta^n_{2\beta}$ are simplices of the triangulations of the first and second ball and the homology class $h(\varphi+\psi)$ is represented by the cycle $\sum_\alpha \varphi(\Delta^n_{1\alpha})+\sum_\beta \varphi(\Delta^n_{2\beta})$. Therefore, h is a group homomorphism. It is called the *Hurewicz homomorphism.*

The homomorphism h, as well as the group $\pi_n(K,x_0)$ itself, depends on the choice of the base point x_0. However, this dependence is not very essential in the sense that if γ is any path from x_0 to x_1, then the following diagram is commutative:

$$\begin{array}{ccc} \pi_n(K,x_0) & \xrightarrow{\ \gamma_*\ } & \pi_n(K,x_1) \\ & {\scriptstyle h}\searrow \quad \swarrow{\scriptstyle h} & \\ & H_n(X). & \end{array}$$

Theorem 3.1 (Poincaré). *If K is a connected simplicial complex, then the homomorphism $h\colon \pi_1(K)\to H_1(K)$ is an epimorphism, and its kernel is the commutator subgroup*[1] $[\pi_1(K),\pi_1(K)]$ *of the group* $\pi_1(K)$.

Proof. We assume that the base point x_0 is a vertex of the complex K. First, we prove surjectivity. Let $\Delta^1_i=[v_{0i},v_{1i}]$, where v_{0i} and v_{1i} are vertices of K and $\sum a_i\Delta^1_i$ is a cycle. Then $0=\partial(\sum a_i\Delta^1_i)=\sum a_i v_{1i}-\sum a_i v_{0i}$. Suppose that γ_α is a path between the base point x_0 and a vertex v_α going along edges of the complex K. This path corresponds to a chain $\hat\gamma_\alpha\in C_1(K)$. The path $\gamma_{0i}\Delta^1_i\gamma^{-1}_{1i}$ is a loop based at x_0. Clearly,

$$h\Big(\prod(\gamma_{0i}\Delta^1_i\gamma^{-1}_{1i})^{a_i}\Big)=\sum a_i(\hat\gamma_{0i}+\Delta^1_i-\hat\gamma_{1i}).$$

The equality $\sum a_i v_{1i}=\sum a_i v_{0i}$ implies $\sum a_i\hat\gamma_{1i}=\sum a_i\hat\gamma_{0i}$ because v_α is a free generator of the group $C_0(K)$. Thus, we have constructed an element of the fundamental group which is mapped to $\sum a_i\Delta^1_i$.

The image of the homomorphism h is an Abelian group; therefore, its kernel is contained in the commutator subgroup of $\pi_1(K)$. Let ω be a loop in K based at x_0 for which $h(\omega)=\partial(\sum a_i\Delta^2_i)$. Any loop in K based at a vertex is homotopic to a loop along edges. Therefore, we can assume that $\omega=\prod\gamma_{j0,j1}$, where $\gamma_{j0,j1}$ is an oriented edge of the complex K. For 1-chains, we have

$$\sum\hat\gamma_{j0,j1}=\sum a_i(\hat\gamma_{i0,i1}+\hat\gamma_{i1,i2}+\hat\gamma_{i2,i0}),\tag{20}$$

where $\hat\gamma_{\alpha,\beta}$ is the oriented 1-simplex $[v_\alpha,v_\beta]$.

Let γ_α be a path from the base point x_0 to a vertex v_α (if $v_\alpha=x_0$, then the path γ_α is constant). Suppose that $\overline{\gamma_\alpha\gamma_{\alpha,\beta}\gamma^{-1}_\beta}$ is the image of the loop

[1]The *commutator subgroup* $[G,G]$ of a group G is the subgroup generated by all elements of the form $aba^{-1}b^{-1}$, $a,b\in G$.

$\overline{\gamma_\alpha\gamma_{\alpha,\beta}\gamma_\beta^{-1}}$ in the commutator quotient of $\pi_1(K)$; this quotient is an Abelian group. Equality (20) holds for the generators $\hat{\gamma}_{\alpha,\beta}$ of the free group; hence a similar equality holds for the elements $\overline{\gamma_\alpha\gamma_{\alpha,\beta}\gamma_\beta^{-1}}$. Let us write it in the multiplicative form:

$$\prod \overline{\gamma_{j0}\gamma_{j0,j1}\gamma_{j1}^{-1}} = \prod \Big(\overline{\gamma_{i0}\gamma_{i0,i1}\gamma_{i1}^{-1}} \cdot \overline{\gamma_{i1}\gamma_{i1,i2}\gamma_{i2}^{-1}} \cdot \overline{\gamma_{i2}\gamma_{i2,i0}\gamma_{i0}^{-1}}\Big)^{a_i}.$$

Since the path ω is closed, we have

$$\prod \overline{\gamma_{j0}\gamma_{j0,j1}\gamma_{j1}^{-1}} = \prod \overline{\gamma_{j0,j1}} = \overline{\omega}.$$

Moreover,

$$\overline{\gamma_{i0}\gamma_{i0,i1}\gamma_{i1}^{-1}} \cdot \overline{\gamma_{i1}\gamma_{i1,i2}\gamma_{i2}^{-1}} \cdot \overline{\gamma_{i2}\gamma_{i2,i0}\gamma_{i0}^{-1}} = \overline{\gamma_{i0}\gamma_{i0,i1}\gamma_{i1,i2}\gamma_{i2,i0}\gamma_{i0}^{-1}} = 1,$$

because the loop $\gamma_{i0,i1}\gamma_{i1,i2}\gamma_{i2,i0}$ is contractible (see Figure 1). Thus, $\overline{\omega} = 1$, as required. □

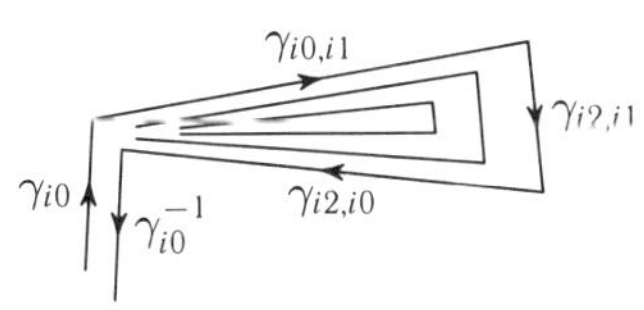

Figure 1. A contraction of the loop

Theorem 3.2 (Hurewicz [**64**]). *If K is a simplicial complex and $\pi_0(K) = \pi_1(K) = \cdots = \pi_{n-1}(K) = 0$ for $n \geq 2$, then the Hurewicz homomorphism $h\colon \pi_n(K) \to H_n(K)$ is an isomorphism.*

Proof. First, we show that h is an epimorphism. Take any cycle $\sum a_i\Delta_i^n \in C_n(K)$. The equalities $\pi_0(K) = \pi_1(K) = \cdots = \pi_{n-1}(K) = 0$ imply that the identity map $K \to K$ is homotopic to a map $f\colon K \to K$ taking the skeleton K^{n-1} to the base point $x_0 \in K$ (see the proof of the Whitehead theorem in Part I). We can assume that x_0 is a vertex of K. Then the restriction of f to K^{n-1} is a simplicial map, hence it has a simplicial approximation $g\colon K^{(m)} \to K$ whose restriction to K^{n-1} coincides with f, i.e., $g(K^{n-1}) = x_0$.

Thus, $g\colon K^{(m)} \to K$ is a simplicial map homotopic to the identity and taking K^{n-1} to x_0. Let φ_i be the restriction of g to Δ_i^n; it can be treated as a spheroid $(D^n, \partial D^n) \to (K, x_0)$. Clearly, $\sum a_i h(\varphi_i) = g_*\big([\sum a_i\Delta_i^n]\big)$, where $[\sum a_i\Delta_i^n]$ is the homology class represented by the cycle $\sum a_i\Delta_i^n$. Since the map g is homotopic to the identity, it follows that $g_*\big([\sum a_i\Delta_i^n]\big) = [\sum a_i\Delta_i^n]$.

Now, we show that h is a monomorphism. Consider a map $\varphi\colon S^n \to K$ for which $\varphi_*([S^n]) = 0$. It is required to prove that φ is the zero element of the group $\pi_n(K)$. We can assume the map φ to be simplicial.

Lemma. *There exists a simplicial complex $L^{n+1} \supset S^n$ such that the cycle $[S^n]$ in this complex is homologous to zero and the map φ can be extended to L^{n+1}.*

Proof. Suppose that at the level of chains, $\varphi_*([S^n]) = \partial(\sum \Delta_i^{n+1})$ (some of the simplices $\Delta_i^{n+1} \subset K$ may coincide). For the first approximation to L^{n+1} we take the disjoint union of the sphere S^n and the simplices $\hat{\Delta}_i^{n+1}$, each identified with the corresponding simplex $\Delta_i^{n+1} \subset K$; the map φ can be extended over $\hat{\Delta}_i^{n+1}$ to the identity map. In the simplicial complex thus obtained, the cycle $z_n = [S^n] - \sum \partial\hat{\Delta}_i^{n+1}$ is homologous to $[S^n]$, and $\varphi_*(z_n) = 0$.

The next approximation to L^{n+1} is constructed so that the cycle z_n becomes homologous to a cycle $w_n = \sum \Delta_j^n$, where $\varphi(\Delta_j^n) \subset K^{n-1}$. Suppose that $z_n = \sum \Delta_\alpha^n$ and $\varphi(\Delta_{\alpha_1}^n) \not\subset K^{n-1}$, i.e., φ maps $\Delta_{\alpha_1}^n$ homeomorphically to a simplex from K. The equality $\varphi_*(z_n) = 0$ implies that some other simplex $\Delta_{\alpha_2}^n$ is mapped homeomorphically onto the same simplex as $\Delta_{\alpha_1}^n$, but the orientations of $\Delta_{\alpha_1}^n$ and $\Delta_{\alpha_2}^n$ are opposite. To $\Delta_{\alpha_1}^n$ and $\Delta_{\alpha_2}^n$ we attach the triangulated prism $\Delta^n \times I$ oriented so that the orientations induced on $\Delta^n \times \{0\} = \Delta_{\alpha_1}^n$ and $\Delta^n \times \{1\} = \Delta_{\alpha_2}^n$ are as required. Consider an extension of φ to $\Delta^n \times I$ with constant value on $\{x\} \times I$. We have $\partial(\Delta^n \times I) = \Delta_{\alpha_1}^n + \Delta_{\alpha_2}^n + c_n$, where the chain c_n contains only simplices satisfying the condition $\varphi(\Delta^n) \subset K^{n-1}$. Let us replace the chain $\Delta_{\alpha_1}^n + \Delta_{\alpha_2}^n$ in the cycle z_n by the homologous chain $-c_n$. After several such operations, we obtain the desired cycle w_n.

Finally, to the simplicial complex obtained at the preceding step, we add the cones over all simplices Δ_j^n in the cycle w_n. Let V be the union of all such simplices, and let CV be the cone over V. This cone is acyclic; therefore, the cycle w_n is now homologous to zero. It remains to extend φ over CV. A map $\varphi\colon V \to K$ can be extended over CV if and only if φ is null-homotopic (that is, homotopic to a constant map). By assumption, we have $\varphi(V) \subset K^{n-1}$. The equalities $\pi_0(K) = \pi_1(K) = \dots = \pi_{n-1}(K) = 0$ imply that the identity map $K \to K$ is homotopic to a map $f\colon K \to K$ taking the skeleton K^{n-1} to x_0. Therefore, the map $\varphi\colon V \to K^{n-1} \subset K$ is homotopic to a constant map to K. □

Suppose that $\varphi\colon L^{n+1} \to K$ is a continuous map, $S^n \subset L^{n+1}$, and $[S^n] = \partial c_{n+1}$ for some chain $c_{n+1} \in C_{n+1}(K)$. Replacing φ by a homotopic map, we can assume that $\varphi(L^{n-1}) = x_0$. Then each simplex $\Delta^n \subset L^{n+1}$

corresponds to a map $(\Delta^n, \partial\Delta^n) \xrightarrow{\varphi} (K, x_0)$, and we obtain a homomorphism $\Phi_*\colon C_n(L^{n+1}) \to \pi_n(K)$.

It remains to show that $\Phi_*([S^n])$ is the homotopy class of the map $S^n \xrightarrow{\varphi} K$. Let $\Delta_1^n, \ldots, \Delta_k^n$ be the simplices in S^n for which $\varphi(\Delta_i^n) \neq x_0$. If $k \leq 2$, then the required assertion follows directly from the definitions. For $k > 2$, we use induction on k.

Thus, we must prove that $\Phi_*([S^n]) = 0$. For any simplex $\Delta^{n+1} \subset L^{n+1}$, the element $\Phi_*(\partial\Delta^{n+1})$ corresponds to the homotopy class of the map $\partial\Delta^{n+1} \xrightarrow{\varphi} K$. This map is null-homotopic because it can be extended over Δ^{n+1}. Therefore, $\Phi_*(\partial\Delta^{n+1}) = 0$. Let $c_{n+1} = \sum a_i \Delta_i^{n+1}$. Then $\Phi_*([S^n]) = \Phi_*(\partial c_{n+1}) = \sum a_i \Phi_*(\partial\Delta_i^{n+1}) = 0$, as required. □

Remark. The proof of the Hurewicz theorem given above follows largely the paper [**111**] by Rokhlin. In the original proof of Hurewicz, more general spaces than simplicial complexes were considered.

Example 23. The *dunce hat* is the CW-complex K obtained by identifying the edges of the triangle as shown in Figure 2. The dunce hat is contractible.

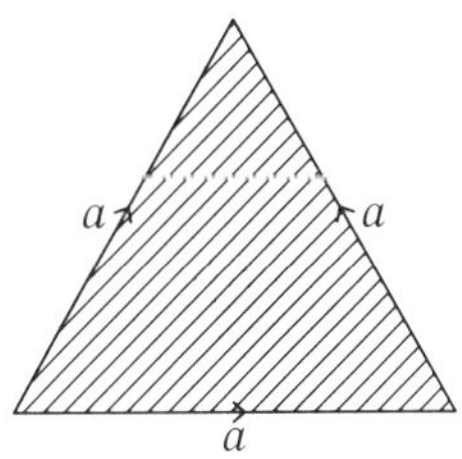

Figure 2. The dunce hat

Proof. The fundamental group $\pi_1(K)$ has one generator a, and this generator satisfies the relation $a^2a^{-1} = 1$, i.e., $a = 1$. Thus, the complex K is simply connected.

The chain complex for calculating the cell homology of the complex K has the form $\cdots \to 0 \to 0 \to C_2 \xrightarrow{\partial} C_1 \to 0$, where $C_2 \cong \mathbb{Z}$, $C_1 \cong \mathbb{Z}$, and ∂ is an isomorphism. Therefore, $H_i(K) = 0$ for $i \geq 1$. It follows from the Hurewicz theorem that $\pi_i(K) = 0$ for all $i \geq 1$. Therefore, according to the Whitehead theorem (see Part I, p. 179), K is contractible. □

Problem 86. (a) Prove that the suspension over any acyclic CW-complex is contractible.

(b) Give an example of a noncontractible space whose suspension is contractible.

1.2. Obstruction Theory. Suppose that X is a path-connected topological space, K is a simplicial complex, L is a subcomplex of K, and K^n is the n-skeleton of K. Let $f\colon \hat{K}^n \to X$ be a map defined on $\hat{K}^n = K^n \cup L$. We want to determine whether this map can be extended to $\hat{K}^{n+1}$. For $n = 0$, the construction of an extension of a map to a path-connected space is obvious; thus, we assume that $n \geq 1$.

We will need to add elements of groups $\pi_n(X, x)$ with different base points $x \in X$; for this reason, we assume that the space X is n-simple.[2] For $n = 1$, this means that the group $\pi_1(X, x)$ is commutative.

We associate with a cochain $c^{n+1}(f) \in C^{n+1}(K; \pi_n(X))$ the map $f\colon \hat{K}^n \to X$ as follows. The consideration of chains and cochains for K assumes that all simplices in K are oriented. An orientation of a simplex $\Delta^{n+1} \subset K$ induces an orientation of the boundary $\partial\Delta^{n+1} \approx S^n$; hence the restriction of f to $\partial\Delta^{n+1}$ determines an element of the group $\pi_n(X)$. We assume that the value of the cochain $c^{n+1}(f)$ at the simplex Δ^{n+1} is equal to this element of $\pi_n(X)$.

It follows directly from the definition that the map f can be extended to a simplex $\Delta^{n+1} \subset K$ if and only if $c^{n+1}(f)(\Delta^{n+1}) = 0$. In particular, if $\Delta^{n+1} \subset L$, then $c^{n+1}(f)(\Delta^{n+1}) = 0$ because f is defined on L. Therefore, $c^{n+1}(f) \in C^{n+1}(K, L; \pi_n(X)) \subset C^{n+1}(K; \pi_n(X))$. In what follows, we treat the cochain $c^{n+1}(f)$ as a relative cochain. It is called an *obstruction* to extending f over $\hat{K}^{n+1}$.

Problem 87. Prove that $S^p \times S^q = (S^p \vee S^q) \cup_f D^{p+q}$, where the map $f\colon S^{p+q-1} \to S^p \vee S^q$ is not null-homotopic.

Theorem 3.3. *We have the equality $\delta c^{n+1}(f) = 0$, which means that $c^{n+1}(f) \in Z^{n+1}(K, L; \pi_n(X))$ is a relative cocycle.*

Proof. We must show that $c^{n+1}(f)(\partial\Delta^{n+2}) = 0$ for any $(n+2)$-simplex in K. First, note that the n-skeleton of the simplicial complex $\partial\Delta^{n+2} \approx S^{n+1}$ is $(n-1)$-connected. This skeleton is homotopy equivalent to the punctured sphere S^{n+1} (from each $(n+1)$-face of the simplex Δ^{n+2}, one interior point is removed), but puncturing an $(n+1)$-manifold M^{n+1} does not affect the homotopy groups $\pi_0, \pi_1, \ldots, \pi_{n-1}$. Indeed, for $k \leq n$, any map $S^k \to M^{n+1}$ is homotopic to a map not involving the removed point y_0, and for $k \leq n-1$, any homotopy between two maps $S^k \to M^{n+1} \setminus \{y_0\}$ in M^{n+1} can be replaced by a homotopy that does not involve y_0.

Let $\partial\Delta^{n+2} = \sum \Delta_i^{n+1}$. By definition, the element $c^{n+1}(f)(\Delta_i^{n+1})$ of $\pi_n(X)$ corresponds to the map $\partial\Delta_i^{n+1} \xrightarrow{f} X$. Consider the simplicial complex B^n that is the n-skeleton of $\partial\Delta^{n+2}$. Take the element α_i of $\pi_n(B^n)$

[2]The definition of n-simple spaces was given in Part I, p. 168.

that corresponds to an orientation-preserving homeomorphism $S^n \to \partial\Delta_i^{n+1}$. Clearly, the homomorphism $f_*\colon \pi_n(B^n) \to \pi_n(X)$ takes α_i to $c^{n+1}(f)(\Delta_i^{n+1})$. This means that the composition

$$\text{the cycle } \partial\Delta_i^{n+1} \longrightarrow \text{a class in } H_n(B^n) \xrightarrow{h^{-1}} \pi_n(B^n) \xrightarrow{f_*} \pi_n(X),$$

where h is the Hurewicz homomorphism, yields the element $c^{n+1}(f)(\Delta_i^{n+1})$. Indeed, since the space B^n is $(n-1)$-connected, it follows that h is an isomorphism for $n \geq 2$. If $n = 1$, then f_*h^{-1} is well defined since, by assumption, the group $\pi_1(X)$ is Abelian and, therefore, f_* maps the commutator subgroup of $\pi_1(B^2)$ to 0. Thus, the element $h^{-1}(z)$ is determined up to the addition of $[\pi_1(B^2), \pi_1(B^2)]$, and the element $f_*h^{-1}(z)$ is determined uniquely.

The element $c^{n+1}(f)(\partial\Delta^{n+2})$ is equal to the image of the homology class of the cycle $\sum \partial\Delta_i^{n+1} = \partial\partial\Delta^{n+2} = 0$ under the homomorphism f_*h^{-1}; therefore, $c^{n+1}(f)(\partial\Delta^{n+2}) = 0$. □

We denote the cohomology class of the cocycle $c^{n+1}(f)$ by $C^{n+1}(f)$.

Theorem 3.4 (Eilenberg [**31**]). *The equality $C^{n+1}(f) = 0$ holds if and only if the map $f\colon \hat{K}^n \to X$ can be extended over $\hat{K}^{n+1}$ after having been changed on $\hat{K}^n \setminus \hat{K}^{n-1}$.*

Proof. Let $f, g\colon \hat{K}^n \to X$ be two maps coinciding on $\hat{K}^{n-1}$. We define the *difference cochain* $d^n(f,g) \in C^n(K, L; \pi_n(X))$ as follows. We take two copies of the simplex $\Delta^n \subset \hat{K}^n$ and glue them together along the boundary $\partial\Delta^n \subset \hat{K}^{n-1}$. Then we identify the first copy with the upper hemisphere and the second copy with the lower hemisphere; the orientation of the sphere is assumed to be compatible with that of the second simplex. We apply f to the first copy of Δ^n and g to the second. The maps f and g coincide on $\partial\Delta^n$; thus, we obtain a continuous map $S^n \to X$. The corresponding element of the group $\pi_n(X)$ is the value of the cochain $d^n(f,g)$ on Δ^n.

The equality $d^n(f,g)(\Delta^n) = 0$ holds if and only if the map $\partial(\Delta^n \times I) \to X$, which coincides with f on $\Delta^n \times \{1\}$, with g on $\Delta^n \times \{0\}$, and with $f|_{\partial\Delta^n} = g|_{\partial\Delta^n}$ on $\partial\Delta^n \times \{t\}$ for all t, can be extended to $\Delta^n \times I$, i.e., the maps $\Delta^n \xrightarrow{f} X$ and $\Delta^n \xrightarrow{g} X$ can be connected by a homotopy which is the identity on $\partial\Delta^n$. In particular, we have $d^n(f,g)(\Delta^n) = 0$ if $\Delta^n \subset L$; thus, $d^n(f,g)(\Delta^n)$ is indeed a relative cochain. Moreover, $d^n(f,g)(\Delta^n) = 0$ if and only if the maps f and g can be connected by a homotopy which is the identity on $\hat{K}^{n-1}$.

Lemma 1. *For any map $f\colon \hat{K}^n \to X$ and cochain $d^n \in C^n(K, L; \pi_n(X))$, there exists a map $g\colon \hat{K}^n \to X$ such that it coincides with f on $\hat{K}^{n-1}$ and $d^n(f,g) = d^n$.*

Proof. For each simplex $\Delta^n \subset \hat{K}^n$, there exists a map g such that the map $S^n \to X$ which coincides with f on the upper hemisphere and with g on the lower hemisphere represents a given element of $\pi_n(X)$; such a map for the simplex $\Delta^n \subset L$ represents the zero element of $\pi_n(X)$, so that $g = f$. The desired map g can always be constructed because $\pi_n(X)$ is a group. □

Lemma 2. $\delta d^n(f, g) = c^{n+1}(g) - c^{n+1}(f)$.

Proof. Any element of the form $(c^{n+1}(g) - c^{n+1}(f))(\Delta^{n+1})$ is equal to the difference of the homotopy classes of the maps $\partial\Delta^{n+1} \xrightarrow{g} X$ and $\partial\Delta^{n+1} \xrightarrow{f} X$. Let $\partial\Delta^{n+1} = \sum \Delta_i^n$. Then the element $d^n(f,g)(\partial\Delta^{n+1})$ equals the sum of $n+2$ homotopy classes of maps $S^n \to X$ each of which coincides with $\Delta_i^n \xrightarrow{f} X$ on the upper hemisphere and with $\Delta_i^n \xrightarrow{g} X$ on the lower hemisphere ($i = 1, \dots, n+2$).

For instance, if $n = 1$, then one element is represented by the sum of the homotopy classes of the two curves shown in Figure 3a and the other,

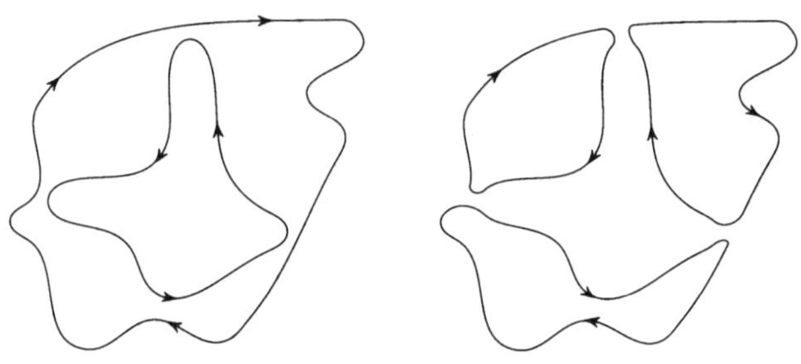

a b

Figure 3. Equal homotopy classes

by the sum of the homotopy classes of the three curves shown in Figure 3b. These elements are equal because, by assumption, the sum of the homotopy classes of two curves does not depend on the base point. For an arbitrary n, the homotopy classes coincide as well. □

Now we can proceed to the proof of the theorem. Suppose $C^{n+1}(f) = 0$, i.e., $c^{n+1}(f) = \delta d^n$ for some cochain d^n. According to Lemma 1, there exists a map $g\colon \hat{K}^n \to X$ such that it coincides with f on $\hat{K}^{n-1}$ and $d^n(f,g) = -d^n$. By Lemma 2, we have

$$c^{n+1}(g) = \delta d^n(f,g) + c^{n+1}(f) = -\delta d^n + \delta d^n = 0;$$

thus, the map g can be extended to $\hat{K}^{n+1}$.

Now suppose that there exists a map $g\colon \hat{K}^n \to X$ such that it coincides with f on $\hat{K}^{n-1}$ and can be extended to $\hat{K}^{n+1}$. Then $c^{n+1}(g) = 0$ and $c^{n+1}(f) = c^{n+1}(f) - c^{n+1}(g) = -\delta d^n(f,g)$, i.e., $C^{n+1}(f) = 0$. □

Remark. In what follows, we use the equality $d^n(f,g)+d^n(g,h)+d^n(h,f) = 0$. It is proved by the same argument as Lemma 2.

1.3. The Hopf–Whitney Theorem. In Part I, p. 231, we proved the Hopf theorem about a one-to-one correspondence between the set of homotopy classes of maps $M^n \to S^n$ and the group $\mathbb{Z}$ (to each map $f\colon M^n \to S^n$ corresponds the integer $\deg f \in \mathbb{Z}$). This theorem can be substantially generalized by methods of obstruction theory. We use the notation $[X, Y]$ for the set of classes of homotopy equivalent maps $X \to Y$. For spaces with base points, we consider only maps that take base points to base points and study homotopies in the class of such maps.

The following theorem was proved by Hopf [**59**], but Hopf considered homology rather than cohomology. For this reason, in the case where the $(n-1)$-dimensional homology group of the complex K has torsion, the statement was fairly intricate. Whitney [**150**] noticed that, in terms of cohomology, not only the statement is simpler but also the correspondence acquires a more natural description.

Theorem 3.5 (Hopf–Whitney)**.** *For any n dimensional simplicial complex K (with a base vertex) and any $(n-1)$-connected space X (with a base point x_0), there is a one-to-one correspondence*

$$[K, X] \longleftrightarrow H^n(K; \pi_n(X))$$

Proof. Consider the set of maps of pairs $(K, K^{n-1}) \to (X, x_0)$. We declare two such maps to be equivalent if they are joined by a homotopy fixed on K^{n-2}. We denote the set of equivalence classes by $[\![K, X]\!]$. Let us show that the natural map $\theta\colon [\![K, X]\!] \to [K, X]$ is one-to-one.

For an arbitrary map $f\colon K \to X$, we construct a homotopy between the maps $K^0 \to x_0$ and $K^0 \xrightarrow{f} X$ and extend it to a homotopy between $K^{n-1} \to x_0$ and $K^{n-1} \xrightarrow{f} X$ using the $(n-1)$-connectedness of X. According to the Borsuk lemma (see Part I, p. 164), this homotopy can be extended to a homotopy of the map $f\colon K \to X$. As a result, we obtain a map $g\colon K \to X$ which is homotopic to f and takes K^{n-1} to x_0. Therefore, the map θ is surjective.

Now consider two maps $f, g\colon (K, K^{n-1}) \to (X, x_0)$ joined by a homotopy $H_0\colon K \times I \to X$. We must prove that they can be joined by a homotopy fixed on K^{n-2}. By assumption, we have $H_0(K^{n-1} \times \partial I) = x_0$; this means that $H_0(K^{n-2} \times \partial I) = x_0$. Let us extend the map $K^{n-2} \times \partial I \times I \to x_0 \in X$ to a map $K^{n-2} \times I \times I \to X$ which coincides with H_0 on $K^{n-2} \times I \times \{0\}$ and is constant on $K^{n-2} \times I \times \{1\}$. This can be done by induction on the dimension of the skeleton because the dimension of the CW-complex $K^{n-2} \times I \times I$ is at most n and the space X is $(n-1)$-connected. The homotopy constructed on $K^{n-2} \times I$ can be extended to a homotopy H_t of the map H_0. For $f'(y) = H_1(y, 0)$ and $g'(y) = H_1(y, 1)$, we have $f \sim f' \sim g' \sim g$, and all the homotopies involved are fixed on K^{n-2}.

To each map $f\colon (K, K^{n-1}) \to (X, x_0)$ we assign the cohomology class $k(f) \in H^n(K; \pi_n(X))$ corresponding to the cocycle $d^n(f, *)$, where $*$ is the map of the complex K to x_0 (the difference cochain $d^n(f, *)$ is a cocycle because $\dim K = n$). Let us show that the map

$$k\colon [\![K, X]\!] \to H^n(K; \pi_n(X))$$

thus defined is one-to-one.

Let $f, g\colon (K, K^{n-1}) \to (X, x_0)$ be maps for which $k(f) = k(g)$. Then the cocycle $d^n(f, g) = d^n(f, *) - d^n(g, *)$ is a coboundary, i.e., f and g are joined by a homotopy fixed on K^{n-2}. Therefore, the map k is injective. Consider any cocycle $d^n \in Z^n(K; \pi_n(X))$. It can be realized as the difference cochain $d^n(f, *)$ for some map $f\colon (K, K^{n-1}) \to (X, x_0)$. Moreover, $k(f)$ is the cohomology class of the cocycle $d^n = d^n(f, *)$. Therefore, the map k is surjective. □

For $X = S^n$, the correspondence $[K, S^n] \longleftrightarrow H^n(K; \mathbb{Z})$ can be defined explicitly: corresponding to each map $f\colon K \to S^n$ is the element $f^*\alpha \in H^n(K; \mathbb{Z})$, where α is the generator of the group $H^n(S^n; \mathbb{Z})$ (determined by orientation). Indeed, we can assume that f takes K^{n-1} to the base point $*$. By definition, the value of the difference cochain $d^n(f, *)$ at each simplex $\Delta^n \subset K^n = K$ is equal to the degree of the map $\Delta^n/\partial\Delta^n \xrightarrow{f} S^n$. The cocycle representing the cohomology class $f^*\alpha$ takes the same value at this simplex. Thus, the Hopf–Whitney theorem has the following corollary.

Theorem 3.6. *If K is an n-dimensional simplicial complex, then any element of the group $H^n(K; \mathbb{Z})$ can be represented in the form $f^*\alpha$, where α is the generator of $H^n(S^n; \mathbb{Z})$ and $f\colon K \to S^n$ is a map determined uniquely up to homotopy.*

1.4. Algebraically Trivial Maps. A map $f\colon X \to Y$ is said to be *algebraically trivial* if the maps $f_*\colon H_i(X) \to H_i(Y)$ and $f^*\colon H^i(Y) \to H^i(X)$ are trivial for all $i \geq 1$.

An example of an algebraically trivial map is an arbitrary map $f\colon S^n \to S^m$, where $n \neq m$. In particular, the Hopf fibration $p\colon S^3 \to S^2$ is algebraically trivial. We use this observation to obtain a homotopy classification of algebraically trivial maps $f\colon X \to S^2$, where X is a three-dimensional simplicial complex; to be more precise, we establish a one-to-one correspondence between the homotopy classes of algebraically trivial maps $X \to S^2$ and the elements of $H^3(X)$. Namely, according to the Hopf–Whitney theorem, we have $[X, S^3] \cong H^3(X, \pi_3(S^3)) = H^3(X)$. Therefore, each cohomology class

$\alpha \in H^3(X)$ corresponds to a homotopy class of maps $F\colon X \to S^3$. We associate a class $\alpha \in H^3(X)$ to the map $f = pF$, where $p\colon S^3 \to S^2$ is the Hopf fibration. The map f is algebraically trivial because so is p.

A classification of algebraically trivial maps $f\colon X \to S^3$ for a three-dimensional simplicial complex X is obtained from a more general classification for an arbitrary simplicial complex X.

Theorem 3.7 (Pontryagin). *For any simplicial complex X, $[F] \mapsto [pF]$ is a one-to-one correspondence between the homotopy classes of maps $X \to S^3$ and the algebraically trivial maps $X \to S^2$.*

Proof (see [**63**]). As usual, the proof that the correspondence is one-to-one consists of two steps: (1) any map belongs to the image; (2) if the images of two maps are homotopic, then so are the initial maps.

Step 1. For any algebraically trivial map $f\colon X \to S^2$, there exists a map $F\colon X \to S^3$ for which $pF = f$.

The map $p\colon S^3 \to S^2$ is a locally trivial fibration; hence there exists an open cover $\{U_\alpha\}$ over which the fibration is trivial. Consider a refinement of the triangulation of X such that the image of each simplex under f is contained in some U_α.

We set $f_k = f|_{X^k}$. Let us construct maps $F_k\colon X^k \to S^3$ such that $pF_k = f_k$ for $k = 2, 3, \ldots$. We start with F_2. The map f_2 is algebraically trivial because it is a composition of the form $X^2 \to X \xrightarrow{f} S^2$ and f is algebraically trivial. It follows that f_2 is null-homotopic. Indeed, the homotopy class of the map $f_2\colon X^2 \to S^2$ corresponds to the element $f_2^*\alpha \in H^2(X^2)$, where α is the generator of the group $H^2(S^2)$, and this element is zero because the map f_2 is algebraically trivial.

Consider a homotopy H between f_2 and a constant map. Obviously, the constant map admits a lifting; therefore, we can construct a lifting of the homotopy H. As a result, we obtain a map $F_2\colon X^2 \to S^3$ for which $pF_2 = f_2$.

Suppose that the map F_{k-1} is constructed for some $k > 2$. We construct F_k as follows. Let $\sigma = \Delta^k$ be a k-simplex in X. On its boundary $\partial\Delta^k \approx S^{k-1}$, the map F_{k-1} is defined; hence we can consider the map

$$\partial\Delta^k \xrightarrow{F_{k-1}} p^{-1}(U_\sigma) \xrightarrow{h_\sigma} U_\sigma \times S^1 \xrightarrow{\pi_\sigma} S^1.$$

Here U_σ is the set from the cover that contains the image of Δ^k, h_σ is the homeomorphism from the definition of a locally trivial fibration, and π_σ is the projection onto the second factor. We have constructed a map $S^{k-1} \to S^1$, where $k - 1 \geq 2$. Any such map is null-homotopic; therefore, it can be extended to a map $g_\sigma \colon \Delta^k \to S^1$. Now we can define F_k as follows:

$$F_k(x) = \begin{cases} F_{k-1}(x) & \text{if } x \in X^{k-1}, \\ h_\sigma^{-1}(f_k(x), g_\sigma(x)) & \text{if } x \in \Delta^k \subset X^k. \end{cases}$$

Finally, the map $F \colon X \to S^3$ is defined by the condition that its restriction to X^k coincides with F_k for any $k \geq 2$.

Step 2. If maps $F, G \colon X \to S^3$ are such that $pF \simeq pG$, then $F \simeq G$.

If $pF \simeq pG$, then, by the covering homotopy theorem, we can lift the homotopy between these maps, obtaining a map $F' \colon X \to S^3$ such that it is homotopic to F and $pF' = pG$. So, we can assume from the very beginning that $pF = pG$.

Consider the sphere S^3 as the group of unit quaternions. Then the Hopf fibration is the map taking each unit quaternion to the corresponding coset modulo the subgroup S^1 of complex numbers with modulus 1. Let $H \colon X \to S^3$ be the map defined by

$$H(x) = F(x) \cdot [G(x)]^{-1}.$$

Since $pF = pG$, the quaternions $F(x)$ and $G(x)$ belong to the same coset modulo S^1. Therefore, the map H takes X to a proper subset $S^1 \subset S^3$, which means that H is null-homotopic, i.e., there exists a homotopy $H_t \colon X \to S^3$, where $H_0 = H$ and $H_1(X) = 1 \in S^3$.

Let $\Phi_t(x) = H_t(x) \cdot G(x)$. Then $\Phi_0 = F$ and $\Phi_1 = G$, and Φ_t is the required homotopy. □

1.5. The Eilenberg–MacLane Spaces. Suppose that n is a positive integer and π is a group; for $n \geq 2$, we assume that π is Abelian. A *CW*-complex X is called a *$K(\pi, n)$ space*, or an *Eilenberg–MacLane space*, if

$$\pi_k(X) = \begin{cases} \pi & \text{for } k = n, \\ 0 & \text{for } k \neq n. \end{cases}$$

These spaces were introduced in [**32**, **34**].

Example 24. The circle S^1 is a $K(\mathbb{Z}, 1)$ space.

Example 25. Any closed 2-manifold M^2 different from S^2 and $\mathbb{R}P^2$ is a $K(\pi, 1)$ space for $\pi = \pi_1(M^2)$.

Example 26. The space $\mathbb{R}P^\infty$ is $K(\mathbb{Z}_2, 1)$.

Example 27. The space $\mathbb{C}P^\infty$ is $K(\mathbb{Z}, 2)$.

Example 28. Let X_n be the set of all points of $\mathbb{C}^n$ with pairwise different coordinates. Then X_n is a $K(P_n, 1)$ space for some group P_n. (This group P_n is called the *colored braid group* on n strands.)

Proof. Consider the map $X_n \to X_{n-1}$ forgetting the last coordinate. This map is a locally trivial fibration. Ifs fiber F is $\mathbb{C}$ with $n-1$ points deleted. Thus, it is homotopy equivalent to the wedge of $n-1$ circles. In particular, $\pi_k(F) = 0$ for $k \geq 2$. The exact sequence $\pi_k(F) \to \pi_k(X_n) \to \pi_k(X_{n-1})$ shows that if $k \geq 2$ and $\pi_k(X_{n-1}) = 0$, then $\pi_k(X_n) = 0$. If remains to note that the space $X_1 = \mathbb{C}$ is contractible. □

Example 29. Let Y_n be the quotient of X_n by the action of the group S_n (of permutations of coordinates). Then Y_n is a $K(B_n, 1)$ space for some group B_n. (This group B_n is called the *braid group on n strands*.)

Proof. The map $X_n \to X_n/S_n = Y_n$ is a covering. Therefore, $\pi_k(X_n) \cong \pi_k(Y_n)$ for $k \geq 2$. □

Remark. It can be proved that the group B_n is defined by generators b_1, $\dots$, b_{n-1} and the relations

$$b_i b_{i+1} b_i = b_{i+1} b_i b_{i+1} \quad \text{for} \quad 1 \leq i \leq n-1$$

and

$$b_i b_j = b_j b_i \quad \text{for} \quad |i-j| \geq 2$$

(the latter relations are called the far commutativity relations).

We regard S^∞ as the subset of $\mathbb{C}^\infty$ consisting of the points $(z_1, z_2, \dots)$ for which $\sum |z_i|^2 = 1$. On S^∞, we introduce the equivalence relation defined by $(z_1, z_2, \dots) \sim (\varepsilon z_1, \varepsilon z_2, \dots)$, where $\varepsilon = \exp(2\pi i/m)$. The space $L_m^\infty = S^\infty/\sim$ is called the *infinite-dimensional lens space*.

Example 30. The space L_m^∞ is $K(\mathbb{Z}_m, 1)$.

Proof. The covering $S^\infty \to L_m^\infty$ is universal because the space S^∞ is contractible (see Part I, Problem 35). This covering is a regular covering with automorphism group $\mathbb{Z}_m$; hence $\pi_1(L_m^\infty) = \mathbb{Z}_m$. Moreover, $\pi_i(L_m^\infty) = \pi_i(S^\infty) = 0$ for $i \geq 2$. □

Problem 88. Prove that

$$H_n(L_m^\infty) = \begin{cases} \mathbb{Z}_m & \text{for odd } n, \\ 0 & \text{for even } n > 0. \end{cases}$$

Problem 89. Given a 3-manifold M^3 with infinite fundamental group for which $\pi_2(M^3) = 0$, prove that M^3 is a $K(\pi, 1)$ space.

Theorem 3.8. *For any n and π, there exists a $K(\pi, n)$ simplicial complex.*

Proof. Consider a free resolution $0 \to R \to F \to \pi \to 0$, where R and F are free groups (for $n = 1$) or free Abelian groups (for $n \geq 2$). In other words, let π be a group (Abelian for $n \geq 2$) with generators $\{b_j \mid j \in J\}$

and relations $\{a_i \mid i \in I\}$; here $\{a_i \mid i \in I\}$ and $\{b_j \mid j \in J\}$ are bases of the groups R and F. Consider the wedge K^n of triangulated n-spheres S_j^n, where $j \in J$. The group $\pi_n(K^n)$ is canonically isomorphic to F. For each basis relation $a_i \in R \subset F = \pi_n(K^n)$, take a continuous map $f_i\colon S_i^n \to K^n$ representing the homotopy class $a_i \in \pi_n(K^n)$ and consider its simplicial approximation φ_i. The triangulation of S_i^n can be naturally extended to a triangulation of $D_i^{n+1} \supset S_i^n$ (to each n-simplex we add one vertex at the center of the ball). Let K^{n+1} be the simplicial complex obtained from K^n by attaching the balls D_i^{n+1} via the maps φ_i. If $n = 1$, then, according to Theorem 6.1 in Part I, we have $\pi_1(K^2) \cong \pi$. If $n > 1$, then the Hurewicz theorem can be applied. Indeed, it follows from the cellular approximation theorem that the space K^{n+1} is $(n-1)$-connected. Clearly, $H_n(K^{n+1}) \cong \pi$. Therefore, by the Hurewicz theorem, $\pi_n(K^{n+1}) \cong \pi$.

Now, we choose a system of generators in $\pi_{n+1}(K^{n+1})$, consider the simplicial map $\psi\colon S^{n+1} \to K^{n+1}$ corresponding to each generator, and attach D^{n+2} to K^{n+1} via the map ψ. As a result, we obtain a simplicial complex K^{n+2} whose homotopy groups up to dimension $n+1$ are as required. This is implied by the following lemma.

Lemma. *Suppose that a CW-complex X' is obtained from a CW-complex X by attaching an $(n+2)$-cell via a map $\psi\colon S^{n+1} \to X^{n+1}$. Then the inclusion $i\colon X \hookrightarrow X'$ induces an isomorphism $i_*\colon \pi_k(X) \to \pi_k(X')$ for $k \le n$; for $k = n+1$, it induces an epimorphism $i_*\colon \pi_k(X) \to \pi_k(X')$ whose kernel contains the subgroup generated by the homotopy class of ψ.*

Proof. According to the cellular approximation theorem, for $k \le n+1$, any map from the sphere S^k into a CW-complex Y is homotopic to a map into the $(n+1)$-skeleton Y^{n+1}, and for $k \le n$, any homotopy in Y between two maps $S^k \to Y^{n+1}$ can be replaced by a homotopy in Y^{n+1}. This implies that i_* is an isomorphisms for $k \le n$ and an epimorphism for $k = n+1$.

The composition $S^{n+1} \xrightarrow{\psi} X \xrightarrow{i} X'$ is null-homotopic; the contraction is performed along the cell D^{n+2}. □

Then, we kill the group $\pi_{n+2}(K^{n+2})$ in the same way as the $(n+1)$-dimensional homotopy group of K^{n+1}, and so on. □

Theorem 3.9. *Any two $K(\pi, n)$ spaces (with the same π and n) are homotopy equivalent.*

Proof. It is sufficient to verify that any $K(\pi, n)$ space X is homotopy equivalent to the simplicial complex K constructed in the proof of Theorem 3.8. Let us construct a continuous map $K \to X$ inducing isomorphisms $\pi_k(K) \to \pi_k(X)$ for all k. First, we define this map on $K^n = \bigvee S_j^n$.

Each sphere S_j^n corresponds to a generator b_j of the group $\pi_n(X)$. Choose a map $S_j^n \to X$ representing the element $b_j \in \pi_n(X)$. As a result, we obtain a map $K^n \to X$. We extend it over K^{n+1} as follows. For each $(n+1)$-cell D_i^{n+1}, the characteristic map $S_j^n \to K^n$ corresponds to a basis relation $a_i \in R \subset F = \pi_n(K^n)$. The map $K^n \to X$ induces an epimorphism $\pi_n(K^n) \to \pi_n(X)$ with kernel R. Therefore, the composition $S_i^n \to K^n \to X$ is homotopic to a constant map and can therefore be extended over the entire cell D_i^{n+1}. We have obtained a map $K^{n+1} \to X$. Consider the composition $K^n \hookrightarrow K^{n+1} \to X$ and the induced group homomorphisms $\pi_n(K^n) \to \pi_n(K^{n+1}) \to \pi_n(X)$. We know that both homomorphisms $\pi_n(K^n) \to \pi_n(K^{n+1})$ and $\pi_n(K^n) \to \pi_n(X)$ are epimorphisms with kernel R. It follows that $\pi_n(K^{n+1}) \to \pi_n(X)$ is an isomorphism.

There are no obstructions to extending the map $K^{n+1} \to X$ over cells of dimension higher than $n+1$ because $\pi_n(X) = 0$ for $k \geq n+1$. The extension $K \to X$ induces group isomorphisms $\pi_k(K) \to \pi_k(X)$ for all k; for $k = n$, this is so because the inclusion $K^{n+1} \hookrightarrow K$ induces an isomorphism $\pi_n(K^{n+1}) \to \pi_n(K)$, and for $k \neq n$, because $\pi_k(K) = \pi_k(X) = 0$. By assumption, both spaces K and X are CW-complexes; thus, we can apply the Whitehead theorem (Theorem 4.21 in Part I). □

Problem 90. Let X be a finite-dimensional $K(\pi, 1)$ simplicial complex. Prove that the group π contains no elements of finite order.

1.6. Cohomology and Maps to $K(\pi, n)$ Spaces. Let K be an $(n-1)$-connected simplicial complex, where $n \geq 2$. Then the group $\mathrm{Hom}(H_n(K), \pi_n(K))$ contains the homomorphism h^{-1} inverse to the Hurewicz homomorphism $h\colon \pi_n(K) \to H_n(K)$. Moreover, $H_{n-1}(K) = 0$; hence the exact sequence

$$0 \longrightarrow \mathrm{Ext}(H_{n-1}(K), G) \longrightarrow H^n(K; G) \longrightarrow \mathrm{Hom}(H_n(K), G) \longrightarrow 0$$

implies the isomorphism $H^n(K; G) \cong \mathrm{Hom}(H_n(K), G)$. The cohomology class $F_K \in H^n(K; \pi_n(K))$ corresponding to h^{-1} is called the *fundamental cohomology class* of the space K. The most important case is where K is a $K(\pi, n)$ simplicial complex. In this case, the fundamental cohomology class is denoted by $F_\pi \in H^n(K; \pi) \cong \mathrm{Hom}(H_n(K), \pi)$; note that $H_n(K) \cong \pi$ and F_π corresponds to the identity map $\pi \to \pi$.

For the $K(\pi, n)$ simplicial complex K constructed in the proof of Theorem 3.8, the fundamental cohomology class F_π is represented by the cocycle c^n defined as follows. Recall that $K^n = \bigvee S_i^n$, where each sphere S_i^n corresponds to a generator $a_i \in \pi$. To the fundamental class of S_i^n the cocycle c^n assigns the value a_i because the Hurewicz homomorphism assigns to this fundamental class the element $a_i \in \pi_n(K)$.

Theorem 3.10. *Let L be a $K(\pi, n)$ simplicial complex. Then, for any simplicial complex K, the map $[K, L] \to H^n(K, \pi)$ that takes $f\colon K \to L$ to $f^*(F_\pi) \in H^n(K; \pi)$ is one-to-one.*

Proof. According to the Hopf–Whitney theorem, there is a one-to-one correspondence

$$[K^n, L] \longleftrightarrow H^n(K^n; \pi_n(L)) = H^n(K^n; \pi).$$

We must only verify that this correspondence has the required form and is determined by the n-skeleton K^n. Consider the diagram

$$\begin{array}{ccc} [K, L] & \xrightarrow{\ i^\#\ } & [K^n, L] \\ & & \big\downarrow k\theta^{-1} \\ H^n(K; \pi) & \xrightarrow{\ i^*\ } & H^n(K^n; \pi). \end{array} \tag{21}$$

In this diagram both horizontal arrows are induced by the natural embedding $i\colon K^n \to K$, and $k\theta^{-1}$ is the map from the Hopf–Whitney theorem. The map $i^*\colon H^n(K; \pi) \to H^n(K^n; \pi)$ is injective because the cochains of dimension at most n in K are identified with the cochains in K^n and the equality $z^n = \delta w^{n-1}$ carries over from K^n to K. The map $i^\#\colon [K, L] \to [K^n, L]$ is injective as well. Indeed, if maps $f, g\colon K \to L$ have homotopic restrictions to K^n, then they are themselves homotopic because all obstructions to extending the homotopies lie in cocycles with values in $\pi_m(L)$ with $m > n$, whereas, by assumption, $\pi_m(L) = 0$ for $m > n$.

To prove that the diagram (21) defines a one-to-one map $\varphi\colon [K, L] \to H^n(K; \pi)$, it remains to show that $k\theta^{-1}$ maps the image of $i^\#$ to the image of i^*. By the properties of θ, it suffices to consider maps $f\colon (K^n, K^{n-1}) \to (L, x_0)$ up to a homotopy fixed on K^{n-2}. The homotopy class of such a map f belongs to $\operatorname{Im} i^\#$ if and only if f can be extended to the entire space K, and the latter condition holds if and only if f can be extended to K^{n+1} because further obstructions to the extension lie in the trivial groups ($\pi_m(L) = 0$ for $m > n$). If such an extension exists, then $c^{n+1}(f) = 0$; here $c^{n+1} \in C^{n+1}(K; \pi)$ is the obstruction to extending the map f. The equality $c^{n+1}(f) = 0$ is equivalent to $\delta d^n(f, *) = 0$ because $\delta d^n(f, *) = c^{n+1}(*) - c^{n+1}(f)$ and $c^{n+1}(*) = 0$. By definition, $k(f)$ is the cohomology class of the cocycle $d^n(f, *)$ in K^n. The cochain $d^n(f, *)$ is also a cocycle in K; therefore, $k(f) \in \operatorname{Im} i^\#$, as required.

The map φ is natural in the sense that for any $h\colon K \to K'$, the diagram

$$\begin{array}{ccc} [K', L] & \xrightarrow{h^{\#}} & [K, L] \\ \downarrow{\scriptstyle\varphi} & & \downarrow{\scriptstyle\varphi} \\ H^n(K';\pi) & \xrightarrow{h^*} & H^n(K;\pi) \end{array}$$

is commutative. We can assume that L is the $K(\pi, n)$ simplicial complex that was constructed in the proof of Theorem 3.8. Then the skeleton L^{n-1} consists of one point, and it follows directly from the definitions that $\varphi(\mathrm{id}_L) = F_\pi$. We set $K' = L$ and $h = f$. Then $\varphi(f) = \varphi(f^{\#}(\mathrm{id}_L)) = f^*(\varphi(\mathrm{id}_L)) = f^*(F_\pi)$, i.e., the map φ does have the required form. □

Corollary. *There is a one-to-one correspondence*

$$[K(\pi, n), K(\pi', n)] \leftrightarrow \mathrm{Hom}(\pi, \pi'),$$

under which the homotopy class of any map $f\colon K(\pi, n) \to K(\pi', n)$ corresponds to the homomorphism $f_*\colon \pi \to \pi'$ induced by this map.

Proof. According to Theorem 3.10, we have

$$[K(\pi, n), K(\pi', n)] \leftrightarrow H^n(K(\pi, n); \pi').$$

The universal coefficient theorem implies the isomorphism $H^n(K(\pi, n); \pi') \cong \mathrm{Hom}(H_n(K(\pi, n)), \pi')$. Finally, from the Hurewicz theorem, it follows that $H_n(K(\pi, n)) \cong \pi$. □

Theorem 3.10 allows us to describe all *cohomology operations*, which are defined as follows. Take Abelian groups G and H and positive integers m and n. A cohomology operation of type (m, n, G, H) is a family of maps (not necessarily homomorphisms) $\Theta_K\colon H^m(K; G) \to H^n(K; H)$ that are defined for each simplicial complex K so that the diagram

$$\begin{array}{ccc} H^m(L; G) & \xrightarrow{\Theta_L} & H^n(L; H) \\ \downarrow{\scriptstyle f^*} & & \downarrow{\scriptstyle f^*} \\ H^m(K; G) & \xrightarrow{\Theta_K} & H^n(K; H) \end{array}$$

is commutative for any continuous map $f\colon K \to L$.

Example 31. A homomorphism of groups $G \to H$ induces a homomorphism $H^m(K; G) \to H^m(K; H)$, which is a dimension-preserving cohomology operation.

Example 32. If R is the additive group of a unital ring, then for any k, the map $\alpha \mapsto \alpha^k$ is a cohomology operation of type (m, mk, R, R).

Example 33. The Bockstein homomorphism

$$\beta^*\colon H^k(K;G'') \to H^{k+1}(K;G')$$

corresponding to an exact sequence

$$0 \longrightarrow G' \longrightarrow G \longrightarrow G'' \longrightarrow 0$$

is a cohomology operation that increases the dimension by 1.

Theorem 3.11. *There exists a one-to-one correspondence between the cohomology operations of type (m, n, G, H) and the elements of the group $H^n(K(G,m);H)$. This correspondence is defined by $\Theta \mapsto \Theta(F_G)$, where $F_G \in H^m(K(G,m);G)$ is the fundamental cohomology class.*

Proof. An element $\alpha \in H^m(K;G)$ can be represented in the form $\alpha = f^*(F_G)$, where $f\colon K \to K(G;m)$ is a map. Therefore, $\Theta(\alpha) = \Theta(f^*(F_G)) = f^*(\Theta(F_G))$, i.e., the cohomology operation Θ is completely determined by the element $\Theta(F_G)$.

It remains to show that any element $\theta \in H^n(K(G,m);H)$ can be represented in the form $\theta = \Theta(F_G)$ for some cohomology operation Θ. The element θ corresponds to a map $\varphi\colon K(G,m) \to K(H,n)$ for which $\theta = \varphi^*(F_H)$. Taking the composition of every $g\colon K \to K(G,m)$ with φ, we obtain a map $[K, K(G,m)] \to [K, K(H,n)]$. This map corresponds to the cohomology operation $\Theta\colon H^m(K;G) \to H^n(K;H)$ that takes $g^*(F_G)$ to $g^*\varphi^*(F_H)$. In particular, $\Theta(F_G) = \varphi^*(F_H) = \theta$. □

Corollary 1. *Cohomology operations cannot decrease dimension.*

Proof. If $n < m$, then $H^n(K(G,m);H) = 0$ because the space $K(G,m)$ is $(m-1)$-connected. □

Corollary 2. *All dimension-preserving cohomology operations are induced by homomorphisms $G \to H$.*

Proof. In the proof of the corollary of Theorem 3.10, it was mentioned that

$$H^n(K(G,n);H) \cong \operatorname{Hom}(H_n(K(G,n)), H) \cong \operatorname{Hom}(G,H).$$ □

1.7. The Moore Spaces. The definition of the Moore spaces is similar to that of the Eilenberg–MacLane spaces, with the only difference that homotopy groups are replaced by homology groups. Namely, for an Abelian group G and a positive integer n, a *Moore space* $M(G,n)$ is a simplicial complex X for which

$$\tilde{H}_k(X) = \begin{cases} G & \text{if } k = n, \\ 0 & \text{if } k \neq n. \end{cases}$$

For $n > 1$, it is also assumed that $\pi_1(X) = 0$.

Let us show that, for any Abelian group G and any positive integer n, there exists a Moore space $M(G,n)$. For $M(\mathbb{Z},n)$ we take the sphere S^n, and for $M(\mathbb{Z}_m,n)$ we can take the sphere S^n to which the $(n+1)$-cell D^{n+1} is attached by a map $\partial D^{n+1} \to S^n$ of degree m. In this case, the chain complex for calculating the cellular homology has the form $\mathbb{Z} \xrightarrow{\times m} \mathbb{Z} \to 0 \to \cdots$. If the group G is finitely generated, then we can take the wedge of several spaces $M(\mathbb{Z},n)$ and $M(\mathbb{Z}_m,n)$ for $M(G,n)$.

For an arbitrary Abelian group G, the space $M(G,n)$ can be constructed as follows. There exists an epimorphism $F \to G$, where F is a free Abelian group. Let H be its kernel. The group H, being a subgroup of the free Abelian group F, is itself a free Abelian group. Let $\{f_\alpha\}$ be a basis of the group F, and let $\{h_\beta\}$ be a basis of H. Then $h_\beta = \sum_\alpha n_{\alpha\beta} f_\alpha$. We set $X^n = \bigvee_\alpha S^n_\alpha$. To X^n we attach cells $\{D^{n+1}_\beta\}$ via the maps $\chi_\beta \colon \partial D^{n+1}_\beta \to X^n$ defined as follows. First, ∂D^{n+1}_β is mapped to the wedge of k_β copies of the n-sphere, where k_β is the number of nonzero coefficients $n_{\alpha\beta}$, by contracting $k_\beta - 1$ $(n-1)$-spheres in ∂D^{n+1}_β. Then each of the spheres in the wedge is mapped to the corresponding sphere S^n_α by a map of degree $n_{\alpha\beta}$. The resulting CW-complex can easily be made into a simplicial complex X; this is the Moore space $M(G,n)$. Indeed, the chain complex for calculating the cellular homology of the complex X has the form $H \xrightarrow{i} F \to 0 \to \cdots$, where i is the inclusion of the subgroup H into the group F.

If $G_1, G_2, \dots$ are any Abelian groups, then there exists a simplicial complex X for which $H_i(X) = G_i$; for X we can take the wedge $\bigvee_i M(G_i, i)$.

Surprisingly, for cohomology, a similar assertion is false. An example was constructed by Kan and Whitehead [**67**].

Theorem 3.12 (Kan–Whitehead). *For no positive integer n, there exists a simplicial complex X for which $H^{n-1}(X;\mathbb{Z}) = 0$ and $H^n(X;\mathbb{Z}) = \mathbb{Q}$, where $\mathbb{Q}$ is the additive group of rational numbers.*

Proof. Suppose there exists a simplicial complex X for which $H^{n-1}(X;\mathbb{Z}) = 0$ and $H^n(X;\mathbb{Z}) = \mathbb{Q}$. Then the universal coefficient formulas yield

$$0 = H^{n-1}(X) \cong \operatorname{Hom}(H_{n-1}(X),\mathbb{Z}) \oplus \operatorname{Ext}(H_{n-2}(X),\mathbb{Z}) \tag{22}$$

and

$$\mathbb{Q} = H^n(X) \cong \operatorname{Hom}(H_n(X),\mathbb{Z}) \oplus \operatorname{Ext}(H_{n-1}(X),\mathbb{Z}). \tag{23}$$

Relations (22) imply, in particular,that $\operatorname{Hom}(H_{n-1}(X),\mathbb{Z}) = 0$.

Lemma. *For any Abelian group A such that* $\mathrm{Hom}(A,\mathbb{Z})=0$, *the following assertions are valid.*

(a) *The group* $\mathrm{Ext}(A,\mathbb{Z})$ *is divisible if and only if A is torsion-free;*

(b) *The group* $\mathrm{Ext}(A,\mathbb{Z})$ *is torsion-free if and only if A is divisible.*

Proof. First, we introduce the notation

$$mA=\{b\in A\,|\,b=ma, a\in A\},$$
$$_mA=\{a\in A\,|\,ma=0\},$$
$$A_m=A/mA.$$

Note that if B is an Abelian group such that $mB=0$, then $\mathrm{Hom}(B,\mathbb{Z})=0$. Indeed, $\varphi(b)=n$ implies $0=\varphi(mb)=mn$. Both groups $_mA$ and A_m have this property; hence $\mathrm{Hom}(_mA,\mathbb{Z})=0$ and $\mathrm{Hom}(A_m,\mathbb{Z})=0$. Taking into account this observation and applying Problem 22 to the exact sequences

$$0\longrightarrow {}_mA\longrightarrow A\xrightarrow{\times m} mA\longrightarrow 0,\qquad 0\longrightarrow mA\xrightarrow{\subset} A\longrightarrow A_m\longrightarrow 0,$$

we obtain the exact sequences

$$0\longrightarrow \mathrm{Ext}(mA,\mathbb{Z})\longrightarrow \mathrm{Ext}(A,\mathbb{Z})\longrightarrow \mathrm{Ext}(_mA,\mathbb{Z})\longrightarrow 0$$

and

$$0\longrightarrow \mathrm{Hom}(A,\mathbb{Z})\longrightarrow \mathrm{Hom}(mA,\mathbb{Z})$$
$$\longrightarrow \mathrm{Ext}(A_m,\mathbb{Z})\longrightarrow \mathrm{Ext}(A,\mathbb{Z})\longrightarrow \mathrm{Ext}(mA,\mathbb{Z})\longrightarrow 0.$$

By assumption, $\mathrm{Hom}(A,\mathbb{Z})=0$. Moreover, according to Problem 14, we have $\mathrm{Hom}(mA,\mathbb{Z})\cong\mathrm{Hom}(A,\mathbb{Z})$. Thus, the second exact sequence becomes

$$0\longrightarrow \mathrm{Ext}(A_m,\mathbb{Z})\longrightarrow \mathrm{Ext}(A,\mathbb{Z})\longrightarrow \mathrm{Ext}(mA,\mathbb{Z})\longrightarrow 0.$$

Using these exact sequences, we construct the exact sequence

$$0\longrightarrow \mathrm{Ext}(A_m,\mathbb{Z})\longrightarrow \mathrm{Ext}(A,\mathbb{Z})\xrightarrow{\varphi} \mathrm{Ext}(A,\mathbb{Z})\longrightarrow \mathrm{Ext}(_mA,\mathbb{Z})\longrightarrow 0,$$

in which φ is the composition of the maps $\mathrm{Ext}(A,\mathbb{Z})\to\mathrm{Ext}(mA,\mathbb{Z})\to\mathrm{Ext}(A,\mathbb{Z})$ induced by the inclusion $mA\subset A$ and by the map $A\xrightarrow{\times m}mA$. Hence, φ is the multiplication by m, which implies the formulas $_m\mathrm{Ext}(A,\mathbb{Z})=\mathrm{Ext}(A_m,\mathbb{Z})$ and $\mathrm{Ext}(A,\mathbb{Z})_m=\mathrm{Ext}(_mA,\mathbb{Z})$.

For a periodic group T, the equality $\mathrm{Ext}(T,\mathbb{Z})=0$ holds if and only if $T=0$ (see Problem 18). Hence $_mA=0$ if and only if $\mathrm{Ext}(_mA,\mathbb{Z})=\mathrm{Ext}(A,\mathbb{Z})_m=0$, and $A_m=0$ if and only if $\mathrm{Ext}(A_m,\mathbb{Z})={}_m\mathrm{Ext}(A,\mathbb{Z})=0$. To prove the lemma, it remains to note that, first, an Abelian group B has no torsion if and only if $_mB=0$ for all $m>1$ and, second, an Abelian group B is divisible if and only if $B_m=0$ for all $m>1$. □

We continue the proof of Theorem 3.12. Consider equality (23). Note that the group $\mathbb{Q}$ cannot be represented as a direct sum of two nonzero groups. Indeed, suppose that $\mathbb{Q} = A \oplus B$ and a and b are nonzero elements of A and B. Then $pa \neq qb$ for any $p, q \in \mathbb{Z}$ such that $pa \neq 0$ and $qb \neq 0$. But no rational numbers a and b have this property. Therefore, either $\operatorname{Hom}(H_n(X), \mathbb{Z}) = \mathbb{Q}$ or $\operatorname{Ext}(H_{n-1}(X), \mathbb{Z}) = \mathbb{Q}$. The equality $\operatorname{Hom}(H_n(X), \mathbb{Z}) = \mathbb{Q}$ cannot hold because, for no Abelian group A, the group $\operatorname{Hom}(A, \mathbb{Z})$ is divisible . Indeed, suppose that $\varphi(a) = k \neq 0$. Choose a positive integer m so that k/m is not an integer. If $m\psi = \varphi$ for some $\psi \in \operatorname{Hom}(A, \mathbb{Z})$, then $m\psi(a) = \varphi(a) = k$, i.e., $\psi(a) = k/m$; hence $m\psi$ never equals φ. Thus, $\operatorname{Ext}(H_{n-1}(X), \mathbb{Z}) = \mathbb{Q}$ is a torsion-free divisible group. According to the lemma, the group $H_{n-1}(X)$ is torsion-free and divisible. Such a group necessarily contains $\mathbb{Q}$ as a subgroup, i.e., there exists a monomorphism $0 \to \mathbb{Q} \to H_{n-1}(X)$. This implies the existence of an epimorphism $\operatorname{Ext}(H_{n-1}(X), \mathbb{Z}) \to \operatorname{Ext}(\mathbb{Q}, \mathbb{Z}) \to 0$.

To obtain a contradiction, it remains to prove that the group $\operatorname{Ext}(\mathbb{Q}, \mathbb{Z})$ is uncountable, and an epimorphism $\mathbb{Q} = \operatorname{Ext}(H_{n-1}(X), \mathbb{Z}) \to \operatorname{Ext}(\mathbb{Q}, \mathbb{Z}) \to 0$ cannot exist. The group $\operatorname{Ext}(A, \mathbb{Z})$ can be calculated by using an injective resolution $0 \to \mathbb{Z} \to \mathbb{Q} \to \mathbb{Q}/\mathbb{Z} \to 0$. As a result, for $A = \mathbb{Q}$, we obtain an exact sequence

$$0 \longrightarrow \operatorname{Hom}(\mathbb{Q}, \mathbb{Q}) \longrightarrow \operatorname{Hom}(\mathbb{Q}, \mathbb{Q}/\mathbb{Z}) \longrightarrow \operatorname{Ext}(\mathbb{Q}, \mathbb{Z}) \longrightarrow 0.$$

The group $\operatorname{Hom}(\mathbb{Q}, \mathbb{Q}) \cong \mathbb{Q}$ is countable, whereas $\operatorname{Hom}(\mathbb{Q}, \mathbb{Q}/\mathbb{Z})$ is uncountable (see Problem 15). Therefore, the group $\operatorname{Ext}(\mathbb{Q}, \mathbb{Z})$ is uncountable. □

2. Characteristic Classes

2.1. Vector Bundles. A locally trivial fiber bundle $p\colon E \to B$ is called an n-dimensional *vector bundle* if its fiber F is the linear space $\mathbb{R}^n$, each set $p^{-1}(b)$ (where $b \in B$) is endowed with the structure of an n-dimensional linear space, and the homeomorphisms $h\colon U \times F \to p^{-1}(U)$ have the property that for each point $b \in U$, the map $x \mapsto h(b, x) \in p^{-1}(b)$ is a linear space isomorphism between $F = \mathbb{R}^n$ and $p^{-1}(b)$.

A vector bundle is said to be *smooth* if E and B are smooth manifolds, p is a smooth map, and the homeomorphisms h are diffeomorphisms.

Example 34. Let TM^n be the tangent bundle[3] of a closed manifold M^n. Then the natural projection $p\colon TM^n \to M^n$ is a smooth vector bundle. We denote this bundle by τ_{M^n}.

A map $s\colon B \to E$ is called a *section* of the bundle if $ps = \mathrm{id}_B$; the sections of the tangent bundle are vector fields on the manifold. For any

[3]The definition of the tangent bundle is given in Part I on p. 202.

vector bundle, the *zero section* is defined, which takes each point $b \in B$ to the zero vector of the space $p^{-1}(b)$.

Vector bundles $p_1 \colon E_1 \to B$ and $p_2 \colon E_2 \to B$ are *isomorphic* if there exists a homeomorphism $\varphi \colon E_1 \to E_2$ which isomorphically maps $p_1^{-1}(b)$ onto $p_2^{-1}(b)$ for all $b \in B$. Isomorphic vector bundles are also called *equivalent*.

Note that, for isomorphic vector bundles, the images of the zero sections have homeomorphic complements in the total spaces of the bundles.

Theorem 3.13. *Suppose that $p_1 \colon E_1 \to B$ and $p_2 \colon E_2 \to B$ are vector bundles and $f \colon E_1 \to E_2$ is a continuous map which induces an isomorphism of linear spaces $p_1^{-1}(b)$ and $p_2^{-1}(b)$ for each $b \in B$. Then f is a homeomorphism, i.e., the bundles p_1 and p_2 are isomorphic.*

Proof. For each point $b \in B$, we choose neighborhoods U_i $(i = 1, 2)$ and homeomorphisms $h_i \colon U_i \times \mathbb{R}^n \to p_i^{-1}(U_i)$. It is sufficient to verify that the map

$$h_2^{-1} f h_1 \colon (U_1 \cap U_2) \times \mathbb{R}^n \to (U_1 \cap U_2) \times \mathbb{R}^n$$

is a homeomorphism. By assumption, this map takes (b, v) to $(b, A(b)v)$, where $A(b)$ is a nondegenerate matrix whose elements continuously depend on b. The inverse map has the form $(b, w) \mapsto (b, A^{-1}(b)w)$. The elements of the matrix $A^{-1}(b)$ continuously depend on b as well; therefore, the inverse map is continuous. □

The natural projection $B \times \mathbb{R}^n \to B$ is an n-dimensional vector bundle. It is called a *trivial*, or a *product* vector bundle. Any vector bundle isomorphic to a trivial bundle is also said to be trivial.

Problem 91. Prove that an n-dimensional vector bundle $p \colon E \to B$ is trivial if and only if there exists a continuous map $\pi \colon E \to \mathbb{R}^n$ whose restriction to each fiber $p^{-1}(b)$, $b \in B$, is a linear space isomorphism.

We say that a closed manifold M^n is *parallelizable* if its tangent bundle $p \colon TM^n \to M^n$ is trivial.

Theorem 3.14. *A manifold M^n is parallelizable if and only if there exist n continuous vector fields $v_1(x), \dots, v_n(x)$ on M^n that are linearly independent at each point $x \in M^n$.*

Proof. Let $\varphi \colon TM^n \to M^n \times \mathbb{R}^n$ be a homeomorphism that isomorphically maps T_xM^n to $\{x\} \times \mathbb{R}^n$. We choose a basis $e_1, \dots, e_n$ in $\mathbb{R}^n$ and put $v_i(x) = \varphi^{-1}(x, e_i)$.

Now suppose that $v_1(x), \dots, v_n(x)$ are linearly independent vector fields. Consider the map $M^n \times \mathbb{R}^n \to TM^n$ defined by $(x, a) \mapsto a_1 v_1(x) + \dots + a_n v_n(x)$. According to Theorem 3.13, this map is a homeomorphism. □

Example 35. The spheres S^1, S^3, and S^7 are parallelizable manifolds.

The construction of linearly independent vector fields on these spheres is discussed in detail in [**104**, Section 41].

Problem 92. Prove that the tangent bundle TS^3 has infinitely many pairwise nonhomotopic[4] trivializations.

Problem 93. Prove that there exist infinitely many pairwise nonhomotopic[5] vector fields without singular points on S^3.

Example 36. The torus T^n is a parallelizable manifold.

Now we discuss some constructions related to vector bundles.

Pullbacks. Let $p\colon E \to B$ be a vector bundle, and let $f\colon B' \to B$ be a continuous map. Then we can construct a *pullback*, or an *induced bundle* $p' = f^*p\colon E' \to B'$, where

$$E' = \{(e, b') \in E \times B' \mid p(e) = f(b')\}$$

and $p'(e, b') = b'$. For the pullback, the diagram

$$\begin{array}{ccc} E' & \xrightarrow{\tilde{f}} & E \\ \downarrow{\scriptstyle p'} & & \downarrow{\scriptstyle p} \\ B' & \xrightarrow{f} & B, \end{array}$$

where $\tilde{f}(e, b') = e$, is commutative.

Let us show that the pullback is indeed a vector bundle. The linear space structure on the fiber $(p')^{-1}(b)$ is determined by

$$\lambda_1(e_1, b') + \lambda_2(e_2, b') = (\lambda_1 e_1 + \lambda_2 e_2, b');$$

here $p(e_1) = p(e_2) = f(b')$, i.e., e_1 and e_2 belong to the same linear space $p^{-1}(f(b'))$.

To a neighborhood $U \subset B$ and a homeomorphism $h\colon U \times \mathbb{R}^n \to p^{-1}(U)$ we assign the neighborhood $U' = f^{-1}(U) \subset B'$ and the homeomorphism $h'\colon U' \times \mathbb{R}^n \to (p')^{-1}(U')$ defined by $(b', v) \mapsto (e, b')$, where $e = h(f(b'), v)$. The condition $p(e) = f(b')$ does hold.

Exercise. Prove that a vector bundle is trivial if and only if it is a pullback of a bundle over a point.

[4]Two trivializations of a bundle are *homotopic* if they can be joined by a continuous family of trivializations.

[5]Two vector fields without singular points are said to be *homotopic* if they can be joined by a continuous family of vector fields without singular points.

Direct Products of Bundles. The *direct product* of vector bundles $p_1\colon E_1 \to B_1$ and $p_2\colon E_2 \to B_2$ is defined as

$$p_1 \times p_2\colon E_1 \times E_2 \to B_1 \times B_2;$$

the fiber $(p_1 \times p_2)^{-1}(b_1, b_2) = p_1^{-1}(b_1) \times p_2^{-1}(b_2)$ has the natural vector space structure.

Example 37. Let M_1 and M_2 be closed manifolds, and let $M = M_1 \times M_2$. Then the tangent bundle τ_M is isomorphic to the direct product $\tau_{M_1} \times \tau_{M_2}$.

The Whitney Sum. The Whitney sum of two bundles can be defined as the bundle whose fibers are direct sums of the fibers of these bundles. The formal definition is as follows. Suppose that $p_1\colon E_1 \to B$ and $p_2\colon E_2 \to B$ are two vector bundles over the same base. Their *Whitney*, or *direct, sum* is the bundle $d^*(p_1 \times p_2)$, where $d\colon B \to B \times B$ is the diagonal map defined by $d(b) = (b, b)$. It is often convenient to denote the bundle $p\colon E \to B$ by one letter ξ; the Whitney sum of bundles ξ_1 and ξ_2 is denoted by $\xi_1 \oplus \xi_2$. This notation is used because the fiber of the bundle $\xi_1 \oplus \xi_2$ over a point b is canonically isomorphic to $p_1^{-1}(b) \oplus p_2^{-1}(b)$.

Example 38. The Whitney sum of the tangent bundle over the sphere S^n and the one-dimensional trivial bundle is the $(n+1)$-dimensional trivial bundle.

Proof. The trivial $(n+1)$-dimensional vector bundle η over S^n can be represented as follows. Consider the standard embedding $S^n \to \mathbb{R}^{n+1}$ and assume that each point $x \in S^n$ is supplied with the linear space $\mathbb{R}^{n+1}$ obtained by translating the origin to x. Under this interpretation, the total space $E(\eta)$ of the bundle η contains the total space $E(\tau_{S^n})$ of the tangent bundle.

Consider the one-dimensional vector bundle ν for which the total space $E(\nu) \subset E(\eta)$ consists of all pairs (x, v), where $x \in S^n$ and $v \perp T_x S^n$ (i.e., v is the normal vector to S^n at x). It is easy to verify that ν is trivial. Indeed, treating x as a unit vector, we can write v in the form $v = \lambda x$, where $\lambda \in \mathbb{R}$; then the map $(x, v) \mapsto (x, \lambda)$ is an isomorphism between ν and the trivial bundle.

Consider the map $E(\tau \oplus \nu) \to E(\eta)$ defined by $(x, v_1, v_2) \mapsto (x, v_1 + v_2)$. According to Theorem 3.13, this is an isomorphism of bundles. □

Problem 94. Prove that $S^{n_1} \times S^{n_2} \times \dots \times S^{n_k}$ can be embedded in the Euclidean space of dimension $n_1 + \dots + n_k + 1$.

Remark. In [**3**], an assertion similar to Problem 94 for embeddings of products of 2-manifolds in the Euclidean space was proved.

Example 39. For any closed manifolds M_1 and M_2, $\tau_{M_1 \times M_2} \cong \pi_1^* \tau_{M_1} \oplus \pi_2^* \tau_{M_2}$, where the $\pi_i\colon M_1 \times M_2 \to M_i$ are natural projections.

Proof. The fiber of the bundle $\pi_1^*\tau_{M_1} \oplus \pi_2^*\tau_{M_2}$ over the point $(x, y) \in M_1 \times M_2$ is canonically isomorphic to $T_xM_1 \oplus T_yM_2 \cong T_{(x,y)}(M_1 \times M_2)$. $\square$

Problem 95*. Prove that the manifold $S^{n_1} \times \cdots \times S^{n_k}$, where $k \geq 2$, is parallelizable if and only if at least one of the numbers $n_1, \ldots, n_k$ is odd.

The Bundle $\mathrm{Hom}(\xi_1, \xi_2)$. Let ξ_1 and ξ_2 be two vector bundles over B. They determine the bundle $\mathrm{Hom}(\xi_1, \xi_2)$ whose fiber over each point $b \in B$ consists of homomorphisms (linear maps) from the fiber of ξ_1 over b to that of ξ_2. To construct this bundle, we must endow the total space with a topology. Let U be an open set over which the bundles ξ_1 and ξ_2 are trivial, and let $h_i \colon U \times \mathbb{R}^{n_i} \to p_i^{-1}(U)$ be the coordinate homeomorphisms. When a point $b \in U$ is fixed, the map h_i identifies $\mathbb{R}^{n_i}$ with the fiber of ξ_i over b. Thus, we have a one-to-one map $h \colon U \times \mathrm{Hom}(\mathbb{R}^{n_1}, \mathbb{R}^{n_2}) \to p^{-1}(U)$. It is required that such maps be homeomorphisms for all U and that all the sets $p^{-1}(U)$ be open. These conditions uniquely determine a topology on E. We must verify only that it is consistent, i.e., that the map

$$(U \cap U') \times \mathrm{Hom}(\mathbb{R}^{n_1}, \mathbb{R}^{n_2}) \xrightarrow{h^{-1} \circ h'} (U \cap U') \times \mathrm{Hom}(\mathbb{R}^{n_1}, \mathbb{R}^{n_2})$$

is continuous for any two intersecting coordinate neighborhoods U and U'. The continuity of this map follows from the continuity of the maps

$$(U \cap U') \times \mathbb{R}^{n_i} \xrightarrow{h_i^{-1} \circ h_i'} (U \cap U') \times \mathbb{R}^{n_i}.$$

A similar construction can be applied to obtain the bundle $\xi_1 \otimes \xi_2$. Applying it to copies of the same bundle, we obtain the bundle $\Lambda^k \xi$ (the exterior power of ξ).

2.2. Cohomology with Local Coefficients. The constructions we used in obstruction theory (Section 1.2) show that it is sometimes natural to consider cohomology the coefficients of which are not in a fixed group but rather each point has its own coefficient group; the groups corresponding to different points are isomorphic, but the isomorphisms are not canonical. Using this kind of cohomology with local coefficients, we can develop obstruction theory that applies to a larger class of spaces. We use it to construct linearly independent sections of vector bundles.

Remark. In fact, the construction of characteristic classes of vector bundles involves only trivial systems of local coefficients, which can be handled without using cohomology with local coefficients. But the reason for this is that some characteristic classes of vector bundles are obtained by reducing nontrivial systems of local coefficients modulo 2. To elucidate the relationship between obstruction theory and the characteristic classes of vector bundles, we present the theory of cohomology with local coefficients.

Suppose that associated with each point x of a topological space X is an Abelian group G_x (all the groups G_x are isomorphic), and with each path γ from x to y, an isomorphism $\gamma_*\colon G_x \to G_y$ so that the same isomorphisms correspond to homotopic paths, and to any composition of paths corresponds the composition of the respective isomorphisms. Then we say that X is endowed with a *local system* of Abelian groups.

A local system of Abelian groups on X determines an action of the group $\pi_1(X, x)$ on G_x because each loop γ based at x is associated with an isomorphism $\gamma\colon G_x \to G_x$. If the space X is path-connected, then the local system of Abelian groups is uniquely determined (up to a natural equivalence) by the group G_x associated with one point x and by the action of $\pi_1(X, x)$ on this group G_x. Indeed, for each point $y \in X$, we can fix a path ω_y from x to y and assign the same group $G = G_x$ to all points of X. To a path γ_{yz} from y to z we assign the isomorphism $G_x \to G_x$ induced by the loop $\omega_y\gamma_{yz}\omega_z^{-1}$.

Example 40. The groups $\pi_n(X, x)$, where $n \geq 2$, form a local system of Abelian groups. A path from x to y induces an isomorphism $\pi_n(X, x) \to \pi_n(X, y)$ in a standard way.

Example 41. On a manifold M^n, consider a local system of Abelian groups $G_x \cong \mathbb{Z}$ on which every element of the fundamental group acts as either the map $a \mapsto -a$ or the identity map, depending on whether or not the orientation changes when going along the representative path. This local system of Abelian groups is called an *orientation system* of groups and denoted by $\mathbb{Z}_{\operatorname{Or} M^n}$.

If the space X is locally path-connected and locally simply connected, then there exists a canonical isomorphism between groups from a local system of Abelian groups in sufficiently small neighborhoods. Thus, for a simplicial complex, such isomorphisms exist on each simplex. It follows that a local system of Abelian groups on a simplicial complex can be specified by choosing a group G_v for each vertex v and defining an isomorphism $\gamma_{uv}\colon G_v \to G_u$ for any vertices u and v belonging to the same simplex so that $\gamma_{uv}\gamma_{vw} = \gamma_{uw}$ whenever u, v, and w are vertices of the same simplex.

Now we can define the *homology* and *cohomology with coefficients in a local system* $\mathfrak{G}$ of Abelian groups on a simplicial complex K. Let Δ be a simplex in K. The groups G_x are canonically isomorphic; therefore, we can use the notation G_Δ. The group $C_k(K; \mathfrak{G})$ of k-chains consists of finite formal sums $\sum a_i \Delta_i^k$, where $a_i \in G_{\Delta_i^k}$. The boundary homomorphism $\partial_k\colon C_k \to C_{k-1}$ is given by

$$\partial[v_0, \dots, v_k] = \sum (-1)^i [v_0, \dots, \hat{v}_i, \dots, v_k].$$

This definition makes sense because if $\Delta' \subset \partial\Delta \subset \Delta$, then the group $G_{\Delta'}$ is canonically isomorphic to G_Δ. The proof of the equality $\partial\partial = 0$ is completely similar to that for usual simplicial chains. The homology group $H_k(K; \mathfrak{G})$ is defined to be the quotient group $\operatorname{Ker}\partial_k / \operatorname{Im}\partial_{k+1}$.

The group $C^k(K; \mathfrak{G})$ of k-cochains consists of maps taking each simplex Δ^k to an element of the group G_{Δ^k}. The coboundary homomorphism δ is given by

$$(\delta c^k)[v_0, \dots, v_{k+1}] = \sum (-1)^i c^k([v_0, \dots, \hat{v}_i, \dots, v_{k+1}]).$$

The definition of the cohomology group $H^k(K; \mathfrak{G})$ is standard.

If a simplicial complex is endowed with the structure of a CW-complex (the cells are unions of simplices), then its cellular homology and cohomology can be defined in the standard manner. It must be taken into account that if $\chi\colon D^k \to X$ is a characteristic map of a cell, then the groups G_x are canonically isomorphic for all $x \in D^k$, but different points $x, y \in D^k$ may be mapped to the same point. In this case, the two copies of the group G associated with the points x and y are identified by the action on G of the element of the fundamental group represented by the image of the interval xy.

Example 42. Let $G_x \cong \mathbb{Z}$ be a local system of Abelian groups on $\mathbb{R}P^n$ with $n \geq 2$ on which the generator of the fundamental group acts as $a \mapsto -a$. We denote this local system of Abelian groups by $\mathbb{Z}_T$ (for *twisted*). We have

$$H_k(\mathbb{R}P^n; \mathbb{Z}_T) = \begin{cases} \mathbb{Z} & \text{if } k = n \text{ and } n \text{ is even,} \\ \mathbb{Z}_2 & \text{if } k < n \text{ and } k \text{ is even,} \\ 0 & \text{otherwise} \end{cases}$$

and

$$H^k(\mathbb{R}P^n; \mathbb{Z}_T) = \begin{cases} \mathbb{Z} & \text{if } k = n \text{ and } n \text{ even,} \\ \mathbb{Z}_2 & \text{if } k \leq n \text{ and } k \text{ odd,} \\ 0 & \text{otherwise.} \end{cases}$$

Proof. The boundary homomorphism $\partial\colon C_k \to C_{k-1}$ acts as multiplication by $1 + (-1)^k(-1) = 1 + (-1)^{k+1}$. The coboundary homomorphism $\delta\colon C^{k-1} \to C^k$ has the same form because it is dual to ∂. □

Remark. The local system $\mathbb{Z}_T$ is defined for any space with fundamental group $\mathbb{Z}_2$.

Poincaré Duality for Local Coefficients. Recall that the usual Poincaré duality for an oriented manifold M^n with two dual cell decompositions K and K^* is established by an isomorphism $C_k(K; \mathbb{Z}) \to C^{n-k}(K^*; \mathbb{Z})$, which induces the isomorphism $H_k(M^n) \cong H^{n-k}(M^n;\mathbb{Z})$. Namely, each chain $c_k = \sum a_i\sigma_i$ is associated with the cochain c^{n-k} for which $\langle c^{n-k}, \sigma_i^*\rangle = a_i$. The orientation of the cell σ_i^* depends on that of the manifold M^n; for the same manifold with opposite orientation, σ_i^* is replaced by $-\sigma_i^*$.

In a similar way, for an arbitrary manifold M^n (orientable or not), we can construct isomorphisms $C_k(K;\mathrm{Or}) \to C^{n-k}(K^*;\mathbb{Z})$ and $C_k(K;\mathbb{Z}) \to C^{n-k}(K^*;\mathrm{Or})$, where $\mathrm{Or} = \mathbb{Z}_{\mathrm{Or}\, M^n}$. The first isomorphism is constructed as follows. The elements of the groups $C_k(K;\mathrm{Or})$ are the sums $\sum(\varepsilon_i a_i)\sigma_i$, where $a_i \in \mathbb{Z}$ and each ε_i is the orientation of M^n at the points of the k-dimensional cell σ_i. To such a sum we associate the cochain $c^{n-k} \in C^{n-k}(K^*;\mathbb{Z})$ for which $\langle c^{n-k}, \sigma_i^*\rangle = a_i$. This definition makes sense because the choice of ε_i determines both the number a_i and the orientation of the cell σ_i^*. Replacing ε_i by $-\varepsilon_i$, we obtain $-a_i$ instead of a_i and $-\sigma_i^*$ instead of σ_i^*.

The second isomorphism is constructed as follows. Each cochain in the group $C^{n-k}(K^*;\mathrm{Or})$ takes an $(n-k)$-cell σ_i^* to $a_i\varepsilon_i$, where $a_i \in \mathbb{Z}$ and ε_i is the orientation of M^n at the points of σ_i^*. To the chain $c_k = \sum a_i\sigma_i \in C_k(K;\mathbb{Z})$ we assign the cochain $c^{n-k} \in C^{n-k}(K^*;\mathrm{Or})$ that takes every σ_i^* to $a_i\varepsilon_i$. This definition makes sense because replacing ε_i by $-\varepsilon_i$, we obtain the cochain taking $-\sigma_i^*$ to $-a_i\varepsilon_i$.

Thus, we have proved the following theorem.

Theorem 3.15 (the Poincaré duality isomorphism). *For any closed manifold M^n, the Poincaré isomorphisms with local coefficients $H_k(M^n;\mathrm{Or}) \cong H^{n-k}(M^n;\mathbb{Z})$ and $H_k(M^n;\mathbb{Z}) \cong H^{n-k}(M^n;\mathrm{Or})$ hold.*

Obstructions to Extending Sections. Consider a locally trivial bundle $p\colon E \to B$ with fiber F. We assume that B is a simplicial complex. Suppose that on B^n (the n-skeleton of the complex B), a section $s\colon B^n \to E$ of the bundle p is given. We want to determine whether or not this section can be extended to B^{n+1}, and if not, whether it can be changed on $B^n \setminus B^{n+1}$ so that the new section does admit such an extension. This problem closely resembles the usual problem of obstruction theory. Indeed, the bundle is trivial over each simplex $\Delta^{n+1} \subset B^{n+1}$, so its section over Δ^{n+1} is completely determined by a map to the fiber F, and it is required to extend a map $\partial\Delta^{n+1} \to F$ to a map $\Delta^{n+1} \to F$. However, this argument is valid only if a trivialization $p^{-1}(\Delta^{n+1}) \to F \times \Delta^{n+1}$ is fixed.

The obstructions $c^{n+1}(s) \in C^{n+1}(B;\pi_n(p^{-1}(b)))$ are defined as in the usual theory of obstructions, but the cochains have coefficients in a local system of Abelian groups $\{\pi_n(p^{-1}(b))\}$ on the space B. Indeed, under the assumption that the space F is n-simple (which is a standard assumption in obstruction theory), each path between $b_1 \in B$ and $b_2 \in B$ determines an isomorphism $\pi_n(p^{-1}(b_1)) \to \pi_n(p^{-1}(b_2))$; this isomorphism is constructed by using the covering homotopy theorem.

All assertions of Section 1.2 carry over to obstructions to extending sections without substantial changes except that cohomology with coefficients in the local system of homotopy groups of the fiber should be considered. As mentioned before, the fiber F must be simple.

2.3. The Stiefel–Whitney Characteristic Classes. Suppose a bundle $p\colon E \to B$ has fiber F such that $\pi_0(F) = \dots = \pi_{n-1}(F) = 0$ but $\pi_n(F) \neq 0$ for $n \geq 1$; for $n = 1$, we assume in addition that the group $\pi_1(F)$ is Abelian. Then the fiber F is n-simple.

There are no obstructions to extending sections over the k-skeletons for $k \leq n$. The first nontrivial obstruction lies in the group $H^{n+1}(B; \mathfrak{G})$, where $\mathfrak{G}$ is the local system $\{\pi_n(p^{-1}(b))\}$. This obstruction does not depend on the choice of a section on the n-skeleton. Indeed, if s_1 and s_2 are two sections on the n-skeleton, and $c^{n+1}(s_1), c^{n+1}(s_2) \in Z^{n+1}(B; \mathfrak{G})$ are obstructions to extending these sections over B^{n+1}, then there exists a cochain $d^n(s_1, s_2) \in C^n(B; \mathfrak{G})$ for which $\delta d^n(s_1, s_2) = c^{n+1}(s_1) - c^{n+1}(s_2)$ (see Lemma 2 on p. 118).

Thus, we can associate the bundle with the first nontrivial obstruction to extending sections.

For vector bundles, this construction cannot be used directly to define characteristic classes because the fiber of a vector bundle is a contractible space. But to each n-dimensional vector bundle we can assign $n-1$ bundles over the same base whose fibers are sets of k orthonormal vectors, where $1 \leq k \leq n-1$. Indeed, let ξ be a vector bundle over a finite simplicial complex B. The construction used in Part I, p. 204, to define a Riemannian metric on a manifold yields a Riemannian metric on ξ. Clearly, any two Riemannian metrics are homotopic in the class of Riemannian metrics. Indeed, if $(\cdot,\cdot)_0$ and $(\cdot,\cdot)_1$ are two inner products on a vector space, then $t(\cdot,\cdot)_0 + (1-t)(\cdot,\cdot)_1$ is an inner product for any $t \in [0,1]$. Consider the bundle $p_k\colon E_k \to B$, where E_k consists of all pairs

$$(b \in B, \text{ an ordered set of } k \text{ orthonormal vectors in the fiber of } \xi \text{ over } b).$$

The space E_k is endowed with the natural topology.

The fiber of the bundle ξ_k constructed above is known as the *Stiefel manifold* $V(n,k)$; it consists of k-tuples of orthonormal vectors in $\mathbb{R}^n$. For the bundle ξ_k, a characteristic class is defined. We start by calculating the first nontrivial homotopy group of the space $V(n,k)$.

Problem 96. Prove that $V(n,k)$ is a manifold of dimension $nk - \frac{k(k+1)}{2}$.

Exercise. Prove that $V(n+1, 2)$ is the manifold of unit tangent vectors to the sphere S^n.

Theorem 3.16. *If $i < n-k$, then $\pi_i(V(n,k)) = 0$. The first nontrivial homotopy group $\pi_{n-k}(V(n,k))$ coincides with $\mathbb{Z}$ (if $n-k$ is even or $k=1$) or $\mathbb{Z}_2$ (if $n-k$ is odd and $k>1$).*

Proof. First, let us show (by induction on k) that $\pi_i(V(n,k)) = 0$ for $i < n-k$. For $k=1$, the manifold $V(n,1)$ is the sphere S^{n-1}. Clearly, $\pi_i(S^{n-1}) = 0$ for $i < n-1$. The induction step is as follows. Consider the map $V(n, k+1) \to V(n,k)$ deleting the last vectors from each $(k+1)$-tuple of orthonormal vectors. This map is a locally trivial bundle with fiber S^{n-k-1}. The exact homotopy sequence for this bundle is

$$\pi_i(S^{n-k-1}) \longrightarrow \pi_i(V(n,k+1)) \longrightarrow \pi_i(V(n,k)) \longrightarrow \pi_{i-1}(S^{n-k-1}).$$

For $i < n-k-1$, this fragment of the exact sequence has the form $0 \to \pi_i(V(n,k+1)) \to 0 \to 0$; therefore, $\pi_i(V(n,k+1)) = 0$.

Now, let us calculate $\pi_{n-k}(V(n,k))$. The case $k=1$ is handled separately: $V(n,1) \approx S^{n-1}$. In what follows, we assume that $1 < k < n$. Consider the map $V(n+1, k+1) \to V(n+1, 1) = S^n$ assigning v_{k+1} to orthonormal vectors $v_1, \dots, v_{k+1}$. This map is a locally trivial bundle with fiber $V(n,k)$. The exact homotopy sequence for this bundle is

$$\pi_{i+1}(S^n) \longrightarrow \pi_i(V(n,k)) \longrightarrow \pi_i(V(n+1,k+1)) \longrightarrow \pi_i(S^n).$$

If $i \le n-2$, then $\pi_i(V(n,k)) \cong \pi_i(V(n+1,k+1))$. Successively applying such isomorphisms, we obtain $\pi_{n-k}(V(n,k)) \cong \pi_{n-k}(V(n-k+2,2))$; the isomorphism $\pi_{n-k}(V(n-k+3,3)) \cong \pi_{n-k}(V(n-k+2,2))$ follows from the inequality $n-k \le (n-k+2)-2$.

It remains to verify that the group $\pi_{n-2}(V(n,2))$ coincides with $\mathbb{Z}$ if n is even and with $\mathbb{Z}_2$ if n is odd. Consider the bundle $V(n,2) \to V(n,1) = S^{n-1}$ with fiber $V(n-1,1) = S^{n-2}$. The exact homotopy sequence for this bundle is

$$\pi_{n-1}(S^{n-1}) \xrightarrow{\partial_*} \pi_{n-2}(S^{n-2}) \longrightarrow \pi_i(V(n,2)) \longrightarrow \pi_{n-2}(S^{n-1}) = 0.$$

The homomorphism ∂_* maps $\mathbb{Z}$ to $\mathbb{Z}$; therefore, it suffices to calculate $\partial_*(\mathrm{id}_{S^{n-1}})$. By definition, the element $\partial_*(\mathrm{id}_{S^{n-1}}) \in \pi_{n-2}(S^{n-2})$ is constructed as follows. On the sphere S^{n-1}, consider the one-parameter family of spheres S_t^{n-2}, where $t \in (0,1)$, which pass through a fixed point $e_0 \in S^{n-1}$ and have a common tangent $(n-2)$-plane at this point. We lift the point e_0 to the total space $V(n,2)$ of the bundle and extend the lifting to S_t^{n-2} first for small t and then for all $t<1$. Letting t tend to 1, we obtain a map from the sphere S_t^{n-2} to the fiber. The homotopy class of this map is the required element of $\pi_{n-2}(S^{n-2})$.

To lift the point $e_0 \in S^{n-1}$ (treated as a unit vector) to $V(n,2)$, we have to choose a unit vector e_1 orthogonal to e_0. The vector e_1 can be regarded as a vector tangent to S^{n-1} at e_0.

The lifting can be constructed using a vector field with one isolated singular point. It is required that the vectors be parallel in a half of a neighborhood of the singular point and all the rotation of the vectors take place in the second half (the first half of the neighborhood corresponds to the spheres S_t^{n-2} with $t < 1/2$ and the second, to the spheres with $t > 1/2$). In Figure 4, the dashed line shows the circle S_t^1 with $t > 1/2$ for $n = 3$. As

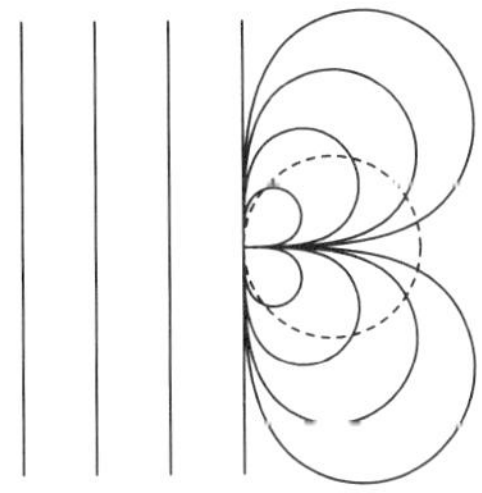

Figure 4. The trajectories of the vector field on S^2

$t \to 1$, the degree of the map $S_t^{n-2} \to S^{n-2}$ taking each point to the tangent vector at this point equals the index of the singular point, i.e., the Euler characteristic of S^{n-1}. Thus, $\partial_*(\mathrm{id}_{S^{n-1}}) = \chi(S^{n-1}) = 1 + (-1)^{n-1}$. This means that if n is even, then $\operatorname{Im}\partial_* = 0$, and if n is odd, then $\operatorname{Im}\partial_* = 2\mathbb{Z}$. □

Remark. Many other homotopy groups of Stiefel manifolds, which are very important for homotopy theory, were calculated in the series of papers [**99**].

Problem 97. Calculate the homology groups with coefficients in $\mathbb{Z}$ for the Stiefel manifold $V(n+2,2)$, where $n \geq 1$.

The characteristic class of the bundle p_k, which is contained in the group $H^{n-k+1}(B, \{\pi_{n-k}(V(n,k))\})$, is an obstruction to extending k linearly independent sections over the $(n-k+1)$-skeleton. If $n-k$ is odd and $k > 1$, then this class is contained in the ordinary cohomology groups $H^{n-k+1}(B;\mathbb{Z}_2)$ because there are no nontrivial automorphisms $\mathbb{Z}_2 \to \mathbb{Z}_2$; thus, in this case, the system of local coefficients is trivial. The corresponding cohomology class $w_{2j} \in H^{2j}(B;\mathbb{Z}_2)$ is called the *Stiefel–Whitney characteristic class*. This class is an obstruction to extending $n+1-2j$ linearly independent sections over the $2j$-skeleton.

Problem 98. A manifold M^n is said to be *almost parallelizable* if the manifold $M^n \setminus \{x\}$, where $x \in M^n$, is parallelizable. Prove that a simply connected 4-manifold M^4 is almost parallelizable if and only if the class w_2 for its tangent bundle vanishes.

The odd-dimensional Stiefel–Whitney classes (and the class w_n for even n) are defined by reducing the integer coefficients modulo 2. As a result, the Stiefel–Whitney classes $w_{2j+1} \in H^{2j+1}(B; \mathbb{Z}_2)$ are obtained. We emphasize that they do not coincide with the obstructions

$$\mathfrak{o}_{2j+1} \in H^{2j+1}(B, \{\pi_{2j}(V(n, n-2j))\}),$$

although they are very close to them (see Problem 100).

Steenrod showed that the obstruction $\mathfrak{o}_{2j+1}$, where $j \geq 1$, can be expressed in terms of the characteristic class $w_{2j} = \mathfrak{o}_{2j}$ by means of the following construction. Consider the locally trivial bundle $V(n, n-2j+1) \xrightarrow{S^{2j-1}} V(n, n-2j)$. The exact homotopy sequence for this bundle is

$$(24) \quad \pi_{2j}(V(n, n-2j)) \xrightarrow{\partial_*} \pi_{2j-1}(S^{2j-1}) \xrightarrow{i_*} \pi_{2j-1}(V(n, n-2j+1)) \to 0.$$

On the right-hand side, we have 0 instead of $\pi_{2j-1}(V(n, n-2j))$ because $2j-1 < n-(n-2j)$. Thus, we have an exact sequence $\mathbb{Z} \xrightarrow{\partial_*} \mathbb{Z} \xrightarrow{i_*} \mathbb{Z}_2 \to 0$; therefore, $\operatorname{Im} \partial_* = \operatorname{Ker} i_* = 2\mathbb{Z}$, and $\operatorname{Ker} \partial_* = 0$. Adding $0 \to$ to the exact sequence (24) on the left, we obtain a short exact sequence of groups. Moreover, in fact, we obtain a short exact sequence of local systems of Abelian groups (this is easy to check). As for ordinary cohomology groups, this exact sequence induces the Bockstein homomorphism

$$\beta^*\colon H^{2j}(B, \{\pi_{2j-1}(V(n, n-2j+1))\}) \to H^{2j+1}(B, \{\pi_{2j}(V(n, n-2j))\}).$$

Problem 99. Prove that $\mathfrak{o}_{2j+1} = \beta^* \mathfrak{o}_{2j}$.

Remark. Problem 113 shows that if $w_1 = 0$, then $w_{2j+1} = \operatorname{Sq}^1 w_{2j}$.

Problem 100. Prove that the obstructions $\mathfrak{o}_k$, where $k < \dim \xi$, have order 2. Moreover, if the number $n = \dim \xi$ is odd, then the class $\mathfrak{o}_n$ has order 2 as well.

Problem 100 shows that almost all classes $w_k(\xi)$ are true obstructions, i.e., extensions of sections exist if and only if these classes vanish. The only exception is the class $w_n(\xi)$ with even $n = \dim \xi$. Only this class is obtained by the reduction modulo 2 of a class which may be of order different from 2.

2.4. Properties of the Stiefel–Whitney Classes. We have defined the classes $w_2, \ldots, w_n$ for n-dimensional vector bundles. We assume that $w_0 = 1$ and $w_i = 0$ for $i > n$.

The Class w_1. The class w_1 must be considered separately because it must correspond to an obstruction to extending n linearly independent sections from the 0-skeleton to the 1-skeleton. Applying the standard construction of obstruction theory, we see that the corresponding cocycle must take values in $\pi_0(V(n, n))$, and π_0 is not a group. We have defined $V(n, k)$ only for

$k \leq n-1$; applying the same definition to $k = n$, we obtain a disconnected manifold diffeomorphic to $\mathrm{O}(n)$. Thus, the set $\pi_0(V(n,n))$ consists of two elements. This violates uniformity too because for the obstructions $\mathfrak{o}_i$ with odd numbers, the local coefficient group is $\mathbb{Z}$. For this reason, instead of $\pi_0(V(n,n))$, we take the reduced homology group $\tilde{H}_0(V(n,n);\mathbb{Z}) \cong \mathbb{Z}$ and obtain the local coefficient system $a\varepsilon$, where $a \in \mathbb{Z}$ and ε is the orientation of the fiber. Indeed, the connected components of the space $V(n,n)$ are identified with orientations of the fiber, and hence the elements of $\tilde{H}_0(V(n,n);\mathbb{Z}) \cong \mathbb{Z}$ have the form $a\,\mathrm{Or}_1 - a\,\mathrm{Or}_2$; to each orientation we assign its coefficient.

The orientation of the fiber can be transferred along any path in the base; thus, we indeed obtain a local system of coefficients.

We now construct obstructions to extending n linearly independent sections from the 0-skeleton to the 1-skeleton. The connected components of the space $V(n,n)$ are orientations. So, we are to specify orientations in the fibers over the 1-skeleton. Suppose that at each vertex v_i, an orientation ε_i is given. Consider the cochain $c^1 \in C^1(B;\{\tilde{H}_0(V(n,n);\mathbb{Z})\})$ for which $\langle c^1, [v_iv_j]\rangle = \varepsilon_j - \varepsilon_i$. This cochain is a cocycle because

$$\begin{aligned}\langle \delta c^1, [v_iv_jv_k]\rangle &= \langle c^1, [v_jv_k]\rangle - \langle c^1, [v_iv_k]\rangle + \langle c^1, [v_iv_j]\rangle \\ &= (\varepsilon_k - \varepsilon_j) - (\varepsilon_k - \varepsilon_i) + (\varepsilon_j - \varepsilon_i) = 0.\end{aligned}$$

The orientations ε_i extend to the 1-skeleton if and only if $c^1 = 0$. Changing orientations at some vertices corresponds to adding some coboundary δc^0 to c^1. Therefore, the orientations extend to the 1-skeleton if and only if the cohomology class of the cocycle c^1, which we denote by $\mathfrak{o}_1$, vanishes.

The class $\mathfrak{o}_1$ does not depend on the choice of the section over the 0-skeleton. Indeed, as mentioned above, changing the orientation at vertices corresponds to adding a coboundary. This implies, in particular, that $2\mathfrak{o}_1 = 0$. Indeed, if we replace the orientations ε_i by $-\varepsilon_i$ at all vertices v_i, then the cocycle c^1 will change sign, as well as the class $\mathfrak{o}_1$. But the class $\mathfrak{o}_1$ does not depend on the choice of orientations; therefore, $\mathfrak{o}_1 = -\mathfrak{o}_1$, i.e., $2\mathfrak{o}_1 = 0$.

Reducing the local coefficient system $\tilde{H}_0(V(n,n);\mathbb{Z}) \cong \mathbb{Z}$ modulo 2, we obtain the class w_1 from the class $\mathfrak{o}_1$. The equalities $\mathfrak{o}_1 = 0$ and $w_1 = 0$ are equivalent because $2\mathfrak{o}_1 = 0$. Thus, $w_1 = 0$ if and only if the fibers of the bundle over the 1-skeleton admit compatible orientations. This property has the following geometric meaning.

Let ξ be an n-dimensional vector bundle over a linearly connected base. Consider the set of unit vectors in the one-dimensional bundle $\Lambda^n\xi$ endowed with a Riemannian metric (these vectors correspond to orientations). If the space thus obtained is connected, then the bundle ξ is said to be *nonorientable*, and if it consists of two connected components, then ξ is *orientable*.

Choosing one of these two components, we specify orientations of all fibers of the bundle.

Example 43. A one-dimensional bundle is orientable if and only if it is trivial.

Theorem 3.17. *A vector bundle over a simplicial complex is orientable if and only if its restriction to the* 1*-skeleton is orientable.*

Proof. The space $\{x \in \Lambda^n \xi \mid \|x\| = 1\}$ doubly covers the base. This space is a CW-complex, and the connectedness of a CW-complex is equivalent to that of its 1-skeleton. □

Thus, $w_1(\xi) = 0$ if and only if the bundle ξ is orientable. We apply this property to calculate the class w_1 of the bundle γ_n^1 over $\mathbb{R}P^n$, which is very important for various applications of characteristic classes. The total space $E(\gamma_n^1)$ of γ_n^1 consists of pairs $(x \in \mathbb{R}P^n, v = \lambda x)$; it is assumed that each point $x \in \mathbb{R}P^n$ is represented by a nonzero vector in $\mathbb{R}^{n+1}$, and the fiber over this point consists of all vectors proportional to this vector.

Theorem 3.18. *The class $w_1(\gamma_n^1)$ is equal to the generator of the group $H^1(\mathbb{R}P^n; \mathbb{Z}_2)$.*

Proof. The group $H^1(\mathbb{R}P^n; \mathbb{Z}_2)$ has only one nonzero element; therefore, it suffices to verify that $w_1(\gamma_n^1) \neq 0$. The bundle γ_n^1 is one-dimensional; hence we must show that it is nontrivial.

For every section $s\colon \mathbb{R}P^n \to E(\gamma_n^1)$, consider the composition $S^n \to \mathbb{R}P^n \xrightarrow{s} E(\gamma_n^1)$, where the first map is the canonical double covering. We obtain a map $S^n \to \mathbb{R}$, which is defined by $x \mapsto \lambda(x)$. We have $\lambda(x)x = \lambda(-x)(-x)$, i.e., $\lambda(-x) = -\lambda(x)$. Thus, the values of the function λ at the endpoints of any path in S^n between x and $-x$ are of different signs; therefore, this function vanishes at some point of any such path. It follows that the bundle γ_n^1 has no nowhere vanishing sections, i.e., is nontrivial. □

The Class w_n and the Euler Class e. By definition, the element $\mathfrak{o}_n \in H^n(B; \{\pi_{n-1}(V(n,1))\})$ is an obstruction to extending a single nowhere vanishing section to the n-skeleton. Note that $V(n,1) \approx S^{n-1}$.

If the bundle is oriented, then each point $b \in B$ is associated with an isomorphism $\varphi\colon \pi_{n-1}(S^{n-1}) \to \mathbb{Z}$. (There exist two isomorphisms $\mathbb{Z} \to \mathbb{Z}$; changing the orientation corresponds to changing the isomorphism.) Applying the isomorphism φ, we obtain an isomorphism $H^n(B; \{\pi_{n-1}(V(n,1))\}) \to H^n(B; \mathbb{Z})$. The image of the class $\mathfrak{o}_n$ under this isomorphism is called the *Euler class* and denoted by e. The class w_n is obtained from e by reduction modulo 2.

For the bundle with opposite orientation, e has the opposite sign; i.e., $e(-\xi) = -e(\xi)$.

Theorem 3.19. *If the dimension of the bundle ξ is odd, then $2e(\xi) = 0$.*

Proof. The map $v \mapsto -v$ induces a fiberwise self-homeomorphism of the total space of the bundle. Therefore, if the dimension is odd, then the bundles ξ and $-\xi$ are isomorphic. (If the dimension is even, then the map $v \mapsto -v$ preserves orientation.) $\square$

Thus, for vector bundles of odd dimension, the classes $w_n(\xi)$ and $e(\xi)$ do not differ substantially, although they belong to different groups. Both of them are true obstructions, i.e., sections extend to the n-skeleton if and only if $w_n(\xi) = 0$ or $e(\xi) = 0$ (these conditions are equivalent). But for even-dimensional orientable bundles, only the class $e(\xi)$ is a true obstruction, while $w_n(\xi)$ is not: the equality $w_n(\xi) = 0$ does not imply the existence of a nowhere vanishing section over the n-skeleton. Such a section exists only if $e(\xi) = 0$. For example, $w_2(S^2) = 0$ but $e(S^2) = 2 \neq 0$; therefore, no vector field on the sphere S^2 is nowhere vanishing.

Let M^n be a closed oriented manifold, and let ξ^n be an oriented n-dimensional bundle over M^n. Using the interpretation of the class $e(\xi^n) \in H^n(M^n; \mathbb{Z})$ as an obstruction, we calculate it as follows. Locally, the total space E of the bundle ξ^n is the product of the oriented manifold M^n and another oriented manifold, the fiber of ξ^n; hence this space can be oriented. (Note that for odd n, the order of factors in the product is essential; for transposed factors, the orientation is opposite.) Consider any section $s\colon M^n \to E$ transversally intersecting the zero section $M_0^n \subset E$ in isolated points. We have $\langle e(\xi^n), [M^n]\rangle = \langle\langle M_0^n, s(M^n)\rangle\rangle$. Indeed, for a representative of the cohomology class $e(\xi)$ we can take the cocycle that is an obstruction to extending the section s from the $(n-1)$-skeleton to the n-skeleton; to each n-simplex Δ^n in the triangulation of the manifold M^n this cocycle assigns the sum of the indices of the singular points of s in Δ^n. On the other hand, this sum is equal to the intersection number of Δ^n with $s(M^n)$ in M_0^n.

Example 44. If M^n is a closed oriented manifold, then $\langle e(\tau_{M^n}), [M^n]\rangle = \chi(M^n)$.

Note that under a change of orientation, both the Euler class $e(\tau_{M^n})$ and the fundamental class $[M^n]$ change sign.

Example 45. The Euler characteristic of the sphere S^n is even; therefore, $w_n(\tau_{S^n}) = 0$ for all $n \geq 1$.

For the sphere S^2, the tangent bundle is nontrivial but $w_1 = 0$ and $w_2 = 0$. Therefore, even the complete set of Stiefel–Whitney classes does not distinguish between trivial and nontrivial bundles.

Theorem 3.20. *If* $\dim \xi = n$ *and* $\dim \eta = m$, *then* $w_{n+m}(\xi \times \eta) = w_n(\xi) \times w_m(\eta)$.

Proof. Take an n-cell σ in the base of the bundle ξ. Let us construct a (nonvanishing) section of ξ over $\partial\sigma$ and extend it inside σ so that all zeros of the extension are nondegenerate. Let $a(\sigma)$ be the number of these zeros. Then the cohomology class $w_n(\xi)$ is represented by the cocycle which assigns the element $a(\sigma) \pmod 2 \in \mathbb{Z}_2$ to the cell σ. Over each m-cell τ of the base of η, we similarly construct a section with $b(\tau)$ nondegenerate zeros. The direct sum of these sections is a section of the bundle $\xi \times \eta$ over the cell $\sigma \times \tau$. Let $c(\sigma \times \tau)$ be the number of zeros of this section. Clearly, $c(\sigma \times \tau) = a(\sigma)b(\tau)$ because the direct sum of two sections vanishes precisely at the common zeros of these sections. Thus, to the cell $\sigma \times \tau$ the cocycle representing the class $w_{n+m}(\xi \times \eta)$ assigns the product of the values of the cocycles representing $w_n(\xi)$ and $w_m(\eta)$ for the cells σ and τ. According to the definition of cohomology cross product, this means that the representatives of the classes $w_{n+m}(\xi \times \eta)$ and $w_n(\xi) \times w_m(\eta)$ take equal values at the cell $\sigma \times \tau$. It remains to verify that if $\dim \sigma' + \dim \tau' = m + n$ but $\dim \sigma' \neq n$, then the representatives of both classes vanish at $\sigma' \times \tau'$. For the class $w_n(\xi) \times w_m(\eta)$, this follows directly from the definition. Suppose that, say, $\dim \sigma' < n$. Then there exists a nowhere vanishing section over σ'. The direct sum of this section and any section over τ' is nowhere vanishing. Therefore, the representative of the class $w_{n+m}(\xi \times \eta)$ vanishes at the cell $\sigma' \times \tau'$. □

The following theorem is a refinement of Theorem 3.20.

Theorem 3.21. *If* $\dim \xi = n$ *and* $\dim \eta = m$, *then* $e(\xi \times \eta) = e(\xi) \times e(\eta)$.

Proof. The main argument is the same as in the proof of Theorem 3.20. We must only trace the orientations. Namely, we need to prove that if vector fields v and w on the spaces V and W have nondegenerate singular points with indices $(-1)^k$ and $(-1)^l$ at the origin, then the vector field $v \oplus w$ on the space $V \oplus W$ has a singular point of index $(-1)^{k+l}$ at the origin. It suffices to prove this for the linear vector fields $v(x) = Ax$ and $w(y) = By$. The vector field $v \oplus w$ is determined by the matrix $\left(\begin{smallmatrix} A & 0 \\ 0 & B \end{smallmatrix}\right)$, which has determinant $\det A \det B$. □

Naturality Property of the Stiefel–Whitney Classes. The Stiefel–Whitney classes have the following *naturality property*: If ξ is a bundle over

B and $f\colon B' \to B$ is a continuous map, then $w_i(f^*(\xi)) = f^*(w_i(\xi))$, i.e., the characteristic classes of the pullback $f^*(\xi)$ are obtained from those of the bundle ξ by means of the induced map $f^*\colon H^*(B;\mathbb{Z}_2) \to H^*(B';\mathbb{Z}_2)$ of cohomology groups.

The naturality property is fairly obvious. Indeed, let E' and E be the total spaces of the bundles $f^*(\xi)$ and ξ. By definition, we have

$$E' = \{(b', v) \mid v \in F_{f(b')}\};$$

i.e., the fibers over $b' \in B'$ and $f(b') \subset B$ are canonically isomorphic. Therefore, the map f induces a cochain map

$$f^{\#}\colon C^{n-k+1}(B; \{\pi_{n-k}(V(n,k))\}) \to C^{n-k+1}(B'; \{\pi_{n-k}(V(n,k))\}).$$

This cochain map takes the obstructing cocycle of the bundle ξ to that of the bundle $f^*(\xi)$.

The Whitney Formula. Let ξ and η be vector bundles over a simplicial complex B. Then

$$(25) \qquad w_k(\xi \oplus \eta) = \sum_{i+j=k} w_i(\xi) \smile w_j(\eta) \quad \text{(the Whitney formula)}.$$

Instead of the Whitney formula, we shall prove the equivalent relation

$$(26) \qquad w_k(\xi \times \eta) = \sum_{i+j=k} w_i(\xi) \times w_j(\eta),$$

where ξ and η are bundles over different bases. Relation (25) is derived from (26) as follows. Suppose that ξ and η are bundles over the same base B, $p_1\colon B \times B \to B$ and $p_2\colon B \times B \to B$ are the projections to the first and second factors respectively, and $d\colon B \to B \times B$ is the diagonal map. Then $p_1 d = p_2 d = \mathrm{id}_B$ and $\xi \times \eta = (p_1^*\xi) \oplus (p_2^*\eta)$. Therefore,

$$w_k((p_1^*\xi) \oplus (p_2^*\eta)) = w_k(\xi \times \eta) = \sum_{i+j=k} w_i(\xi) \times w_j(\eta).$$

Applying the map d^* to both parts in this equality and using naturality of the characteristic classes, we obtain

$$d^*(w_k((p_1^*\xi) \oplus (p_2^*\eta))) = w_k((d^*p_1^*\xi) \oplus (d^*p_2^*\eta)) = w_k(\xi \oplus \eta).$$

Moreover, $d^*(w_i(\xi) \times w_j(\eta)) = w_i(\xi) \smile w_j(\eta)$. Now, let us derive (26) from (25). First, note that $\xi \times \eta = (p_X^*\xi) \oplus (p_Y^*\eta)$, where p_X and p_Y are the natural projections of $X \times Y$ to X and Y. Hence $w_k(\xi \times \eta) = w_k((p_X^*\xi) \oplus (p_Y^*\eta)) = \sum_{i+j=k} w_i(p_X^*\xi) \smile w_j(p_Y^*\eta) = \sum_{i+j=k} p_X^* w_i(\xi) \smile p_Y^* w_j(\eta)$ (the last equality follows from naturality of the characteristic classes). Finally, the corollary of Theorem 2.27 on p. 107 implies $p_X^* w_i(\xi) \smile p_Y^* w_j(\eta) = w_i(\xi) \times w_j(\eta)$.

Vector bundles ξ and η over the same base are said to be *stably equivalent* if $\xi \oplus \varepsilon^m \cong \eta \oplus \varepsilon^p$, where ε^m and ε^p are the trivial bundles of dimensions m and p, respectively. We start by proving the following very special case of the Whitney formula.

Lemma. *Stably equivalent bundles have the same Stiefel–Whitney characteristic classes.*

Proof. We show that obstructions to extending the sections of the bundles ξ and $\xi \oplus \varepsilon^m$ can be chosen so as to coincide even at the level of cocycles, or, to be more precise, after the groups of cochains have been identified.

The obstructing cocycle $\mathfrak{o}_k(\xi)$ belongs to $C^{n-k+1}(B; \{\pi_{n-k}(V(n,k))\})$, and the obstructing cocycle $\mathfrak{o}_k(\xi \oplus \varepsilon^m)$, to $C^{n-k+1}(B; \{\pi_{n-k}(V(n+m, k+m))\})$; here $n = \dim \xi$. The coefficient groups $\pi_{n-k}(V(n,k))$ and $\pi_{n-k}(V(n+m, k+m))$ reduced modulo 2 are identified by the map $V(n,k) \to V(n+m, k+m)$ that takes any k vectors in $\mathbb{R}^n$ to the same k vectors in $\mathbb{R}^n \oplus \mathbb{R}^m$ plus a fixed orthonormal basis of $\mathbb{R}^m$.

Take k linearly independent sections over the $(n-k)$-skeleton of the base of ξ. They can be naturally supplemented to $k+m$ sections of the bundle $\xi \oplus \varepsilon^m$. After the identification of the groups, the coefficients of the obstructing cocycles $\mathfrak{o}_k(\xi)$ and $\mathfrak{o}_k(\xi \oplus \varepsilon^m)$ in these sections coincide. □

We prove (26) by using this lemma and Theorem 3.20 on p. 146. Let ξ and η be bundles over simplicial complexes X and Y. Consider the inclusions $i_X \colon X^i \hookrightarrow X$ and $j_Y \colon Y^j \hookrightarrow Y$ and the restrictions $\xi|_{X^i}$ and $\eta|_{Y^j}$. The first bundle has $\dim \xi - i$ linearly independent sections; therefore, it is stably equivalent to an i-dimensional bundle ξ_0. Similarly, the second bundle is stably equivalent to a j-dimensional bundle η_0. According to Theorem 3.20, we have $w_{i+j}(\xi_0 \times \eta_0) = w_i(\xi_0) \times w_j(\eta_0)$. The naturality property of the characteristic classes implies

$$\begin{aligned}(i_X \times j_Y)^*(w_{i+j}(\xi \times \eta)) &= w_{i+j}(\xi|_{X^i} \times \eta|_{Y^j}) \\ &= w_{i+j}(\xi_0 \times \eta_0) = w_i(\xi_0) \times w_j(\eta_0) \\ &= w_i(i_X^*\xi) \times w_j(j_Y^*\eta) = (i_X \times j_Y)^*(w_i(\xi) \times w_j(\eta)).\end{aligned}$$

As is easily seen, if $\alpha + \beta = i + j$ and $\alpha \neq i$, then $(i_X \times j_Y)^*(w_{\alpha+\beta}(\xi \times \eta)) = w_\alpha(\xi_0) \times w_\beta(\eta_0) = 0$. Indeed, if $\alpha > i$, then $w_\alpha(\xi_0) = 0$ because $\dim \xi_0 = i < \alpha$, and if $\beta > j$, then $w_\beta(\eta_0) = 0$.

Let $\omega_k = w_k(\xi \times \eta) - \sum_{\alpha+\beta=k} w_\alpha(\xi) \times w_\beta(\eta)$. We have proved that ω_k belongs to the kernel of the homomorphism $(i_X \times j_Y)^* \colon H^k(X \times Y; \mathbb{Z}_2) \to H^k(X^i \times Y^j; \mathbb{Z}_2)$ for all dimensions i and j such that $i + j = k$. Let us prove that $\omega_k = 0$. To simplify the formulas, we omit $\mathbb{Z}_2$ from the notation. According to the Künneth theorem, we have $H^k(X \times Y) \cong$

$\bigoplus_{\alpha+\beta=k} H^\alpha(X)\otimes H^\beta(Y)$ and $H^k(X^i\times Y^j)\cong\bigoplus_{\alpha+\beta=k} H^\alpha(X^i)\otimes H^\beta(Y^j) = H^i(X^i)\otimes H^j(Y^j)$ because $H^\alpha(X^i)=0$ for $\alpha>i$ and $H^\beta(Y^j)=0$ for $\beta>j$. Therefore, the direct sum of the maps $(i_X\times j_Y)^*$ over all i and j such that $i+j=k$ is a homomorphism

$$\bigoplus_{\alpha+\beta=k} H^\alpha(X)\otimes H^\beta(Y)\to\bigoplus_{\alpha+\beta=k} H^\alpha(X^\alpha)\otimes H^\beta(Y^\beta).$$

This homomorphism is injective because the maps $H^\alpha(X)\to H^\alpha(X^\alpha)$ and $H^\beta(Y)\to H^\beta(Y^\beta)$ are monomorphisms (they are dual to the epimorphisms $H_\alpha(X^\alpha)\to H_\alpha(X)$ and $H_\beta(Y^\beta)\to H_\beta(Y)$).

It is convenient to write the Whitney formula in terms of the *total Stiefel–Whitney class* $w(\xi)=1+w_1(\xi)+w_2(\xi)+\dots+w_n(\xi)$, where $n=\dim\xi$. In this notation, the Whitney formula becomes $w(\xi\oplus\eta)=w(\xi)\smile w(\eta)$.

2.5. Applications of Stiefel–Whitney Classes. The total Stiefel–Whitney class $w=1+w_1+w_2+\cdots$ is invertible in the sense that there exists an element $\bar w=1+\bar w_1+\bar w_2+\cdots\in H^*(B;\mathbb{Z}_2)$ for which $w\bar w=1$ (for brevity, instead of cups, we use the standard notation for the product of ring elements). Indeed, if a is a nilpotent element of the ring, then $(1+a)^{-1}=1-a+a^2-a^3+\cdots$. Therefore,

$$\begin{aligned}(1&+w_1+w_2+\cdots)^{-1}\\ &=1-(w_1+w_2+\cdots)+(w_1+w_2+\cdots)^2-(w_1+w_2+\cdots)^3+\cdots\\ &=1-w_1+(w_1^2-w_2)+(-w_1^3+2w_1w_2-w_3)+\cdots.\end{aligned}$$

All elements of the ring $H^*(B;\mathbb{Z}_2)$ have additive order 2; hence

$$\begin{aligned}\bar w_1&=w_1,\\ \bar w_2&=w_1^2+w_2,\\ \bar w_3&=w_1^3+w_3,\\ \bar w_4&=w_1^4+w_1^2w_2+w_2^2+w_4,\\ &\dots\dots\dots\dots\end{aligned}$$

The classes $\bar w_i(\xi)$ are called the *dual Stiefel–Whitney classes*.

The dual Stiefel–Whitney classes are important because of the following *Whitney duality theorem*.

Theorem 3.22 (Whitney). *Suppose that a manifold M^n is immersed in $\mathbb{R}^N$ and ν_{M^n} is the normal bundle over M^n. Then $w_i(\nu_{M^n})=\bar w_i(\tau_{M^n})$. In particular, $w_i(\nu_{M^n})$ does not depend on the immersion.*

Proof. The bundle $\nu_{M^n}\oplus\tau_{M^n}$ is isomorphic to the trivial bundle ε^N over M^n; the isomorphism takes each pair of vectors in the fibers of these bundles over a point $x \in M^n$ to their sum (one of the vectors is normal to the tangent space and the other belongs to this space). Thus, $w(\nu_{M^n})w(\tau_{M^n}) = w(\nu_{M^n} \oplus \tau_{M^n}) = 1$, i.e., $w(\nu_{M^n}) = \bar{w}(\tau_{M^n})$. □

Corollary. $w(\tau_{S^n}) = 1$.

Proof. For the standard embedding $S^n \subset \mathbb{R}^{n+1}$, the normal bundle is trivial. □

Problem 101. Given a Riemannian manifold M, prove that the normal bundle to $d(M)$ in $M \times M$ endowed with the structure of a Riemannian manifold is isomorphic to the tangent bundle of the manifold M; here $d(x) = (x, x)$ is the diagonal map.

For a manifold M^n, the class $w(\tau_{M^n})$ is called the *Stiefel–Whitney class of the manifold* M^n and is denoted by $w(M^n)$.

Problem 102. Calculate the classes w_1 and w_2 for a sphere with g handles.

Problem 103. Calculate the classes w_1 and w_2 for the sphere with m Möbius bands attached and prove that $w_1^2 = w_2$.

Problem 104. Prove that if a manifold M^n is immersed in $\mathbb{R}^{n+1}$, then $w_k = w_1^k$ for all k (here $w_k = w_k(M^n)$).

The Stiefel–Whitney classes of all spheres are trivial. A more interesting example is the Stiefel–Whitney classes of projective spaces $\mathbb{R}P^n$. They are calculated using the following assertion.

Theorem 3.23. *If* ε^1 *is the trivial one-dimensional bundle over* $\mathbb{R}P^n$, *then* $\tau_{\mathbb{R}P^n} \oplus \varepsilon^1 \cong \underbrace{\gamma_n^1 \oplus \dots \oplus \gamma_n^1}_{n+1}$.

Proof. The total spaces of the bundles involved in the statement of the theorem can be described as follows:

$$\begin{aligned}
\tau(\mathbb{R}P^n) &\text{ consists of the pairs } \pm(x, v), x \in S^n, v \in \mathbb{R}^{n+1}, x \perp v;\\
\varepsilon^1 &\text{ consists of the pairs } (\pm x, \lambda), \lambda \in \mathbb{R};\\
\gamma_n^1 &\text{ consists of the pairs } (\pm x, \lambda(x)), \lambda(-x) = -\lambda(x);\\
(\gamma_n^1)^{n+1} &\text{ consists of the pairs } (\pm x, v(x)), v(-x) = -v(x).
\end{aligned}$$

The map $\tau_{\mathbb{R}P^n} \oplus \varepsilon^1 \to (\gamma_n^1)^{n+1}$ takes pairs $\pm(x, v)$ and $(\pm x, \lambda)$ to the pair $(\pm x, v + \lambda x)$. The inverse map acts as follows. Let us represent a vector $v(x)$ in the form $v(x) = v_\tau + v_\nu$, where $v_\nu = \lambda x$ and $v_\tau \perp x$. The pair $(\pm x, v(x))$ is mapped to the pairs $\pm(x, v_\tau)$ and $\pm(x, \varepsilon v_\nu)$, where $\varepsilon = 1$ if the directions of the vectors x and v_ν coincide, and $\varepsilon = -1$ if they are opposite. □

Corollary. $w(\mathbb{R}P^n) = (1+\alpha)^{n+1} = 1 + \binom{n+1}{1}\alpha + \cdots + \binom{n+1}{n}\alpha^n$.

Exercise. Prove that $w_1(\mathbb{R}P^n) = 0$ if and only if n is odd.

Theorem 3.24. *The equality $w(\mathbb{R}R^n) = 1$ holds if and only if $n = 2^k - 1$ for some k.*

Proof. Clearly, $(a+b)^2 \equiv a^2 + b^2 \pmod 2$. Therefore, $(1+\alpha)^{2^k} = 1 + \alpha^{2^k}$. Suppose that $n = 2^k - 1$. Then $w(\mathbb{R}R^n) = (1+\alpha)^{2^k} = 1 + \alpha^{2^k} = 1$ because $2^k > n$.

Now, suppose that $n + 1 = 2^k m$, where m is an odd number larger than 1. Then

$$(1+\alpha)^{n+1} = (1+\alpha^{2^k})^m = 1 + m\alpha^{2^k} + \frac{m(m-1)}{2}\alpha^{2 \cdot 2^k} + \cdots,$$

where $\alpha^{2^k} \neq 0$ and $m \not\equiv 0 \pmod 2$. □

Corollary 1. *If $\mathbb{R}P^n$ is parallelizable, then $n = 2^k - 1$ for some k.*

Corollary 2. *If $n+1 = 2^k m$, where m is an odd number, then $w_{2^k}(\mathbb{R}P^n) = \alpha^{2^k} \neq 0$; hence there exist no 2^k linearly independent vector fields on $\mathbb{R}P^n$.*

Problem 105. Prove that if $\mathbb{R}P^n$ is immersed in $\mathbb{R}^{n+1}$, then n has the form $2^r - 1$ or $2^r - 2$.

Theorem 3.25. *Suppose that $\mu\colon \mathbb{R}^n \times \mathbb{R}^n \to \mathbb{R}^n$ is a bilinear multiplication without divisors of zero. Suppose also that there exists a left identity, i.e., an element e such that $\mu(e, x) = x$ for all $x \in \mathbb{R}^n$ (the multiplication is not assumed to be associative). Then $n = 2^k$ for some k.*

Proof. Consider a basis $e_1 = e, e_2, \dots, e_n$ in $\mathbb{R}^n$. At each point $x \in S^{n-1} \subset \mathbb{R}^n$, the vectors $x = \mu(e, x), \mu(e_2, x), \dots, \mu(e_n, x)$ are linearly independent because otherwise $\mu(y, x) = 0$ for some nonzero vectors $y = \lambda_1 e_1 + \cdots + \lambda_n e_n$ and x. The projections of $\mu(e_2, x), \dots, \mu(e_n, x)$ to the tangent space to S^{n-1} at x are linearly independent. These projections form linearly independent vector fields on the sphere. Since $\mu(e_i, -x) = -\mu(e_i, x)$, it follows that these vector fields can be transferred to $\mathbb{R}P^{n-1}$. Thus, the manifold $\mathbb{R}P^{n-1}$ is parallelizable. It remains to apply Corollary 1 of Theorem 3.24. □

Theorem 3.26. *If $n = 2^k$, then the manifold $\mathbb{R}P^n$ cannot be immersed in $\mathbb{R}^{2n-2}$.*

Proof. If $n = 2^k$, then $w(\mathbb{R}P^n) = (1+\alpha)(1+\alpha)^{2^k} = (1+\alpha)(1+\alpha^{2^k}) = 1 + \alpha + \alpha^{2^k} + \alpha^{2^k+1} = 1 + \alpha + \alpha^n$. By assumption, we have $\alpha^{n+1} = 0$; therefore, $(1 + \alpha + \alpha^n)(1 + \alpha + \alpha^2 + \cdots + \alpha^{n-1}) = 1$. Thus, $\bar{w}(\mathbb{R}P^n) = 1 + \alpha + \alpha^2 + \cdots + \alpha^{n-1}$.

Suppose that $\mathbb{R}P^n$ is immersed in $\mathbb{R}^{n+m}$. Then the dimension of the normal bundle cannot be lower than the maximal dimension of a nonzero dual Stiefel–Whitney class. In the case under consideration, we have $m \geq n-1$, i.e., $\mathbb{R}P^n$ cannot be immersed in $\mathbb{R}^{2n-2}$. □

Theorem 3.27. *If a number n has the binary representation $2^{k_1}+\cdots+2^{k_s}$, then the manifold $\mathbb{R}P^{2^{k_1}} \times \cdots \times \mathbb{R}P^{2^{k_s}}$ cannot be immersed in $\mathbb{R}^{2n-s-1}$.*

Proof. To avoid cumbersome notation, we assume that $s=2$. We set $2^{k_1} = p$ and $2^{k_2} = q$. Let π_p and π_q be the projections of the product $\mathbb{R}P^p \times \mathbb{R}P^q$ onto the first and second factors. Then $w(\mathbb{R}P^p \times \mathbb{R}P^q) = w(\pi_p^*(\tau_{\mathbb{R}P^p}) \oplus \pi_p^*(\tau_{\mathbb{R}P^q})) = \pi_p^*(w(\mathbb{R}P^p)) \smile \pi_q^*(w(\mathbb{R}P^q)) = \pi_p^*(1+\alpha_p)^{p+1} \smile \pi_q^*(1+\alpha_p)^{q+1}$. In the proof of Theorem 3.26, we showed that $((1+\alpha_p)^{p+1})^{-1} = 1+\alpha_p+\cdots+\alpha_p^{p-1}$. For coefficients in $\mathbb{Z}_2$, we have $(\alpha\times\beta)\smile(\gamma\times\delta) = (\alpha\smile\gamma)\times(\beta\smile\delta)$; thus, $\bar{w}(\mathbb{R}P^p \times \mathbb{R}P^q) = (1+\beta_p+\cdots+\beta_p^{p-1}) \times (1+\beta_q+\cdots+\beta_q^{q-1})$, where $\beta_p = \pi_p^*\alpha_p$ and $\beta_q = \pi_q^*\alpha_q$. The leading dimension of a nonzero dual class $\bar{w}(\mathbb{R}P^p \times \mathbb{R}P^q)$ equals $(p-1)+(q-1) = n-2$. For an arbitrary s, the calculations are similar. □

The Stiefel–Whitney Numbers. Let M^n be a closed manifold (possibly disconnected), and let $[M^n] \in H_n(M;\mathbb{Z}_2)$ be its fundamental class. Then, for any cohomology class $\alpha \in H^n(M;\mathbb{Z}_2)$, the number $\langle \alpha, [M^n]\rangle \in \mathbb{Z}_2$ is defined. The number

$$\langle w_1(\tau_{M^n})^{r_1} \cdots w_n(\tau_{M^n})^{r_n}, [M^n]\rangle \in \mathbb{Z}_2$$

obtained for the class $\alpha = w_1(\tau_{M^n})^{r_1} \cdots w_n(\tau_{M^n})^{r_n}$, where $r_1, \dots, r_n$ are nonnegative integers for which $r_1 + 2r_2 + \cdots + nr_n = n$, is called the *Stiefel–Whitney number* corresponding to the cohomology class $w_1^{r_1} \cdots w_n^{r_n}$. This number is denoted by $w_1^{r_1} \cdots w_n^{r_n}[M^n]$.

Problem 106. Prove that if n is even, then $w_n[\mathbb{R}P^n] \neq 0$ and $w_1^n[\mathbb{R}P^n] \neq 0$.

Problem 107. Prove that if n is odd, then all the Stiefel–Whitney numbers of the manifold $\mathbb{R}P^n$ are zero.

Theorem 3.28 (Pontryagin). *All the Stiefel–Whitney numbers of the manifold $M^n = \partial W^{n+1}$, where W^{n+1} is a compact manifold, are zero.*

Proof. We consider homology and cohomology with coefficients in $\mathbb{Z}_2$. Let $[W^{n+1}] \in H_{n+1}(W^{n+1}, M^n)$ be the fundamental class of W^{n+1}. In the proof of the Lefschetz duality theorem, we showed that the homomorphism $\partial_*\colon H_{n+1}(W^{n+1}, M^n) \to H_n(M^n)$ takes $[W^{n+1}]$ to the fundamental class $[M^n] \in H_n(M^n)$. Hence, if $\alpha \in H^n(M^n)$, then $\langle \alpha, [M^n]\rangle = \langle \alpha, \partial_*[W^{n+1}]\rangle = \langle \delta^*\alpha, [W^{n+1}]\rangle$, where $\delta^*\colon H^n(M^n) \to H^{n+1}(W^{n+1}, M^n)$ is the homomorphism dual to ∂_*.

The collar theorem (for smooth manifolds) readily implies $\tau_{W^{n+1}}|_{M^n} \cong \tau_{M^n} \oplus \varepsilon^1$, where ε^1 is the trivial one-dimensional bundle over M^n. Indeed, we are interested only in the part of W^{n+1} near $\partial W^{n+1} = M^n$. Therefore, we can assume that $W^{n+1} = M^n \times I$; in this case, $w_j(M^n) = w_j(\tau_{W^{n+1}}|_{M^n})$ for all j. We have $w_j(\tau_{W^{n+1}}|_{M^n}) = i^* w_j(W^{n+1})$, where $i^*\colon H^j(W^{n+1}) \to H^j(M^n)$ is the homomorphism induced by the inclusion $i\colon M^n \hookrightarrow W^{n+1}$. Hence $w_1^{r_1}(M^n)\cdots w_n^{r_n}(M^n) = i^*\alpha$, where $\alpha = w_1^{r_1}(W^{n+1})\cdots w_n^{r_n}(W^{n+1})$.

The exact cohomology sequence of a pair

$$H^n(W^{n+1}) \xrightarrow{i^*} H^n(M^n) \xrightarrow{\delta^*} H^{n+1}(W^{n+1}, M^n)$$

implies $\delta^* i^* = 0$. Consequently, $\delta^*(w_1^{r_1}(M^n)\cdots w_n^{r_n}(M^n)) = \delta^* i^*\alpha = 0$, and therefore

$$\langle w_1^{r_1}(M^n)\cdots w_n^{r_n}(M^n), [M^n]\rangle = \langle \delta^*(w_1^{r_1}(M^n)\cdots w_n^{r_n}(M^n)), [W^{n+1}]\rangle = 0,$$

i.e., $w_1^{r_1}\cdots w_n^{r_n}[M^n] = 0$. $\square$

2.6. The Universal Bundle. There exists another important approach to constructing characteristic classes, which is based on the fact that any vector bundle of a given dimension is a pullback of a fixed bundle over the Grassmann manifold. Under this approach, the Stiefel–Whitney characteristic classes are obtained from a certain set of multiplicative generators of the cohomology ring of the Grassmann manifold by means of the induced map in cohomology.

We recall that the points of the Grassmann manifold $G(n,k)$ are k-dimensional subspaces of $\mathbb{R}^n$. The natural embedding $\mathbb{R}^n \hookrightarrow \mathbb{R}^{n+1}$ taking every point with coordinates $(x_1,\ldots,x_n)$ to the point with coordinates $(x_1,\ldots,x_n,0)$, induces an embedding $G(n,k) \hookrightarrow G(n+1,k)$. Thus, we can consider the *infinite Grassmann manifold* $G(\infty,k) = \bigcup_{n\geq k} G(n,k)$, which is endowed with the direct limit topology; a set $U \subset G(\infty,k)$ is open in this topology if and only if all the sets $U \cap G(n,k)$, $n \geq k$, are open.

In terms of the infinite Grassmann manifold, the main results of this section (Theorems 3.29, 3.30, and 3.31) can be formulated as follows: There is a one-to-one correspondence between the k-dimensional vector bundles over a compact Hausdorff space B considered up to isomorphism and the homotopy classes of maps $[B, G(\infty,k)]$. This assertion can also be stated and proved in terms of the ordinary Grassmann manifolds; this is what we do below.

There is a *canonical* k-dimensional vector bundle γ_n^k over the Grassmann manifold $G(n,k)$. The space $E(\gamma_n^k)$ of this bundle consists of the pairs

$$(\text{a } k\text{-dimensional subspace } \Pi^k \subset \mathbb{R}^n,\ \text{a vector } v \in \Pi^k);$$

the projection $p\colon E(\gamma_n^k) \to G(n,k)$ assigns to every such pair the subspace Π^k viewed as a point of the manifold $G(n,k)$. The topology of the space $E(\gamma_n^k)$ is induced from $G(n,k) \times \mathbb{R}^n$. The fiber over a point $\Pi^k \in G(n,k)$ is naturally identified with the k-space Π^k; thus, is has a linear space structure. It remains to verify that γ_n^k satisfies the local triviality condition.

We use the coordinate neighborhoods U_I constructed in the proof that $G(n,k)$ is a smooth manifold (see Part I, Chapter 5, Section 1.5). Here I denotes a set of numbers $\{i_1, \dots, i_k\}$ satisfying the inequalities $1 \le i_1 < \dots < i_k \le n$; the set $U_I \subset G(n,k)$ consists of k-dimensional subspaces of $\mathbb{R}^n$ transversal[6] to the orthogonal complement of the space $\mathbb{R}_I^k$ generated by the vectors $e_{i_1}, \dots, e_{i_k}$. The orthogonal projection of any k-subspace $\Pi^k \in U_I$ to $\mathbb{R}_I^k$ is an isomorphism of linear spaces. There is a homeomorphism $h_i\colon U_I \times \mathbb{R}_I^k \to p^{-1}(U_I)$; it maps each pair $(\Pi^k, v) \in U_I \times \mathbb{R}_I^k$ to the pair $(\Pi^k, v') \in E(\gamma_n^k)$, where $v' \in \Pi^k$ is the vector with orthogonal projection $v \in \mathbb{R}_I^k$.

Theorem 3.29. *Any k-dimensional vector bundle ξ over a compact Hausdorff base B is a pullback of the canonical bundle γ_n^k for some n.*

Proof. Choose open sets $U_1, \dots, U_m$ so that they cover B and the bundle ξ is trivial over each of them. Any compact Hausdorff space is paracompact and, therefore, normal. Hence there exists a partition of unity $f_1, \dots, f_m$ subordinate to the cover $U_1, \dots, U_m$.

We use the homeomorphism $h_i\colon U_i \times \mathbb{R}^k \to p^{-1}(U_i)$ to define a map $\varphi_i\colon p^{-1}(U_i) \to \mathbb{R}^k$ as the composition

$$p^{-1}(U_i) \xrightarrow{h_i^{-1}} U_i \times \mathbb{R}^k \longrightarrow \mathbb{R}^k$$

and use the latter to define a map $g_i\colon E(\xi) \to \mathbb{R}^k$ as

$$g_i(x) = \begin{cases} f_i(x)\varphi_i(x) & \text{if } x \in U_i, \\ 0 & \text{if } x \notin U_i. \end{cases}$$

Finally, we set $g(x) = (g_1(x), \dots, g_m(x))$. Each fiber of the map $g\colon E(\xi) \to \mathbb{R}^{mk}$ is isomorphic to some k-dimensional subspace $\Pi^k \subset \mathbb{R}^{mk}$. Thus, we have obtained a map $\hat{g}\colon B \to G(mk, k)$.

It is easy to verify that $\hat{g}^*\gamma_{mk}^k \cong \xi$. Indeed, the total space of the bundle $\hat{g}^*\gamma_{mk}^k$ consists of all triples $(\Pi^k, v \in \Pi^k, b)$, where $\hat{g}(b) = \Pi^k$. Therefore, the fibers of the bundles ξ and $\hat{g}^*\gamma_{mk}^k$ over each point $b \in B$ are canonically isomorphic. □

[6]Two linear subspaces of dimensions k and $n-k$ in $\mathbb{R}^n$ are called transversal if their intersection consists of the zero vector only.

For a compact Hausdorff space B, two homotopic maps $f_0, f_1\colon B \to G(n,k)$ induce isomorphic bundles, i.e., $f_0^*\gamma_n^k \cong f_1^*\gamma_n^k$. This assertion is valid even in the following more general situation.

Theorem 3.30. *Suppose that B is a compact Hausdorff space, ξ is a bundle over X, and $f_0, f_1\colon B \to X$ are homotopic maps. Then $f_0^*\xi \cong f_1^*\xi$.*

Proof. Let $F\colon B \times I \to X$ be a homotopy between f_0 and f_1, and let f_t be the restriction of F to $B \times \{t\}$. Consider two bundles $F^*\xi$ and $p^*f_t^*\xi$ over $B \times I$, where $p\colon B \times I \to B$ is the natural projection. The restrictions of these bundles to $B \times \{t\}$ are isomorphic. If two bundles ξ_1 and ξ_2 are isomorphic, then the isomorphism between them is a section of the bundle $\operatorname{Hom}(\xi_1, \xi_2)$, for which all maps of fibers are isomorphisms.

Lemma. *Let Y be a closed subset of a compact Hausdorff space Z, and let ξ be a vector bundle over Z. Then any section of the bundle $\xi|_Y$ can be extended to a section of the entire bundle ξ.*

Proof. Let us cover Z by finitely many coordinate neighborhoods. Over each coordinate neighborhood U_i, any section of $\xi|_Y$ can be regarded as a map $f_i\colon U_i \cap Y \to \mathbb{R}^n$. The set $U_i \cap Y$ is closed in the topology of U_i; therefore, f_i can be extended to a map $F_i\colon U_i \to \mathbb{R}^n$. Take a partition of unity $\{\lambda_i\}$ subordinate to the cover $\{U_i\}$. We set

$$s_i(x) = \begin{cases} \lambda_i(x)F_i(x) & \text{if } x \in U_i, \\ 0 & \text{if } x \notin U_i. \end{cases}$$

The sum $\sum s_i(x)$ is a section of ξ extending the given section over Y. □

Take an isomorphism between the restrictions of the bundles $F^*\xi$ and $p^*f_t^*\xi$ to $B\times\{t\}$ and consider an extension of this isomorphism to a section of the bundle $\operatorname{Hom}(F^*\xi, p^*f_t^*\xi)$. In the space $\operatorname{Hom}(\mathbb{R}^n, \mathbb{R}^n)$, isomorphisms form an open subset. Since B is compact, the restrictions of the bundles $F^*\xi$ and $p^*f_t^*\xi$ to $B\times\{V(t)\}$, where $V(t)$ is an open subset of the interval I containing t, are isomorphic. Consequently, the bundles $p^*f_\tau^*\xi$ are isomorphic for all $\tau \in V(t)$, and $f_0^*\xi \cong f_1^*\xi$ because the interval I is compact (and connected). □

The embedding $\mathbb{R}^n \hookrightarrow \mathbb{R}^{n+a}$ induces an embedding $i_a\colon G(n,k) \to G(n+a,k)$. Moreover, $i_a^*(\gamma_{n+a}^k) = \gamma_n^k$. Thus, any pullback of γ_n^k is also a pullback of γ_{n+a}^k for all $a \in \mathbb{N}$.

Theorem 3.31. *Let B be a compact Hausdorff space, and let f_0 and f_1 be maps from B to $G(n,k)$. Then the bundles $f_0^*\gamma_n^k$ and $f_1^*\gamma_n^k$ are isomorphic if and only if the maps $i_a f_0$ and $i_a f_1$ are homotopic for some a.*

Proof. We have just proved that homotopic maps induce isomorphic bundles. Thus, we must only show that if bundles $f_0^*\gamma_n^k$ and $f_1^*\gamma_n^k$ are isomorphic, then the maps $i_a f_0$ and $i_a f_1$ are homotopic for some a.

We identify the isomorphic bundles $f_0^*\gamma_n^k$ and $f_1^*\gamma_n^k$ and assume that the commutative diagrams

$$\begin{array}{ccc} E & \xrightarrow{\varphi_i} & E(\gamma_n^k) \subset G(n,k) \times \mathbb{R}^n \\ \downarrow p & & \downarrow p' \\ B & \xrightarrow{f_i} & G(n,k) \end{array}$$

are given for $i = 0, 1$. Instead of the map φ_i, we consider its composition $\psi_i\colon E \to \mathbb{R}^n$ with the natural projection $G(n,k) \times \mathbb{R}^n \to \mathbb{R}^n$. Note that $f_i(b)$ is a subspace of $\psi_i(p^{-1}(b)) \in G(n,k)$, i.e., the map ψ_i uniquely determines f_i.

Consider the embeddings $j_{n,0}, j_{n,1}\colon \mathbb{R}^n \to \mathbb{R}^n \oplus \mathbb{R}^n$, where $j_{n,0}$ is an isomorphism between $\mathbb{R}^n$ and the first summand and $j_{n,1}$ is an isomorphism between $\mathbb{R}^n$ and the second summand. They induce embeddings $i_{n,0}, i_{n,1}\colon G(n,k) \to G(2n,k)$, where $i_{n,0} = i_n$ and $i_{n,1} \simeq i_n$ (for $n = 1$, the homotopy between $i_{n,0}$ and $i_{n,1}$ is constructed by rotating the coordinate axes; for $n > 1$, it is constructed by applying the same construction to each plane generated by vectors of the form e_α and $e_{n+\alpha}$).

For each t, the map $\psi_{n,t} = (1-t)j_{n,0}\psi_0 + tj_{n,1}\psi_1$ determines a map $f_{n,t}\colon B \to G(2n,k)$. Indeed, since ψ_0 and ψ_1 are monomorphisms from $p^{-1}(b)$ to $\mathbb{R}^n$, the map $\psi_{n,t}$ is also a monomorphism because the vectors $j_{n,0}\psi_0(v)$ and $j_{n,1}\psi_1(v)$ belong to orthogonal subspaces.

Thus, the maps $f_{n,0} = i_{n,0}f_0 = i_n f_0$ and $f_{n,1} = i_{n,1}f_1 \simeq i_n f_1$ are homotopic; therefore, so are the maps $i_n f_0$ and $i_n f_1$. □

Problem 108. Given a k-dimensional vector bundle ξ^k over a compact CW-complex B of dimension less than k and the trivial one-dimensional vector bundle ε^1 over B, prove that the bundle ξ^k is trivial if and only if so is the bundle $\xi^k \oplus \varepsilon^1$.

A manifold M^n is said to be *stably parallelizable* if there exists a trivial k-dimensional bundle ε^k over M^n for which the bundle $\tau_{M^n} \oplus \varepsilon^k$ is trivial.

Problem 109. Prove that a manifold M^n is stably parallelizable if and only if one of the following conditions holds:

(a) the bundle $\tau_{M^n} \oplus \varepsilon^1$ is trivial;

(b) the normal bundle for an embedding of M^n into $\mathbb{R}^N$, where $N \geq 2n+1$, is trivial.

2.7. Stable Cohomology of Grassmann Manifolds. The one-to-one correspondence between the k-dimensional vector bundles over B and the homotopy classes of maps $B \to G(\infty, k)$ implies that the cohomology groups of the manifolds $G(n,k)$ and their behavior under the embeddings $G(n,k) \to G(n, k+a)$ are of great importance in the study of vector bundles. Indeed, every cohomology class stable with respect to such embeddings allows us to associate to any bundle ξ over B a cohomology class of the space B, namely, the characteristic class of the bundle ξ.

In this section, we calculate the homology and cohomology groups with coefficients in $\mathbb{Z}_2$ of the (real) Grassmann manifolds $G(n,k)$. We shall use the coordinate neighborhoods U_I and open Schubert cells $e(\sigma)$ introduced in Part I on p. 196. Any open Schubert cell $e(\sigma)$ is contained entirely in the chart U_I, where $I = \sigma$; moreover, it is determined by a system of equations of the form $x_i = 0$ in this chart. Its closure $\overline{e(\sigma)}$ intersects only those charts U_I for which $i_1 \le \sigma_1, \ldots, i_k \le \sigma_k$. In any such chart, the set $\overline{e(\sigma)} \cap U_I$ is determined by equations of the same form. Therefore, $\overline{e(\sigma)}$ is a submanifold in $G(n,k)$. It is called the *Schubert manifold.*

The open Schubert cells form a cell decomposition of the Grassmann manifold, and their closures are submanifolds. Hence there exists a triangulation of $G(n,k)$ that induces a triangulation on each Schubert manifold. Given such a triangulation, we can calculate cellular homology. First, we must determine the structure of the characteristic maps of Schubert cells.

The dimension of the cell $e(\sigma)$ equals $(\sigma_1 - 1) + (\sigma_2 - 2) + \cdots + (\sigma_k - k)$. We are interested in the coefficients of the cells of dimension lower by 1 in $\partial[e(\sigma)]$. If a cell $e(\sigma')$ is contained in $\overline{e(\sigma)}$, then $\sigma'_1 \le \sigma_1, \ldots, \sigma'_k \le \sigma_k$. Therefore, for such a cell, the equality $\sigma'_1 + \cdots + \sigma'_k = \sigma_1 + \cdots + \sigma_k - 1$ can hold only if $\sigma'_i = \sigma_i - 1$ and $\sigma'_j = \sigma_j$ for $j \ne i$; in this case, $\sigma_{i-1} = \sigma_i - 1$. In an appropriate coordinate system, the subset $e(\sigma')$ of $e(\sigma)$ is determined by the equation $v_{i,\sigma_i} = 0$ (see Figure 5). Depending on the orientations of the

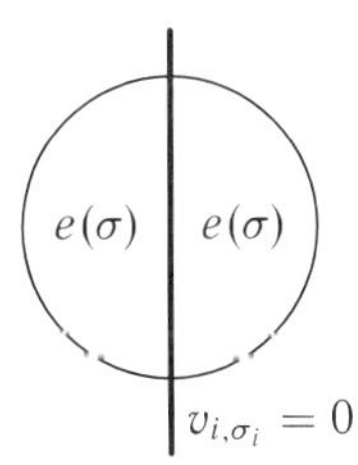

Figure 5. The boundary of a Schubert cell

two parts of $e(\sigma)$ adjacent to $\overline{e(\sigma')}$, the coefficient of the cell $e(\sigma')$ in $\partial[e(\sigma)]$ is equal to ± 2 or 0, which means that $\partial[e(\sigma)] = 0$ for $\mathbb{Z}_2$-homology. Thus, $H_r(G(n,k); \mathbb{Z}_2) \cong \mathbb{Z}_2^{N(r,n,k)}$, where $N(r,n,k)$ is the number of r-dimensional Schubert cells in the manifold $G(n,k)$.

Each Schubert symbol σ can be associated with the sequence of numbers $\sigma_1 - 1, \sigma_2 - 2, \dots, \sigma_k - k$. Removing all zeros from this sequence, we obtain a sequence $i_1, i_2, \dots, i_s$, where $s \le k$ and $1 \le i_1 \le i_2 \le \dots \le i_s \le n-k$ (the inequality $i_\alpha \le i_{\alpha+1}$ follows from $\sigma_\beta < \sigma_{\beta+1}$).

An unordered set of positive integers $i_1, i_2, \dots, i_s$ is called a *partition* of an integer r if $i_1 + \dots + i_s = r$. Every unordered set of numbers uniquely determines the set of the same numbers arranged in increasing order. Therefore, $N(r, n, k)$ is equal to the number of partitions of r into at most k positive integers not exceeding $n-k$. In particular, if both numbers k and $n-k$ are not smaller than r, then $N(r,n,k) = p(r)$ equals the number of all partitions of r.

If $n \ge r + k$, then the embedding $G(n,k) \to G(n+1,k)$ induces an isomorphism $H_r(G(n,k);\mathbb{Z}_2) \to H_r(G(n+1,k);\mathbb{Z}_2)$ of cohomology groups and an isomorphism $H^r(G(n+1,k);\mathbb{Z}_2) \to H^r(G(n,k);\mathbb{Z}_2)$ of dual cohomology spaces; here we have in mind the duality of linear spaces rather than Poincaré duality. For $n \ge r+k$, we denote the group $H^r(G(n,k);\mathbb{Z}_2)$ by $H^r(G(\infty,k);\mathbb{Z}_2)$.

We determine the multiplicative structure of the ring $H^*(G(\infty,k);\mathbb{Z}_2)$. To do this, consider the embedding

$$f\colon \underbrace{\mathbb{R}P^{n-1} \times \dots \times \mathbb{R}P^{n-1}}_{k} \to G(nk,k)$$

that identifies every set of vectors $v_1, \dots, v_k \in \mathbb{R}^n$ with the subspace of $\mathbb{R}^n \times \dots \times \mathbb{R}^n$ spanned by $(v_1, 0, \dots, 0), (0, v_2, 0, \dots, 0), \dots, (0, 0, \dots, v_k)$. This map induces a ring homomorphism

$$H^*(G(nk,k);\mathbb{Z}_2) \to H^*(\mathbb{R}P^{n-1} \times \dots \times \mathbb{R}P^{n-1};\mathbb{Z}_2).$$

The ring $H^*(\mathbb{R}P^{n-1} \times \dots \times \mathbb{R}P^{n-1};\mathbb{Z}_2)$ is isomorphic to the quotient of the polynomial ring $\mathbb{Z}_2[\alpha_1, \dots, \alpha_k]$ by the relations $\alpha_i^n = 0$, where $i = 1, \dots, k$.

For $r \le n$, the embedding $\mathbb{R}P^n \to \mathbb{R}P^{n+1}$ induces a ring isomorphism $H^r(\mathbb{R}P^{n+1};\mathbb{Z}_2) \to H^r(\mathbb{R}P^n;\mathbb{Z}_2)$. Hence we can introduce the ring $H^*(\mathbb{R}P^\infty;\mathbb{Z}_2) \cong \mathbb{Z}_2[\alpha]$; for every fixed r, we simply treat $\mathbb{R}P^\infty$ as $\mathbb{R}P^n$ with sufficiently large n. Similarly, we introduce the ring $H^*(\mathbb{R}P^\infty \times \dots \times \mathbb{R}P^\infty;\mathbb{Z}_2) \cong \mathbb{Z}_2[\alpha_1, \dots, \alpha_k]$ and define the homomorphism

$$f^*\colon H^*(G(\infty,k);\mathbb{Z}_2) \to H^*(\mathbb{R}P^\infty \times \dots \times \mathbb{R}P^\infty;\mathbb{Z}_2) \cong \mathbb{Z}_2[\alpha_1, \dots, \alpha_k].$$

Theorem 3.32. *The homomorphism f^* is an isomorphism onto a subring of symmetric polynomials. Moreover, $f^*(w_i(\gamma_\infty^k)) = \sigma_i(\alpha_1, \dots, \alpha_k)$, where σ_i is the ith elementary symmetric function.*

Proof. First, we show that the dimension of the space $H^r(G(\infty,k);\mathbb{Z}_2)$ coincides with that of the space of degree r symmetric polynomials in k

variables over the field $\mathbb{Z}_2$. To this end, we must establish a one-to-one correspondence between the partitions of r into at most k positive numbers and the polynomials of the form $\sigma_1^{r_1}\cdots\sigma_k^{r_k}$, where $r_1 + 2r_2 + \cdots + kr_k = r$ ($r_1 \geq 0$). Note that for such r_i, the sum of the numbers $r_k \leq r_k + r_{k-1} \leq \cdots \leq r_k + \cdots + r_1$ equals r. Conversely, given $r = s_1 + \cdots + s_k$, where $s_1 \geq s_2 \geq \cdots \geq s_k \geq 0$, we set $r_k = s_k$, $r_{k-1} = s_{k-1} - s_k$, $\ldots$, $r_1 = s_1 - s_2$.

It is easy to verify that $f^*\gamma_\infty^k \cong \gamma_\infty^1 \times \cdots \times \gamma_\infty^1$; to be more precise, $f^*\gamma_{nk}^k \cong \gamma_{n-1}^1 \times \cdots \times \gamma_{n-1}^1$. Indeed, the fiber of the bundle $f^*\gamma_{nk}^k$ over a point $(x_1, \ldots, x_k)$, where $x_i \in \mathbb{R}^n$, consists of vectors $\lambda_1 x_1 + \cdots + \lambda_k x_k$.

Using the naturality property, we obtain $f^*(w(\gamma_\infty^k)) = w(f^*(\gamma_\infty^k)) = w(\gamma_\infty^1 \times \cdots \times \gamma_\infty^1) = (1+\alpha_1)\cdots(1+\alpha_k)$, where α_i is the generator of the cohomology ring of the ith copy of $\mathbb{R}P^\infty$. Thus, $f^*(w(\gamma_\infty^k)) = \sigma_i(\alpha_1, \ldots, \alpha_k)$. It follows that f^* is an epimorphism from the ring $H^*(G(\infty, k); \mathbb{Z}_2)$ onto the ring of symmetric polynomials in $\alpha_1, \ldots, \alpha_k$ over the field $\mathbb{Z}_2$. Therefore, f^* is an isomorphism between $H^r(G(\infty, k); \mathbb{Z}_2)$ and the space of symmetric polynomials of degree r. □

Corollary (an axiomatic approach to the Stiefel–Whitney classes). *Suppose that to any vector bundle ξ over any simplicial complex B we associate cohomology classes $\tilde{w}_i(\xi) \in H^i(B; \mathbb{Z}_2)$ so that $\tilde{w}_0(\xi) = 1$, $\tilde{w}_i(\xi) = 0$ for $i > \dim\xi$, $\tilde{w}_1(\gamma_\infty^1) = \alpha$, the naturality condition holds, and the Whitney formula (see (26) on p. 147) is valid. Then the $\tilde{w}_i(\xi) = w_i(\xi)$ are the Stiefel–Whitney classes.*

Proof. The proof of Theorem 3.32 uses only the properties of Stiefel–Whitney classes mentioned in the statement. Therefore, the same argument as in that proof implies $f^*(\tilde{w}_i(\gamma_\infty^k)) = f^*(w_i(\gamma_\infty^k))$. Since f^* is an isomorphism, it follows that $\tilde{w}_i(\gamma_\infty^k) = w_i(\gamma_\infty^k)$, and since the bundle γ_∞^k is universal, it follows that $\tilde{w}_i(\xi) = w_i(\xi)$ for any k-dimensional bundle ξ. □

Using Theorem 3.32, we can obtain an expression for the characteristic classes of the bundle $\xi^m \otimes \eta^n$ (where m and n are the dimensions of the bundles) in terms of those of the bundles ξ^m and η^n.

Theorem 3.33. *Let $\sigma_1, \ldots, \sigma_m$ and $\sigma'_1, \ldots, \sigma'_n$ be elementary symmetric functions of variables $t_1, \ldots, t_m$ and $t'_1, \ldots, t'_n$ over the field $\mathbb{Z}_2$, and let $p_{m,n}$ be the (unique) polynomial in $m+n$ variables for which*

$$p_{m,n}(\sigma_1, \ldots, \sigma_m, \sigma'_1, \ldots, \sigma'_n) = \prod_{i=1}^{m}\prod_{j=1}^{n}(1 + t_i + t'_j).$$

Then $w(\xi^m \otimes \eta^n) = p_{m,n}(w_1, \ldots, w_m, w'_1, \ldots, w'_n)$, where $w_i = w_i(\xi^m)$ and $w'_j = w_j(\eta^n)$.

Proof. First, note that $w_1(\xi^1 \otimes \eta^1) = w_1(\xi^1) + w_1(\eta^1)$ for one-dimensional bundles. Indeed, for a one-dimensional bundle, the class w_1 is completely determined by the restriction of this bundle to the 1-skeleton; therefore, it is sufficient to check the required equality for one-dimensional circle bundles. For such bundles, $\xi^1 \otimes \eta^1$ is orientable if and only if the bundles ξ^1 and η^1 are both orientable or both nonorientable (if the orientations of the fibers of ξ^1 and η^1 are multiplied by ε_1 and ε_2 when the circle is traversed, then the orientation of the fiber of $\xi^1 \otimes \eta^1$ is multiplied by $\varepsilon_1\varepsilon_2$).

This equality can also be derived from the fact that, for any 1-cell, the value of any cocycle representing, say, the class $w_1(\xi^1)$ is congruent to the number of nondegenerate zeros of a section of ξ^1 over this cell modulo 2. Choose sections of the bundles ξ^1 and η^1 over the 1-skeleton so that they have no common zeros and all of their zeros are nondegenerate. The tensor product of these sections is a section of the bundle $\xi^1 \otimes \eta^1$ whose zeros are those of the sections of ξ^1 and η^1.

Step 1. The required assertion is true for direct sums of one-dimensional bundles, i.e., in the case of $\xi^m = \xi_1 \oplus \dots \oplus \xi_m$ and $\eta^n = \eta_1 \oplus \dots \oplus \eta_n$, where all bundles ξ_i and η_j are one-dimensional.

We set $t_i = w_1(\xi_i)$ and $t'_j = w_1(\eta_i)$. Then $w = \prod(1 + t_i)$ and $w' = \prod(1 + t'_j)$, i.e., $w_i = \sigma_i$ and $w'_j = \sigma'_j$. Moreover,

$$(\xi_1 \oplus \dots \oplus \xi_m) \otimes (\eta_1 \oplus \dots \oplus \eta_n) = \bigoplus_{i,j} (\xi_i \otimes \eta_j).$$

Therefore,

$$w(\xi^m \otimes \eta^n) = \prod_{i,j} w(\xi_i \otimes \eta_j) = \prod_{i,j}^{n} (1 + t_i + t'_j).$$

By assumption,

$$\prod_{i,j}^{n} (1 + t_i + t'_j) = p_{m,n}(\sigma_1, \dots, \sigma_m, \sigma'_1, \dots, \sigma'_n),$$

i.e., $w(\xi^m \otimes \eta^n) = p_{m,n}(w_1, \dots, w_m, w'_1, \dots, w'_n)$.

Step 2. It is sufficient to prove the required assertion for direct sums of one-dimensional bundles.

Let p_1 and p_2 be the projections of $G(\infty, m) \times G(\infty, n)$ to the first and second factors, respectively. We set $\gamma_1^m = p_1^*\gamma^m$ and $\gamma_2^m = p_2^*\gamma^m$. The bundle $\gamma_1^m \otimes \gamma_2^m$ over $G(\infty, m) \times G(\infty, n)$ is universal for bundles of the form $\xi^m \otimes \eta^n$. Indeed, if $\xi^m = f_1^*\gamma^m$ and $\eta^n = f_2^*\gamma^n$, then $\xi^m \otimes \eta^n = (f_1 \times f_2)^*(\gamma_1^m \otimes \gamma_2^m)$; here $(f_1 \times f_2)(b) = (f_1(b), f_2(b))$.

Thus,

$$w(\xi^m \otimes \eta^n) \in (f_1 \times f_2)^* H^*(G(\infty, m) \times G(\infty, n); \mathbb{Z}_2);$$

to shorten the notation, we do not indicate the coefficient groups $\mathbb{Z}_2$. If $\alpha \in H^*(G(\infty, m))$ and $\beta \in H^*(G(\infty, n))$, then $\alpha \times \beta = p_1^*\alpha \smile p_2^*\beta$. According to the Künneth theorem, the map $\alpha \otimes \beta \mapsto \alpha \times \beta$ induces a cohomology isomorphism

$$H^*(G(\infty, m)) \otimes H^*(G(\infty, n)) \to H^*(G(\infty, m) \times G(\infty, n)).$$

Therefore, the ring $H^*(G(\infty, m) \times G(\infty, n))$ is multiplicatively generated by the free generators $w_i(\gamma_1^m) = w_i$ and $w_i(\gamma_2^n) = w_j'$. Hence the total Stiefel–Whitney class of the bundle $\gamma_1^m \otimes \gamma_2^m$ has a *unique* representation in the form $p_{m,n}(w_1, \dots, w_m, w_1', \dots, w_n')$. □

Example 46. $w_1(\xi^m \otimes \eta^n) = nw_1(\xi^m) + mw_1(\eta^n)$.

Proof. We have $\prod_{i=1}^m \prod_{j=1}^n (1 + t_i + t_j') = 1 + n(t_1 + \dots + t_m) + m(t_1' + \dots + t_n') + \cdots$. □

Theorem 3.33 can be applied to calculate the Stiefel–Whitney classes of Grassmann manifolds. Let $\overline{\gamma_n^k}$ be the bundle over $G(n, k)$ whose fiber over $\Pi^k \in G(n, k)$ is the orthogonal complement of the subspace Π^k. Then the bundle $\gamma_n^k \oplus \overline{\gamma_n^k}$ is trivial, and therefore $w(\overline{\gamma_n^k}) = \bar{w}(\gamma_n^k)$.

Theorem 3.34. $\tau_{G(n,k)} \cong \mathrm{Hom}(\gamma_n^k, \overline{\gamma_n^k})$.

Proof. A tangent vector at a point $\Pi^k \in G(n, k)$ is determined by a curve $\Pi^k(t)$ such that $\Pi^k(0) = \Pi^k$. For small t, the subspace $\Pi^k(t)$ is uniquely determined by the linear map $\varphi_t \colon \Pi^k \to (\Pi^k)^\perp$. Indeed, choose a basis $e_1, \dots, e_k$ in Π^k. Let $e_1', \dots, e_k'$ be vectors in $\Pi^k(t)$ which are mapped to $e_1, \dots, e_k$ under the orthogonal projection onto Π^k. Suppose that $e_1'', \dots, e_k''$ are the orthogonal projections of the vectors $e_1', \dots, e_k'$ on $(\Pi^k)^\perp$. The formula $\varphi_t(e_i) = e_i''$ defines the linear maps $\varphi_t \colon \Pi^k \to (\Pi^k)^\perp$; for $t = 0$, we have $\varphi_0 = 0$. Each map φ_t uniquely determines the subspace $\Pi^k(t)$; it is spanned by the vectors $e_1 + \varphi_t(e_1), \dots, e_k + \varphi_t(e_k)$.

Thus, any curve in $G(n, k)$ passing through the point Π^k can be viewed as a curve in $\mathrm{Hom}(\Pi^k, (\Pi^k)^\perp)$ passing through the origin for small t. The tangent vectors to such curves are in one-to-one correspondence with the elements of the space $\mathrm{Hom}(\Pi^k, (\Pi^k)^\perp)$. Thus, we have proved the isomorphism $\tau_{G(n,k)} \cong \mathrm{Hom}(\gamma_n^k, \overline{\gamma_n^k})$. □

It is known from linear algebra (see, e.g., [**104**, p. 120]) that $\mathrm{Hom}(\gamma_n^k, \overline{\gamma_n^k}) \cong (\gamma_n^k)^* \otimes \overline{\gamma_n^k}$, where $*$ denotes the passage to the dual space. Endowing the bundle γ_n^k with a Riemannian metric, we get the isomorphism $(\gamma_n^k)^* \cong \gamma_n^k$. Thus, $\tau_{G(n,k)} \cong \gamma_n^k \otimes \overline{\gamma_n^k}$; therefore,

$$w(G(n,k)) = w(\gamma_n^k \otimes \overline{\gamma_n^k}) = p_{k,n-k}(w_1, \dots, w_k, \bar{w}_1, \dots, \bar{w}_{n-k}),$$

where $w_i = w_i(\gamma_n^k)$ and $\bar{w}_j = w_j(\overline{\gamma_n^k}) = \bar{w}_j(\gamma_n^k)$.

Example 47. The Grassmann manifold $G(n,k)$ is orientable if and only if n is even.

Proof. Applying Example 46, we obtain $w_1(G(n,k)) = (n-k)w_1(\gamma_n^k) + k\bar{w}_1(\gamma_n^k) = nw_1(\gamma_n^k)$ because $\bar{w}_1 = w_1$. Moreover, $w_1(\gamma_n^k) \neq 0$. □

2.8. The Chern Characteristic Classes. By analogy with real vector bundles, we can define *complex* vector bundles, whose fibers are vector spaces over $\mathbb{C}$. Applying the constructions of obstruction theory to complex vector bundles, we obtain characteristic classes. For complex bundles, the situation is simpler in many respects. One of the reasons for this is that the realification of a complex space has a canonical orientation, and therefore the realification of any complex vector bundle is orientable. Below, we explain this in more detail.

Let V be a vector space over $\mathbb{C}$ with basis $e_1, \dots, e_n$. It can be assigned the vector space $V_{\mathbb{R}}$ over $\mathbb{R}$ with basis $e_1, \dots, e_n, ie_1, \dots, ie_n$. If the transition matrix between the bases e and ε has the form $A + iB$, where A and B are real matrices, then the transition matrix from e, ie to $\varepsilon, i\varepsilon$ has the form $\left(\begin{smallmatrix} A & -B \\ B & A \end{smallmatrix}\right)$. The determinant of this matrix is equal to $|\det(A+iB)|^2 > 0$ because

$$\begin{vmatrix} A & -B \\ B & A \end{vmatrix} = \begin{vmatrix} A+iB & -B+iA \\ B & A \end{vmatrix} = \begin{vmatrix} A+iB & 0 \\ B & A-iB \end{vmatrix}.$$

(These equalities of determinants are obtained as follows: first, we add the $(n+k)$th row multiplied by i to the kth row for each $1 \le k \le n$, and then, subtract the $(n+k)$th column multiplied by i from the kth row for each $1 \le k \le n$.) Therefore, the space $V_{\mathbb{R}}$ has a canonical orientation.

Let ω be a complex bundle with realification $\omega_{\mathbb{R}}$; this bundle is oriented. We construct the universal bundle in the complex case in analogy with the real case. The base of this bundle is the complex Grassmann manifold $G_{\mathbb{C}}(\infty, k)$, where $k = \dim_{\mathbb{C}} \omega = \frac{1}{2} \dim_{\mathbb{R}} \omega_{\mathbb{R}}$.

We can endow every fiber of a complex vector bundle ω over a compact base with a Hermitian metric and construct the bundle ω_k whose fiber is the *complex Stiefel manifold* $V_{\mathbb{C}}(n,k)$; the points of this manifold are k-element sets of vectors in $\mathbb{C}^n$ orthonormal with respect to Hermitian inner product. To construct the characteristic class of the bundle ω_k, we must calculate the first nontrivial homotopy group $\pi_i(V_{\mathbb{C}}(n,k))$.

Theorem 3.35. *If $i \leq 2(n-k)$, then $\pi_i(V_{\mathbb{C}}(n,k)) = 0$. The first nontrivial homotopy group $\pi_{2(n-k)+1}(V_{\mathbb{C}}(n,k))$ is isomorphic to $\mathbb{Z}$.*

Proof. The proof of this theorem follows the same scheme as that of Theorem 3.16 (see p. 140). In the complex case, the argument becomes simpler because we do not have to calculate the homomorphism ∂_*.

For $k=1$, the manifold $V_{\mathbb{C}}(n,1)$ is homeomorphic to S^{2n-1}; clearly, $\pi_i(S^{2n-1}) = 0$ for $i \leq 2(n-1)$. The induction step uses the locally trivial bundle $V_{\mathbb{C}}(n,k+1) \to V_{\mathbb{C}}(n,k)$ with fiber $S^{2(n-k)-1}$; as a result, we obtain $\pi_i(V_{\mathbb{C}}(n,k+1)) = 0$ for $i \leq 2(n-k-1)$.

Using the locally trivial bundle $V_{\mathbb{C}}(n+1,k+1) \to V_{\mathbb{C}}(n+1,1) = S^{2n+1}$ with fiber $V_{\mathbb{C}}(n,k)$, we can show that $\pi_i(V_{\mathbb{C}}(n,k)) \cong \pi_i(V_{\mathbb{C}}(n+1,k+1))$ for $i \leq 2n-1$. Therefore, for $i \leq 2(n-k)+1$, we have $\pi_i(V_{\mathbb{C}}(n-k+1,1)) \cong \pi_i(V_{\mathbb{C}}(n-k+2,2)) \cong \cdots$. Thus, we see that $\pi_{2(n-k)+1}(V_{\mathbb{C}}(n,k)) \cong \pi_{2(n-k)+1}(V_{\mathbb{C}}(n-k+1,1)) = \pi_{2(n-k)+1}(S^{2(n-k)+1}) = \mathbb{Z}$. □

The canonical orientation of $\mathbb{C}^n$ determines a canonical isomorphism $\pi_{2(n-k)+1}(V_{\mathbb{C}}(n,k)) \to \mathbb{Z}$. Thus, instead of a cohomology with local coefficients, we obtain the ordinary integral cohomology. The *Chern characteristic class* $c_{i+1}(\omega)$ is defined as the obstruction to extending $n-i$ linearly independent sections of the bundle ω over the $(2i+2)$-dimensional skeleton; this obstruction belongs to the group $H^{2i+2}(B; \{\pi_{2i+1}(V_{\mathbb{C}}(n,n-i))\}) = H^{2i+2}(B;\mathbb{Z})$.

The definition directly implies the following properties of the Chern classes.

(i) If $\dim_{\mathbb{C}} \omega = n$, then $c_n(\omega) = e(\omega_{\mathbb{R}})$; i.e., the highest Chern class coincides with the Euler class of the realification of the bundle ω.

(ii) The homomorphism $H^i(B;\mathbb{Z}) \to H^i(B;\mathbb{Z}_2)$ maps the total Chern class $c(\omega)$ to the total Stiefel–Whitney class $w(\omega_{\mathbb{R}})$. In particular, $w_{2j+1}(\omega_{\mathbb{R}}) = 0$ for all j.

In the same way as for the Stiefel–Whitney classes, the following assertions are proved.

(i) The Chern classes satisfy the naturality condition.

(ii) The Chern classes of stably equivalent bundles coincide.

Let γ_n^1 be the bundle over $\mathbb{C}P^n$ whose total space consists of all pairs $(x \in \mathbb{C}P^n, v = \lambda x)$, where $\lambda \in \mathbb{C}$.

Theorem 3.36. *The Chern class $c_1(\gamma_n^1)$ is a generator α of the group $H^2(\mathbb{C}P^n;\mathbb{Z})$.*

Proof. First, we calculate $c_1(\gamma_1^1)$. To this end, we describe the structure of the bundle γ_1^1 in more detail. Let us represent $\mathbb{C}P^1 = \{(z_1 : z_2)\}$ as the union of the sets U_1 and U_2 determined by the inequalities $|z_1| \leq |z_2|$ and $|z_1| \geq |z_2|$. On U_1 and U_2, we introduce the coordinates $w_1 = z_1/z_2$ and $w_2 = z_2/z_1$. The fiber of the bundle γ_1^1 over a point $(z_1 : z_2)$ consists of all vectors $(\lambda z_1 : \lambda z_2)$, where $\lambda \in \mathbb{C}$. The restrictions of γ_1^1 to U_1 and U_2 are trivial; the trivializations have the forms $(\lambda w_1, \lambda)$ and $(\mu, \mu w_2)$. We identify the fibers over w_1 and $w_2 = w_1^{-1}$ by setting $(\lambda w_1, \lambda) = (\mu, \mu w_2)$, i.e., $\mu = \lambda w_1$. The section determined by $\lambda = 1$ over the circle $|w_1| = 1$ is identified with that determined by $\mu = w_1$ over $w_2 = w_1^{-1}$. We obtain a map $S^1 \to S^1$ of degree -1. Therefore, extending this section inside U_2, we obtain a section with one nondegenerate zero. It follows that $c_1(\gamma_1^1) = e((\gamma_1^1)_{\mathbb{R}})$ is a generator of the group $H^2(\mathbb{C}P^n; \mathbb{Z}) \cong \mathbb{Z}$.

Let $i\colon \mathbb{C}P^1 \to \mathbb{C}P^n$ be the natural embedding. It is seen directly from the definition that $i^*(\gamma_n^1) \cong \gamma_1^1$; therefore, $i^* c_1(\gamma_n^1) = c_1(\gamma_1^1)$. Clearly, i induces an isomorphism $i^*\colon H^2(\mathbb{C}P^n; \mathbb{Z}) \to H^2(\mathbb{C}P^1; \mathbb{Z})$. Hence $c_1(\gamma_n^1)$ is a generator of the group $H^2(\mathbb{C}P^n; \mathbb{Z})$. □

Remark. Since the degree of the map $S^1 \to S^1$ under consideration equals -1, it follows that $c_1(\gamma_n^1)$ is the generator of $H^2(\mathbb{C}P^n; \mathbb{Z})$ that takes the value -1 for $\mathbb{C}P^1$ with canonical orientation (rather than the generator taking the value $+1$).

Theorem 3.37. *If ω and φ are complex vector bundles over simplicial complexes X and Y, then $c_k(\omega \times \varphi) = \sum_{i+j=k} c_i(\omega) \times c_j(\varphi)$.*

Proof. The proof of this theorem closely resembles that of the generalized Whitney formula (see (26) on p. 147), but there are substantial differences. Again, consider the embeddings i_X and j_Y. The bundle $\omega|_{X^i}$ has $\dim \omega - [i/2]$ linearly independent sections; therefore, it is stably equivalent to a bundle ω_0 of dimension $[i/2]$. Applying Theorem 3.21 instead of Theorem 3.20, we see that the element $c_k(\omega \times \varphi) - \sum_{\alpha+\beta=k} c_\alpha(\omega) \times c_\beta(\varphi)$ belongs to the kernel of the homomorphism $(i_X \times j_Y)^*$ for all i and j such that $[i/2] + [j/2] = k$. We prove that this element vanishes. Note that if $i + j = 2k + 1$, then $[i/2] + [j/2] = k$. Indeed, $i = 2i'$ and $j = 2j' + 1$ (or *vice versa*); therefore, $[i/2] + [j/2] = i' + j' = (i + j - 1)/2 = k$. We show that the direct sum of the maps $(i_X \times j_Y)^*$ over all i and j such that $i + j = 2k + 1$ is a monomorphism.

For the coefficient group $\mathbb{Z}$, the Künneth theorem implies that the group $H^{2k}(X \times Y)$ is the direct sum of the groups

$$\bigoplus_{\alpha+\beta=2k} H^\alpha(X) \otimes H^\beta(Y) \quad \text{and} \quad \bigoplus_{\alpha+\beta=2k+1} \operatorname{Tor}(H^\alpha(X), H^\beta(Y)).$$

Similarly, the group $H^{2k}(X^i \times Y^j)$ is the direct sum of the groups

$$\bigoplus_{\alpha+\beta=2k} H^\alpha(X^i) \otimes H^\beta(Y^j) \cong (H^i(X^i) \otimes H^{j-1}(Y^j)) \oplus (H^{i-1}(X^i) \otimes H^j(Y^j))$$

and

$$\bigoplus_{\alpha+\beta=2k+1} \operatorname{Tor}(H^\alpha(X^i), H^\beta(Y^j)) = \operatorname{Tor}(H^i(X^i), H^j(Y^j)).$$

Both maps from $H^\alpha(X) \otimes H^\beta(Y)$ to $H^\alpha(X^\alpha) \otimes H^\beta(Y^{\beta+1})$ and to $H^\alpha(X^{\alpha+1}) \otimes H^\beta(Y^\beta)$ are monomorphisms. For the Tor summand, the corresponding maps are monomorphisms also. □

The properties of Chern classes just proved allow us to prove the following assertion using the scheme of the proof of Theorem 3.32.

Theorem 3.38. *The homomorphism*

$$f^*\colon H^*(G_{\mathbb{C}}(\infty, k); \mathbb{Z}) \to H^*(\mathbb{C}P^\infty \times \cdots \times \mathbb{C}P^\infty; \mathbb{Z}) \cong \mathbb{Z}[\alpha_1, \ldots, \alpha_k]$$

is an isomorphism onto the subring of symmetric polynomials. Moreover, $f^*(c_i(\gamma^k_\infty)) = \sigma_i(\alpha_1, \ldots, \alpha_n)$, where σ_i is the ith elementary symmetric function.

The only essential difference between the complex and real cases is that for the complex Grassmann manifolds, all Schubert cells have even dimension, and therefore the boundary homomorphism for integer coefficients is zero for obvious reasons.

The Chern characteristic classes, as well as the Stiefel–Whitney classes, can be defined axiomatically; the corresponding reformulation of the corollary to Theorem 3.32 is fairly obvious.

The expression for the Chern class $c(\xi^m \otimes \eta^n)$ in terms of the classes $c(\xi^m)$ and $c(\eta^n)$ is precisely the same as for the Stiefel–Whitney classes (Theorem 3.33). The only essential difference in the proof is that the following assertion is needed in the case of Chern classes.

Lemma. *If ω_1^1 and ω_2^1 are one-dimensional complex bundles over the same base, then $c_1(\omega_1^1 \otimes \omega_2^1) = c_1(\omega_1^1) + c_1(\omega_2^1)$.*

Proof. For a one-dimensional bundle, the class c_1 is the obstruction to extending a nonzero section to the 2-skeleton. The value of the cocycle representing this class for a 2-cell is equal to the sum of the indices of the singular points of the section (it is assumed that all singular points are nondegenerate). Choose generic sections (that is, sections with pairwise different zeros) of the bundles ω_1^1 and ω_2^1. The tensor product of these sections vanishes at all points at which one of the sections vanishes. We show that the zeros of the section and those of the tensor product of sections have the same sign over each singular point.

It is sufficient to consider the case where one of the sections is constant and the other is determined by a linear map. Suppose that, in some basis, the section of the bundle ω_1^1 is given by $\left(\begin{smallmatrix} e_1 \\ e_2 \end{smallmatrix}\right) \mapsto \left(\begin{smallmatrix} a_{11}+a_{12}i \\ a_{21}+a_{22}i \end{smallmatrix}\right)$; its value at any vector $xe_1 + ye_2$ is determined by linearity. The section of the bundle ω_2^1 is constant; suppose that it takes all vectors to $\lambda + \mu i$. Then the section of the bundle $\omega_1^1 \otimes \omega_2^1$ is determined by

$$\begin{pmatrix} e_1 \\ e_2 \end{pmatrix} \mapsto \begin{pmatrix} (a_{11}+a_{12}i)\otimes(\lambda+\mu i) \\ (a_{21}+a_{22}i)\otimes(\lambda+\mu i) \end{pmatrix} = \begin{pmatrix} \lambda a_{11} - \mu a_{12} + (\lambda a_{12} + \mu a_{11})i \\ \lambda a_{21} - \mu a_{22} + (\lambda a_{22} + \mu a_{21})i \end{pmatrix}.$$

The determinant of the matrix of this map is equal to $(\lambda^2 + \mu^2)(a_{11}a_{22} - a_{12}a_{21})$. Therefore, the singular point of the section of ω_1^1 has the same index as that of the section of $\omega_1^1 \otimes \omega_2^1$. □

Similarly to real bundles, it can be proved that the complex k-dimensional vector bundles over any compact Hausdorff space B are in one-to-one correspondence with the homotopy classes of the maps $[B, G_{\mathbb{C}}(\infty, k)]$, where $G_{\mathbb{C}}(\infty, k)$ is the infinite complex Grassmann manifold (it is defined by analogy with the real case). For one-dimensional bundles, these two classification theorems (real and complex) can be substantially improved by using the fact that $G(\infty, 1) = \mathbb{R}P^\infty$ and $G_{\mathbb{C}}(\infty, 1) = \mathbb{C}P^\infty$ are $K(\pi, n)$ spaces; namely, $\mathbb{R}P^\infty = K(\mathbb{Z}_2, 1)$ and $\mathbb{C}P^\infty = K(\mathbb{Z}, 2)$. Therefore, according to Theorem 3.10 (see p. 125), if X is a finite simplicial complex, then the elements of the sets $[X, \mathbb{R}P^\infty]$ and $[X, \mathbb{C}P^\infty]$ are in one-to-one correspondence with those of the groups $H^1(X; \mathbb{Z}_2)$ and $H^2(X; \mathbb{Z})$. On the other hand, the elements of the sets $[X, \mathbb{R}P^\infty]$ and $[X, \mathbb{C}P^\infty]$ are in one-to-one correspondence with the one-dimensional real and complex bundles over X considered up to isomorphism.

This correspondence can be described explicitly by using characteristic classes. Namely, suppose that a bundle ξ is determined by a map $f\colon X \to \mathbb{R}P^\infty$ or $f\colon X \to \mathbb{C}P^\infty$. Theorem 3.10 assigns to the map f the element $f^*(F_\pi) \in H^n(X; \pi)$; here $n = 1$ and $\pi = \mathbb{Z}_2$ (in the real case) or $n = 2$ and $\pi = \mathbb{Z}$ (in the complex case). Moreover, the cohomology class F_π is a generator of the group $H^1(\mathbb{R}P^\infty; \mathbb{Z}_2)$ or $H^2(\mathbb{C}P^\infty; \mathbb{Z})$, which is seen from its description. Thus, the one-to-one correspondence between the one-dimensional bundles and the elements of $H^1(\mathbb{R}P^\infty; \mathbb{Z}_2)$ or $H^2(\mathbb{C}P^\infty; \mathbb{Z})$ is defined by $\xi \mapsto w_1(\xi)$ or $\xi \mapsto c_1(\xi)$, respectively, and the addition of cohomology classes corresponds to tensor multiplication of the bundles. As a result, we obtain the following theorem.

Theorem 3.39. *Let X be a finite-dimensional simplicial complex.*

(a) *One-dimensional real bundles ξ and η over X are isomorphic if and only if $w_1(\xi) = w_1(\eta)$; moreover, for each element $\alpha \in H^1(X; \mathbb{Z}_2)$, there exists a one-dimensional real bundle ξ over X such that $w_1(\xi) = \alpha$.*

(b) *One-dimensional complex bundles ξ and η over X are isomorphic if and only if $c_1(\xi) = c_1(\eta)$; moreover, for each element $\alpha \in H^2(X; \mathbb{Z})$, there exists a one-dimensional complex bundle ξ over X such that $c_1(\xi) = \alpha$.*

Note that a one-dimensional complex bundle is the same object as an oriented real two-dimensional bundle with a fixed Riemannian metric. Indeed, to specify a complex structure on the plane $\mathbb{R}^2$, we must define multiplication by i, i.e., rotation through 90°. It is also clear that the Euler class of an oriented two-dimensional real bundle coincides with the first Chern class of the corresponding one-dimensional complex bundle. Recall that the Euler class of a bundle changes sign under the change of orientation. Theorem 3.39 (b) implies the following assertion.

Theorem 3.40. *Oriented two-dimensional real bundles ξ and η over a finite simplicial complex X are isomorphic if and only if $e(\xi) = e(\eta)$; moreover, for any element $\alpha \in H^2(X; \mathbb{Z})$, there exists an oriented two dimensional real bundle ξ such that $e(\xi) = \alpha$.*

2.9. Splitting Maps. Let ξ be a vector bundle over a base B. A continuous map $f \colon B_1 \to B$ is said to be *splitting* for ξ if $f^*(\xi)$ is a direct sum of one-dimensional bundles and $f^* \colon H^*(B) \to H^*(B_1)$ is a monomorphism. Using splitting maps makes it possible to reduce the proofs of many assertions about characteristic classes to the case of direct sums of one-dimensional bundles. To prove the existence of splitting maps, we need the following construction involving vector bundles.

To each vector bundle ξ over a base B we can assign its *projectivization* $P\xi$, which is a locally trivial bundle over B. The fiber of the bundle $P\xi$ over a point $b \in B$ is obtained from the fiber V of ξ by identifying (in $V \setminus \{0\}$) all points of a straight line passing through zero. Thus, the fiber of the bundle $P\xi$ coincides with $\mathbb{R}P^{n-1}$ or $\mathbb{C}P^{n-1}$, where $n = \dim \xi$.

Problem 110. Let $E(P\gamma_n^k) \to G(n,k)$ be the projectivization of the canonical bundle γ_n^k over $G(n,k)$, and let $E(P\bar{\gamma}_n^{k-1}) \to G(n, k-1)$ be the projectivization of the orthogonal complement of the canonical bundle γ_n^{k-1} over $G(n, k-1)$. Prove that $E(P\gamma_n^k) \approx E(P\bar{\gamma}_n^{k-1})$.

Suppose that a bundle $P\xi$ is determined by a map $q \colon E(P\xi) \to B$. Consider the induced vector bundle $q^*(\xi)$ over $E(P\xi)$. The points of the space $E(P\xi)$ are the lines l passing through zero. Therefore, $q^*(\xi)$ has a one-dimensional subbundle λ_ξ for which the total space consists of all pairs (v, l), where $v \in l$. If the base B is compact, then the bundle $q^*(\xi)$

admits a Riemannian (Hermitian in the complex case) metric; therefore, $q^*(\xi) \cong \lambda_\xi \oplus \sigma_\xi$, where σ_ξ is the orthogonal complement of λ_ξ.

The bundle λ_ξ, as well as any other one-dimensional bundle, is obtained from the canonical bundle γ^1 over $\mathbb{R}P^\infty$ (or $\mathbb{C}P^\infty$); i.e., there exists a map $f\colon E(\xi) \to \mathbb{R}P^\infty$ for which $f^*(\gamma^1) \cong \lambda_\xi$. Let α be the generator of the group $H^1(\mathbb{R}P^\infty; \mathbb{Z}_2)$ (in the real case) or $H^1(\mathbb{C}P^\infty; \mathbb{Z})$ (in the complex case). Consider the element $a_\xi = f^*(\alpha) \in H^*(E(P\xi))$; it depends only on the bundle ξ because if two maps $E(P\xi) \to \mathbb{R}P^\infty$ induce isomorphic bundles when applied to γ^1, then these maps are homotopic.

Theorem 3.41 (Leray–Hirsh). *Let ξ be a real or complex vector bundle of dimension n over a compact base $B(\xi)$.*

(a) *In the real case, the linear space $\langle 1, a_\xi, \dots, a_\xi^{n-1} \rangle$ generated by the elements $1, a_\xi, a_\xi^2, \dots, a_\xi^{n-1}$ has dimension n, and*

$$H^*(E(P\xi); \mathbb{Z}_2) \cong H^*(B(\xi); \mathbb{Z}_2) \otimes_{\mathbb{Z}_2} \langle 1, a_\xi, \dots, a_\xi^{n-1} \rangle.$$

The space $H^*(B(\xi); \mathbb{Z}_2)$ can be identified with the image of the homomorphism $q^*\colon H^*(B(\xi); \mathbb{Z}_2) \to H^*(E(P\xi); \mathbb{Z}_2)$ so that each element $\beta \otimes a_\xi^i$ is identified with $\beta \smile a_\xi^i$.

(b) *In the complex case, the free Abelian group $\langle 1, a_\xi, \dots, a_\xi^{n-1} \rangle$ generated by the elements $1, a_\xi, a_\xi^2, \dots, a_\xi^{n-1}$ in $H^*(E(P\xi); \mathbb{Z})$ has rank n, and*

$$H^*(E(P\xi); \mathbb{Z}) \cong H^*(B(\xi); \mathbb{Z}) \otimes \langle 1, a_\xi, \dots, a_\xi^{n-1} \rangle.$$

This decomposition is also compatible with the homomorphism q^*.

Proof. For each point $b \in B$, consider the map $i_b\colon \mathbb{R}P^{n-1} \to E(P\xi)$ that is the inclusion of the fiber in the total space of the bundle (in the complex case, $\mathbb{R}P^{n-1}$ is replaced by $\mathbb{C}P^{n-1}$). It follows directly from the definitions that $i_b^*(\lambda_\xi)$ is the canonical bundle γ_{n-1}^1 over $\mathbb{R}P^{n-1}$ or $\mathbb{C}P^{n-1}$. The bundle $i_b^*(f^*(\gamma^1))$ coincides with γ_{n-1}^1; therefore, the map fi_b is homotopic to the natural embedding $\mathbb{R}P^{n-1} \to \mathbb{R}P^\infty$ or $\mathbb{C}P^{n-1} \to \mathbb{C}P^\infty$. This implies the independence of the elements $i_b^*(1), \dots, i_b^*(a_\xi^{n-1})$.

The required isomorphisms are proved as follows. Let K be a subcomplex in B over which the bundle $P\xi$ is trivial (for example, a simplex), and let $E_K = q^{-1}(K)$. Then there is a homeomorphism $K \times F \to E_K$ compatible with the projection q. An isomorphism $H^*(E_K) \cong H^*(K) \otimes H^*(F)$ compatible with q exists by the Künneth theorem. In the real case, $H^*(K) \otimes H^*(F)$ is understood as the tensor product of vector spaces over $\mathbb{Z}_2$. In the complex case, the proof uses the freeness of the group $H^*(F) = \langle 1, a_\xi, \dots, a_\xi^{n-1} \rangle$.

Formally, an isomorphism between $H^*(E_K)$ and $H^*(K) \otimes H^*(F)$ compatible with the projection can be written as follows. Consider the homomorphism φ_K^m defined by $\sum \beta^k \mapsto \sum q^*(\beta^k) \smile a_\xi^l$, where $\beta^k \in H^k(K)$ and l is such that $k+l=m$ (in the real case) or $k+2l=m$ (in the complex case). In the real case, the inequalities $0 \le m-k \le n-1$ must hold; in the complex case, the inequalities $0 \le m-k \le 2n-2$ must hold and the number $m-k$ must be even. Consider the group $\mathcal{H}^m(K)$, which coincides with $\bigoplus_{k=1-n+m}^{m} H^k(K)$ in the real case and with $\bigoplus_{\substack{k=2-2n+m \\ m-k \text{ even}}}^{m} H^k(K)$ in the complex case. The map $\varphi_K^m \colon \mathcal{H}^m(K) \to H^m(E_K)$ is an isomorphism for all m.

The group $\mathcal{H}^m(L)$ and the homomorphism φ_L^m can be defined for any subcomplex $L \subset B$. We must prove that φ_B^m is an isomorphism for any m. To do this, we use the Mayer–Vietoris exact sequence and the five lemma. The Mayer–Vietoris exact sequence holds for the groups $\mathcal{H}^*$ because any direct sum of exact sequences is an exact sequence. The commutative diagram

$$\begin{array}{ccccc}
\mathcal{H}^{m+1}(K\cap L) & \longrightarrow & \mathcal{H}^m(K)\oplus\mathcal{H}^m(L) & \longrightarrow & \mathcal{H}^m(K\cup L) \\
\Big\downarrow{\scriptstyle \varphi^{m+1}_{K\cap L}} & & \Big\downarrow{\scriptstyle \varphi^m_K\oplus\varphi^m_L} & & \Big\downarrow{\scriptstyle \varphi^m_{K\cup L}} \\
H^{m+1}(E_{K\cap L}) & \longrightarrow & H^m(E_K)\oplus H^m(E_L) & \longrightarrow & H^m(E_{K\cup L})
\end{array}$$

$$\begin{array}{ccccc}
\longrightarrow & \mathcal{H}^m(K\cap L) & \longrightarrow & \mathcal{H}^{m-1}(K)\oplus\mathcal{H}^{m-1}(L) \\
 & \Big\downarrow{\scriptstyle \varphi^m_{K\cap L}} & & \Big\downarrow{\scriptstyle \varphi^{m-1}_K\oplus\varphi^{m-1}_L} \\
\longrightarrow & H^m(E_{K\cap L}) & \longrightarrow & H^{m-1}(E_K)\oplus H^{m-1}(E_L)
\end{array}$$

shows that if φ_K, φ_L, and $\varphi_{K\cap L}$ are isomorphisms, then so is $\varphi_{K\cup L}$. This implies the required assertion. □

The existence of splitting maps follows easily from Theorem 3.41.

Theorem 3.42. *For any vector bundle ξ over a compact simplicial complex B, there exists a splitting map $f \colon B_1 \to B$.*

Proof. We prove the theorem by induction on $n = \dim \xi$. For $n = 1$, the map id_B is splitting. Suppose that for any vector bundle of dimension less than n, a splitting map exists. According to Theorem 3.41, the map $q^* \colon H^*(B) \to H^*(E(P\xi))$ is a monomorphism. Moreover, $q^*(\xi) = \lambda_\xi \oplus \sigma_\xi$, where λ_ξ is a one-dimensional bundle. By the induction hypothesis, for σ_ξ, a splitting map $g \colon B_1 \to E(P\xi)$ exists. We show that the map $f = qg$ is splitting as well. Indeed, $f^* = g^*q^*$ is a monomorphism because g^* and q^* are monomorphisms. Moreover, $f^*(\xi) = g^*(\lambda_\xi) \oplus g^*(\sigma_\xi)$, and, by assumption, $g^*(\sigma_\xi)$ is a direct sum of one-dimensional bundles. □

To illustrate applications of splitting maps, we give a new proof of an assertion which we have proved earlier by a different method. Namely, we show that the existence of splitting maps implies the uniqueness of Stiefel–Whitney and Chern classes (under the axiomatic approach; see p. 159). Indeed, for a one-dimensional bundle, the Stiefel–Whitney or Chern class is uniquely determined by the naturality condition, the condition on $w_1(\gamma^1)$, and the universality of the bundle γ^1. Let $f\colon B_1 \to B$ be a splitting map for the bundle ξ. Then $f^*(\xi) = \xi_1 \oplus \cdots \oplus \xi_n$, where the bundles $\xi_1, \ldots, \xi_n$ are one-dimensional. Therefore, the class $f^*(w(\xi)) = \prod(1 + w_1(\xi_i))$ is determined uniquely. Since the map f^* is a monomorphism, the class $w(\xi)$ is unique.

We have already discussed two approaches to the Stiefel–Whitney and Chern characteristic classes, namely, treating these classes as obstructions to extending sections and the axiomatic approach. Theorem 3.41 suggests yet another construction of the characteristic classes. According to this theorem, the element a_ξ^n can be uniquely written as $\sum_{i=1}^n (-1)^{i+1} x_i(\xi) a_\xi^{n-i}$, where each $x_i(\xi)$ belongs to $H^i(B;\mathbb{Z}_2)$ (in the real case) or to $H^{2i}(B;\mathbb{Z})$ (in the complex case). The Stiefel–Whitney and Chern characteristic classes can be constructed using the following theorem.

Theorem 3.43. *If $x_0(\xi) = 1$ and $x_i(\xi) = 0$ for $i > \dim \xi$, then $x_i(\xi) = w_i(\xi)$ in the real case and $x_i(\xi) = c_i(\xi)$ in the complex case.*

It is most convenient for us to prove Theorem 3.43 using singular cohomology, so we postpone the proof until p. 252.

Any vector space V over $\mathbb{C}$ can be associated with a vector space $\overline{V}$ over $\mathbb{C}$. Each vector $v \in V$ corresponds to $\overline{v} \in \overline{V}$, and multiplication by complex numbers in $\overline{V}$ is defined by $\lambda\overline{v} = \overline{\overline{\lambda} v}$; i.e., multiplication by λ is replaced with multiplication by $\overline{\lambda}$. Applying this operation to every fiber of a complex vector bundle ω, we obtain a bundle $\overline{\omega}$, which is said to be *conjugate* to ω.

Example 48. The tangent bundle τ^1 of $\mathbb{C}P^1$ is not isomorphic to $\overline{\tau^1}$.

Proof. Let $f\colon V \to \overline{V}$ be an isomorphism over $\mathbb{C}$. If $f(v) = \overline{w}$, then $f(\lambda v) = \lambda\overline{w} = \overline{\overline{\lambda} w}$. This means that if we identify corresponding elements of the spaces V and $\overline{V}$, then $f(\lambda v) = \overline{\lambda} f(v)$. If $\dim_{\mathbb{C}} V = 1$, then f reverses the orientation of $V_{\mathbb{R}}$. Thus, geometrically, f is the symmetry with respect to some straight line; this line uniquely determines the isomorphism f.

The realification of the bundle τ^1 is the tangent bundle of the sphere S^2. Therefore, if the bundles τ^1 and $\overline{\tau^1}$ were isomorphic, then we would obtain a continuous line-element field on S^2, which does not exist. Indeed, a closed manifold admits a continuous line-element field if and only if it admits a continuous vector field without singular points (see Problem 84 in Part I). □

The conjugate bundle $\overline{\omega}$ is closely related to the *dual* bundle $\omega^* = \mathrm{Hom}_{\mathbb{C}}(\omega, \mathbb{C})$. If a space V over $\mathbb{C}$ is endowed with a Hermitian inner product for which $(\lambda v, \mu w) = \lambda\overline{\mu}(v, w)$, then the formula $\overline{v} \mapsto \varphi(x) = (x, v)$ defines an isomorphism $\overline{V} \to \mathrm{Hom}_{\mathbb{C}}(V, \mathbb{C})$. Therefore, in the presence of a Hermitian metric on the bundle ω, the conjugate bundle $\overline{\omega}$ is canonically isomorphic to the dual bundle $\mathrm{Hom}_{\mathbb{C}}(\omega, \mathbb{C})$.

Theorem 3.44. *The bundle $\omega^* \otimes \omega$ has a nonzero section. In particular, if $\dim_{\mathbb{C}} \omega = 1$, then the bundle is trivial.*

Proof. There is a canonical isomorphism $V^* \otimes V \to \mathrm{Hom}(V, V)$ (see, for example, [**104**, p. 120]). The space $\mathrm{Hom}(V, V)$ has a distinguished element, the identity map. Hence the bundle $\omega^* \otimes \omega$ has a nonzero section. □

Theorem 3.45. $c_k(\overline{\omega}) = (-1)^k c_k(\omega)$.

Proof. If $\dim \omega = 1$, then $c_1(\omega) = e(\omega_{\mathbb{R}})$ and $c_1(\overline{\omega}) = e(\overline{\omega}_{\mathbb{R}})$. The bundles $\omega_{\mathbb{R}}$ and $\overline{\omega}_{\mathbb{R}}$ have opposite orientations; therefore, $c_1(\overline{\omega}) = -c_1(\omega)$.

Suppose that ω is a complex vector bundle of dimension n and f is a splitting map for ω. Then $f^*(\omega) = \omega_1 \oplus \cdots \oplus \omega_n$; hence

$$\begin{aligned} f^*(c(\overline{\omega})) &= c(f^*(\overline{\omega})) = c(\overline{\omega}_1) \cdots c(\overline{\omega}_n) \\ &= (1 - c_1(\overline{\omega}_1)) \cdots (1 - c_1(\overline{\omega}_n)) \\ &= \sum (-1)^k c_k(f^*(\overline{\omega})) = f^*\Big(\sum (-1)^k c_k(\overline{\omega})\Big). \end{aligned}$$

This implies the required assertion because f^* is a monomorphism. □

Corollary. *If $\overline{\omega} \cong \omega$, then $2c_{2k+1}(\omega) = 0$.*

Proof. If $\overline{\omega} \cong \omega$, we have $c_{2k+1}(\omega) = c_{2k+1}(\overline{\omega}) = (-1)^{2k+1} c_{2k+1}(\omega) = -c_{2k+1}(\omega)$. □

The *Chern classes of a complex manifold* are defined as the Chern classes of its tangent bundles.

The calculation of the Chern classes of complex projective spaces is similar to that of the Stiefel–Whitney classes of real projective spaces (see the corollary to Theorem 3.23).

Example 49. We have $c(\mathbb{C}P^n) = (1 + \alpha)^{n+1}$, where $\alpha = -c_1(\gamma_n^1)$ is a generator of the group $H^2(\mathbb{C}P^n; \mathbb{Z})$.

Proof. The fiber of the bundle γ_n^1 over a point $l \in \mathbb{C}P^n$ is the line l in $\mathbb{C}^{n+1}$. Let ω^n be the bundle over $\mathbb{C}P^n$ whose fiber over the point l is the orthogonal complement $l^\perp$ of the line l in $\mathbb{C}^{n+1}$ with respect to Hermitian product. We claim that $\tau_{\mathbb{C}P^n} \cong \mathrm{Hom}_{\mathbb{C}}(\gamma_n^1, \omega^n)$. Indeed, each tangent vector to $\mathbb{C}P^n$ is determined by a pair (x, v), where $x \in l \setminus \{0\}$ and v is a vector orthogonal to x; the pairs $(\lambda x, \lambda v)$ with $\lambda \in \mathbb{C} \setminus \{0\}$ determine the same tangent vector. Therefore, each tangent vector is uniquely determined by the linear map $l \to l^\perp$ taking x to v.

The bundle $\mathrm{Hom}_{\mathbb{C}}(\gamma_n^1, \gamma_n^1)$ is trivial because it has a zero section, which corresponds to the identity map. Let ε^k be a trivial k-dimensional bundle over $\mathbb{C}P^n$. Then $\tau_{\mathbb{C}P^n} \oplus \varepsilon^1 \cong \mathrm{Hom}_{\mathbb{C}}(\gamma_n^1, \omega^n \oplus \gamma_n^1) \cong \mathrm{Hom}_{\mathbb{C}}(\gamma_n^1, \varepsilon^{n+1})$; therefore, the bundle $\tau_{\mathbb{C}P^n} \oplus \varepsilon^1$ is isomorphic to the direct sum of $n+1$ copies of the bundle $\mathrm{Hom}_{\mathbb{C}}(\gamma_n^1, \varepsilon^1) \cong \overline{\gamma_n^1}$. According to Theorem 3.45, we have $c_1(\overline{\gamma_n^1}) = -c_1(\gamma_n^1)$. Therefore, Theorems 3.37 and 3.36 imply $c(\mathbb{C}P^n) = (1 - c_1(\gamma_n^1))^{n+1} = (1+\alpha)^{n+1}$. □

To each vector space V over $\mathbb{R}$ we can assign its *complexification* $V \otimes_{\mathbb{R}} \mathbb{C}$. As a space over $\mathbb{R}$, it is canonically isomorphic to the space consisting of the vectors $v + iw$, where $v, w \in V$. Multiplication by i is defined as $i(v+iw) = iv - w$.

If ξ is a real bundle, then we can complexify each of its fibers. We denote the bundle thus obtained by $\xi \otimes \mathbb{C}$.

Lemma. *The bundle $\xi \otimes \mathbb{C}$ is isomorphic to the conjugate bundle $\overline{\xi \otimes \mathbb{C}}$.*

Proof. Consider the map f given by $f(v + iw) = v - iw$. It is easy to verify that $f(i(v+iw)) = -if(v+iw)$. Therefore, the map f determines a fiberwise isomorphism of the bundles $\xi \otimes \mathbb{C}$ and $\overline{\xi \otimes \mathbb{C}}$. □

Corollary. $2c_{2k+1}(\xi \otimes \mathbb{C}) = 0$.

The *Pontryagin characteristic classes* of an arbitrary real bundle ξ are defined by

$$p_k(\xi) = (-1)^k c_{2k}(\xi \otimes \mathbb{C}) \in H^{4k}(B; \mathbb{Z}).$$

The elements of order 2 (namely, $c_{2k+1}(\xi \otimes \mathbb{C})$) are ignored. The Pontryagin classes have the naturality property; the Pontryagin classes of stably equivalent bundles coincide. However, the equality $p(\xi \oplus \eta) = p(\xi)p(\eta)$ does not hold because we have discarded the elements of order 2. We can assert only that the relation given in the following theorem is valid.

Theorem 3.46. *The equality $2(p(\xi \oplus \eta) - p(\xi)p(\eta)) = 0$ holds.*

Proof. The bundle $(\xi \oplus \eta) \otimes \mathbb{C}$ is isomorphic $(\xi \otimes \mathbb{C}) \oplus (\eta \otimes \mathbb{C})$. Therefore,

$$c_k((\xi \oplus \eta) \otimes \mathbb{C}) = \sum_{i+j=k} c_i(\xi \otimes \mathbb{C}) c_j(\eta \otimes \mathbb{C}).$$

Discarding the odd-dimensional Chern classes yields

$$2c_{2k}((\xi \oplus \eta) \otimes \mathbb{C}) = 2 \sum_{i+j=k} c_{2i}(\xi \otimes \mathbb{C})c_{2j}(\eta \otimes \mathbb{C}).$$

The required equality is obtained by multiplying both sides by $(-1)^k = (-1)^i(-1)^j$. □

Theorem 3.47. *Suppose that ω is a complex n-dimensional vector bundle, $p_k = p_k(\omega_{\mathbb{R}})$, and $c_k = c_k(\omega)$. Then*

$$\sum_{k=0}^{n}(-1)^k p_k = \left(\sum_{k=0}^{n}(-1)^k c_k\right) \sum_{k=0}^{n} c_k.$$

Proof. By definition, $(-1)^k p_k = c_{2k}(\omega_{\mathbb{R}} \otimes \mathbb{C})$. Therefore, is suffices to verify that $\omega_{\mathbb{R}} \otimes \mathbb{C} \cong \omega \oplus \overline{\omega}$.

Let W be a complex space and $V = W_{\mathbb{R}}$. The space $V \otimes \mathbb{C}$ is canonically isomorphic to $V \oplus V$ under the operation $J(x, y) = (-y, x)$, which corresponds to multiplication by i. In the space W, multiplication by i is given. Thus, the two maps $f_\pm(x) = (x, \mp ix)$ from V to $V \oplus V$ are defined. It is easy to verify that $f_\pm(ix) = \pm J(f_\pm(x))$, i.e., f_+ is complex-linear and f_- is antilinear. Moreover, $V \oplus V = \operatorname{Im} f_+ \oplus \operatorname{Im} f_-$ because $(x, y) = f_+\left(\frac{x+iy}{2}\right) + f_-\left(\frac{x-iy}{2}\right)$. The space $\operatorname{Im} f_+$ is canonically isomorphic to W, and $\operatorname{Im} f_-$ is canonically isomorphic to $\overline{w}$. Having constructed such canonical decompositions for all fibers of the bundle $\omega_{\mathbb{R}} \otimes \mathbb{C}$, we obtain the required isomorphism. □

Corollary. *We have $p(\mathbb{C}P^n) = (1 + \alpha^2)^{n+1}$, where α is a generator of the group $H^2(\mathbb{C}P^n; \mathbb{Z})$.*

Proof. Let τ be the tangent bundle of $\mathbb{C}P^n$ treated as a complex manifold, and $\tau_{\mathbb{R}}$ its realification, i.e., the tangent bundle of $\mathbb{C}P^n$ as a real manifold. We already know that $c(\tau) = (1 + \alpha)^{n+1} = \sum c_k$. Therefore, $\sum(-1)^k c_k = (1 - \alpha)^{n+1}$, whence $\sum(-1)^k p_k(\tau_{\mathbb{R}}) = ((1 - \alpha)(1 + \alpha))^{n+1} = (1 - \alpha^2)^{n+1}$. It follows that $\sum p_k(\tau_{\mathbb{R}}) = (1 + \alpha^2)^{n+1}$. □

3. Group Actions

3.1. Simplicial Actions. A topological space X with an action of a group G is called a *G-space*. A G-space map $f: X \to Y$ is said to be *equivariant* if $f(g(x)) = g(f(x))$ for all $g \in G$ and all $x \in X$.

Let G be a finite group acting on the space $|K|$, where K is a simplicial complex. This action is said to be *simplicial* if the map $g: |K| \to |K|$ is simplicial for any $g \in G$. A simplicial complex K with a fixed simplicial action of a finite group G is called a *simplicial G-complex*.

Any simplicial map of simplicial complexes induces a simplicial map of the barycentric subdivisions of these complexes. Therefore, the barycentric subdivision of a simplicial G-complex is a simplicial G-complex as well.

For a simplicial G-complex K, the simplicial complex K/G is defined. Its vertices are the orbits of the action of G on the vertex set of K; i.e., each vertex of K/G has the form $v^* = G(v)$, where v is a vertex of K. Vertices $v_0^*, \dots, v_n^*$ are the vertices of a simplex in K/G if and only if the orbits $v_0^*, \dots, v_n^*$ contain vertices $v_0, \dots, v_n$ that are the vertices of a simplex in K. In this case, we say that $[v_0, \dots, v_n]$ is a simplex *over* the simplex $[v_0^*, \dots, v_n^*]$. Note that other vertices in the orbits $v_0^*, \dots, v_n^*$ do not necessarily determine a simplex.

Assigning to each vertex v the orbit v^*, we obtain a simplicial map $K \to K/G$ and, therefore, a continuous map $|K| \to |K/G|$. The group G acts on the topological space $|K|$: an element $g \in G$ takes each point $\sum \lambda_i v_i$ to the point $\sum \lambda_i g v_i$ (we use the assumption that g takes simplices to simplices). Thus, the orbit space $|K|/G$ is defined. Assigning the point $\sum \lambda_i G(v_i) = \sum \lambda_i v_i^*$ to each orbit $G(\sum \lambda_i v_i)$, we obtain a map $|K|/G \to |K/G|$. This map is one-to-one if and only if for every two simplices $[v_0, \dots, v_n]$ and $[g_0 v_0, \dots, g_n v_n]$ in K, there exists an element $g \in G$ such that $g v_i = g_i v_i$ for $i = 0, 1, \dots, n$.

Many theorems about simplicial G-complexes can be proved only under additional conditions on the G-complexes. But, as rule, these conditions can be fulfilled by refining the triangulation (e.g., by passing to the first or second barycentric subdivision). They include the properties (R) and (A) specified below.

A simplicial G-complex K is said to be *regular* if any subgroup $H \subset G$ has the following *property* (R):

Suppose that $h_0, \dots, h_n \in H$ and $[v_0, \dots, v_n]$ and $[h_0 v_0, \dots, h_n v_n]$ are two simplices in K. Then there exists an element $h \in H$ such that $h v_i = h_i v_i$ for $i = 0, 1, \dots, n$.

We say that a simplicial G-complex has *property* (A) if it satisfies any of the two conditions specified in the following lemma.

Lemma. *For a simplicial G-complex K, the following conditions are equivalent.*

(i) *For each element $g \in G$ and every simplex Δ from K, all points of the set $\Delta \cap g\Delta$ are fixed under the action of g.*

(ii) *If v and gv are vertices of the same simplex, then $v = gv$.*

Proof. Suppose that (i) holds. Let v and gv be vertices of the same simplex Δ. Then $g^{-1}v$ and v are vertices of the simplex $g^{-1}\Delta$; therefore, they are joined by an edge Δ^1 (or coincide). Clearly, $v \in \Delta^1 \cap g\Delta^1$, whence $gv = v$.

Suppose that (ii) holds. Let $v \in \Delta \cap g\Delta$. Then the vertices v and gv belong to the same simplex Δ, and therefore $v = gv$. □

Theorem 3.48. *The second barycentric subdivision of any simplicial G-complex is a regular G-complex.*

Proof. First, we prove that the barycentric subdivision K' of a simplicial G-complex K has property (A).

Each vertex v of the complex K' determines a simplex $\Delta(v)$ in K for which this point is interior. Moreover, vertices v_1 and v_2 belong to the same simplex of K' if and only if one of the simplices $\Delta(v_1)$ and $\Delta(v_2)$ is a face of the other. We have $\Delta(gv) = g\Delta(v)$; in particular, the simplices $\Delta(v)$ and $\Delta(gv)$ have the same dimension. Therefore, if vertices v and gv belong to the same simplex of K', then one of the simplices $\Delta(v)$ and $\Delta(gv)$ is a face of the other, and the dimensions of these simplices are equal. Therefore, $\Delta(v) = \Delta(gv)$, and $v = gv$.

Now we prove that if a simplicial G-complex K has property (A), then its barycentric subdivision K' has property (R).

The proof is by induction on n. Suppose that $[v_0, \dots, v_n]$ and $[h_0v_0, \dots, h_nv_n]$ are simplices of K' and all $(n-1)$-simplices of K', where $n \geq 1$, have property (R) (zero-dimensional simplices always have property (R)). Renumbering the vertices if necessary, we can assume that $\Delta(v_0) \subset \Delta(v_1) \subset \dots \subset \Delta(v_n)$, i.e., v_0 is a vertex of a simplex of K, v_1 is the midpoint of an edge, and so on. By the induction hypothesis, there exists an element $h \in H$ such that $hv_i = h_iv_i$ for $i = 0, 1, \dots, n-1$. Let us show that $h_nv_i = h_iv_i$ for all i. This equality is obvious for $i = n$; it remains to verify it for $i = 0, 1, \dots, n-1$. Under the action of h^{-1}, the simplex $[h_0v_0, \dots, h_nv_n]$ is mapped to $[v_0, \dots, v_{n-1}, h^{-1}h_nv_n]$. Moreover,

$$\Delta(v_0) \subset \Delta(v_1) \subset \dots \subset \Delta(v_{n-1}) \subset \Delta(h^{-1}h_nv_n) = h^{-1}h_n\Delta(v_n).$$

Therefore, $\Delta(v_{n-1}) \subset \Delta(v_n) \cap h^{-1}h_n\Delta(v_n)$. Property (A) implies that all points of the $(n-1)$-simplex $\Delta(v_{n-1})$ are fixed under the action of $h^{-1}h_n$. Hence $h^{-1}h_nv_i = v_i$ for $i = 0, 1, \dots, n-1$. Therefore, $h_nv_i = hv_i = h_iv_i$, as required. □

Thus, after refining the triangulation of the complex K, we can identify the sets $|K|/G$ and $|K/G|$. The natural topologies on the identified sets coincide, i.e., the map $|K|/G \to |K/G|$ is a homeomorphism. Indeed, since the simplicial map $K \to K/G$ is surjective, it follows that a subset of $|K/G|$ is open if and only if its preimage is open in $|K|$. On the other hand, a subset of the orbit space $|K|/G$ is open if and only if its preimage is open in $|K|$. It remains to note that the composition $|K| \to |K|/G \to |K/G|$ coincides with the map $|K| \to |K/G|$.

3.2. Equivariant Simplicial Approximation. We say that a simplicial G-complex L has *property* (I) if for an arbitrary element $g \in G$ and an arbitrary simplex Δ in L, the equality $g(\Delta) = \Delta$ implies that the restriction of the action of g to Δ is the identity map.

The barycentric subdivision K' of any simplicial G-complex K has property (I). This is a general property of simplicial maps (see Part I, Theorem 3.14).

Theorem 3.49. *Suppose that K and L are simplicial G-complexes and L has property* (I). *Then for any continuous map $f\colon |K| \to |L|$, there exists an equivariant simplicial approximation $\varphi\colon K^{(n)} \to L$.*

Proof. Let $\psi\colon K^{(n)} \to L$ be a simplicial approximation of the map f. We define an equivariant simplicial map $\varphi\colon K^{(n)} \to L$ as follows. For each vertex $v \in K^{(n)}$, consider its orbit $\{g(v) \mid g \in G\}$. The orbits of two vertices either coincide or are disjoint. In each orbit, we choose some vertex v and put $\varphi(v) = \psi(v)$; for all other vertices from the same orbit, we set $\varphi(g(v)) = g(\varphi(v)) = g(\psi(v))$. First, we must verify that the map φ is well defined, i.e., $g_1(\psi(v)) = g_2(\psi(v))$ whenever $g_1(v) = g_2(v)$, or, equivalently, $g(\psi(v)) = \psi(v)$ whenever $g(v) = v$; here $g = g_1^{-1}g_2$. To show this, we use property (I). Let Δ be the simplex of L that contains $f(v)$ as an interior point. The point $g(f(v)) = f(g(v)) = f(v)$ is also an interior point of the simplex Δ; therefore, $g(\Delta) = \Delta$. According to property (I), we have $g|_\Delta = \mathrm{id}_\Delta$. But ψ is a simplicial approximation of f; hence $\psi(v) \in \Delta$, which means that $g(\psi(v)) = \psi(v)$.

It remains to show that φ is a simplicial approximation of the map f, i.e., $f(\operatorname{st} v) \subset \operatorname{st}\varphi(v)$ for any vertex $v \in K^{(n)}$. If v is the distinguished point of its orbit, then $\varphi(v) = \psi(v)$ and $f(\operatorname{st} v) \subset \operatorname{st}\varphi(v)$ because ψ is a simplicial approximation of f. For any element $g \in G$, the simplicial map g is one-to-one; hence $\operatorname{st} g(v) = g(\operatorname{st} v)$, where v is a vertex of $K^{(n)}$ or L. Thus, if $f(\operatorname{st} v) \subset \operatorname{st}\varphi(v)$, then $f(\operatorname{st} g(v)) = f(g(\operatorname{st} v)) = g(f(\operatorname{st} v)) \subset g(\operatorname{st}\varphi(v)) = \operatorname{st}(g\varphi(v)) = \operatorname{st}\varphi(g(v))$. $\square$

Remark. The map φ constructed in the proof of Theorem 3.49 is equivariantly homotopic to f; i.e., these maps are homotopic in the class of equivariant maps. The homotopy is constructed in a standard way (for any x, the points $\varphi(x)$ and $f(x)$ belong to the same simplex and can therefore be joined by a line segment).

3.3. Fixed Points and Fixed Simplices. For a G-space X, let X^G be the set of fixed points of the action of G, i.e., $X^G = \{x \in X \mid gx = x \ \forall g \in G\}$.

For a simplicial G-complex K, by K^G we denote the subcomplex of K formed by the simplices that are pointwise fixed under the action of G. Clearly, $|K^G| \subset |K|^G$. A point x belongs to the set $|K|^G \setminus |K^G|$ if and only if it is interior for a simplex $\Delta = [v_0, \ldots, v_n]$ and $gx = x$ for all $g \in G$ but $gv_i \neq v_i$ for some i and some $g \in G$. If an interior point of a simplex Δ is fixed under the action of an element g, then $g\Delta = \Delta$. In particular, if a simplicial G-complex K has property (A), then $|K^G| = |K|^G$. Recall that the barycentric subdivision of any simplicial G-complex has property (A). Therefore, after refining the subdivision, we can assume that $|K^G| = |K|^G$.

Theorem 3.50. *Let K be a simplicial $\mathbb{Z}_n$-complex with property* (A)[7], *and let g be a generator of the group $\mathbb{Z}_n$. Then $\chi(K^G) = \Lambda(g)$, where $\Lambda(g)$ is the Lefschetz number of the map $g\colon |K| \to |K|$.*

Proof (see [**148**]). We represent $C_k(K;\mathbb{R})$ in the form $V_k \oplus W_k$, where $V_k = C_k(K^G;\mathbb{R})$ and W_k is the space spanned by the simplices not belonging to K^G (such simplices form a basis in this space). By definition, $\chi(K^G) = \sum(-1)^k \dim V_k$ and $\Lambda(g) = \sum(-1)^k \operatorname{tr} g_k$, where $g_k\colon C_k(K;\mathbb{R}) \to C_k(K;\mathbb{R})$ is the map induced by the action of g on k-simplices. Property (A) implies that if $g\Delta = \Delta$, then $\Delta \in K^G$. Therefore, the first $\dim V_k$ diagonal elements in the matrix of g_k in the natural basis of the space $V_k \oplus W_k$ are equal to 1 and the remaining diagonal elements are zeros. Hence $\operatorname{tr} g_k = \dim V_k$. □

The action of a group G on the space X is said to be *effective* if for each nonidentity element $g \in G$, there exists a point $x \in X$ such that $gx \neq x$.

Theorem 3.51 (Minkowski). *Suppose that M is a triangulated sphere with n handles ($n \geq 2$) on which a finite group G acts effectively* (*and simplicially*) *and the G-complex M has property* (A). *Then for any nonidentity element $g \in G$, the map $g_*\colon H_1(M;\mathbb{R}) \to H_1(M;\mathbb{R})$ is not the identity.*

Proof (see [**148**]). We show that the set of points of the manifold M that are fixed under the action of the subgroup $\mathbb{Z}_m \subset G$ generated by g consists of several isolated points and disjoint circles. First, an effective action on M cannot have fixed 2-simplices because any 2-simplex sharing an edge with a fixed simplex must be fixed as well (and a free vertex can be mapped only to itself). Second, if v is a fixed vertex, then it is either isolated in the set of fixed points or incident with precisely two fixed edges. Indeed, let $v_1, \ldots, v_m$ be the vertices neighboring v and numbered in some order (say, clockwise). Suppose that $gv_1 = v_k$. Then $gv_2 = v_{k+1}$ or $gv_2 = v_{k-1}$. We have $gv_i = v_{k-1+i}$ in the former case and $gv_i = v_{k+1-i}$ in the latter. In the former case, the only fixed point inside the polygon $v_1 \ldots v_m$ is v; in

[7]Property (A) was defined on p. 174. Recall that the barycentric subdivision of any simplicial G-complex has property (A).

the latter case, there are two fixed line segments joining v either to some vertex v_j or to the midpoint of some edge $v_j v_{j+1}$. Property (A) eliminates the second possibility.

Suppose that g_* is the identity map. Then $\Lambda(g) = \chi(M) = 2 - 2n < 0$ because $n > 1$. On the other hand, according to Theorem 3.50, we have $\Lambda(g) = \chi(M^{\mathbb{Z}_m})$ and $\chi(M^{\mathbb{Z}_m}) \geq 0$ because $M^{\mathbb{Z}_m}$ consists of several isolated points and pairwise disjoint circles. □

3.4. Transfer Homomorphisms. Suppose that G is a finite group, K is a regular simplicial G-complex, and $C_*(K)$ is a simplicial chain complex. We define an action of the group G on $C_*(K)$ by setting $g[v_0, \dots, v_n] = [gv_0, \dots, gv_n]$. Consider the homomorphism $\sigma\colon C_*(K) \to C_*(K)$ that takes every chain c to $\sum_{g \in G} gc$.

If a subcomplex $L \subset K$ is invariant with respect to the action of G, then we can transfer this action to $C_*(K, L) = C_*(K)/C_*(L)$ and define a similar homomorphism $\sigma\colon C_*(K, L) \to C_*(K, L)$.

Let $p\colon K \to K/G$ be the canonical projection (which is a simplicial map). The map p induces the chain map $p_\#\colon C_*(K, L) \to C_*(K/G, L/G)$.

Theorem 3.52. *If a subcomplex $L \subset K$ is G-invariant, then* $\operatorname{Ker} \sigma = \operatorname{Ker} p_\#$.

Proof. First, we show that the regularity of the complex K implies property (A). Suppose that vertices v and gv belong to the same simplex. Then $[v, v]$ and $[v, gv]$ are simplices in K. By the regularity condition, $v = g'v$ and $gv = g'v$ for some $g' \in G$. Thus, $v = g'v = gv$, as required.

Consider a simplex Δ^k in K/G. Let $\Delta_1^k, \dots, \Delta_n^k$ be the simplices over Δ^k in K. Since K is regular, each of these simplices admits the orientation compatible with that of Δ^k, i.e., such that the projection $p_i\colon \Delta_i^k \to \Delta^k$ is orientation-preserving. Indeed, the regularity condition implies property (A). Therefore, if $g[v_0, \dots, v_n] = [v_0, \dots, v_n]$, then all points of the simplex $[v_0, \dots, v_n]$ are fixed under the action of the element g; in particular, g preserves the orientation of the simplex $[v_0, \dots, v_n]$.

Let $c = \sum_{i=1}^n a_i \Delta_i^k$. Then

$$p_\# c = \left(\sum_{i=1}^n a_i \right) \Delta^k$$

and

$$\sigma c = \frac{|G|}{n} \left(\sum_{i=1}^n a_i \right) \left(\sum_{i=1}^n \Delta_i^k \right);$$

the second equality holds because

$$\sum_{g\in G} g\Delta_j^k = \frac{|G|}{n}\sum_{i=1}^{n}\Delta_i^k.$$

To calculate $\operatorname{Ker}\sigma$ and $\operatorname{Ker}p_\#$, we decompose every chain into summands of the form $\sum_{i=1}^n a_i\Delta_i^k$; a chain belongs to the kernel of σ or $p_\#$ if and only if so does each of the summands. The conditions for such a summand to belong to the kernels of σ and $p_\#$ almost coincide: they are $\frac{|G|}{n}\bigl(\sum_{i=1}^n a_i\bigr) = 0$ and $\sum_{i=1}^n a_i = 0$. For integer coefficients a_i, these conditions coincide exactly. □

Remark. Theorem 3.52 is valid not only for $\mathbb{Z}$ but also for an arbitrary coefficient group A provided that A has no element a for which $|G|a = 0$.

Since $p_\#\colon C_*(K,L) \to C_*(K/G, L/G)$ is an epimorphism, it follows that $C_*(K/G, L/G) \cong C_*(K,L)/\operatorname{Ker}p_\# = C_*(K,L)/\operatorname{Ker}\sigma \cong \sigma C_*(K,L)$; thus, there is a canonical chain isomorphism $\mu\colon C_*(K/G,L/G) \to \sigma C_*(K,L)$. The composition of this homomorphism and the inclusion $\sigma C_*(K,L) \subset C_*(K,L)$ induces a homomorphism $\mu_*\colon H_*(K/G,L/G) \to H_*(K,L)$, which is called the *transfer homomorphism*. Clearly, the map p induces a homomorphism $p_*\colon H_*(K,L) \to H_*(K/G,L/G)$. Consider the compositions $p_*\mu_*$ and μ_*p_*.

It follows directly from the definition that $\mu p_\# c = \sigma c$; therefore, $\mu_* p_* = \sum_{g\in G} g_*$, where $g_*\colon H_*(K,L) \to H_*(K,L)$ is the map induced by the action of g. Moreover, any element $\tilde c \in C_*(K/G,L/G)$ can be represented in the form $\tilde c = p_\# c$, where $c \in C_*(K,L)$. Hence $\mu\tilde c = \mu p_\# c = \sigma c$, which means that $p_\#\mu\tilde c = p_\#\sigma c = |G|p_\# c = |G|\tilde c$. Thus, $p_*\mu_*$ is multiplication by $|G|$.

Any chain $\sigma c = \sum_{g\in G} gc$ is G-invariant. Hence, $\operatorname{Im}\mu \subset C_*(K,L)^G$ and $\operatorname{Im}\mu_* \subset H_*(K,L)^G$. Clearly, the restriction of the map μ_*p_* to $H_*(K,L)^G$ is the sum of $|G|$ identity maps; thus, this restriction is multiplication by $|G|$ as well. Using the remark after Theorem 3.52, we obtain the following result.

Theorem 3.53. *If F is a field whose characteristic is either coprime to $|G|$ or equal to zero, then the maps*

$$p_*\colon H_*(K,L;F)^G \to H_*(K/G,L/G;F)$$

and

$$\mu_*\colon H_*(K/G,L/G;F) \to H_*(K,L;F)^G$$

are isomorphisms. Also, the kernel of the homomorphism $p_*\colon H_*(K,L;F) \to H_*(K/G,L/G;F)$ coincides with that of σ_*.

We now show that in Theorem 3.53, the assumption that the characteristic of the coefficient field F is coprime to $|G|$ cannot be removed. Consider the lens space $L(p,q)$. Let $G = \mathbb{Z}_p$. For $K = S^3$, we have $L(p,q) = K/G$, and K is a regular G-complex. It is easy to derive from the universal coefficient theorem that if p' is a prime, then, for $k = 1$ and 2, we have

$$H_k(L(p,q); \mathbb{Z}_{p'}) \cong \begin{cases} \mathbb{Z}_{p'} & \text{if } p \text{ is divisible by } p', \\ 0 & \text{if } p \text{ is not divisible by } p'. \end{cases}$$

Thus, if $p = |G|$ is divisible by p', then the groups $H_k(L(p,q); \mathbb{Z}_{p'})$ and $H_k(S^3; \mathbb{Z}_{p'})^G = 0$ are not isomorphic.

Theorem 3.53 can be carried over to cohomology groups in a natural way.

3.5. Smith Theory. In this section, we consider $G = \mathbb{Z}_p$ for a prime p. (Note that, for applications, even the simplest case $p = 2$ is interesting.) Moreover, we consider homology and cohomology only with coefficients in $\mathbb{Z}_p$.

Let K be a regular G-complex, and let L be an invariant subcomplex of K. The chains from $C_*(K, L; \mathbb{Z}_p)$ can be multiplied by elements of the *group ring* $\mathbb{Z}_pG$, which consists of the formal sums $\sum a_i g_i$, where $a_i \in \mathbb{Z}_p$ and $g_i \in G = \mathbb{Z}_p$. Let g be a generator (that is, any nonzero element) of the group G. We use multiplicative notation for the group G, that is, assume that G consists of the elements $1, g, \dots, g^{p-1}$. Consider the elements $\sigma = 1 + g + \cdots + g^{p-1}$ and $\tau = 1 - g$ in the group ring $\mathbb{Z}_pG$. The equality $g^p = 1$ implies $\sigma\tau = 0$.

Let us show that $\tau^{p-1} = \sigma$, i.e., $(-1)^k \binom{p-1}{k} \equiv 1 \pmod p$. Indeed, $(1-p)(2-p)\cdots(k-p) \equiv 1 \cdot 2 \cdots k \pmod p$; therefore,

$$\frac{1 \cdot 2 \cdots (p-k-1) \cdot (-1)^k (p-k) \cdots (p-1)}{1 \cdot 2 \cdots (p-k-1) \cdot 1 \cdot 2 \cdots k} \equiv 1 \pmod p.$$

For each element of the form $\rho = \tau^k$, we set $\bar{\rho} = \tau^{p-k}$. In particular, $\tau = \bar{\sigma}$ and $\sigma = \bar{\tau}$. For every $\rho = \tau^k$, where $1 \le k \le p-1$, consider the chain subcomplex $\rho C_*(K, L; \mathbb{Z}_p) \subset C_*(K, L; \mathbb{Z}_p)$. In what follows, we omit the coefficient group $\mathbb{Z}_p$ from the notation.

Theorem 3.54. *For any $\rho = \tau^k$, where $1 \le k \le p-1$, the sequence of chain complexes*

$$0 \longrightarrow \bar{\rho} C_*(K, L) \oplus C_*(K^G, L^G) \xrightarrow{\ i\ } C_*(K, L) \xrightarrow{\rho\times} \rho C_*(K, L) \longrightarrow 0$$

is exact. (Here i is the sum of natural embeddings and $\rho\times$ is multiplication by ρ.)

Proof. First, note that the embedding $C_*(K^G, L^G) \to C_*(K, L)$ is well defined because $L^G = L \cap K^G$.

We are interested only in simplices in $K \setminus L$. Suppose that $\Delta \subset K^G$. Then $\tau\Delta = 0$; therefore, $\rho\Delta = 0$ and $\bar{\rho}\Delta = 0$. This means, in particular, that the map i takes $\bar{\rho}C_*(K, L) \oplus 0$ and $0 \oplus C_*(K^G, L^G)$ to subspaces with trivial intersection. Moreover, $\rho\times$ (multiplication by ρ) maps the subspace $C_*(K^G, L^G) \subset C_*(K, L)$ to zero. Hence it is sufficient to verify exactness only for chains of the form $\sum a_i g^i \Delta$, where $\Delta \not\subset K^G$. For fixed Δ, every chain of this form is uniquely determined by an element $\sum a_i g^i \subset \Lambda$, where $\Lambda = \mathbb{Z}_p G$ is the group ring. It remains to prove the exactness of the sequence

$$0 \longrightarrow \bar{\rho}\Lambda \xrightarrow{\ i\ } \Lambda \xrightarrow{\ \rho\times\ } \rho\Lambda \longrightarrow 0.$$

Clearly, i is a monomorphism, and $\rho\times$ is an epimorphism. Thus, we have to verify that $\dim(\rho\Lambda) + \dim(\bar{\rho}\Lambda) = \dim\Lambda = p$ (the dimension is over the field $\mathbb{Z}_p$). Recall that $\rho = \tau^k$, where $1 \le k \le p-1$, and $\bar{\rho} = \tau^{p-k}$. We show that $\dim(\tau^k\Lambda) = p - k$. The kernel of τ consists of the elements $\sum a_i g^i$ such that $\sum a_i g^i = \sum a_i g^{i+1}$. This means that $a_i = a_{i+1} = \cdots$, i.e., the kernel is generated by the element $\sigma = 1 + g + \cdots + g^{p-1}$. Therefore, $\dim(\tau\Lambda) = p - 1$. Moreover, $\sigma = \tau^{p-1} = \tau^k\tau^{p-k-1}$, whence $\operatorname{Ker}\tau = \langle\sigma\rangle \subset \tau\Lambda$. Thus, $\dim(\tau^{k+1}\Lambda) = \dim\tau(\tau^k\Lambda) = \dim(\tau^k\Lambda) - 1$. ☐

The homology groups $H_*^{\rho}(K, L; \mathbb{Z}_p) = H_*(\rho C_*(K, L; \mathbb{Z}_p))$ are known as the *Smith homology groups.* The short exact sequence from Theorem 3.54 induces the exact sequence of homology groups

$$\begin{aligned} \cdots \longrightarrow H_q^{\bar{\rho}}(K, L) \oplus H_q(K^G, L^G) \xrightarrow{\ i_*\ } H_q(K, L) \xrightarrow{\ (\rho\times)_*\ } H_q^{\rho}(K, L) \\ \xrightarrow{\ d_*\ } H_{q-1}^{\bar{\rho}}(K, L) \oplus H_{q-1}(K^G, L^G) \longrightarrow \cdots. \end{aligned}$$

It is called the *Smith exact sequence.* As a rule, the Smith exact sequence is used for $\rho = \sigma$ and $\rho = \tau$.

For the Smith homology groups, yet another set of exact sequences can be constructed (for $k = 1, \ldots, p-2$). The argument at the end of the proof of Theorem 3.54 shows that the sequence

$$0 \longrightarrow \sigma\Lambda \xrightarrow{\ i\ } \tau^k\Lambda \xrightarrow{\ \tau\times\ } \tau^{k+1}\Lambda \longrightarrow 0 \tag{27}$$

is exact. It follows that $\sigma = \tau^{p-1}$ and $\dim(\tau^k\Lambda) = p - k$. Note that, for $k = p-1$, the sequence (27) degenerates: $0 \to \sigma\Lambda \to \sigma\Lambda \to 0 \to 0$. From the short exact sequence (27) we obtain the exact sequence of homology groups

$$\begin{aligned} \cdots \longrightarrow H_q^{\sigma}(K^G, L^G) \xrightarrow{\ i_*\ } H_q^{\tau^k}(K, L) \\ \xrightarrow{\ (\tau\times)_*\ } H_q^{\tau^{k+1}}(K, L) \xrightarrow{\ d_*\ } H_{q-1}^{\sigma}(K^G, L^G) \longrightarrow \cdots. \end{aligned} \tag{28}$$

Suppose that K is a finite simplicial complex and L is a subcomplex of K. We define the *Euler characteristic* of the pair (K, L) by $\chi(K, L) = \sum(-1)^i \dim H_i(K, L; F)$, where F is a field. Like the ordinary Euler characteristic, the number $\chi(K, L)$ does not depend on the choice of F.

Theorem 3.55. (a) *For an integer $n \geq 0$,*

$$\dim H_n^\rho(K, L) + \sum_{i \geq n} \dim H_n(K^G, L^G) \leq \sum_{i \geq n} \dim H_n(K, L);$$

(b) $\chi(K^G, L^G) \equiv \chi(K, L) \pmod{p}$.

Proof. (a) Suppose that $a_i = \dim H_i(K^G, L^G)$, $b_i = \dim H_i(K, L)$, $c_i = \dim H_i^\rho(K, L)$, and $\bar{c}_i = \dim H_i^{\bar{\rho}}(K, L)$. The Smith exact sequence

$$H_{i+1}^\rho(K, L) \longrightarrow H_i^{\bar{\rho}}(K, L) \oplus H_i(K^G, L^G) \longrightarrow H_i(K, L)$$

implies $\bar{c}_i + a_i \leq c_{i+1} + b_i$. Similarly, writing the Smith sequence for $\bar{\rho}$, we obtain $c_i + a_i \leq \bar{c}_{i+1} + b_i$. Consider the sequence of inequalities

$$\begin{aligned} c_n + a_n &\leq \bar{c}_{n+1} + b_n, \\ \bar{c}_{n+1} + a_{n+1} &\leq c_{n+2} + b_{n+1}, \\ c_{n+2} + a_{n+2} &\leq \bar{c}_{n+3} + b_{n+2}, \\ &\cdots\cdots\cdots\cdots \end{aligned}$$

The number of inequalities is finite (that is, the inequalities with large numbers have the form $0 \leq 0$) because the simplicial complex K is assumed to be finite. Summing up these inequalities, we obtain the desired relation.

(b) Any exact sequence of spaces $0 \to V_0 \to V_1 \to \cdots \to V_n \to 0$ can be represented as the direct sum of the short exact sequences $0 \to \cdots \to 0 \to V_{i-1}^\alpha \to V_i^\alpha \to V_{i+1}^\alpha \to 0 \to \cdots \to 0$. For each of these short exact sequences, we have $\sum(-1)^i \dim V_i^\alpha = 0$; therefore, $\sum(-1)^i \dim V_i = 0$. Thus, if spaces A_i, B_i, and C_i are such that an exact sequence $0 \to A_0 \to B_0 \to C_0 \to A_1 \to B_1 \to \cdots \to C_n \to 0$ holds, then $\sum(-1)^i \dim A_i + \sum(-1)^i \dim C_i = \sum(-1)^i \dim B_i$. Hence the Smith exact sequence implies

$$\chi(\bar{\rho}) + \chi(K^G, L^G) + \chi(\rho) = \chi(K, L), \tag{29}$$

where $\chi(\rho) = \sum(-1)^i \dim H_i^\rho(K, L)$.

Similarly, the exact sequence (28) gives

$$\chi(\sigma) + \chi(\tau^{k+1}) = \chi(\tau^k), \quad k = 1, \ldots, p-2. \tag{30}$$

Summing the equalities (30) over $k = 1, \ldots, p-2$, we obtain $(p-2)\chi(\sigma) + \chi(\tau^{p-1}) = \chi(\tau)$. But $\sigma = \tau^{p-1}$; therefore, $\chi(\tau) = (p-1)\chi(\sigma)$.

Setting $\rho = \tau$ in (29) (so that $\bar{\rho} = \tau^{p-1} = \sigma$), we obtain

$$\chi(\sigma) + \chi(\tau) + \chi(K^G, L^G) = \chi(K, L).$$

Thus, $\chi(K, L) = p\chi(\sigma) + \chi(K^G, L^G) \equiv \chi(K^G, L^G) \pmod{p}$. □

Theorem 3.55 gives strong constraints on the structure of the sets of fixed points for actions of the group $\mathbb{Z}_p$ on spheres and disks. Its proof uses only homological information; so, the same argument applies to homology spheres and disks. A simplicial complex K is called a *$\mathbb{Z}_p$-homology r-sphere* if $H_i(K;\mathbb{Z}_p) \cong H_i(S^r;\mathbb{Z}_p)$ for all i. A pair (K, L) is called a *$\mathbb{Z}_p$-homology r-disk* if $H_i(K, L;\mathbb{Z}_p) \cong H_i(D^r, S^{r-1};\mathbb{Z}_p)$ for all i.

Theorem 3.56 (Smith [**122**]). *Suppose that $G = \mathbb{Z}_p$ (for prime p), K is a regular G-complex, and L is an invariant subcomplex of K.*

(a) *If K is a $\mathbb{Z}_p$-homology n-sphere, then K^G is a $\mathbb{Z}_p$-homology r-sphere, where $-1 \le r \le n$. (It is assumed that the sphere of dimension -1 is the empty set.) If $p > 2$, then $r = n \pmod 2$.*

(b) *If (K, L) is a $\mathbb{Z}_p$-homology n-disk, then (K^G, L^G) is a $\mathbb{Z}_p$-homology r-disk, where $1 \le r \le n$. If $p > 2$, then $r = n \pmod 2$.*

Proof (see [**36**]). (a) Applying Theorem 3.55 to $L = \varnothing$, we obtain $\sum_{i\ge 0} \dim H_i(K^G) \le \sum_{i\ge 0} \dim H_i(K) = 2$ and $\chi(K^G) \equiv \chi(K) \pmod{p}$. But $\chi(K)$ is equal to 0 or 2. Therefore, the equality $\sum_{i\ge 0} \dim H_i(K^G) = 1$ cannot hold. Thus, $\sum_{i\ge 0} \dim H_i(K^G) = 0$ or 2. In the former case, we have $K^G = \varnothing$, and in the latter case, K^G is a $\mathbb{Z}_p$-homology sphere of some nonnegative dimension r.

The inequality $r \le n$ follows from

$$\sum_{i\ge n+1} \dim H_i(K^G) \le \sum_{i\ge n+1} \dim H_i(K) = 0.$$

If $p > 2$, then the congruence $\chi(K^G) \equiv \chi(K) \pmod{p}$ (i.e., $(-1)^r + 1 \equiv (-1)^n + 1 \pmod{p}$) implies $r \equiv n \pmod 2$.

(b) For homology disks, the proof is even simpler: it suffices to apply the relations

$$\sum_{i\ge 0} \dim H_i(K^G, L^G) \le \sum_{i\ge 0} \dim H_i(K, L) = 1,$$

$$\sum_{i\ge n+1} \dim H_i(K^G, L^G) \le \sum_{i\ge n+1} \dim H_i(K, L) = 0.$$ □

The action of a group G on a topological space X is said to be *free* if $gx \ne x$ for any point $x \in X$ and any nonidentity element $g \in G$.

Problem 111 (Smith). Prove that finite groups cannot act freely on $\mathbb{R}^n$.

Exercise. Prove that if $G = \mathbb{Z}_p$ (for prime p) and K is an acyclic regular G-complex, then K^G is an acyclic complex. (In particular, $K^G \neq \varnothing$.)

4. Steenrod Squares

In this section, we construct the Steenrod squares and prove some of their properties. We use them in Section 3.2 to derive the Thom and Wu formulas and prove the Stiefel theorem on the parallelizability of orientable 3-manifolds.

4.1. Construction of the Steenrod Squares. The Steenrod squares are additional algebraic operations on cohomology rings, which generalize cup product. Recall that cup product can be constructed using any diagonal approximation $D_0\colon C_*(K) \to C_*(K) \otimes C_*(K)$ that takes every vertex v to $v \otimes v$. Such a diagonal approximation is not unique. Two diagonal approximations D_0 and TD_0 for a 1-simplex K are shown in Figure 6. The diagonal approximation TD_0 is symmetric to D_0 with respect to the diagonal. Formally, symmetry with respect to the diagonal can be defined as the chain transformation T of $C_*(K) \otimes C_*(K)$ given by $T(\sigma^p \otimes \tau^q) = (-1)^{pq}\tau^q \otimes \sigma^p$; the sign is chosen so as to make T the chain map. The diagonal approximation TD_0 is the composition of the diagonal approximation D_0 and the chain transformation T.

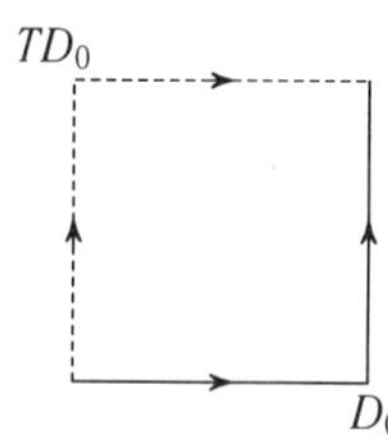

Figure 6. Two diagonal approximations

The diagonal map $d\colon |K| \to |K \times K|$ is not cellular; therefore, $TD_0 \neq D_0$, and we can consider the difference $D_0 - TD_0$. It turns out that the maps D_0 and TD_0 are chain homotopic, i.e., $D_0 - TD_0 = \partial D_1 + D_1\partial$ for some homomorphism $D_1\colon C_*(K) \to C_*(K) \otimes C_*(K)$ increasing the dimension of the chain by 1. Moreover, extending this construction, we obtain a sequence of homomorphisms $D_k\colon C_*(K) \to (C_*(K) \otimes C_*(K))_{i+k}$. The Steenrod squares are constructed using the homomorphisms D_k. Let us study these homomorphisms in more detail.

Consider the chain complex W corresponding to the CW-complex S^∞ having two cells of each dimension from 0 to ∞ (the k-skeleton of the complex S^∞ is the sphere S^k). The chain transformation T can be treated as an action of the generator T of the group $\mathbb{Z}_2$. The generator T of $\mathbb{Z}_2$ acts on S^∞ as the symmetry about the center of the sphere. Therefore, the chain complex W has two generators, w_k and Tw_k, in each dimension $k \geq 0$. The generators can be chosen so that $\partial w_1 = w_0 - Tw_0$, $\partial w_2 = w_1 + Tw_1$, and in general $\partial w_k = w_{k-1} + (-1)^k Tw_{k-1}$ (T preserves the orientation of cells for odd k and reverses it for even k).

Let K be an arbitrary simplicial complex. We define the action of the group $\mathbb{Z}_2$ on the chain complex $W \otimes C_*(K)$ by $T(w_k \otimes \sigma) = (Tw_k \otimes \sigma)$; its action on $C_*(K) \otimes C_*(K)$ has already been defined. Let $C_*(K \times K)$ be a chain complex for calculating cellular homology. It can be identified with $C_*(K) \otimes C_*(K)$. The chain subcomplex $C_*(\sigma \times \sigma)$ is defined for any cell (simplex) σ in K; we denote by the same symbol the corresponding subcomplex in $C_*(K) \otimes C_*(K)$.

We set $\mathcal{L}(w_k \otimes \sigma) = C_*(\sigma \times \sigma) \subset C_*(K) \otimes C_*(K)$. The subcomplex $\mathcal{L}(w_k \otimes \sigma)$ is acyclic, and $T\mathcal{L}(w_k \otimes \sigma) = \mathcal{L}(w_k \otimes \sigma)$. This allows us to construct a chain map $\varphi\colon W \otimes C_*(K) \to C_*(K) \otimes C_*(K)$ which is equivariant (that is, $\varphi T = T\varphi$) and is supported on the image of $\mathcal{L}$, i.e., $\varphi(w_k \otimes \sigma) \in \mathcal{L}(w_k \otimes \sigma)$. The map φ is constructed by induction on $\dim(w_k \otimes \sigma)$. For a vertex v, we set $\varphi(w_0 \otimes v) = v \otimes v$. Note that

$$\begin{aligned}
\partial(\varphi(w_k \otimes \sigma)) &= \varphi(\partial(w_k \otimes \sigma)) = \varphi(\partial w_k \otimes \sigma \pm w_k \otimes \partial\sigma) \\
&= \varphi((w_{k-1} \pm Tw_{k-1}) \otimes \sigma \pm w_k \otimes \partial\sigma) \\
&\subset \mathcal{L}(w_{k-1} \otimes \sigma) + \mathcal{L}(w_k \otimes \partial\sigma) \\
&= C_*(\sigma \times \sigma) + C_*(\partial\sigma \times \partial\sigma) = C_*(\sigma \times \sigma).
\end{aligned}$$

The chain $\partial(\varphi(w_k \otimes \sigma))$ is a cycle, and it is contained in the acyclic chain complex $C_*(\sigma \times \sigma)$. Therefore, $C_*(\sigma \times \sigma) = \mathcal{L}(w_k \otimes \partial\sigma)$ contains a chain c for which $\partial(\varphi(w_k \otimes \sigma)) = \partial c$. We set $\varphi(w_k \otimes \sigma) = c$ and $\varphi(Tw_k \otimes \sigma) = Tc$.

Now, we can construct the homomorphisms D_k. Set $D_k(c) = \varphi(w_k \otimes c)$ for $c \in C_*(K)$. Clearly, D_k increases the dimension of each chain c by k, and D_0 is a diagonal approximation.

Theorem 3.57. $D_{k-1} + (-1)^k TD_{k-1} = \partial D_k + (-1)^{k+1} D_k \partial$.

Proof. Since φ is a chain map, we have

$$\partial D_k(c) = \partial(\varphi(w_k \otimes c)) = \varphi(\partial(w_k \otimes c)) = \varphi(\partial w_k \otimes c + (-1)^k w_k \otimes \partial c).$$

Taking into account the fact that $D_k \partial c = \varphi(w_k \otimes \partial c)$, we obtain

$$\begin{aligned}
\partial D_k(c) + (-1)^{k+1} D_k \partial c &= \varphi(\partial w_k \otimes c) = \varphi((w_{k-1} + (-1)^k Tw_{k-1}) \otimes c) \\
&= D_{k-1}c + (-1)^k TD_{k-1}c,
\end{aligned}$$

as required. □

Corollary. *$D_0 - TD_0 = \partial D_1 + D_1\partial$. Thus, D_1 is a chain homotopy between D_0 and TD_0.*

For $k \geq 0$, we define the $\smile_k$-product $c^p \smile_k c^q$ of $c^p \in C^p(K)$ and $c^q \in C^q(K)$ to be the cochain that assigns

$$\langle c^p \otimes c^q, D_k c\rangle = \langle c^p \otimes c^q, \varphi(w_k \otimes c)\rangle$$

to every chain $c \in C_{p+q-k}(K)$. It is convenient to set $c^p \smile_{-1} c^q = 0$.

Note that $\smile_k$-product depends on the choice of the map φ. Thus, we have to verify that the operation Sq_k, which is defined in terms of $\smile_k$, does not depend on φ.

Theorem 3.58 (the coboundary formula). *The following equality holds:*

$$\begin{aligned}\delta(c^p \smile_k c^q) = (-1)^k \delta c^p \smile_k c^q + (-1)^{p+k} c^p \smile_k \delta c^q \\ -(-1)^k c^p \smile_{k-1} c^q - (-1)^{pq} c^q \smile_{k-1} c^p.\end{aligned}$$

Proof. Let $c \in C_{p+q-k+1}(K)$. Then

$$\langle \delta(c^p \smile_k c^q), c\rangle = \langle c^p \smile_k c^q, \partial c\rangle = \langle c^p \otimes c^q, \varphi(w_k \otimes c)\rangle.$$

By definition, $\partial(w_k \otimes c) = \partial w_k \otimes c + (-1)^k w_k \otimes \partial c$. Therefore,

$$\begin{aligned}
&\langle \delta(c^p \smile_k c^q), c\rangle \\
&\quad = (-1)^k \langle c^p \otimes c^q, \varphi(\partial(w_k \otimes c))\rangle - (-1)^k \langle c^p \otimes c^q, \varphi(\partial w_k \otimes c)\rangle \\
&\quad = (-1)^k \langle c^p \otimes c^q, \partial(\varphi(w_k \otimes c))\rangle - (-1)^k \langle c^p \otimes c^q, \varphi(w_{k-1} \otimes c)\rangle \\
&\qquad - (-1)^{2k} \langle c^p \otimes c^q, \varphi(T w_{k-1} \otimes c)\rangle \\
&\quad = (-1)^k \langle \delta(c^p \otimes c^q), \varphi(w_k \otimes c)\rangle - (-1)^k \langle c^p \otimes c^q, \varphi(w_{k-1} \otimes c)\rangle \\
&\qquad - \langle c^p \otimes c^q, T\varphi(w_{k-1} \otimes c)\rangle \\
&\quad = (-1)^k \langle \delta(c^p \otimes c^q), \varphi(w_k \otimes c)\rangle - (-1)^k \langle c^p \otimes c^q, \varphi(w_{k-1} \otimes c)\rangle \\
&\qquad - (-1)^{pq} \langle c^q \otimes c^p, \varphi(w_{k-1} \otimes c)\rangle.
\end{aligned}$$

By definition, $\delta(c^p \otimes c^q) = (\delta c^p) \otimes c^q + (-1)^p c^p \otimes \delta c^q$. Hence the first term in the obtained expression is equal to

$$(-1)^k \langle \delta c^p \otimes c^q), \varphi(w_k \otimes c)\rangle + (-1)^{p+k} \langle c^p \otimes \delta c^q), \varphi(w_k \otimes c)\rangle.$$

This completes the proof of the coboundary formula. □

Corollary. *If a cochain $c^p \in C^p(K)$ is a cocycle modulo 2, i.e., $\delta c^p = 2c^{p+1}$, where $c^{p+1} \in C^{p+1}(K)$, then the cochain $c^p \smile_k c^p \in C^{2p-k}(K)$ is a cocycle modulo 2 as well.*

Proof. According to the coboundary formula, we have

$$\begin{aligned}\delta(c^p \smile_k c^p) &= (-1)^k 2c^{p+1} \smile_k c^p + (-1)^{p+k} c^p \smile_k 2c^{p+1}\\ &\quad - (-1)^k c^p \smile_{k-1} c^p - (-1)^{p^2} c^p \smile_{k-1} c^p.\end{aligned}$$

It remains to note that $(-1)^k + (-1)^{p^2} = \pm 1 \pm 1 \equiv 0 \pmod 2$. □

Thus, the formula $\mathrm{Sq}_k\, c^p = c^p \smile_k c^p$ defines a map $\mathrm{Sq}_k \colon Z^p(K;\mathbb{Z}_2) \to Z^{2p-k}(K;\mathbb{Z}_2)$. The natural projection $Z^{2p-k}(K;\mathbb{Z}_2) \to H^{2p-k}(K;\mathbb{Z}_2)$ then gives a map $\mathrm{Sq}_k \colon Z^p(K;\mathbb{Z}_2) \to H^{2p-k}(K;\mathbb{Z}_2)$

Theorem 3.59. *For any $k \geq 0$, the map $c^p \mapsto \mathrm{Sq}_k\, c^p$ induces a group homomorphism $\mathrm{Sq}_k \colon H^p(K;\mathbb{Z}_2) \to H^{2p-k}(K;\mathbb{Z}_2)$.*

Proof. First, we verify that if $c^p \equiv \delta c \pmod 2$, then $\mathrm{Sq}_k\, c^p$ is a coboundary modulo 2. Indeed, $c^p \smile_k c^p \equiv \delta(c \smile_k \delta c + c \smile_{k-1} c) \pmod 2$ because

$$\begin{aligned}\delta(c \smile_k \delta c + c \smile_{k-1} c) &\equiv \delta c \smile_k \delta c + c \smile_{k-1} \delta c + \delta c \smile_{k-1} c\\ &\quad + \delta c \smile_{k-1} c + c \smile_{k-1} \delta c + c \smile_{k-2} c + c \smile_{k-2} c\\ &\equiv \delta c \smile_k \delta c \equiv c^p \smile_k c^p \pmod 2.\end{aligned}$$

Now, we show that the map Sq_k is a homomorphism. Clearly,

$$\mathrm{Sq}_k(c^p + d^p) = \mathrm{Sq}_k\, c^p + \mathrm{Sq}_k\, d^p + c^p \smile_k d^p + d^p \smile_k c^p.$$

Moreover, if $\delta c^p \equiv 0 \pmod 2$ and $\delta d^p \equiv 0 \pmod 2$, then

$$\begin{aligned}\delta(c^p \smile_{k+1} d^p) &\equiv \delta c^p \smile_{k+1} d^p + c^p \smile_{k+1} \delta d^p + c^p \smile_k d^p + d^p \smile_k c^p\\ &\equiv c^p \smile_k d^p + d^p \smile_k c^p \pmod 2.\end{aligned}$$ □

Theorem 3.60. *The homomorphism $f^* \colon H^*(L,\mathbb{Z}_2) \to H^*(K;\mathbb{Z}_2)$ commutes with Sq_k for any simplicial map $f \colon K \to L$.*

Proof. By definition, we have

$$\langle f^*(\mathrm{Sq}_k\, c), c' \rangle = \langle c \otimes c, \varphi_L(w_k \otimes f(c')) \rangle = \langle c \otimes c, \varphi_L(1 \otimes f)(w_k \otimes c') \rangle,$$
$$\langle \mathrm{Sq}_k(f^* c), c' \rangle = \langle f^* c \otimes f^* c, \varphi_K(w_k \otimes c') \rangle = \langle c \otimes c, (f \otimes f)\varphi_K(w_k \otimes c') \rangle.$$

The equivariant chain maps $\varphi_L(1 \otimes f)$ and $(f \otimes f)\varphi_K$ have the common support $\mathcal{L}(w_k \otimes \sigma) = C_*(f\sigma \times f\sigma)$. This support is equivariant in the sense that $T\mathcal{L}(w_k \otimes \sigma) = \mathcal{L}(w_k \otimes \sigma)$. The cyclicity and equivariance of the support allow us to apply a standard construction (see pp. 185 and 7) and obtain a chain homotopy D between the maps $\varphi_L(1 \otimes f)$ and $(f \otimes f)\varphi_K$ for which $DT = D$ (if $D(w_k \otimes \sigma) = c$, then we set $D(Tw_k \otimes \sigma) = c$). Thus,

$$\begin{aligned}\langle (f^* \mathrm{Sq}_k - \mathrm{Sq}_k f^*)c, c' \rangle &= \langle c \otimes c, (\partial D + D\partial)(w_k \otimes c') \rangle\\ &= \langle \delta(c \otimes c), D(w_k \otimes c') \rangle + \langle c \otimes c, D(\partial w_k \otimes c' \pm w_k \otimes \partial c') \rangle.\end{aligned}$$

If $\delta c \equiv 0 \pmod 2$, then $\delta(c \otimes c) \equiv 0 \pmod 2$. Moreover,

$$\begin{aligned} D(\partial w_k \otimes c') &= D((w_{k-1} \pm T w_{k-1}) \otimes c') \\ &= D(w_{k-1} \otimes c') \pm D(w_{k-1} \otimes c') \equiv 0 \pmod 2. \end{aligned}$$

Therefore,

$$(f^* \operatorname{Sq}_k - \operatorname{Sq}_k f^*)c \equiv \delta d \pmod 2,$$

where $\langle d, c'' \rangle = \langle c \otimes c, D(w_k \otimes c') \rangle$. □

Corollary. *The operation* Sq_k *does not depend on the choice of the equivariant chain map* φ.

Proof. For $K = L$ and $f = \mathrm{id}_K$, there are two possible choices of φ, φ_K and φ_L. If Sq_k and Sq'_k are the operations constructed for φ_K and φ_L, then $\operatorname{Sq}_k - \operatorname{Sq}'_k = \mathrm{id}^* \operatorname{Sq}_k - \operatorname{Sq}'_k \mathrm{id}^* = 0$. □

For certain reasons, it is more convenient to use the notation $\operatorname{Sq}^k = \operatorname{Sq}_{p-k} \colon H^p(K; \mathbb{Z}_2) \to H^{p+k}(K; \mathbb{Z}_2)$. It is assumed that $0 \le k \le p$. For $k > p$, Sq^k vanishes identically on $H^p(K; \mathbb{Z}_2)$.

It follows directly from the definition that for the k-dimensional cohomology group, $\operatorname{Sq}^k = \operatorname{Sq}_0$ is the cohomology square map. For this reason, the operations Sq^k, which were introduced by Steenrod in [**125**, **128**, **129**], are called the *Steenrod squares*.

The Steenrod squares can also be defined for relative cohomology. Let $L \subset K$ be a simplicial subcomplex. Then we have the short exact sequence

$$0 \longrightarrow C^*(K, L) \xrightarrow{p^*} C^*(K) \xrightarrow{i^*} C^*(L) \longrightarrow 0.$$

By construction, $\varphi_K(w_k \otimes \sigma) \in C_*(\sigma \times \sigma) \subset C_*(L \times L)$ for all $\sigma \subset L$. Therefore, we can assume that $\varphi_L = \varphi_K|_{W \otimes C_*(L)}$. For cochains $c, c' \in C^*(K)$, we have $i^*(c \smile_k c') = i^*c \smile_k i^*c'$; in particular, $i^*(p^*c \smile_k p^*c') = i^*p^*c \smile_k i^*p^*c' = 0$ for any two cochains $c, c' \in C^*(K, L)$ because $i^*p^* = 0$. Since the sequence is exact, $p^*c \smile_k p^*c' = p^*d$ for some cochain $d \in C^*(K, L)$; this cochain d is determined uniquely because p^* is a monomorphism. We set $c \smile_k c' = d$. By definition, $p^*(c \smile_k c') = p^*c \smile_k p^*c'$. It easy to show that the coboundary formula remains valid for the relative $\smile_k$-product. Thus, we can define a map $\operatorname{Sq}^k \colon H^p(K, L; \mathbb{Z}_2) \to H^{p+k}(K, L; \mathbb{Z}_2)$ similar to that in the absolute case. It can be seen from the definition that $p^* \operatorname{Sq}^k = \operatorname{Sq}^k p^*$.

4.2. Properties of the Steenrod Squares. We have already proved the following properties of the Steenrod squares.

1. The Steenrod squares satisfy the naturality condition $f^*\operatorname{Sq}^k = \operatorname{Sq}^k f^*$ for any map $f \colon (K, L) \to (K_1, L_1)$ of simplicial pairs.

2. If $k > p$, then $\operatorname{Sq}^k \alpha = 0$ for all $\alpha \in H^p(K, L; \mathbb{Z}_2)$.

3. For any $\alpha \in H^k(K, L; \mathbb{Z}_2)$, $\mathrm{Sq}^k\alpha = \alpha \smile \alpha$.

Other properties of the Steenrod squares are given in the following theorem. This section is devoted to proving it; for brevity, we omit the coefficient group $\mathbb{Z}_2$ from the notation.

Theorem 3.61. (a) *The operation* Sq^k *commutes with the connecting homomorphism* $\delta\colon H^*(L) \to H^*(K, L)$.

(b) *The operation* Sq^k *commutes with the suspension isomorphism* $\Sigma^* : \tilde{H}^p(K) \to \tilde{H}^{p+1}(\Sigma K)$.

(c) *The operation* Sq^0 *is the identity map.*

(d) *The operation* Sq^1 *is the Bockstein homomorphism for the exact sequence* $0 \to \mathbb{Z}_2 \to \mathbb{Z}_4 \to \mathbb{Z}_2 \to 0$.

(e) *The* Cartan formula $\mathrm{Sq}^k(\alpha \smile \beta) = \sum_i (\mathrm{Sq}^i\alpha) \smile (\mathrm{Sq}^{k-i}\beta)$ *holds.*

The Steenrod squares have yet another very important property; namely, they satisfy the so-called Adem relations. The formulation and the proof of the Adem relations is beyond the scope of this book.

Commutation with δ. The connecting homomorphism $\delta\colon H^*(L) \to H^*(K, L)$ can be described as follows. Take $c^q \in Z^q(L)$ and let $[c^q]$ be the corresponding cohomology class. The homomorphism $i^*\colon C^*(K) \to C^*(L)$ is an epimorphism; hence there exists a cochain $\tilde{c}^q \in C^*(K)$ for which $i^*\tilde{c}^q = c^q$. But $i^*(\delta\tilde{c}^q) = \delta c^q = 0$; thus, there exists a cochain $d^{q+1} \in C^{q+1}(K, L)$ for which $\delta\tilde{c}^q = p^*d^{q+1}$. The homomorphism $p^*\colon C^*(K, L) \to C^*(K)$ is a monomorphism; therefore, $\delta d^{q+1} = 0$ because $p^*\delta d^{q+1} = \delta^2\tilde{c}^q = 0$. For the cohomology class $\delta^*[c^q]$ we take the class $[d^{q+1}]$.

We must prove that the cocycles $\delta^*\mathrm{Sq}^k c^q$ and $\mathrm{Sq}^k d^{q+1}$ are cohomologous. By definition,

$$\begin{aligned} p^*(\mathrm{Sq}^k d^{q+1}) &= \mathrm{Sq}^k(p^*d^{q+1}) = \mathrm{Sq}^k(\delta\tilde{c}^q) \\ &= \delta\tilde{c}^q \smile_{q+1-k} \delta\tilde{c}^q = \delta(\tilde{c}^q \smile_{q+1-k} \delta\tilde{c}^q + \tilde{c}^q \smile_{q-k} \tilde{c}^q). \end{aligned}$$

Moreover, $i^*(\tilde{c}^q \smile_{q+1-k} \delta\tilde{c}^q + \tilde{c}^q \smile_{q-k} \tilde{c}^q) = c^q \smile_{q-k} c^q$ because $i^*(\delta\tilde{c}^q) = 0$ and $i^*(\tilde{c}^q) = c^q$. Consider $d' = \mathrm{Sq}^k d^{q+1}$, $c' = c^q \smile_{q-k} c^q$, and $\tilde{c}' = \tilde{c}^q \smile_{q+1-k} \delta\tilde{c}^q + \tilde{c}^q \smile_{q-k} \tilde{c}^q$. We have $p^*d' = \delta\tilde{c}'$ and $i^*\tilde{c}' = c'$. This means that $\delta^*[c'] = [d']$. But $c' = c^q \smile_{q-k} c^q$; this implies the required assertion.

Commutation with the Suspension Isomorphism. The proof of (a) does not depend on any special properties of the short exact sequence used to construct the connecting homomorphism δ^*. The same argument shows that the operation Sq^k commutes with the connecting homomorphism from the Mayer–Vietoris cohomology sequence. The suspension isomorphism is the connecting homomorphism from the Mayer–Vietoris sequence for K_0 and K_1, where K_0 and K_1 are copies of the cone over K and $K_0 \cup K_1 = \Sigma K$.

The Operations Sq^0 and Sq^1. Let $\beta\colon H^p(K,L;\mathbb{Z}) \to H^{p+1}(K,L;\mathbb{Z})$ be the Bockstein homomorphism for the short exact sequence $0 \to \mathbb{Z}_2 \to \mathbb{Z}_4 \to \mathbb{Z}_2 \to 0$. To calculate Sq^0 and Sq^1, we need the following auxiliary assertion.

Lemma. $\beta \mathrm{Sq}^k = \begin{cases} 0 & \textit{if } k \textit{ is odd,} \\ \mathrm{Sq}^{k+1} & \textit{if } k \textit{ is even.} \end{cases}$

Proof. Suppose that $c \in H^p(K,L;\mathbb{Z}_2)$ and $\tilde{c}$ is an integer cochain such that reducing it modulo 2 yields a cocycle representing c. By definition, $\mathrm{Sq}^k c$ is the cohomology class of the cochain $\tilde{c} \smile_{p-k} \tilde{c}$ reduced modulo 2. By assumption, $\tilde{c} = 2c'$ for some integer cochain c'; therefore, setting $q = p - k$, we obtain $\delta(\tilde{c} \smile_{p-k} \tilde{c}) = (-1)^q 2c' \smile_q \tilde{c} + (-1)^p \tilde{c} \smile_q 2c' - (-1)^q \tilde{c} \smile_{q-1} \tilde{c} - (-1)^p \tilde{c} \smile_{q-1} \tilde{c}$. Thus, the cohomology class $\beta(\mathrm{Sq}^k c)$ is determined by the cocycle

$$c' \smile_q \tilde{c} + \tilde{c} \smile_q c' + \varepsilon(\tilde{c} \smile_{q-1} \tilde{c}) \pmod 2,$$

where $\varepsilon = 0$ for odd $k = q - p$ and $\varepsilon = 1$ for even k. But $c' \smile_q \tilde{c} + \tilde{c} \smile_q c' \equiv \delta(\tilde{c} \smile_{q+1} c') \pmod 2$ because $\delta\tilde{c} = 2c' \equiv 0 \pmod 2$ and $2\delta c' = \delta^2 \tilde{c} = 0$, i.e., $\delta c' = 0$. Thus, $\beta(\mathrm{Sq}^k c) = 0$ for odd k and $\beta(\mathrm{Sq}^k c) = [\tilde{c} \smile_{q-1} \tilde{c}] = \mathrm{Sq}^{k+1} c$ for even k. $\square$

The lemma shows that (d) follows from (c), because if $\mathrm{Sq}^0 = \mathrm{id}$, then $\mathrm{Sq}^1 = \beta \mathrm{Sq}^0 = \beta$. In turn, (c) can be proved by using the lemma. To be more precise, the lemma is used to prove (c) for $K = \mathbb{R}P^2$, from which (c) is deduced for all simplicial pairs (K, L).

Thus, we show that (c) holds for $\mathbb{R}P^2$. Let α be the generator of the group $H^1(\mathbb{R}P^2;\mathbb{Z}_2)$. According to the lemma, we have $\beta\mathrm{Sq}^0\alpha = \mathrm{Sq}^1\alpha = \alpha \smile \alpha \neq 0$; hence $\mathrm{Sq}^0\alpha \neq 0$. Therefore, $\mathrm{Sq}^0\alpha = \alpha$ because $H^1(\mathbb{R}P^2;\mathbb{Z}_2)$ has only one nonzero element.

The natural embedding $f\colon \mathbb{R}P^1 \to \mathbb{R}P^2$ induces the homomorphism $f^*\colon H^1(\mathbb{R}P^2;\mathbb{Z}_2) \to H^1(\mathbb{R}P^1;\mathbb{Z}_2)$ that takes α to the generator α' of the group $H^1(\mathbb{R}P^1;\mathbb{Z}_2)$. The naturality condition implies $\mathrm{Sq}^0\alpha' = \mathrm{Sq}(f^*\alpha) = f^*(\mathrm{Sq}^0\alpha) = f^*\alpha = \alpha'$. Thus, (c) holds also for $\mathbb{R}P^1 \approx S^1$. Since the operation Sq^0 commutes with the suspension isomorphism, (c) holds for S^n.

Let K be an n-dimensional simplicial complex. According to Theorem 3.6, any element of the group $H^n(K;\mathbb{Z})$ can be realized by means of a map $K \to S^n$. This is also true for the group $H^n(K;\mathbb{Z}_2)$ because the reduction of coefficients modulo 2 induces an epimorphism $H^n(K;\mathbb{Z}) \to H^n(K;\mathbb{Z}_2)$. Note that only the cohomology groups have this property; it does not hold for homology. The map $H_2(\mathbb{R}P^2;\mathbb{Z}) \to H_2(\mathbb{R}P^2;\mathbb{Z}_2)$, say, cannot be an epimorphism because $H_2(\mathbb{R}P^2;\mathbb{Z}) = 0$ whereas $H_2(\mathbb{R}P^2;\mathbb{Z}_2) \neq 0$. The surjectivity of $H^n(K;\mathbb{Z}) \to H^n(K;\mathbb{Z}_2)$ follows from the fact that the group $H_n(K)$ is torsion-free for each n-dimensional simplicial complex; it is proved as follows. According to the universal coefficient theorem, the group $H^n(K;G)$ is the direct sum of the groups $\mathrm{Hom}(H_n(K), G)$ and $\mathrm{Ext}(H_{n-1}(K), G)$. The map $\mathrm{Hom}(H_n(K),\mathbb{Z}) \to \mathrm{Hom}(H_n(K),\mathbb{Z}_2)$ is an epimorphism because the group $H_n(K)$ is torsion-free, and $\mathrm{Ext}(H_{n-1}(K),\mathbb{Z}) \to \mathrm{Ext}(H_{n-1}(K),\mathbb{Z}_2)$ is an epimorphism in any case (this follows from the calculation of $\mathrm{Ext}(A,B)$ for cyclic groups A and B). The naturality of the operation Sq^0 implies (c) for K.

If K is an infinite-dimensional simplicial complex, then the natural embedding $K^n \to K$ induces a monomorphism $H^n(K;\mathbb{Z}_2) \to H^n(K^n;\mathbb{Z}_2)$. Therefore, the naturality of Sq^0 implies (c) for an arbitrary K.

Now, consider the relative cohomology groups $H^n(K,L;\mathbb{Z}_2)$. Let CL be the cone over L. Then $H^n(K,L) \cong H^n(K\cup CL, CL) \cong \tilde{H}^n(K\cup CL)$. The first isomorphism is obvious at the level of (co)chains, and the second follows from the exact sequence for the pair $(K\cup CL, CL)$ because the space CL is contractible. Both isomorphisms commute with Sq^0; therefore, (c) holds for any pair (K,L) of simplicial complexes.

The Cartan Formula. It is sufficient to show that $\mathrm{Sq}^k(\alpha\times\beta) = \sum_i \mathrm{Sq}^i\alpha \times \mathrm{Sq}^{k-i}\beta$. Indeed, we have $\alpha \smile \beta = d^*(\alpha\times\beta)$, where $d\colon |K| \to |K\times K|$ is the diagonal map. Therefore, $\mathrm{Sq}^k d^*(\alpha\times\beta) = d^*\mathrm{Sq}^k(\alpha\times\beta) = d^*\sum_i \mathrm{Sq}^i\alpha\times \mathrm{Sq}^{k-i}\beta = \sum_i \mathrm{Sq}^i\alpha \smile \mathrm{Sq}^{k-i}\beta$.

Let $r\colon W \to W\otimes W$ be the linear extension of the map defined by $r(w_k) = \sum(-1)^{i(k-i)} w_i \otimes T^i w_{k-i}$. This is a chain map. Indeed, we have

$$\begin{aligned} r(\partial w_k) &= r(w_{k-1} + (-1)^k T w_{k-1}) \\ &= \sum(-1)^{i(k-1-i)} w_i \otimes T^i w_{k-1-i} \\ &\quad + \sum(-1)^{i(k-1-i)+k} T w_i \otimes T^i w_{k-1-i} \end{aligned}$$

and

$$\begin{aligned} \partial r(w_k) &= \sum(-1)^{i(k-i)}\partial w_i \otimes T^i w_{k-i} + \sum(-1)^{i(k-i)+i} w_i \otimes T^i \partial w_{k-i} \\ &= \sum(-1)^{i(k-i)} w_{i-1} \otimes T^i w_{k-i} + \sum(-1)^{i(k-i)+i} T w_{i-1} \otimes T^i w_{k-i} \\ &\quad + \sum(-1)^{i(k-i)+i} w_i \otimes T^i w_{k-i-1} \\ &\quad + \sum(-1)^{i(k-i)+k} w_i \otimes T^{i+1} w_{k-i-1}. \end{aligned}$$

In the expression for $\partial r(w_k)$, the first and fourth sums cancel each other because $i(k-i)$ is replaced by $i(k-i)+k-1$ when i is replaced with $i+1$.

Now it is easy to see that $\partial r(w_k) = r(\partial w_k)$ (the first sum in the expression for $r(\partial w_k)$ coincides with the third sum in the expression for $\partial r(w_k)$, and the coincidence of the second sums in these expressions is proved by replacing i with $i+1$).

We apply r to define the map $\varphi_{K\times L}$ as the composition

$$\begin{aligned} &W\otimes C_*(K)\otimes C_*(L) \xrightarrow{r\otimes 1} W\otimes W\otimes C_*(K)\otimes C_*(L)\\ &\xrightarrow{T} W\otimes C_*(K)\otimes W\otimes C_*(L) \xrightarrow{\varphi_K\otimes\varphi_L} C_*(K)\otimes C_*(K)\otimes C_*(L)\otimes C_*(L)\\ &\qquad\qquad\qquad\qquad\qquad\xrightarrow{T} C_*(K)\otimes C_*(L)\otimes C_*(K)\otimes C_*(L), \end{aligned}$$

where T transposes the second and third factors in both cases. The map $\varphi_{K\times L}$ is equivariant, and is supported on the image of the corresponding map $\mathcal{L}$; hence we can use it to calculate the $\smile_k$-product in the chain complex $C_*(K\times L)\cong C_*(K)\otimes C_*(L)$, and therefore, to calculate the operations Sq^k.

Suppose that $\alpha\in Z^p(K;\mathbb{Z}_2)$, $\beta\in Z^p(L;\mathbb{Z}_2)$, $a\in C_*(K;\mathbb{Z}_2)$, and $b\in C_*(L;\mathbb{Z}_2)$. We are interested in the action of the operation $\mathrm{Sq}^k=\mathrm{Sq}_{p+q-k}$ on the $(p+q)$-cocycle $\alpha\times\beta$. For brevity, we set $p+q-k=n$. Using the definition of $\varphi_{K\times L}$ and discarding signs (because we deal with $\mathbb{Z}_2$ coefficients), we obtain

$$\begin{aligned} \langle \mathrm{Sq}^k(\alpha\otimes\beta), a\otimes b\rangle &= \langle(\alpha\otimes\beta)\smile_n(\alpha\otimes\beta), a\otimes b\rangle\\ &= \langle \alpha\otimes\beta\otimes\alpha\otimes\beta, \varphi_{K\times L}(w_n\otimes a\otimes b)\rangle\\ &= \Big\langle \alpha\otimes\alpha\otimes\beta\otimes\beta, \sum \varphi_K(w_i\otimes a)\otimes T^i\varphi_L(w_{n-i}\otimes b)\Big\rangle\\ &= \sum\langle\alpha\smile_i\alpha, a\rangle\langle\beta\smile_{n-i}\beta, b\rangle\\ &= \sum\langle \mathrm{Sq}^{p-i}\alpha, a\rangle\langle\mathrm{Sq}^{q-n+i}\beta, b\rangle\\ &= \sum\langle \mathrm{Sq}^{p-i}\alpha\times\mathrm{Sq}^{q-n+i}\beta, a\otimes b\rangle, \end{aligned}$$

where the summation is over all i from 0 to n. Thus, $\mathrm{Sq}^k(\alpha\times\beta) = \sum_{i=0}^n \mathrm{Sq}^{p-i}\alpha\times\mathrm{Sq}^{q-n+i}\beta = \sum_{j=p-n}^p \mathrm{Sq}^j\alpha\times\mathrm{Sq}^{k-j}\beta$. It remains to show that the summation limits are as required. The terms in $\mathrm{Sq}^k(\alpha\times\beta) = \sum_i \mathrm{Sq}^i\alpha\times\mathrm{Sq}^{k-i}\beta$ are nonzero only for $0\le i\le p$ and $0\le k-i\le q$, i.e., for $k-q\le i\le k$. Therefore, the summation is over i from $k-q=p-n$ to p.

The Cartan formula has the following interpretation. For each element $\alpha\in H^*(K;\mathbb{Z}_2)$, we set $\mathrm{Sq}\,\alpha=\sum_k\mathrm{Sq}^k\alpha$ (this sum contains only finitely many nonzero terms). The map $\mathrm{Sq}\colon H^*(K;\mathbb{Z}_2)\to H^*(K;\mathbb{Z}_2)$ is a ring homomorphism. Indeed, if $\dim\alpha=p$ and $\dim\beta=q$, then the component of dimension $(p+q+k)$ of the element $\mathrm{Sq}\,\alpha\smile\mathrm{Sq}\,\beta = \left(\sum_i\mathrm{Sq}\,\alpha\right)\smile\left(\sum_j\mathrm{Sq}^j\beta\right)$ has the form $\sum\mathrm{Sq}^i\alpha\smile\mathrm{Sq}^j\beta$, where the summation is over all i and j for which $(i+p)+(j+q)=p+q+k$, i.e., $i+j=k$. Therefore, it has the form

$\sum_i \mathrm{Sq}^i\alpha \smile \mathrm{Sq}^{k-i}\beta$. According to the Cartan formula, this sum is equal to $\mathrm{Sq}^k(\alpha \smile \beta)$; thus, it is equal to the $(p+q+k)$-dimensional component of $\mathrm{Sq}(\alpha \smile \beta)$.

Using the homomorphism Sq, we can calculate $\mathrm{Sq}^k(\alpha^n)$ for any $\alpha \in H^1(K;\mathbb{Z}_2)$.

Theorem 3.62. *If $\alpha \in H^1(K;\mathbb{Z}_2)$, then $\mathrm{Sq}^k(\alpha^n) = \binom{n}{k}\alpha^{n+k}$.*

Proof. The properties of the Steenrod squares imply that $\mathrm{Sq}^k\alpha = 0$ for $k \geq 2$, $\mathrm{Sq}^0\alpha = \alpha$, and $\mathrm{Sq}^1\alpha = \alpha^2$. Therefore, $\mathrm{Sq}\,\alpha = \alpha + \alpha^2$. Taking into account the fact that Sq is a homomorphism, we obtain $\mathrm{Sq}(\alpha^n) = (\mathrm{Sq}\,\alpha)^n = (\alpha+\alpha^2)^n = \alpha\sum_k \binom{n}{k}\alpha^k$. Hence $\mathrm{Sq}^k(\alpha^n) = \binom{n}{k}\alpha^{n+k}$. □

Problem 112. Prove that if the cohomology class α is one-dimensional, then

$$\mathrm{Sq}(1+\alpha+\alpha^2+\alpha^4+\cdots) = 1+\alpha.$$

Problem 113 (Wu formula). Prove that

$$\mathrm{Sq}^k w_m = w_k w_m + \binom{m-k}{1} w_{k-1}w_{m+1} + \binom{m-k+1}{2} w_{k-2}w_{m+2} + \cdots + \binom{m-1}{k} w_0 w_{m+k}.$$

Chapter 4

Singular Homology

1. Basic Definitions and Properties

Simplicial homology groups are defined only for simplicial complexes, and their topological invariance is far from being obvious. There is an approach to define homology for an arbitrary topological space X, and the topological invariance of homology thus defined is obvious. This approach is as follows. Consider all continuous maps $f\colon \Delta^k \to X$, where Δ^k is the simplex $[0, 1, \dots, k]$; such maps are called *singular k-simplices.* By $C_k(X; G)$ we denote the group of all finite sums $\sum a_i f_i$, where $a_i \in G$ and f_i is a singular k-simplex.

To define the boundary homomorphism ∂, we introduce maps $\varepsilon_j = \varepsilon_j^k\colon \Delta^{k-1} \to \Delta^k$ for $j = 0, 1, \dots, k$ by setting $\varepsilon_j^k[i] = [i]$ for $i < j$ and $\varepsilon_j^k[i] = [i+1]$ for $i \geq j$; these maps are extended to the entire simplex by linearity. For a singular simplex $f\colon \Delta^k \to X$, we set $\partial f = \sum_{j=0}^k (-1)^j f\varepsilon_j^k \in C_{k-1}(X; G)$. For a zero-dimensional simplex f, we set $\partial f = 0$.

Theorem 4.1. $\partial\partial = 0$.

Proof. First, note that if $i < j$, then $\varepsilon_j^{k+1}\varepsilon_i^k = \varepsilon_i^{k+1}\varepsilon_{j-1}^k$. Indeed, both maps $\varepsilon_j^{k+1}\varepsilon_i^k$ and $\varepsilon_i^{k+1}\varepsilon_{j-1}^k$ take $[a]$ to $[a]$ (if $a < i$), to $[a+1]$ (if $i \leq a < j-1$), or to $[a+2]$ (if $a \geq j-1$). Therefore,

$$\begin{aligned}\partial\partial f &= \partial\Big(\sum_j (-1)^j f\varepsilon_j\Big) = \sum_{i,j} (-1)^{i+j} f\varepsilon_j\varepsilon_i \\ &= \sum_{i \geq j} (-1)^{i+j} f\varepsilon_j\varepsilon_i + \sum_{i<j} (-1)^{i+j} f\varepsilon_i\varepsilon_{j-1}.\end{aligned}$$

Let us replace i by j and j by $i+1$ in the second sum. After such a change, the second sum cancels the first one. $\square$

Again, consider the groups $Z_k = \operatorname{Ker} \partial_k$ and $B_k = \operatorname{Im} \partial_{k+1}$. According to Theorem 4.1, we have $B_k \subset Z_k$; therefore, the quotient $H_k(X; G) = Z_k/B_k$ is defined. This group is called the *singular homology group* of the space X with coefficients in G. We shall use the same notation for singular and simplicial homology; we shall show in Section 1.4 that the singular homology of any simplicial complex is isomorphic to its simplicial homology.

Theorem 4.2. *If a space X has n path-connected components, then we have* $H_0(X; G) = \underbrace{G \oplus \cdots \oplus G}_{n}$.

Proof. Choose points $x_1, \dots, x_n$, one in each path-connected component of X. If a point y_i belongs to the same component as x_i, then there exists a path from x_i to y_i. This path is a singular simplex $f\colon \Delta^1 \to X$; we have $\partial(af) = a[y_i] - a[x_i]$. Thus, the singular chain $a[y_i]$ coincides with $a[x_i]$ up to a boundary. Now suppose that $a[x_i] = \partial c$, where $c = \sum a_i f_i$ and f_i is a path from z_i to w_i. Then $\partial c = \sum a_i[w_i] - \sum a_i[z_i]$. In this expression, the sum of all coefficients vanishes; comparing the sum of coefficients on the left- and right-hand sides of the equality $a[x_i] = \sum a_i[w_i] - \sum a_i[z_i]$, we obtain $a = 0$. $\square$

Consider $C_k = C_k(X)$ and $C'_k = C_k(Y)$. Any continuous map $\varphi\colon X \to Y$ induces the map $\varphi_k\colon C_k \to C'_k$ that assigns to a singular simplex $f\colon \Delta^k \to X$ the singular simplex $\varphi f\colon \Delta^k \to Y$. Moreover, $\varphi_{k-1}\partial_k f = \sum (-1)^j \varphi f \varepsilon_j^k = \partial'_k \varphi_k f$. The equality $\varphi_{k-1}\partial_k = \partial'_k \varphi_k$ implies $\varphi_k(\operatorname{Ker} \partial_k) \subset \operatorname{Ker} \partial'_k$ and $\varphi_k(\operatorname{Im} \partial_{k+1}) \subset \operatorname{Im} \partial'_{k+1}$. We obtain a homomorphism $\varphi_*\colon H_k(X) \to H_k(Y)$.

The identity map $X \to X$ induces the identity map $H_k(X) \to H_k(X)$; hence a homeomorphism $X \to Y$ induces an isomorphism $H_k(X) \to H_k(Y)$. This assertion can be generalized as follows: any homotopy equivalence $X \to Y$ induces an isomorphism $H_k(X) \to H_k(Y)$. The following theorem is even more general.

Theorem 4.3. *If maps $\varphi, \psi\colon X \to Y$ are homotopic, then the homomorphisms $\varphi_*, \psi_*\colon H_k(X) \to H_k(Y)$ coincide.*

Proof. As in the case of simplicial homology, we define a *chain homotopy* as a family of homomorphisms $D_k\colon C_k(X) \to C_{k+1}(Y)$ for which

$$D_{k-1}\partial_k + \partial'_{k+1} D_k = \varphi_k - \psi_k. \tag{31}$$

The proof of the equality $\varphi_* = \psi_*$ remains the same.

Let $H\colon T \times I \to Y$ be a homotopy between φ and ψ. From this homotopy we construct a chain homotopy $D_k\colon C_k(X) \to C_{k+1}(Y)$ as follows. Let

$f\colon \Delta^k \to X$ be a singular simplex. Consider the map $F\colon \Delta^k \times I \to Y$ defined by $F(x,t) = H(f(x),t)$. We set $D_k f = \sum_{i=0}^{k}(-1)^i F\delta_i^{k+1}$, where δ_i^{k+1} is the linear map from the simplex $[0,1,\dots,k+1]$ to the simplex $[0_0,\dots,i_0,i^1,\dots,k^1] \subset \Delta^k \times I$; here $0_0,\dots,i_0$ are the vertices of $\Delta^k \times \{0\}$ and $i^1,\dots,k^1$ are the vertices of $\Delta^k \times \{1\}$.

The singular chains $D_{k-1}\partial_k f$ and $\partial'_{k+1} D_k f$ are linear combinations of singular simplices of the form $F\alpha\colon \Delta^k \to Y$, where α is a map from $[0,\dots,k]$ to a simplex in $\delta^k \times I$. To prove (31), is suffices to determine which simplices in $\delta^k \times I$ are obtained in this way (with signs taken into account). First, consider the case $k=2$. The map D_2 takes the simplex $[0,1,2]$ to

$$+0_0 0^1 1^1 2^1 - 0_0 1_0 1^1 2^1 + 0_0 1_0 2_0 2^1,$$

and the map ∂'_3 takes these simplices to

$$\begin{array}{ccc}
+0^1 1^1 2^1 & -1_0 1^1 2^1 & +1_0 2_0 2^1 \\
\underline{-0_0 1^1 2^1} & \underline{+0_0 1^1 2^1} & -0_0 2_0 2^1 \\
+0_0 0^1 2^1 & \underline{\underline{-0_0 1_0 2^1}} & \underline{\underline{+0_0 1_0 2^1}} \\
-0_0 0^1 1^1 & +0_0 1_0 1^1 & -0_0 1_0 2_0.
\end{array}$$

Note that the simplices underlined likewise cancel each other. For an arbitrary k, the picture is similar: for $i \neq 0, k$, the map ∂'_{k+1} deletes the vertex $i+2$ from the simplices $(-1)^i[0_0,\dots,i_0,i^1,(i+1)^1,\dots,k^1]$ and $(-1)^{i+1}[0_0,\dots,i_0,(i+1)_0,(i+1)^1,\dots,k^1]$, and the simplex $[0_0,\dots,i_0,(i+1)^1,\dots,k^1]$ occurs twice with opposite signs.

The map ∂_2 takes the simplex $[0,1,2]$ to $[1,2]-[0,2]+[0,1]$, and D_1 takes these simplices to

$$\begin{array}{ccc}
+1_0 1^1 2^1 & -0_0 0^1 2^1 & +0_0 0^1 1^1 \\
-1_0 2_0 2^1 & +0_0 2_0 2^1 & -0_0 1_0 1^1.
\end{array}$$

In the sum $\partial'_3 D_2 + D_1\partial_2$, all simplices except $+0^1 1^1 2^1$ and $-0_1 1_1 2_1$ are canceled. The remaining simplices correspond to the maps φf and ψf. It is required to show that, for any k, each of the simplices $[\dots\hat{\jmath}\dots i_0 i^1\dots]$ (or $[\dots i_0 i^1\dots\hat{\jmath}\dots]$) is encountered precisely once in $\partial'_{k+1}D_k$ and in $D_{k-1}\partial_k$ with opposite signs. This is easy. □

Example 50. Suppose X is a contractible space. Then $H_0(X;G) = G$ and $H_k(X;G) = 0$ for $k > 0$.

Proof. Any contractible space is homotopy equivalent to a singleton $*$. Therefore, it suffices to show that $H_0(*;G) = G$ and $H_k(*;G) = 0$ for $k > 0$. For any $k \geq 0$, there exists a unique map $\Delta^k \to *$; let us denote it by f_k. Clearly, $\partial_k f_k = \sum_{i=0}^{k}(-1)^i f_{k-1}$ for $k \geq 1$ and $\partial_0 f_0 = 0$. Hence

$\partial_{2k-1} f_{2k-1} = 0$ and $\partial_{2k} f_{2k} = f_{2k-1}$ ($k \geq 1$). Thus, we have $Z_0 = G$ and $B_0 = 0$, $Z_{2k-1} = G$ and $B_{2k-1} = G$, and $Z_{2k} = 0$ and $B_{2k} = 0$. □

1.1. Excision Theorem and the Mayer–Vietoris Exact Sequence. The definitions of relative and reduced singular homology groups, as well as of singular cohomology groups, are the same as in the simplicial case. The construction of the exact homology sequence of a pair does not require any changes either. But when constructing the Mayer–Vietoris exact sequence, one encounters problems because not every singular chain in $X \cup Y$ can be represented as the sum of two singular chains with supports in X and Y. This problem is essential; the Mayer–Vietoris exact sequence for singular homology exists only under certain constraints. Its existence is closely related to the following theorem.

Theorem 4.4 (on excision). *Suppose $U \subset A \subset X$ and $\overline{U} \subset \operatorname{int} A$. Then the inclusion of pairs $(X \setminus U, A \setminus U) \subset (X, A)$ induces an isomorphism $H_*(X \setminus U, A \setminus U) \to H_*(X, A)$.*

Proof. Consider $X_1 = A$ and $X_2 = X \setminus U$. We have $\operatorname{int}(X \setminus U) = X \setminus \overline{U} \supset X \setminus \operatorname{int} A$; therefore, $\operatorname{int} X_1 \cup \operatorname{int} X_2 = X$.

Lemma 1. *Let $X = \operatorname{int} X_1 \cup \operatorname{int} X_2$. Then, for any singular simplex $f\colon \Delta^k \to X$, there exists a positive integer $m = m(f)$ such that the image of any simplex from the mth barycentric subdivision of Δ^k is contained entirely either in X_1 or in X_2.*

Proof. The sets $f^{-1}(\operatorname{int} X_1)$ and $f^{-1}(\operatorname{int} X_2)$ form an open cover of the compact set $\Delta^k \subset \mathbb{R}^k$. Let δ be the Lebesgue number (see Part I, p. 59) of this cover. If d is the diameter of the simplex Δ^k, then the diameter of any simplex in its mth barycentric subdivision is at most $\left(\frac{k}{k+1}\right)^m d$. It remains to choose m so that $\left(\frac{k}{k+1}\right)^m d < \delta$. □

To each singular simplex $f\colon \Delta^k \to X$ we can assign its barycentric subdivision $\operatorname{sd} f = \sum f_\alpha$, where f_α is the restriction of f to a simplex Δ^k_α from the barycentric subdivision of Δ^k. Formally, the singular chain $\operatorname{sd} f$ is defined as follows. On p. 8, we defined a chain map $i_k\colon C_k(K) \to C_k(K')$ for which $i_k(\Delta^k) = [b, i_{k-1}\partial\Delta^k] = \sum \pm[b, b', b'', \dots, v_i]$, where b is the barycenter of the simplex Δ^k, b' is the barycenter of its $(k-1)$-face $\Delta' \subset \Delta^k$, b'' is the barycenter of a $(k-2)$-face $\Delta'' \subset \Delta' \subset \Delta^k$, $\dots$, and v_i is a vertex of Δ^k. We set $\operatorname{sd} f = \sum \pm g_\alpha$, where g_α is the composition

$$\Delta^k = [v_0, v_1, \dots, v_k] \to [b, b', b'', \dots, v_i] \xrightarrow{f} X$$

(here the first arrow is a homeomorphism of simplices treated as ordered sets of vertices).

The map sd can be extended to a homomorphism $\mathrm{sd}\colon C_*(X) \to C_*(X)$ by linearity. This is a chain map: the equality $\partial\,\mathrm{sd} = \mathrm{sd}\,\partial$ follows from $i\partial = \partial i$.

Lemma 2. *For each positive integer m, there exists a family of homomorphisms $D_k\colon C_k(X) \to C_{k+1}(X)$ such that*

(i) $\partial_{k+1} D_k c + D_{k-1}\partial_k c = \mathrm{sd}^m\, c - c$;

(ii) *the support*[1] *of the chain $D_k c$ is contained in that of the chain c.*

Proof. We set $D_0 = 0$ and construct the homomorphism D_k assuming that $D_0, \ldots, D_{k-1}$ are already constructed. Consider the singular simplex $\mathrm{id}_k\colon \Delta^k \to \Delta^k$ in $C_k(\Delta^k)$. It is easy to show that the chain $z_k = \mathrm{sd}^m \mathrm{id}_k - \mathrm{id}_k - D_{k-1}\partial_k \mathrm{id}_k$ is a cycle. Indeed, by assumption, for the chain $j_{k-1} = \partial_k \mathrm{id}_k$, we have

$$\partial_k D_{k-1} j_{k-1} + D_{k-2}\partial_{k-1} j_{k-1} = \mathrm{sd}^m\, j_{k-1} - j_{k-1},$$

and $D_{k-2}\partial_{k-1} j_{k-1} = D_{k-2}\partial_{k-1}\partial_k \mathrm{id}_k = 0$. Therefore,

$$\partial_k z_k = \partial_k \mathrm{sd}^m \mathrm{id}_k - \partial_k \mathrm{id}_k - \mathrm{sd}^m \partial_k \mathrm{id}_k + \partial_k \mathrm{id}_k = 0$$

because $\partial_k \mathrm{sd}^m = \mathrm{sd}^m \partial_k$.

We already know that $H_k(\Delta^k) = 0$ for $k \geq 1$ because the simplex Δ^k is a contractible space. Hence there exists a singular chain $b_{k+1} \in C_{k+1}(\Delta^k)$ for which $z_k = \partial b_{k+1}$. Suppose that $b_{k+1} = \sum a_i f_i$, where $f_i\colon \Delta^{k+1} \to \Delta^k$. For any singular simplex $f\colon \Delta^k \to X$, we set $D_k f = \sum a_i (f \circ f_i)$ and extend D_k by linearity. The required equality

$$\partial_{k+1} D_k f + D_{k-1}\partial_k f = \mathrm{sd}^m\, f - f$$

follows from $\sum a_i f_i + D_{k-1}\partial_k = \mathrm{sd}^m \mathrm{id}_k - \mathrm{id}_k$ in view of $\partial_{k+1} D_k f = \sum a_k \partial_{k+1}(f \circ f_i)$. $\square$

We continue the proof of Theorem 4.4. According to Lemma 2, any relative cycle $z_k \in C_k(X, A)$ is homologous to the relative cycle $\mathrm{sd}^m\, z_k$ because $D_{k-1}\partial_k z_k \in C_k(A)$. Any cycle contains only finitely many singular simplices; therefore, by Lemma 1, we can choose a number m for each cycle z_k so that the support of any singular simplex included in $\mathrm{sd}^m\, z_k$ is contained entirely in $X_1 = A$ or $X_2 = X \setminus U$. Thus, the map $H(X \setminus U, A \setminus U) \to H_*(X, A)$ is an epimorphism. It remains to show that if $z_k \in C_k(X \setminus U)$ and $z_k - \partial b \in C_k(A)$ for some $b \in C_{k+1}(X)$, then $z_k - \partial b' \in C_k(A \setminus U)$ for some $b' \in C_k(X \setminus U)$.

Let $C_k^{\{X_1, X_2\}}(X)$ be the group of sums of chains from $C_k(X_1)$ and $C_k(X_2)$. By assumption, $\partial b \in C_k^{\{X_1, X_2\}}(X)$. Choose a number m so that

[1] By the support of a chain we mean here the union of the images of all singular simplices in this chain.

$\operatorname{sd}^m b \in C_{k+1}^{\{X_1,X_2\}}(X)$. We have

$$b + \partial D b = \operatorname{sd}^m b - D\partial b \in C_{k+1}^{\{X_1,X_2\}}(X).$$

Let $\operatorname{sd}^m b - D\partial b = b' + b''$, where $b' \in C_{k+1}(X_2)$ and $b'' \in C_{k+1}(X_1)$. Then

$$z_k - \partial b' - \partial b'' = z_k - \partial(b + \partial D b) = z_k - \partial b \in C_k(X_1).$$

Moreover, $z_k, \partial b' \in C_k(X_2)$ and $\partial b'' \in C_k(X_1)$. Hence

$$z_k - \partial b' \in C_k(X_1) \cap C_k(X_2) = C_k(A) \cap C_k(X \setminus U) = C_k(A \setminus U),$$

as required. □

Now, consider the Mayer–Vietoris exact sequence. For any two sets $X_1, X_2 \subset X$, we have the short exact sequence

$$0 \longrightarrow C_*(X_1 \cap X_2) \xrightarrow{(j_1,-j_2)} C_*(X_1) \oplus C_*(X_2) \xrightarrow{(i_1,i_2)} C_*^{\{X_1,X_2\}}(X_1 \cup X_2) \longrightarrow 0,$$

where $C_*^{\{X_1,X_2\}}(X_1 \cup X_2) = C_*(X_1) + C_*(X_2)$ is the group of sums of chains. This short exact sequence induces the exact sequence

$$\cdots \longrightarrow H_k(X_1 \cap X_2) \xrightarrow{j_*} H_k(X_1) \oplus H_k(X_2)$$
$$\xrightarrow{i_*} H_k(C_*^{\{X_1,X_2\}}(X_1 \cup X_2)) \xrightarrow{\partial_*} H_{k-1}(X_1 \cap X_2) \longrightarrow \cdots.$$

If the embedding $C_*^{\{X_1,X_2\}}(X_1 \cup X_2) \to C_*(X_1 \cup X_2)$ induces an isomorphism of homology groups, then we say that the pair $\{X_1, X_2\}$ satisfies the *excision axiom*. In this case, the group $H_k(C_*^{\{X_1,X_2\}}(X_1 \cup X_2))$ can be replaced by $H_*(X_1 \cup X_2)$; the exact homology sequence is then called the *Mayer–Vietoris sequence*.

Example 51. Suppose that X_1 and X_2 are two semicircles, X_1 contains both boundary points, and X_2 contains no boundary points (see Figure 1). Then the Mayer–Vietoris exact sequence does not hold for the pair $\{X_1, X_2\}$.

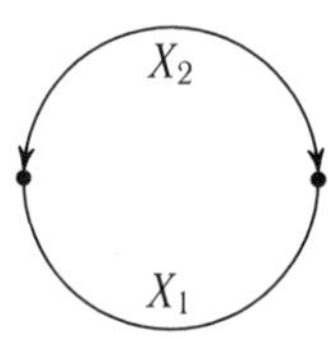

Figure 1. The semicircles X_1 and X_2

Proof. By assumption, $X_1 \cup X_2 = S^1$ and $X_1 \cap X_2 = \varnothing$. Therefore, the fragment $H_1(X_1) \oplus H_1(X_2) \to H_1(X_1 \cup X_2) \to H_1(X_1 \cap X_2)$ has the form $0 \to \mathbb{Z} \to 0$. □

In the proof of the excision theorem, we showed that if $X_1 \cup X_2 = \operatorname{int} X_1 \cup \operatorname{int} X_2$, then the pair $\{X_1, X_2\}$ satisfies the excision axiom. The surjectivity of the homomorphism $H_*(C_*^{\{X_1,X_2\}}(X)) \to H_*(X)$, where $X = X_1 \cup X_2$, was proved explicitly, and its injectivity follows from the existence of a chain $d = Db \in C_{k+2}(X)$ such that $b + \partial d \in C_{k+1}^{\{X_1,X_2\}}(X)$ for any $b \in C_{k+1}(X)$ with $\partial b \in C_k^{\{X_1,X_2\}}(X)$; such a chain exists because ∂b is the boundary of the chain $b + \partial d$.

Example 52. $\tilde{H}_k(S^n) = \begin{cases} \mathbb{Z} & \text{if } k = n, \\ 0 & \text{if } k \neq n. \end{cases}$

Proof. Suppose that x_1 and x_2 are two different points on the sphere S^n, $X_1 = S^n \setminus \{x_1\}$, and $X_2 = S^n \setminus \{x_2\}$. The spaces X_1 and X_2 are contractible; therefore, $\tilde{H}_k(X_i) = 0$ for all k.

The pair $\{X_1, X_2\}$ satisfies the excision axiom because the sets X_1 and X_2 are open in S^n. The Mayer–Vietoris exact sequence gives the isomorphism

$$\tilde{H}_k(S^n) \cong \tilde{H}_{k-1}(S^n \setminus \{x_1, x_2\}).$$

Moreover, $S^n \setminus \{x_1, x_2\} \sim S^{n-1}$. Thus, $\tilde{H}_k(S^n) \cong \tilde{H}_{k-1}(S^{n-1})$. It remains to note that the required assertion is obviously true for $n = 0$. □

Example 53. $H_k(D^n, S^{n-1}) = \begin{cases} \mathbb{Z} & \text{if } k = n, \\ 0 & \text{if } k \neq n. \end{cases}$

Proof. The exact sequence of the pair (D^n, S^{n-1}) shows that

$$H_k(D^n, S^{n-1}) \cong \tilde{H}_{k-1}(S^{n-1}).$$ □

Exercise. Prove the *suspension isomorphism* $H_k(\Sigma X) \cong \tilde{H}_{k-1}(X)$ for all $k \geq 1$.

Problem 114. Prove that $H_i(X, Y) \cong H_i(X \cup CY, CY)$.

Problem 115. Prove that $H_i(X, Y) \cong H_i(X \cup CY)$ for $i \geq 1$.

Problem 116. Given a connected CW-complex X and its subcomplex Y, prove that $H_i(X, Y) \cong \tilde{H}_i(X/Y)$.

Suppose that (X_1, A_1) and (X_2, A_2) are two pairs of topological spaces, $A_i \subset X_i \subset X$ for $i = 1, 2$, and the pairs of sets $\{X_1, X_2\}$ and $\{A_1, A_2\}$ satisfy the excision axiom. Then we have the exact sequence

$$\longrightarrow H_k(X_1 \cap X_2, A_1 \cap A_2) \xrightarrow{j_*} H_k(X_1, A_1) \oplus H_k(X_2, A_2)$$
$$\xrightarrow{i_*} H_k(X_1 \cup X_2, A_1 \cup A_2) \xrightarrow{\partial_*} H_{k-1}(X_1 \cap X_2, A_1 \cap A_2) \longrightarrow;$$

it is called the *relative Mayer–Vietoris sequence.* To construct this sequence, note that the map induced in relative homology by the embedding

$$\frac{C_*(X_1)+C_*(X_2)}{C_*(A_1)+C_*(A_2)} \to \frac{C_*(X_1\cup X_2)}{C_*(A_1\cup A_2)}$$

is an isomorphism. Indeed, consider the exact homology sequence of the pair $(X_1 \cup X_2, A_1 \cup A_2)$ and the corresponding exact sequence for homology of the form $H_*(C_*^{\{X_1,X_2\}}(X_1\cup X_2))$, which arises from the obvious short exact sequence. The natural embeddings of chain complexes induce a map from the latter exact sequence to the former, and the maps of absolute homology groups are isomorphisms. It follows by the five lemma that the maps of relative homology groups are isomorphisms as well.

It remains to construct a short exact sequence

$$0 \longrightarrow \frac{C_*(X_1\cap X_2)}{C_*(A_1\cap A_2)} \longrightarrow \frac{C_*(X_1)}{C_*(A_1)} \oplus \frac{C_*(X_2)}{C_*(A_2)} \longrightarrow \frac{C_*(X_1)+C_*(X_2)}{C_*(A_1)+C_*(A_2)} \longrightarrow 0,$$

i.e., to prove that the quotient of a short exact sequence by an exact subsequence is an exact sequence (this is true for any, not necessarily short, exact sequences). Consider the exact sequence as a chain complex C and the subsequence as a chain subcomplex C'. The exactness of these sequences means that $H_*(C) = 0$ and $H_*(C') = 0$. Let us write the exact sequence induced in homology by the short exact sequence $0 \to C' \to C \to C/C' \to 0$:

$$H_k(C) \xrightarrow{p_*} H_k(C/C') \xrightarrow{\partial_*} H_{k-1}(C').$$

We obtain $H_k(C/C') = 0$, as required.

Singular Cohomology. *Singular cohomology* groups are defined in a natural way. Let $C_k(X)$ be the group of singular chains with coefficients in $\mathbb{Z}$. Then $C^k(X;G) = \operatorname{Hom}(C_k(X), G)$ is the group of singular cochains with coefficients in G. The coboundary operator $\delta\colon C^k(X;G) \to C^{k+1}(X;G)$ is the dual of the boundary operator $\partial\colon C_{k+1}(X) \to C_k(X)$. The group Z^k of cocycles, the group B^k of coboundaries, and the cohomology group Z^k/B^k are defined in a standard way.

For singular cohomology, a theorem similar to the excision theorem for singular homology is valid. Indeed, according to the universal coefficient theorem, any chain map inducing an isomorphism in integral homology induces also an isomorphism in cohomology.

1.2. Axioms of (Co)homology Theory. Singular (co)homology theory is only one of the possible theories of (co)homology defined for all topological spaces. Another example is Čech cohomology. For simplicial complexes, these theories coincide, but for more general topological spaces, they may give different results. There are also other (co)homology theories.

Steenrod and Eilenberg [**131**] suggested axioms which any (co)homology theory must satisfy and proved that if two (co)homology theories satisfy these axioms and have the same coefficient group, then they coincide for simplicial complexes.

We say that a class $\mathcal{A}$ of pairs (X, A) of topological spaces is *admissible* if the following conditions hold:

(i) if a pair (X, A) belongs to $\mathcal{A}$, then the pairs (X, X), $(X, \varnothing)$, (A, A), and $(A, \varnothing)$ belong to $\mathcal{A}$;

(ii) if a pair (X, A) belongs to $\mathcal{A}$, then the pair $(X \times I, A \times I)$ belongs to $\mathcal{A}$;

(iii) the pair $(*, \varnothing)$, where $*$ is a singleton, belongs to $\mathcal{A}$.

Any homology theory is defined for an admissible class of pairs. To every pair (X, A) from the given admissible class and any integer n it assigns an Abelian group $H_n(X, A)$. To any map of pairs $f\colon (X, A) \to (Y, B)$ and any integer n a homomorphism $f_*\colon H_n(X, A) \to H_n(Y, B)$ is assigned. Moreover, boundary homomorphisms $\partial_*\colon H_n(X, A) \to H_{n-1}(A)$, where $H_{n-1}(A) = H_{n-1}(A, \varnothing)$, are defined. Every homology theory must satisfy the following axioms (the *Steenrod–Eilenberg axioms*).

1. If f is the identity map, then so is f_*;
2. $(gf)_* = g_* f_*$;
3. $\partial_* f_* = (f|_A)_* \partial_*$, i.e., the following diagram is commutative:

$$\begin{array}{ccc} H_n(X, A) & \xrightarrow{f_*} & H_n(Y, B) \\ \Big\downarrow \partial_* & & \Big\downarrow \partial_* \\ H_{n-1}(A) & \xrightarrow{(f|_A)_*} & H_{n-1}(B); \end{array}$$

4 (the exactness axiom). If $i\colon A \to X$ and $j\colon X \to (X, A)$ are natural embeddings, then the following sequence is exact:

$$\cdots \xrightarrow{\partial_*} H_n(A) \xrightarrow{i_*} H_n(X) \xrightarrow{j_*} H_n(X, A) \xrightarrow{\partial_*} H_{n-1}(A) \xrightarrow{i_*} \cdots;$$

5 (the homotopy axiom). If maps $f, g\colon (X, A) \to (Y, B)$ are homotopic, then the homomorphisms f_* and g_* coincide for all n;

6 (the excision axiom). If $U \subset X$ is an open set whose closure is contained in $\operatorname{int} A$ and the natural embedding $(X \setminus U, A \setminus U) \to (X, A)$ is an admissible map, then, for any n, this embedding induces an isomorphism $H_n(X \setminus U, A \setminus U) \to H_n(X, A)$;

7 (the dimension axiom). $H_n(*) = 0$ for $n \neq 0$. The group $H_0(*)$ is called the *coefficient group*.

Any cohomology theory is defined for an admissible class of pairs. To every pair (X, A) from the given admissible class and any integer n it assigns an Abelian group $H^n(X, A)$. Any map of pairs $f\colon (X, A) \to (Y, B)$ and integer n are assigned a homomorphism $f^*\colon H^n(Y, B) \to H^n(X, A)$. Moreover, coboundary homomorphisms $\delta^*\colon H^{n-1}(A) \to H^n(X, A)$, where $H^{n-1}(A) = H^{n-1}(A, \varnothing)$, are defined. Any cohomology theory must satisfy the following axioms (*the Steenrod–Eilenberg axioms*):

1. If f is the identity map, then so is f^*;
2. $(gf)^* = f^* g^*$;
3. $\delta^* f^* = (f|_A)^* \delta^*$, i.e., the following diagram is commutative:

$$\begin{array}{ccc} H^n(X,A) & \xleftarrow{f^*} & H^n(Y,B) \\ \downarrow{\scriptstyle\delta^*} & & \downarrow{\scriptstyle\delta^*} \\ H^{n-1}(A) & \xleftarrow{(f|_A)^*} & H^{n-1}(B); \end{array}$$

4 (the exactness axiom). If $i\colon A \to X$ and $j\colon X \to (X, A)$ are natural embeddings, then the following sequence is exact:

$$\cdots \xrightarrow{i^*} H^{n-1}(A) \xrightarrow{\delta^*} H^n(X,A) \xrightarrow{j^*} H^n(X) \xrightarrow{i^*} H^n(A) \xrightarrow{\delta^*} \cdots;$$

5 (the homotopy axiom). If maps $f, g\colon (X, A) \to (Y, B)$ are homotopic, then the homomorphisms f^* and g^* coincide for all n;

6 (the excision axiom). If $U \subset X$ is an open set whose closure is contained in $\operatorname{int} A$ and the embedding $(X \setminus U, A \setminus U) \to (X, A)$ is admissible, then, for any n, this embedding induces an isomorphism $H^n(X, A) \to H^n(X \setminus U, A \setminus U)$;

7 (the dimension axiom). $H^n(*) = 0$ for $n \neq 0$. The group $H^0(*)$ is called the *coefficient group*.

Shortly after Steenrod and Eilenberg published their book, it turned out that the dimension axiom plays a special role; some naturally arising theories satisfy all Steenrod–Eilenberg axioms except the dimension axiom. Such (co)homology theories are said to be *extraordinary*, or *generalized*. They include, e.g., cobordism theory and K-theory.

1.3. The Jordan–Brouwer Theorem. Brouwer [**14–18**] proved that if a subspace $B \subset \mathbb{R}^n$ is homeomorphic to S^{n-1}, then $\mathbb{R}^n \setminus B$ has two connected components. In the special case of $n = 2$, this theorem was proved earlier by Jordan.[2] Generalizing the proof of the Jordan–Brouwer theorem, Alexander obtained the following result.

[2]The proof of Jordan was not quite rigorous. The first complete proof of Jordan's theorem was suggested by Veblen in 1905.

Theorem 4.5 (Alexander [**5**]). *If $A, B \subset \mathbb{R}^n$ are homeomorphic closed subspaces, then $H_k(\mathbb{R}^n \setminus A) \cong H_k(\mathbb{R}^n \setminus B)$ for all $k \geq 0$.*

Proof (see [**30**]). Let $\mathbb{R}^{2n} = \mathbb{R}^n \times \mathbb{R}^n$. Suppose that $A \subset \mathbb{R}^n \times \{0\}$ and $B \subset \{0\} \times \mathbb{R}^n$. Then $(\mathbb{R}^{2n} \setminus A) \approx (\mathbb{R}^{2n} \setminus B)$ (see Part I, Theorem 2.5). If $A \neq \mathbb{R}^n$ and $B \neq \mathbb{R}^n$, then the required assertion is easily derived from the following lemma.

Lemma. *If $A \subset \mathbb{R}^n$ is a closed set and $A \neq \mathbb{R}^n$, then $\tilde{H}_{k+i}(\mathbb{R}^{n+i} \setminus A) \cong \tilde{H}_k(\mathbb{R}^n \setminus A)$ for any $i \geq 0$ (it is assumed that $\mathbb{R}^n = \mathbb{R}^n \times \{0\} \subset \mathbb{R}^{n+i}$).*

Proof. We first consider the case $i = 1$ and then use induction. We set $X = \mathbb{R}^{n+1} \setminus A$,

$$X_+ = \{(x,t) \in \mathbb{R}^n \times \mathbb{R} \mid t > 0 \text{ or } x \in \mathbb{R}^n \setminus A\},$$
$$X_- = \{(x,t) \in \mathbb{R}^n \times \mathbb{R} \mid t < 0 \text{ or } x \in \mathbb{R}^n \setminus A\}.$$

The sets X_+ and X_- are open in X, $X_+ \cup X_- = X$, and $X_+ \cap X_- = (\mathbb{R}^n \setminus A) \times \mathbb{R}$. Let us write the exact Mayer–Vietoris sequence for reduced homology:

$$\cdots \longrightarrow \tilde{H}_{k+1}(X_+) \oplus \tilde{H}_{k+1}(X_-) \longrightarrow \tilde{H}_{k+1}(X_+ \cup X_-)$$
$$\longrightarrow \tilde{H}_k(X_+ \cap X_-) \longrightarrow \tilde{H}_k(X_+) \oplus \tilde{H}_{k+1}(X_-) \longrightarrow \cdots.$$

The spaces X_+ and X_- are contractible; hence they have trivial reduced homology groups, and therefore $\tilde{H}_{k+1}(X_+ \cup X_-) \cong \tilde{H}_k(X_+ \cap X_-)$, i.e., $\tilde{H}_{k+1}(\mathbb{R}^{n+1} \setminus A) \cong \tilde{H}_k((\mathbb{R}^n \setminus A) \times \mathbb{R}) \cong \tilde{H}_k(\mathbb{R}^n \setminus A)$. □

If $A \neq \mathbb{R}^n$, then, according to the lemma, we have $\tilde{H}_k(\mathbb{R}^n \setminus A) \cong \tilde{H}_{k+n}(\mathbb{R}^{2n} \setminus A)$; if $A = \mathbb{R}^n$, then we obtain the isomorphism $\tilde{H}_k(\mathbb{R}^{n+1} \setminus A) \cong \tilde{H}_{k+n-1}(\mathbb{R}^{2n} \setminus A)$. Thus, if $A \neq \mathbb{R}^n$ and $B \neq \mathbb{R}^n$, then $\tilde{H}_k(\mathbb{R}^n \setminus A) \cong \tilde{H}_k(\mathbb{R}^n \setminus B)$ and $H_k(\mathbb{R}^n \setminus A) \cong H_k(\mathbb{R}^n \setminus B)$. If $A = \mathbb{R}^n$, then $\tilde{H}_k(\mathbb{R}^{n+1} \setminus A) \cong \tilde{H}_k(\mathbb{R}^{n+1} \setminus B)$. In particular, $\tilde{H}_0(\mathbb{R}^{n+1} \setminus A) \cong \tilde{H}_0(\mathbb{R}^{n+1} \setminus B)$. Since the set $\mathbb{R}^{n+1} \setminus A = \mathbb{R}^{n+1} \setminus \mathbb{R}^n$ has precisely two connected components, it follows that so does $\mathbb{R}^{n+1} \setminus B$. But this is possible only if $B = \mathbb{R}^n$. □

The Alexander theorem readily implies the following assertion, which was proved by Brouwer in [**17**].

Theorem 4.6 (invariance of domain). *If U and V are homeomorphic subspaces of $\mathbb{R}^n$ and U is open, then V is open as well.*

Proof. Take a point $v_0 \in V$. We must prove that V has a subset that is open in $\mathbb{R}^n$ and contains the point v_0. Consider a homeomorphism $h\colon U \to V$ and let $u_0 = h^{-1}(v_0)$. Choose $\varepsilon > 0$ so that $D^n = \{x \in \mathbb{R}^n \mid \|x - u_0\| \leq \varepsilon\} \subset U$. According to Alexander's theorem, the set $\mathbb{R}^n \setminus h(D^n)$ is connected, while the set $\mathbb{R}^n \setminus h(S^{n-1})$, where $S^{n-1} = \partial D^n$, has two connected components. The sets $\mathbb{R}^n \setminus h(D^n)$ and $h(D^n) \setminus h(S^{n-1}) = h(\operatorname{int} D^n)$ are connected, and their union is $\mathbb{R}^n \setminus h(S^{n-1})$. Therefore, it is these sets that are the connected components. In particular, they are open subsets of the space $\mathbb{R}^n \setminus h(S^{n-1})$, which, in turn, is an open subset in $\mathbb{R}^n$. Hence the set $h(\operatorname{int} D^n) \ni h(u_0) = v_0$ is open in $\mathbb{R}^n$. □

Corollary (invariance of boundary). *A homeomorphism $D^n \to D^n$ cannot take a point $x \in \partial D^n$ to an interior point.*

Example 54. Suppose that $X = (\mathbb{R} \times \{0\}) \cup (\{0\} \times \mathbb{R}) \subset \mathbb{R}^2$, $U = \{(0,t) \mid 2 < t < 3\}$, and $V = \{(0,t) \mid -1 < t < 1\}$ (see Figure 2). Then the sets U

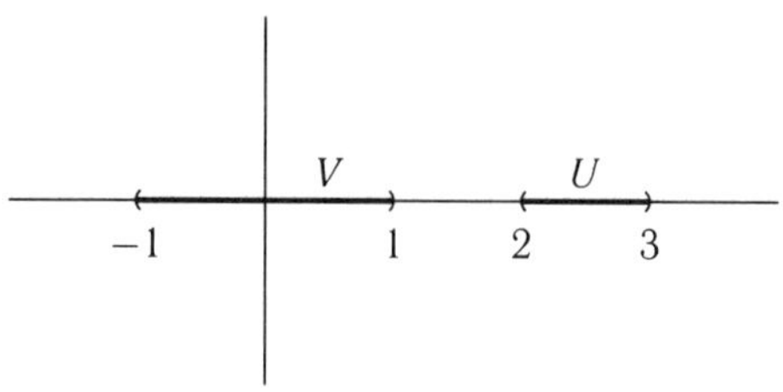

Figure 2. Noninvariance of domains

and V are homeomorphic, U is open in X, but V is not open in X.

Problem 117. Suppose that U and V are homeomorphic subspaces of $\mathbb{R}^n$ and U is closed. Is it true that V must be closed?

1.4. Isomorphism between Simplicial Homology and Singular Homology. For any simplicial complex K, both simplicial and singular homology groups are defined. In this section, we prove that these groups are isomorphic. It is more convenient to construct a chain map that induces this isomorphism by using the total (ordered) chain complex $\hat{C}_*(K)$, whose homology groups are isomorphic to the simplicial homology groups (see Theorem 2.1 on p. 60), rather than the simplicial chain complex.

To any ordered k-simplex $(v_0, \dots, v_k)$ there corresponds the singular simplex $f\colon [0, \dots, k] \to [v_0, \dots, v_k]$, where f is the linear map of simplices taking each point i to v_i. This correspondence, when extended by linearity, determines a homomorphism $j_k\colon \hat{C}_k(K) \to C_k(K)$, where $C_k(K)$ is the singular chain complex. It follows directly from the definition of boundary homomorphisms in $\hat{C}_*$ and C_* that j is a chain map. Therefore, j induces a homomorphism $j_*\colon H_k(\hat{C}_*(K)) \to H_k(K)$ of homology groups.

Theorem 4.7. *The homomorphism j_* is an isomorphism.*

Proof. First, suppose that the simplicial complex K is finite. In this case, the theorem is proved by induction on the number n of all simplices of all dimensions in K. If $n = 1$, then K consists of one point. For a one-point space, the isomorphism is obvious even at the level of chains.

Now suppose that j_* is an isomorphism for any simplicial complex containing fewer than n simplices. Let K be a simplicial complex containing precisely n simplices, and let Δ be a simplex of maximal dimension in K. The simplex Δ is not a face; therefore, $K_1 = K \setminus \Delta$ is a subcomplex of K containing fewer than n simplices.

To simplify notation, we set $\hat{H}_k(K) = H_k(\ddot{C}_*(K))$. For singular homology, we consider the Mayer–Vietoris sequence for Δ and K_1', where $K_1' = K_1 \cup (\operatorname{int}\Delta \setminus \{x_0\})$ for $x_0 \in \operatorname{int}\Delta$ because the pair Δ, K_1 does not satisfy the excision axiom. As a result, we obtain the following commutative diagram with exact rows:

$$
\begin{array}{ccccccccc}
\cdots & \longrightarrow & \hat{H}_{k+1}(\partial\Delta) & \longrightarrow & \hat{H}_k(K) & \longrightarrow & \hat{H}_k(\Delta) \oplus \hat{H}_k(K_1) & \longrightarrow & \cdots \\
 & & \Big\downarrow{\scriptstyle \text{iso}} & & \Big\downarrow{\scriptstyle ?} & & \Big\downarrow{\scriptstyle \text{iso}} & & \\
 & & H_{k+1}(\partial\Delta) & & H_k(K) & & H_k(\Delta) \oplus H_k(K_1) & & \\
 & & \Big\downarrow{\scriptstyle \text{iso}} & & \Big\downarrow{\scriptstyle \text{iso}} & & \Big\downarrow{\scriptstyle \text{iso}} & & \\
\cdots & \longrightarrow & H_{k+1}(\Delta \setminus \{x_0\}) & \longrightarrow & H_k(K) & \longrightarrow & H_k(\Delta) \oplus H_k(K_1') & \longrightarrow & \cdots .
\end{array}
$$

The upper vertical arrows coincide with j_*, and the lower vertical arrows are the isomorphisms induced by homotopy equivalences (or the identity maps). It follows from the five lemma that the map $j_* \colon \hat{H}_k(K) \to H_k(K)$ is an isomorphism (to apply this lemma, we must add two terms, on the left and on the right, to the above fragment of the diagram).

Now suppose that the simplicial complex K is infinite. For any $\alpha \in H_k(K)$, there exists a finite subcomplex L of K such that α belongs to the image of $H_k(L)$ under the homomorphism induced by the embedding $L \to K$. Therefore, $j_* \colon \hat{H}_k(K) \to H_k(K)$ is an epimorphism. Suppose that $\hat{\alpha} \in \hat{H}_k(K)$ and $j_*\hat{\alpha} = 0$. At the level of chains, this means that a singular chain $c_k \in C_k(K)$ corresponding to a representative of the homology class $\hat{\alpha}$ is the boundary of some singular chain $\alpha_{k+1} \in C_{k+1}(K)$. Choose a finite subcomplex L in K so that it is a support for the chain α_{k+1}. The chain c_k represents the zero homology class in $H_k(L)$; therefore, $\hat{\alpha} = 0$, i.e., j_* is a monomorphism. □

Theorem 4.8. *If U is an open subset of $\mathbb{R}^n$, then $H_k(U) = 0$ for $k \geq n$.*

Proof. Consider a singular cycle $z \in C_k(U)$, where $k \geq n$. We must prove that $z = \partial w$, where w is a singular chain in U. Let X be the union of all images of singular simplices in z. The set $X \subset U$ is compact; therefore, the distance ε between X and $\mathbb{R}^n \setminus U$ is positive. Let K be a simplex in $\mathbb{R}^n$ which contains U and is triangulated so that the diameter of any simplex in its triangulation is less than ε. Consider the subcomplex L of K that consists of all simplices intersecting X. We have $|L| \subset U$ because the triangulation is sufficiently fine.

Let us write the exact sequence of simplicial homology groups for the pair (K, L):

$$\cdots \longrightarrow H_k(K, L) \longrightarrow H_k(L) \longrightarrow H_k(K) \longrightarrow \cdots.$$

Here $H_k(K) = 0$ because $K \approx \Delta^n$, and $H_{k+1}(K, L) = 0$ because $k + 1 \geq n + 1 > \dim K$ (by assumption). Therefore, $H_k(L) = 0$.

Now, consider singular homology. We have $H_k(|L|) = 0$. By construction, $X \subset |L|$; hence z is a singular cycle in $|L|$, and therefore $z = \partial w$, where w is a singular chain in $|L| \subset U$. □

Remark. The analog of Theorem 4.8 for closed sets does not hold. In [**10**], an example of a closed subset in $\mathbb{R}^3$ with nontrivial homology groups of arbitrarily high dimensions was constructed.

Using the properties of nerves of covers described in Part I, we can easily derive from Theorem 4.8 the following well-known theorem of Helly.

Theorem 4.9 (Helly [**48**]). *If $U_1, \dots, U_{n+1}$ are open convex sets in $\mathbb{R}^n$ and any n of them have a common point, then all these sets have a common point.*

Proof. We set $X = U_1 \cup \cdots \cup U_{n+1}$ and consider the cover $\mathcal{U} = \{U_i\}$ $(i = 1, \dots, n+1)$ of X. Let N be the nerve of the cover $\mathcal{U}$. According to Theorem 3.21 from Part I, the space X is homotopy equivalent to $|N|$. Suppose that $U_1 \cap \cdots \cap U_{n+1} = \varnothing$. Then N consists of the n-faces of the simplex Δ^{n+1}. Therefore, $|N| \sim S^n$; in particular, $H_n(X) \simeq H_n(|N|) \cong H_n(S^n) \neq 0$. This contradicts Theorem 4.8. □

Helly's theorem (and even a more general assertion, which was also proved by Helly) can be derived from Theorem 4.8 without using the properties of nerves of covers. For this purpose, the following lemma is needed.

Lemma 4.1. *Let $X_1, \dots, X_m$ be acyclic open subsets of a topological space X and any r of them $(1 \leq r \leq m-1)$ have acyclic (in particular, nonempty) intersection.*

(a) *If $X_1 \cap \cdots \cap X_m = \varnothing$, then the group $\tilde{H}_q(X_1 \cup \cdots \cup X_m)$ is nontrivial precisely when $q = m - 2$.*

(b) *If $X_1 \cap \cdots \cap X_m \neq \varnothing$, then $\tilde{H}_q(X_1 \cup \cdots \cup X_m) \cong \tilde{H}_{q-m+1}(X_1 \cap \cdots \cap X_m)$ for all q.*

Proof. We prove the lemma by induction on m. First, consider the case $m = 2$. If $X_1 \cap X_2 = \varnothing$, then $X_1 \cup X_2$ is a union of two disjoint acyclic spaces. Therefore, $\tilde{H}_0(X_1 \cup X_2) \neq 0$ and $\tilde{H}_q(X_1 \cup X_2) = 0$ for $q \neq 0$. If $X_1 \cap X_2 \neq \varnothing$, then the reduced Mayer–Vietoris sequence

$$\tilde{H}_q(X_1) \oplus \tilde{H}_q(X_2) \longrightarrow \tilde{H}_q(X_1 \cup X_2) \\ \longrightarrow \tilde{H}_{q-1}(X_1 \cap X_2) \longrightarrow \tilde{H}_{q-1}(X_1) \oplus \tilde{H}_{q-1}(X_2)$$

implies $\tilde{H}_q(X_1 \cup X_2) \cong \tilde{H}_{q-1}(X_1 \cap X_2)$ because $\tilde{H}_{q-1}(X_i) = 0$ for all q.

Consider an arbitrary m. By the induction hypothesis, the space $U_1 = X_1 \cup \cdots \cup X_{m-1}$ is acyclic because $X_1 \cap \cdots \cap X_{m-1} \neq \varnothing$. The reduced Mayer–Vietoris sequence implies $\tilde{H}_q(U_1 \cup X_m) \cong \tilde{H}_{q-1}(U_1 \cap X_m)$. Moreover, $U_1 \cap X_m = (X_1 \cap X_m) \cup \cdots \cup (X_{m-1} \cap X_m)$. Let us apply the induction hypothesis to the sets $X_1 \cap X_m, \ldots, X_{m-1} \cap X_m$. Clearly, the intersection of r such sets is the intersection of $r+1$ of the sets $X_1, \ldots, X_m$. Suppose that $X_1 \cap \cdots \cap X_m = \varnothing$ (or, equivalently, the intersection of the sets $X_1 \cap X_m, \ldots, X_{m-1} \cap X_m$ is empty). By the induction hypothesis, the group $\tilde{H}_{q-1}(U_1 \cap X_m)$ is nontrivial if and only if $q-1 = (m-1)-2$, i.e., $q = m-2$. If $X_1 \cap \cdots \cap X_m \neq \varnothing$, then $\tilde{H}_{q-1}(U_1 \cap X_m) \cong \tilde{H}_{(q-1)-(m-1)+1}(X_1 \cap \cdots \cap X_m)$. But $(q-1)-(m-1)+1 = q-m+1$. □

Now, we can prove the following theorem; the classical Helly's theorem is its special case.

Theorem 4.10 (Helly [**49**]). *Suppose that $X_1, \ldots, X_m$ is a finite family of open subsets of $\mathbb{R}^n$ and the intersection of any r of these sets is nonempty for $r \leq n+1$ and acyclic for $r \leq n$. Then the space $X_1 \cap \cdots \cap X_m$ is acyclic (in particular, it is nonempty).*

Proof (see [**29**]). Suppose that the required assertion does not hold for sets $X_1, \ldots, X_m$ and the number m is minimal. The minimality of m implies that the intersection of any $r \leq m-1$ of the sets $X_1, \ldots, X_m$ is acyclic.

Suppose that $X_1 \cap \cdots \cap X_m = \varnothing$. This can happen only if $m > n+1$. Applying assertion (a) of Lemma 4.1, we obtain $\tilde{H}_{m-2}(X_1 \cup \cdots \cup X_m) \neq 0$. On the other hand, $m-2 \geq n$; therefore, according to Theorem 4.8, we have $\tilde{H}_{m-2}(X_1 \cup \cdots \cup X_m) = 0$. We have arrived at a contradiction.

Now, suppose that $X_1 \cap \cdots \cap X_m \neq \varnothing$. By assumption, the space $X_1 \cap \cdots \cap X_m$ is not acyclic. This can happen only if $m > n$. There exists a number $p \geq 0$ for which $\tilde{H}_p(X_1 \cup \cdots \cup X_m) \neq 0$. Let us write p in the form $p = q - m + 1$. According to assertion (b) of Lemma 4.1, we have $\tilde{H}_q(X_1 \cap \cdots \cap X_m) \neq 0$. Note that $q = p + m - 1 \geq n$. This contradicts Theorem 4.8. □

Cellular Homology. For any CW-complex X, we can define cellular homology in analogy with the case of simplicial complexes with the only difference that simplicial homology should be replaced by singular homology. Using Problem 116, we obtain $H_i(X^k, X^{k-1}) \cong H_i(X^k/X^{k-1}) \cong H_i(\bigvee_\alpha S^k_\alpha)$ for $i \geq 1$. Therefore, $H_i(X^k, X^{k-1}) = 0$ for $i \neq k$; for $i = k$, this is a free[3] Abelian group whose generators are in one-to-one correspondence with the k-cells of the complex X. This implies, in particular, that $H_{k+1}(X^k) = 0$. Indeed, the exact sequence of a pair

$$\cdots \longrightarrow H_{i+1}(X^k, X^{k-1}) \longrightarrow H_i(X^{k-1}) \longrightarrow H_i(X^k) \longrightarrow H_i(X^k, X^{k-1}) \longrightarrow \cdots$$

shows that $H_i(X^k) \cong H_i(X^{k-1})$ for $i \geq k+1$. Therefore, $H_{k+1}(X^k) \cong H_{k+1}(X^{k-1}) \cong \cdots \cong H_{k+1}(X^0) = 0$.

As in the simplicial case, we set $C_k = H_i(X^k, X^{k-1})$ and define a homomorphism $\partial_k \colon C_k \to C_{k-1}$. Then, we prove that $H_{k-1} \cong \operatorname{Ker} \partial_{k-1} / \operatorname{Im} \partial_k$ and $H_{k-1}(X) \cong H_{k-1}(X^k)$. The proof of the latter isomorphism uses the fact that any singular chain is contained in some finite-dimensional skeleton X^N.

When cellular homology is used, the following assertion becomes obvious. *Let X be a CW-complex containing m_k cells of dimension k. Then the rank of the group $H_k(X)$ is at most m_k.* We need it to prove the Morse inequalities.

1.5. The Morse Inequalities. Suppose that M^n is a closed manifold and f is a Morse function on M^n with pairwise distinct critical values. For each $k = 0, 1, \ldots, n$, we define r_k to be the rank of $H_k(M^n)$ and m_k to be the number of index k critical points of f. These numbers are related by certain equalities and inequalities. First, note that $\sum(-1)^k r_k = \chi(M^n) = \sum(-1)^k m_k$ (see Part I, p. 244). Moreover, it is easy to see that $r_k \leq m_k$. Indeed, in the complex for calculating the cellular homology of M^n, the chain group C_k coincides with $\mathbb{Z}^{m_k}$ (see Part I, Theorem 5.44). Clearly, $\mathbb{Z}^{m_k}$ has a quotient of rank at most m_k. A more detailed study of this chain complex leads to the following sharper inequality.

Theorem 4.11. *Let t_k be the minimum number of generators in the torsion subgroup T_k of $H_k(M^n)$. Then*

$$r_k + t_k + t_{k-1} \leq m_k.$$

[3]This group contains only finite sums of cells because the image of any simplex under a continuous map is compact and any compact subset of a CW-complex intersects only finitely many open cells.

Proof. The groups T_k and T_{k-1} are quotients of free Abelian subgroups F_k and F_{k-1} of ranks t_k and t_{k-1} in C_k and C_{k-1}. The group F_{k-1} is the epimorphic image under the map ∂_k of a rank-t_{k-1} free Abelian group $F'_{k-1} \subset C_k$, and the group F_k is the image of a similar group $F'_k \subset C_{k+1}$.

Thus, the group C_k contains the subgroups F_k and F'_{k-1}; it has also a free subgroup S_k of rank r_k, which is contained in $\operatorname{Ker} \partial_k$ and has trivial intersection with $\operatorname{Im} \partial_{k+1}$; this subgroup corresponds to a free summand of $H_k(M^n)$. The group F_k is contained in $\operatorname{Im} \partial_{k+1}$; therefore, the sum $F_k + S_k$ is direct. Moreover, we have $\partial_k(F_k \oplus S_k) = 0$ and $\partial_k F'_{k-1} = F_{k-1}$, where F'_{k-1} and F_{k-1} are free Abelian groups of the same rank. Hence the sum $F'_{k-1} + F_k \oplus S_k$ is direct as well. Thus, the group C_k, whose rank equals m_k, contains a free Abelian subgroup of rank $t_{k-1} + t_k + r_k$; therefore, $t_{k-1} + t_k + r_k \le m_k$. $\square$

The numbers r_k and m_k are also related by the following series of inequalities (Morse inequalities).

Theorem 4.12. (a) *For every $k = 0, 1, \dots, n$,*

$$r_k - r_{k-1} + r_{k-2} - \dots \pm r_0 \le m_k - m_{k-1} + m_{k-2} - \dots \pm m_0$$

(for $k = n$, this inequality becomes an equality).

(b) *Suppose that $M(t) = \sum m_k t^k$ and $R(t) = \sum r_k t^k$. Then the polynomial $M(t) - R(t)$ is divisible by $1 + t$ and the coefficients of the polynomial $\frac{M(t)-R(t)}{1+t}$ are nonnegative.*

Proof. It is easy to show that (a) follows from (b). Indeed, suppose that $\sum (m_k - r_k)t^k = \sum (1+t) d_k t^k$, where $d_k \ge 0$. Then $m_0 - r_0 - d_0 \ge 0$, $m_1 - r_1 - (m_0 - r_0) = d_1 \ge 0$, $m_2 - r_2 - (m_1 - r_1) + (m_0 - r_0) = d_2 \ge 0$, etc. Clearly, $d_m = 0$. Thus, it suffices to prove (b).

Let $a_1 < a_2 < \dots < a_m$ be the critical values of the function f. Choose numbers $b_1, \dots, b_{m+1}$ so that $-\infty < b_1 < a_1 < b_2 < \dots < a_m < b_{m+1} < \infty$ and consider the manifold $M_p = \{x \in M^n \mid f(x) \le b_p\}$, where $p = 1, 2, \dots, m+1$. Consider also the polynomials $R_p(t) = \sum \operatorname{rk} H_k(M_p) t^k$ and $R_{pq}(t) = \sum \operatorname{rk} H_k(M_p, M_q) t^k$ for $p > q$. Let us write the exact sequence for the pair (M_p, M_q):

$$H_{k+1}(M_p, M_q) \xrightarrow{\partial_{k+1}} H_k(M^n) \xrightarrow{i} H_k(M_p) \xrightarrow{\pi} H_k(M_p, M_q) \xrightarrow{\partial_k} H_{k-1}(M_q).$$

Clearly,

$$\operatorname{rk} H_k(M_p, M_q) = \operatorname{rk} \operatorname{Im} \pi + \operatorname{rk} \operatorname{Im} \partial_k$$

and

$$\operatorname{rk} \operatorname{Im} \pi = \operatorname{rk} H_k(M_p) - \operatorname{rk} \operatorname{Im} i = \operatorname{rk} H_k(M_p) - (\operatorname{rk} H_k(M_q) - \operatorname{rk} \operatorname{Im} \partial_k).$$

Therefore,

$$\operatorname{rk} H_k(M_p, M_q) - (\operatorname{rk} H_k(M_p) - \operatorname{rk} H_k(M_q)) = \operatorname{rk} \operatorname{Im} \partial_k + \operatorname{rk} \operatorname{Im} \partial_{k+1},$$

and hence

$$R_{pq}(t) - (R_p(t) - R_q(t)) = (1+t)d_{pq}(t),$$

where d_{pq} is a polynomial with nonnegative coefficients. Obviously,

$$\sum_{p=1}^{m} (R_{p+1}(t) - R_p(t)) = R_{m+1}(t) = R(t);$$

thus, it remains to prove that $\sum_{p=1}^{m} R_{p+1,p}(t) = M(f)$. To this end, it suffices to verify that $R_{p+1,p}(t) = t^\mu$, where μ is the index of the critical point between the levels $f(x) = b_{p+1}$ and $f(x) = b_p$. We have $M_{p+1} \approx M_p \cup (S^\mu \times D^{n-\mu})$; therefore, $H_k(M_{p+1}, M_p) \cong H_k(M_p \cup (S^\mu \times D^{n-\mu}), M_p) \cong \tilde{H}_k(S^\mu \times D^{n-\mu}) \cong \tilde{H}_k(S^\mu)$. □

1.6. Multiplications.

The Eilenberg–Zilber Theorem. The Eilenberg–Zilber theorem claims that $H_k(X \times Y) \cong H_k(C_*(X) \otimes C_*(Y))$ for any topological spaces X and Y. Its proof, which is based on the acyclic model theorem, is presented below.

Let $\mathcal{T}$ be the category of topological spaces and continuous maps, and let $\mathcal{T} \times \mathcal{T}$ be the category whose objects are ordered pairs of topological spaces (X, Y) and morphisms are ordered pairs of continuous maps $f\colon X \to X'$, $g\colon Y \to Y'$. Let us define two functors, T and T', from the category $\mathcal{T} \times \mathcal{T}$ to the category of nonnegative chain complexes.

(i) The functor T takes each pair (X, Y) to the singular chain complex $C_*(X \times Y)$;

(ii) the functor T' takes each pair (X, Y) to the tensor product $C_*(X) \otimes C_*(Y)$.

Recall that the tensor product of chain complexes is defined by

$$(C_*(X) \otimes C_*(Y))_k = \bigoplus_{p+q=k} C_p(X) \otimes C_q(Y),$$

$$\partial(c_p \otimes c_q) = \partial c_p \otimes c_q + (-1)^p c_p \otimes \partial c_q.$$

The functors T and T' are acyclic[4] with respect to the models from $\mathcal{M} = \{(\Delta^p, \Delta^q)\}$, where p and q range over all nonnegative integers. Indeed, the acyclicity of T is obvious because the space $\Delta^p \times \Delta^q$ is contractible. To prove the acyclicity of T', is suffices to show that if chain complexes C'_* and C''_* are acyclic, then so is the chain complex $C'_* \otimes C''_*$. This follows easily from the algebraic Künneth theorem.

[4] The definition of a functor acyclic with respect to models was given on p. 103.

The functor T' is free[5] with respect to the models from $\mathcal{M}$. Indeed, the group $C_p(X) \otimes C_q(Y)$ is freely generated by the singular chains of the form $c_p \otimes c_q$, which are in one-to-one correspondence with the pairs of maps $f\colon \Delta^p \to X$, $g\colon \Delta^q \to Y$. Therefore, for the element $e_{p,q} \in C_*(\Delta^p) \otimes C_*(\Delta^q)$ we can take the singular chain $\Delta^p \times \Delta^q$.

The functor T is free with respect to the models from $\{(\Delta^k, \Delta^k)\} \subset \mathcal{M}$. Indeed, the group $C_k(X \times Y)$ is freely generated by singular chains that are in one-to-one correspondence with the maps $F\colon \Delta^k \to X \times Y$. Any such map is determined by a pair of maps $f\colon \Delta^k \to X$, $g\colon \Delta^k \to Y$. Let $d_k\colon \Delta^k \to \Delta^k \times \Delta^k$ be the diagonal map. Then F can be represented as the composition $\Delta^k \xrightarrow{d_k} \Delta^k \times \Delta^k \xrightarrow{(f,g)} X \times Y$. Therefore, for the element $e_k \in C_*(\Delta^k \times \Delta^k)$ we can take the singular chain corresponding to the map d_k.

The algebraic Künneth theorem implies the existence of a canonical isomorphism $H_0(C_*(X) \otimes C_*(Y)) \to H_0(X) \otimes H_0(Y)$. Clearly, the path-connected components of $X \times Y$ are the products of those of X and Y. Hence there exists a natural isomorphism $\varphi\colon H_0(X) \otimes H_0(Y) \to H_0(C_*(X) \otimes C_*(Y))$. Applying the acyclic model theorem to the maps φ and φ^{-1}, we obtain chain maps $\tau\colon C_*(X \times Y) \to C_*(X) \otimes C_*(Y)$ and $\bar{\tau}\colon C_*(X) \otimes C_*(Y) \to C_*(X \times Y)$, which induce the maps φ and φ^{-1}, respectively, in the 0-dimensional homology groups.

The chain map $\bar{\tau} \circ \tau$ induces the identity map of the group $H_0(X \times Y)$. Hence, according to the acyclic model theorem, the map $\bar{\tau} \circ \tau$ is chain homotopic to the identity. Similarly, $\tau \circ \bar{\tau}$ is chain homotopic to the identity as well. Thus, the map $\tau_*\colon H_*(X \times Y) \to H_*(C_*(X) \otimes C_*(Y))$ is an isomorphism. We have proved the following theorem.

Theorem 4.13 (Eilenberg–Zilber [**35**]). *For any topological spaces X and Y, the groups $H_k(X \times Y)$ and $H_k(C_*(X) \otimes C_*(Y))$ are isomorphic.*

Combining the Eilenberg–Zilber theorem with the algebraic Künneth theorem, we obtain the *Künneth theorem for singular homology.*

Theorem 4.14. *For any topological spaces X and Y,*

$$H_k(X \times Y) \cong (H_*(X) \otimes H_*(Y))_k \oplus (\mathrm{Tor}(H_*(X), H_*(Y)))_{k-1}.$$

The Alexander–Whitney Diagonal Approximation. Suppose X is a topological space, $d\colon X \to X \times X$ is a diagonal map, $d_*\colon C_*(X) \to C_*(X \times X)$ is the induced chain map, and $\tau\colon C_*(X \times X) \to C_*(X) \otimes C_*(X)$ is the functorial chain map inducing an isomorphism of homology groups. Consider

[5] The definition of a functor free with respect to models was given on p. 103.

the composition

$$C_*(X) \xrightarrow{d_*} C_*(X \times X) \xrightarrow{\tau} C_*(X) \otimes C_*(X).$$

The chain map $\tau d_*\colon C_*(X) \to C_*(X) \otimes C_*(X)$ is functorial and takes every zero-dimensional simplex v to $v \otimes v$. We refer to any functorial chain map $D_0\colon C_*(X) \to C_*(X) \otimes C_*(X)$ that takes v to $v \otimes v$ as a *diagonal approximation*. The acyclic model theorem implies that any diagonal approximation is chain homotopic to τd_*. Indeed, the functor $C_*(X)$ is free with respect to the models from $\mathcal{M} = \{\Delta^k\}$, and the functor $C_*(X) \otimes C_*(X)$ is acyclic with respect to these models.

For singular cohomology, the cup product is defined as follows. If $c^p \in C^p(X; R)$ and $c^q \in C^q(X; R)$, where R is a commutative ring with identity, then the cochain $c^p \smile c^q$ assigns $\langle c^p \otimes c^q, D_0 c^{p+q}\rangle$ to the chain $c^{p+q} \in C^{p+q}(X; R)$; here

$$\langle c^p \otimes c^q, c_i \otimes c_j\rangle = \delta_{pi}\delta_{qj}\langle c^p, c_i\rangle\langle c^q, c_j\rangle.$$

At the level of cochains, the cup product depends on the choice of the diagonal approximation D_0, but at the level of cohomology, it is invariant.

For calculations, it is convenient to have a diagonal approximation specified by some simple formula. Best suited is the *Alexander–Whitney diagonal approximation*

$$[0, 1, \ldots, n] \mapsto \sum_{i=0}^{n} [0, \ldots, i] \otimes [i, \ldots, n];$$

here $[0, 1, \ldots, n]$ is a singular simplex $f\colon [0, 1, \ldots, n] \to X$, and $[0, \ldots, i]$ and $[i, \ldots, n]$ are the restrictions of f to the corresponding faces. The proof that this is a chain map is the same as in the simplicial case (see p. 105).

At the level of cochains, the cup product determined by the Alexander–Whitney diagonal approximation has the form

$$\langle c^p \smile c^q, [0, 1, \ldots, p+q]\rangle = \langle c^p, [0, \ldots, p]\rangle\langle c^q, [p, \ldots, p+q]\rangle.$$

Now let us define the cap product. Any cochain $c^p \in C^p(X; R)$ can be regarded as a homomorphism $c^p\colon C_*(X) \to R$ (for $i \neq p$, this homomorphism vanishes at the i-dimensional chains). Consider the composition

$$C_*(X) \xrightarrow{D_0} C_*(X) \otimes C_*(X) \xrightarrow{\mathrm{id} \otimes c^p} C_*(X) \otimes R.$$

We take the tensor product of the groups involved in this composition and R and apply the multiplication $\mu\colon R \otimes R \to R$ in the ring R to the result:

$$C_*(X) \otimes R \xrightarrow{D_0 \otimes \mathrm{id}} C_*(X) \otimes C_*(X) \otimes R$$
$$\xrightarrow{\mathrm{id} \otimes c^p \otimes \mathrm{id}} C_*(X) \otimes R \otimes R \xrightarrow{\mathrm{id} \otimes \mu} C_*(X) \otimes R.$$

We denote the image of a chain $c_{p+q} \in C_{p+q}(X;R) = C_{p+q}(X) \otimes R$ by $c^p \frown c_{p+q}$. At the level of (co)homology, the cap product does not depend on the choice of the diagonal approximation D_0. For the Alexander–Whitney diagonal approximation, we obtain the usual cap product:

$$c^p \frown [0,1,\dots,p+q] = \langle c^p, [q,\dots,p+q]\rangle [0,\dots,q].$$

The Relative Case. In the relative case, the functorial chain equivalence between $C_*(X\times Y)$ and $C_*(X)\otimes C_*(Y)$ is replaced by that between $C_*(X\times Y, (X\times B)\cup(A\times Y))$ and $C_*(X,A)\otimes C_*(Y,B)$ (the spaces $\{X\times B, A\times Y\}$ must satisfy the excision axiom). This chain equivalence is constructed as follows. If the spaces $\{X\times B, A\times Y\}$ satisfy the excision axiom, then the natural embedding

$$C_*(X\times B) + C_*(A\times Y) \to C_*((X\times B)\cup(A\times Y))$$

induces an isomorphism of homology groups, and therefore so does the natural map

$$\frac{C_*(X\times Y)}{C_*(X\times B)+C_*(A\times Y)} \to \frac{C_*(X\times Y)}{C_*((X\times B)\cup(A\times Y))}.$$

The functorial chain equivalence $C_*(X)\otimes C_*(Y)\to C_*(X\times Y)$ maps $C_*(X)\otimes C_*(B)$ to $C_*(X\times B)$ and $C_*(A)\otimes C_*(Y)$ to $C_*(A\times Y)$. We have obtained a chain equivalence between $\frac{C_*(X\times Y)}{C_*(X\times B)+C_*(A\times Y)}$ and

$$\frac{C_*(X)\otimes C_*(Y)}{C_*(X)\otimes C_*(B)+C_*(A)\otimes C_*(Y)} \cong \frac{C_*(X)}{C_*(A)} \oplus \frac{C_*(Y)}{C_*(B)}.$$

The composition of these two chain equivalences is the required chain equivalence.

This chain equivalence implies the *Künneth theorem* for relative homology:

$$\begin{aligned} &H_k(X\times Y, (X\times B)\cup(A\times Y)) \\ &\qquad \cong (H_*(X,A)\otimes H_*(Y,B))_k \oplus (\operatorname{Tor}(H_*(X,A), H_*(Y,B)))_{k-1}. \end{aligned}$$

Moreover, it allows us to define the homology cross product

$$H_p(X,A)\otimes H_q(Y,B) \to H_{p+q}(X\times Y, (X\times B)\cup(A\times Y))$$

and the cohomology cross product

$$H^p(X,A)\otimes H^q(Y,B) \to H^{p+q}(X\times Y, (X\times B)\cup(A\times Y)).$$

The cup product construction in the relative case, i.e., for $H^p(X,A_1)$ and $H^q(X,A_2)$, involves the cohomology of the complex of cochains vanishing on $C_*(A_1)+C_*(A_2)$, and the cap product construction involves the homology of the complex $\frac{C_*(X)}{C_*(A_1)+C_*(A_2)}$. Therefore, if sets $\{A_1, A_2\}$ in X satisfy the excision axiom, then the cup product

$$H^p(X, A_1) \otimes H^q(X, A_2) \to H^{p+q}(X, A_1 \cup A_2)$$

and the cap product

$$H^p(X, A_1) \otimes H_{p+q}(X, A_1 \cup A_2) \to H_q(X, A_2)$$

are defined.

Cohomology of Products: A Special Case. The proof of the Künneth theorem for cohomology is more involved than for homology (moreover, the cohomology theorem is valid only under certain finiteness conditions). We shall not prove the Künneth theorem in the general form; instead, we shall prove a special case of this theorem, namely, the expression for the cohomology of the pair $(X \times \mathbb{R}^n, X \times (\mathbb{R}^n \setminus \{0\}))$ in terms of the cohomology of X and of the pair $(\mathbb{R}^n, \mathbb{R}^n \setminus \{0\})$, which plays an important role in what follows (in the proof of the Thom isomorphism). First, we calculate the cohomology $H^*(\mathbb{R}^n, \mathbb{R}^n \setminus \{0\})$.

The cohomology sequence for the pair $(\mathbb{R}^n, \mathbb{R}^n \setminus \{0\})$ implies that $H^k(\mathbb{R}^n, \mathbb{R}^n \setminus \{0\}) \cong \tilde{H}^{k-1}(\mathbb{R}^n \setminus \{0\}) \cong \tilde{H}^{k-1}(S^{n-1})$. Therefore, the group $H^n(\mathbb{R}^n, \mathbb{R}^n \setminus \{0\})$ is isomorphic to the additive group of the coefficient ring, and all of the other groups $H^k(\mathbb{R}^n, \mathbb{R}^n \setminus \{0\}), k \neq n$ are trivial.

To simplify notation, we set $\mathbb{R}^n_0 = \mathbb{R}^n \setminus \{0\}$, $\mathbb{R}_+ = \{x \in \mathbb{R} \mid x > 0\}$, and $\mathbb{R}_- = \{x \in \mathbb{R} \mid x < 0\}$. The excision isomorphism $H^0(\mathbb{R}_+) \cong H^0(\mathbb{R}_0, \mathbb{R}_-)$ holds. Let us write the exact sequence for the triple $(\mathbb{R}, \mathbb{R}_0, \mathbb{R}_-)$:

$$H^0(\mathbb{R}, \mathbb{R}_-) \longrightarrow H^0(\mathbb{R}_0, \mathbb{R}_-) \xrightarrow{\delta} H^1(\mathbb{R}, \mathbb{R}_0) \longrightarrow H^1(\mathbb{R}, \mathbb{R}_-).$$

We claim that $H^0(\mathbb{R}, \mathbb{R}_-) = H^1(\mathbb{R}, \mathbb{R}_-) = 0$, and therefore δ is an isomorphism. Since $\mathbb{R}$ is connected, $H^0(\mathbb{R}, \mathbb{R}_-) = 0$, and since the sequence

$$H^k(\mathbb{R}) \longrightarrow H^k(\mathbb{R}, \mathbb{R}_-) \longrightarrow H^{k+1}(\mathbb{R}_-)$$

for the pair $(\mathbb{R}, \mathbb{R}_-)$ is exact, $H^k(\mathbb{R}, \mathbb{R}_-) = 0$ for $k > 0$. Let e be the image of the element $1 \in H^0(\mathbb{R}_+)$ corresponding to the identity element of the coefficient ring under the composition of isomorphisms

$$H^0(\mathbb{R}_+) \cong H^0(\mathbb{R}_0, \mathbb{R}_-) \xrightarrow{\delta} H^1(\mathbb{R}, \mathbb{R}_0).$$

The group $H^1(\mathbb{R}, \mathbb{R}_0)$ is isomorphic to the additive group of the coefficient ring, and e is the identity element of this ring.

The cohomology cross product of the cohomology classes of the pairs (X, A) and (Y, B) belongs to the cohomology of the pair $(X \times Y, (X \times B) \cup (A \times Y))$. For brevity, we denote this pair by $(X, A) \times (Y, B)$.

It is easy to verify that

$$(\mathbb{R}^n, \mathbb{R}^n_0) \times (\mathbb{R}^m, \mathbb{R}^m_0) = (\mathbb{R}^{n+m}, \mathbb{R}^{n+m}_0).$$

Indeed, take $u \in \mathbb{R}^n$ and $v \in \mathbb{R}^m$. Then $\mathbb{R}^n \times \mathbb{R}^m_0$ consists of all pairs (u, v) in which $v \neq 0$, and $\mathbb{R}^n_0 \times \mathbb{R}^m$ consists of all pairs (u, v) in which $u \neq 0$. Clearly, the union of these sets consists of all nonzero vectors in $\mathbb{R}^{n+m}$.

Consider the element

$$e^n = \underbrace{e \times \cdots \times e}_{n},$$

which belongs to $H^n(\mathbb{R}^n, \mathbb{R}^n_0)$.

Theorem 4.15. *If $A \subset X$ is an open set, then the map*

$$H^k(X, A) \to H^{k+n}((X, A) \times (\mathbb{R}^n, \mathbb{R}^n_0))$$

defined by $\alpha \mapsto \alpha \times e^n$ is an isomorphism.

Proof. First, suppose that $n = 1$ and $A = \varnothing$. Take an element $\alpha \in H^k(X)$ and consider the diagram

$$\begin{array}{ccc}
H^0(\mathbb{R}_+) & \xleftarrow{\ \cong\ } & H^0(\mathbb{R}_0, \mathbb{R}_-) \\
\downarrow{\scriptstyle \alpha\times} & & \downarrow{\scriptstyle \alpha\times} \\
H^k(X) \cong H^k(X \times \mathbb{R}_+) & \xleftarrow{\ \cong\ } & H^k(X \times \mathbb{R}_0, X \times \mathbb{R}_-)
\end{array}$$

$$\begin{array}{rl}
\xrightarrow{\ \ \delta\ \ } & H^1(\mathbb{R}, \mathbb{R}^0) \\
& \downarrow{\scriptstyle \alpha\times} \\
\xrightarrow{\ \ \delta'\ \ } & H^{k+1}(X \times \mathbb{R}, X \times \mathbb{R}^0),
\end{array}$$

where the horizontal arrows in the left square are the excision isomorphisms and δ' is the homomorphism from the cohomology sequence of the triple $(X \times \mathbb{R}, X \times \mathbb{R}_0, X \times \mathbb{R}_-)$. The left square of the diagram is commutative, and the right one is commutative up to sign because $\delta'(\alpha \times x) = (-1)^k \alpha \times \delta x$. The homomorphism δ' is an isomorphism because

$$H^i(X \times \mathbb{R}, X \times \mathbb{R}_-) = 0$$

for $i > 0$ (both spaces $X \times \mathbb{R}$ and $X \times \mathbb{R}_-$ are deformation retracts of the same space $X \times \{-1\}$). Thus, the element $\alpha \times e \in H^{k+1}(X \times \mathbb{R}, X \times \mathbb{R}_0)$ is the image of $\alpha \in H^k(X)$ under the composition of isomorphisms.

Now, suppose that $n = 1$ but $A \neq \varnothing$. Let $z \in Z^1(\mathbb{R}, \mathbb{R}_0)$ be a cocycle representing the cohomology class e. Consider the commutative diagram

$$\begin{array}{ccccc} 0 \longrightarrow & C^k(X,A) & \longrightarrow & C^k(X) & \\ & \downarrow{\times z} & & \downarrow{\times z} & \\ 0 \to & C^{k+1}((X,A)\times(\mathbb{R},\mathbb{R}_0)) & \to & C^{k+1}(X\times\mathbb{R}, X\times\mathbb{R}_0) & \end{array}$$

$$\begin{array}{ccc} \longrightarrow C^k(A) & \longrightarrow & 0 \\ \downarrow{\times z} & & \\ \longrightarrow C^{k+1}(A\times\mathbb{R}, A\times\mathbb{R}_0) & \longrightarrow & 0 \end{array}$$

with exact rows. The vertical arrows are cochain maps because $\delta(c \times z) = (\delta c)\times z$. Thus, we obtain the following commutative diagram for cohomology groups:

$$\begin{array}{ccccc} \longrightarrow & H^k(X,A) & \longrightarrow & H^k(X) & \\ & \downarrow{\times e} & & \downarrow{\times e} & \\ \to & H^{k+1}((X,A)\times(\mathbb{R},\mathbb{R}_0)) & \to & H^{k+1}(X\times\mathbb{R}, X\times\mathbb{R}_0) & \end{array}$$

$$\begin{array}{ccc} \longrightarrow H^k(A) & \longrightarrow & \\ \downarrow{\times e} & & \\ \to H^{k+1}(A\times\mathbb{R}, A\times\mathbb{R}_0) & \to \cdot & \end{array}$$

We have already proved that the vertical maps are isomorphisms for absolute cohomology. By the five lemma, they remain isomorphisms for relative cohomology.

The required assertion for an arbitrary n follows from its validity for $n = 1$ because the associativity of the cross product implies that $a \times e^n = (a \times e^{n-1}) \times e$. □

1.7. The Hopf Invariant. The interest in the homotopy groups of spheres was caused by Hopf's example of a map $S^3 \to S^2$ not homotopic to a constant [**58**]. To prove that this map is nontrivial, Hopf defined a homotopy invariant for maps $f\colon S^3 \to S^2$, which is now known as the Hopf invariant. In [**60**], Hopf defined a similar invariant for maps $f\colon S^{2n-1} \to S^n$ with any n. In the original definition, Hopf used the linking number. Steenrod [**126**] suggested a definition in terms of multiplication in cohomology.

Let $n \geq 2$. Consider the oriented n-sphere S^n and a continuous map $f\colon S^{2n-1} \to S^n$. We regard the oriented sphere S^{2n-1} as the boundary of the oriented disk D^{2n}. Thus, the space $X_f = S^n \cup_f D^{2n}$ is defined. The homotopy type of this space depends only on the homotopy class of the map f (see Part I, p. 248). Throughout this section, we consider only integral cohomology. Let us show that $H^k(X_f) = \mathbb{Z}$ for $k = 0$, n, and $2n$; all of the other cohomology groups are trivial. Consider the cohomology sequence of the pair (X_f, S^n):

$$\cdots \longrightarrow H^k(X_f, S^n) \longrightarrow H^k(X_f) \longrightarrow H^k(S^n) \longrightarrow \cdots .$$

The excision theorem implies $H^k(X_f, S^n) \cong H^k(S^{2n})$ for $k > 0$. Therefore, for $k \neq 0, n, 2n$, the group $H^k(X_f)$ is surrounded by trivial groups; for $k = n$ and $2n$, we obtain isomorphisms $H^n(X_f) \xrightarrow{\cong} H^n(S^n)$ and $H^{2n}(S^{2n}) \xrightarrow{\cong} H^{2n}(X_f)$.

Let α and β be the generators of the groups $H^n(X_f)$ and $H^{2n}(X_f)$ corresponding to the orientations of the sphere S^n and the disk D^{2n}. Then $\alpha \smile \alpha = H(f)\beta$, where $H(f)$ is an integer. This integer is called the *Hopf invariant* of the map f.

It follows directly from the definition that the change of the orientation of S^n does not affect the Hopf invariant, whereas the change of the orientation of D^{2n} changes its sign.

Exercise. Given maps $g\colon S^n \to S^n$ and $h\colon S^{2n} \to S^{2n}$, prove that $H(gf) = (\deg g)^2 H(f)$ and $H(fh) = (\deg h) H(f)$.

If n is odd, then $\alpha \smile \alpha = -\alpha \smile \alpha$, and this element belongs to $\mathbb{Z}$. Hence $\alpha \smile \alpha = 0$, and therefore $H(f) = 0$ for any map f.

For the Hopf fibration $f\colon S^3 \to S^2$, we have $X_f = \mathbb{C}P^2$. In this case, $\alpha \smile \alpha = \beta$ and $H(f) = 1$.

Exercise. Using the quaternions and the Cayley numbers, construct maps $f\colon S^7 \to S^4$ and $f\colon S^{15} \to S^8$ with Hopf invariant equal to 1.

Remark. Adams [**2**] proved that the maps $f\colon S^{2n-1} \to S^n$ with Hopf invariant equal to 1 exist only for $n = 2, 4, 8$.

The Hopf invariant can be defined in many different ways. Below, we give yet another cohomological definition of the Hopf invariant (it essentially coincides with that given above but is formulated in a different language). Let $[a]$ and $[b]$ be the generators of the groups $H^n(S^n)$ and $H^{2n-1}(S^{2n-1})$ corresponding to the orientations of the spheres S^n and S^{2n-1}, and let a and b be cocycles representing them. The cohomology class of the cocycle $a \smile a$ belongs to $H^{2n}(S^n) = 0$; hence there exists a cochain $c \in C^{2n-1}(S^n)$ for which $\delta c = a \smile a$. The cohomology class of the cocycle $f^{\#}(a)$ belongs to $H^n(S^{2n-1}) = 0$; hence there exists a cochain $e \in C^n(S^{2n-1})$ for which

$\delta e = f^{\#}(a)$. It is easy to verify that the cochain $f^{\#}(c) - e \smile f^{\#}(a)$ is a cocycle. Indeed,

$$\begin{aligned}\delta(f^{\#}(c) - e \smile f^{\#}(a)) &= f^{\#}(a \smile a) - \delta(e \smile \delta e)\\ &= f^{\#}(a) \smile f^{\#}(a) - \delta e \smile \delta e = 0.\end{aligned}$$

Therefore, the cohomology class of $f^{\#}(c) - e \smile f^{\#}(a)$ is defined. This class belongs to the group $H^{2n-1}(S^{2n-1})$; hence it is proportional to the class $[b]$. The coefficient of proportionality is the Hopf invariant $H(f)$.

Let us prove the equivalence of the two definitions of the Hopf invariant given above. Consider the auxiliary space $S_f = S^n \cup_f (S^{2n-1} \times I)$ (the sphere S^n attached to the cylinder $S^{2n-1} \times I$ along one of its bases) and the pair (S_f, S^{2n-1}), which contains the other base of the cylinder. The excision theorem implies $H^k(S_f, S^{2n-1}) \cong H^k(X_f, *) \cong H^k(X_f)$ for $k \geq 1$.

Consider the diagram

$$\begin{aligned}&H^n(S^n) \xrightarrow[\cong]{p^*} H^n(S_f) \xleftarrow[\cong]{} H^n(S_f, S^{2n-1})\\ &\qquad\qquad\qquad\qquad \xrightarrow{\text{sq}} H^{2n}(S_f, S^{2n-1}) \xleftarrow[\cong]{} H^{2n-1}(S^{2n-1}).\end{aligned}$$

Here $\text{sq}(x) = x \smile x$, and the remaining maps are natural isomorphisms (the first is induced by the projection $p\colon S_f \to S^n$, which is a homotopy equivalence, and the two others are taken from the cohomology sequence of the pair). It is fairly easy to extract the first definition of the Hopf invariant from this diagram: the class $[a]$ is identified with a class α by means of the first two isomorphisms; applying the map sq to α ($\text{sq}(\alpha) = \alpha \smile \alpha$), we obtain $H(f)\beta$, where β is the image of the class $[b]$ under the natural isomorphism. Let us show how to obtain the second definition from this diagram considered at the level of cocycles.

At the level of cocycles, the isomorphism p^* takes the cocycle a to $a_f = p^{\#}(a)$. Moreover, $f^{\#}(a)$ is the restriction of the cocycle a_f to the set of singular simplices S^{2n-1}, i.e., $f^{\#}(a) = i(a_f)$, where $i\colon C^*(S^{2n-1}) \to C^*(S_f)$ is the natural embedding (the restriction map to the subset S^{2n-1}). Take a cochain $c \in C^*(S^{2n})$ such that $\delta c = a \smile a$ and consider $c_f = p^{\#}(c)$ and $f^{\#}(c) = i(c_f)$. By assumption, we have $i(a_f) = f^{\#}(a) = \delta e$ for some $e \in C^*(S^{2n-1})$. Let e' be an extension of e to S_f, i.e., a cochain such that $e = i(e')$. The difference $a_f - \delta e'$ is a cocycle in $C^n(S_f)$, and it vanishes on $C^n(S^{2n-1})$. It can be considered as a cocycle in $C^n(S_f, S^{2n-1})$; in $H^n(S_f, S^{2n-1})$, this cocycle represents the class corresponding to $[a]$. Let us apply the map sq to this cocycle. For a representative of the resulting class we can take the cocycle $(a_f - \delta e') \smile a_f$ because this product vanishes on $C^n(S^{2n-1})$ and $\delta e'$ is a coboundary.

Clearly, $f^{\#}(c) - e \smile f^{\#}(a) = i(c_f) - ie' \smile i(a_f) = i(c_f - e' \smile a_f)$. Therefore, the class of the cocycle $f^{\#}(c) - e \smile f^{\#}(a)$ is identified with the class of the cocycle $\delta(c_f - e' \smile a_f) = a_f \smile a_f - (\delta e') \smile a_f = (a_f - \delta e') \smile a_f$, as required.

1.8. Simplicial Volume (the Gromov Norm). Let M^n be a closed orientable manifold. Each singular chain $\sum \alpha_i f_i \in C_k(M^n; \mathbb{R})$ is assigned its norm $\|\sum \alpha_i f_i\| = \sum |\alpha_i|$. This norm carries over to the homology groups $H_k(M^n; \mathbb{R})$; namely, for a homology class $\zeta \in H_k(M^n; \mathbb{R})$,

$$\|\zeta\| = \inf\{\|z\| \mid z \text{ is a cycle in the class } \zeta\}.$$

The *simplicial volume*, or *Gromov norm*, of a manifold M^n is defined as $\|[M^n]\|$, where $[M^n] \in H_n(M^n; \mathbb{R})$ is the fundamental class of the manifold M^n. We denote the simplicial volume of M^n by $\|M^n\|$; its properties are discussed in detail in [**45**].

Example 55. $\|S^1\| = 0$.

Proof. Let $f_n\colon [0,1] \to S^1$ be the map defined by $f_n(t) = e^{2\pi i n t}$. Then f_n is a cycle representing the homology class $n[S^1]$. Therefore, the class $[S^1]$ is represented by the cycle $\frac{1}{n} f_n$, whose norm equals $\frac{1}{n}$. □

The construction from Example 55 applies to any closed orientable manifold M^n admitting maps $M^n \to M^n$ of arbitrarily large degrees. Such maps are easy to construct, e.g., for the sphere S^n and the torus T^n (it suffices to construct a map of degree larger than 1). Therefore, $\|S^n\| = 0$ and $\|T^n\| = 0$.

Example 56. If M_g^2 is a sphere with g handles, where $g \geq 2$, then $\|M_g^2\| = 2|\chi(M_g^2)| = 2(2g-2)$.

Proof. The manifold M_g^2 can be obtained by gluing together sides of a $4g$-gon. This $4g$-gon can be cut into $4g-2$ triangles. Therefore,

$$\|M_g^2\| \leq 2|\chi(M_g^2)| + 2. \tag{32}$$

Applying (32) to a manifold N^2 which covers M_g^2 n-to-one, we obtain $n\|M_g^2\| \leq \|N^2\| \leq 2|\chi(N^2)| + 2 = 2|\chi(N^2)| + 2 = 2|n\chi(M_g^2)| + 2$. Letting n tend to ∞, we see that $\|M_g^2\| \leq 2|\chi(M_g^2)|$.

Now let us prove the inequality $\|M_g^2\| \geq 2|\chi(M_g^2)|$. Consider a singular 2-simplex $f\colon \Delta^2 \to M_g^2$ and the universal covering $p\colon H^2 \to M_g^2$, where H^2 is the hyperbolic plane; we assume that M_g^2 is a Riemannian manifold endowed with a metric compatible with that on the hyperbolic plane. The map f admits a lifting $\tilde{f}\colon \Delta^2 \to H^2$ for which the diagram

$$\begin{array}{ccc} & & H^2 \\ & \overset{\tilde{f}}{\nearrow} & \downarrow p \\ \Delta^2 & \xrightarrow{f} & M_g^2 \end{array}$$

is commutative. Let us replace the singular simplex $\tilde{f}$ by a singular simplex $\tilde{f}'$ whose image is the triangle in the hyperbolic plane with the same vertices. If a cycle $\sum \alpha_i f_i$ represents the fundamental class $[M_g^2]$, then so does the cycle $\sum \alpha_i p(\tilde{f}_i')$. Moreover, $2\pi|\chi(M_g^2)| = \operatorname{vol}(M_g^2) \le \sum |\alpha_i| \operatorname{vol}(p(\tilde{f}_i'(\Delta^2)))$, where vol denotes area in the hyperbolic plane. The inequality arises because some overlapping parts of simplices may cancel each other; if there are no such cancellations, we obtain an equality. Finally, since the area of any triangle in the hyperbolic plane does not exceed π, we have $\operatorname{vol}(p(\tilde{f}_i'(\Delta^2))) \le \pi$. Therefore, $2|\chi(M_g^2)| \le \sum |\alpha_i| \le \|M_g^2\|$. □

Example 56 admits the following generalization. Let M^n be a closed orientable *hyperbolic* manifold, that is, a Riemannian manifold such that the universal covering over it is the hyperbolic space H^n. Then $\|M^n\| = \operatorname{vol}(M^n)/\operatorname{vol}(\Delta^n)$, where Δ^n is a simplex of maximum volume in H^n. This was proved by Thurston [**139**]; see also [**95**].

1.9. Cohomology with Noncommutative Coefficients and the van Kampen Theorem. Zero- and one-dimensional cohomology groups with coefficients in a non-Abelian group G can be defined as follows [**98**]. A cochain $c^n \in C^n(X;G)$ (of any dimension) is defined as a function taking each singular simplex $\Delta^n \to X$ to an element of G (a relative cochain $c^n \in C^n(X,Y;G)$ must vanish at the simplices $\Delta^n \to Y$). As previously, we use additive notation for the group operation, although it is no longer commutative. We call a one-dimensional cochain $z^1 \in C^1(X,Y;G)$ a *cocycle* if, for any singular simplex $f\colon \Delta^2 \to X$,

$$z^1(f\varepsilon_0^2) - z^1(f\varepsilon_1^2) + z^1(f\varepsilon_2^2) = 0,$$

where $\varepsilon_j^k\colon \Delta^{k-1} \to \Delta^k$ is the map defined on p. 195. A cochain $z^0 \in C^0(X,Y;G)$ is called a *cocycle* if $z^0(f\varepsilon_0^1) = z^0(f\varepsilon_1^1)$ for any singular simplex $f\colon \Delta^1 \to X$.

Two one-dimensional cocycles z^1 and $\bar{z}^1$ are said to be *cohomologous* if there exists a 0-cochain c^0 such that

$$\bar{z}^1(f) = -c^0(f\varepsilon_1^1) + z^1(f) + c^0(f\varepsilon_0^1)$$

for all singular 1-simplices $f\colon \Delta^1 \to X$. (Note that we cannot take $z^1(f)$ to the left-hand side because the group operation is noncommutative.)

It is easy to show that the relation of being cohomologous is an equivalence. The set of equivalence classes is $H^1(X,Y;G)$. It is not a group, but it contains a distinguished (zero) element, namely, the equivalence class of the trivial cocycle, which vanishes at each singular 1-simplex. We put $H^0(X,Y;G) = Z^0(X,Y;G)$; this set is a group.

Let X_1 and X_2 be subsets of $X = X_1 \cup X_2$. Define $H^q(C_*^{\{X_1,X_2\}}(X,Y);G)$ for $q = 0$ and 1 by considering only singular simplices that are contained entirely in X_1 or X_2. To each cochain there corresponds its restriction, that is, a cochain defined only for some of the simplices. Let us show that if $X = \operatorname{int} X_1 \cup \operatorname{int} X_2$, then this correspondence induces a one-to-one map

$$i^*: H^q(X,Y;G) \to H^q(C_*^{\{X_1,X_2\}}(X,Y);G) \tag{33}$$

that takes the zero element to the zero element ($q = 0, 1$).

For $q = 0$, this assertion is obvious. Indeed, on the left-hand side, the cocycles take equal values at the endpoints of any curve, and on the right-hand side, they take equal values at the endpoints of any curve contained entirely in X_1 or X_2. But any curve can be divided into parts contained entirely in $\operatorname{int} X_1$ or $\operatorname{int} X_2$.

To prove the required assertion for $q = 1$, consider the map

$$k: Z^1(C_*^{\{X_1,X_2\}}(X,Y);G) \to Z^1(X,Y;G)$$

defined as follows. Suppose that z^1 is a cocycle defined only for simplices in X_1 and X_2, and $f: \Delta^1 \to X$ is an arbitrary singular simplex. To define the value of $k(z^1)$ at f, we divide the interval Δ^1 into $M = 2^m$ equal intervals $I_1, \dots, I_M$ (the intervals are numbered in the natural order). If m is sufficiently large, then any interval I_α is contained in either X_1 or X_2. In this case, we set $k(z^1)(f) = z^1(f|_{I_1}) + \dots + z^1(f|_{I_M})$. The map $k(z_1)$ is well defined because z^1 is a cocycle. Indeed, if an interval I contained in X_1 or X_2 is partitioned into subintervals I' and I'', then $z^1(f|_{I'}) - z^1(f|_I) + z^1(f|_{I''}) = 0$.

Let us show that $k(z^1)$ is a cocycle. Consider any singular simplex $f: \Delta^2 \to X$. Choose m so that the image of the restriction of f to any simplex from the mth barycentric subdivision is contained entirely in X_1 or X_2. Then, for every simplex from the mth barycentric subdivision, the cocyclicity condition holds. Let us show that this implies cocyclicity for the simplices from the $(m-1)$th barycentric subdivision. Summing the cocyclicity conditions for the simplices $1, 2, \dots, 6$ (see Figure 3), we obtain $x + a_1 + a_2 + b_1 + b_2 + c_1 + c_2 - x = 0$ (here a_i is the value of the cocycle at the simplex a_i, etc.). Any element conjugate to the identity element in a group is itself the identity. Therefore, $a_1 + a_2 + b_1 + b_2 + c_1 + c_2 = 0$. Moreover, by

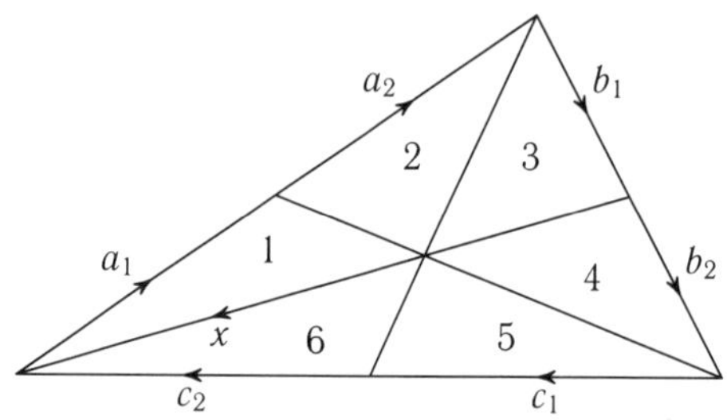

Figure 3. Summation of the cocyclicity conditions

construction, the value of the cocycle $k(z^1)$ at an interval I partitioned into two equal intervals I' and I'' is equal to the sum of its values at I' and I''.

Clearly, k takes cohomologous cocycles to cohomologous cocycles. Thus, we have obtained a map

$$k^*\colon H^1(C_*^{\{X_1,X_2\}}(X,Y);G) \to H^1(X,Y;G).$$

It follows directly from the definition that $k^*i^* = \mathrm{id}$ and $i^*k^* = \mathrm{id}$.

The statement that the map (33) is one-to-one is an analog of the excision axiom for noncommutative cohomology. Let us construct an analog of the Mayer–Vietoris sequence. Suppose that X_1 and X_2 are sets, $X = X_1 \cup X_2$, and $x_0 \in X_1 \cap X_2$. We assume that the natural map

$$i^*\colon H^q(X,Y;G) \to H^q(C_*^{\{X_1,X_2\}}(X,Y);G)$$

is one-to-one. This is so if, e.g., $X = \operatorname{int} X_1 \cup \operatorname{int} X_2$.

As in the commutative case, any continuous map $f\colon (X,Y) \to (X',Y')$ induces a map $f^*\colon H^q(X',Y';G) \to H^q(X,Y;G)$ for $q = 0, 1$, which takes the distinguished element to the distinguished element. This yields the maps

$$\begin{array}{ccc} H^1(C_*^{\{X_1,X_2\}}(X,X_2);G) & \xrightarrow{l_1^*} & H^1(X_1, X_1 \cap X_2; G) \\ \Big\downarrow{\scriptstyle l_2^*} & & \\ H^1(C_*^{\{X_1,X_2\}}(X,x_0);G)\,. & & \end{array}$$

The map l_1^* is one-to-one because there is no essential difference between cochains taking nonzero values at simplices in X_1 and X_2 and vanishing at all simplices in X_2 and those taking nonzero values at simplices in X_1 and vanishing at all simplices in $X_1 \cap X_2$.

For any triple of spaces $Z \subset Y \subset X$, we define a map $\delta\colon H^0(Y,Z;G) \to H^1(X,Y;G)$ as follows. Given a cohomology class $\alpha^0 \in H^0(Y,Z;G)$, we define a cochain $c^0 \in C^0(X,Y;G)$ by setting $c^0(x) = \alpha^0(x)$ for $x \in Y$ and

$c^0(x) = 0$ for $x \notin Y$. Then, we define a cocycle $z^1 \in Z^1(X, Y; G)$ by setting $z^1(f) = -c^0(f\varepsilon_1^1) + c^0(f\varepsilon_0^1)$ for any singular simplex $f\colon \Delta^1 \to X$.

Consider the diagram

(34)

$$\begin{array}{ccccc} H^0(X_1 \cap X_2, x_0; G) & \xrightarrow{\Delta} & H^1(X_1 \cup X_2, x_0; G) & \xrightarrow{i_1^*} & H^1(X_1, x_0; G) \\ & & \downarrow{\scriptstyle i_1^*} & & \downarrow{\scriptstyle j_1^*} \\ & & H^1(X_2, x_0; G) & \xrightarrow{j_2^*} & H^1(X_1 \cap X_2, x_0; G), \end{array}$$

where the maps i_α^* and j_α^* are induced by inclusions and Δ is the composition

$$\begin{array}{ccccc} H^0(X_1 \cap X_2, x_0; G) & & H^1(C_*^{\{X_1,X_2\}}(X, x_0; G)) & \xrightarrow{\cong} & H^1(X, x_0; G) \\ \downarrow{\scriptstyle \delta} & & \uparrow{\scriptstyle l_2^*} & & \\ H^1(X_1, X_1 \cap X_2; G) & \underset{\cong}{\xrightarrow{(l_1^*)^{-1}}} & H^1(C_*^{\{X_1,X_2\}}(X, X_2; G)). & & \end{array}$$

By the kernel of a map to a set with a distinguished element we mean the preimage of the distinguished element.

Theorem 4.16. (a) $\operatorname{Ker} i_1^* \cap \operatorname{Ker} i_2^* = \operatorname{Im} \Delta$.

(b) *For* $a \in H^1(X_1, x_0; G)$ *and* $b \in H^1(X_2, x_0; G)$, $j_1^*(a) = j_2^*(b)$ *if and only if there exists* $c \in H^1(X_1 \cup X_2, x_0; G)$ *for which* $i_1^*(c) = a$ *and* $i_2^*(c) = b$. *Moreover, if* $H^0(X_1 \cap X_2, x_0; G) = 0$, *then such an element* c *is determined uniquely.*

Proof. Consider the map $\Delta' = l_2^*(l_1^*)^{-1}\delta$ (it is obtained from Δ by removing the last isomorphism). It is sufficient to prove the required assertion for the diagram

(35)

$$\begin{array}{ccccc} H^0(X_1 \cap X_2, x_0) & \xrightarrow{\Delta'} & H^1(C_*^{\{X_1,X_2\}}(X, x_0)) & \xrightarrow{(i_1')^*} & H^1(X_1, x_0) \\ & & \downarrow{\scriptstyle (i_2')^*} & & \downarrow{\scriptstyle j_1^*} \\ & & H^1(X_2, x_0) & \xrightarrow{j_2^*} & H^1(X_1 \cap X_2, x_0) \end{array}$$

(to simplify notation, we do not indicate the coefficient group G).

Both sets $\operatorname{Im} \Delta'$ and $\operatorname{Ker}(i_1')^* \cap \operatorname{Ker}(i_2')^*$ consist of equivalence classes of cocycles vanishing at the 1-simplices contained in $X_1 \cap X_2$. The equality $j_1^*(a) = j_2^*(b)$ is equivalent to the existence of a $c' \in H^1(C_*^{\{X_1,X_2\}}(X, x_0))$ for which $(i_1')^*(c') = a$ and $(i_2')^*(c') = b$; this is verified directly. The equality $H^0(X_1 \cap X_2, x_0) = 0$ means that the space $X_1 \cap X_2$ is path-connected (provided that $G \neq 0$). The uniqueness of c' follows from the path-connectedness of $X_1 \cap X_2$. □

Theorem 4.16 allows us to obtain the most general version of van Kampen's theorem about the fundamental group of a union of two sets. For this purpose, we need a dual definition of an amalgam of groups. Namely, a group G is the amalgam of two groups G_1 and G_2 over a group G_0 with respect to homomorphisms $\varphi_1\colon G_0 \to G_1$ and $\varphi_2\colon G_0 \to G_2$ if the commutative diagram

$$\begin{array}{ccccc} 0 & \longrightarrow & \operatorname{Hom}(G, G') & \xrightarrow{\psi_1^*} & \operatorname{Hom}(G_1, G') \\ & & {\scriptstyle\psi_2^*}\big\downarrow & & \big\downarrow{\scriptstyle\varphi_1^*} \\ & & \operatorname{Hom}(G_2, G') & \xrightarrow{\varphi_2^*} & \operatorname{Hom}(G_0, G) \end{array}$$

is exact for any group G'. This definition of an amalgam is equivalent to that given in Part I.

Theorem 4.17 (Olum [**98**]). *Suppose that spaces U_1, U_2, and $U_1 \cap U_2$ are path-connected and $x_0 \in U_1 \cap U_2$. The group $\pi_1(U_1 \cup U_2, x_0)$ is the amalgam of the groups $\pi_1(U_1, x_0)$ and $\pi_1(U_2, x_0)$ over $\pi_1(U_1 \cup U_2, x_0)$ with respect to the homomorphisms induced by the inclusions $U_1 \cap U_2 \subset U_1$ and $U_1 \cap U_2 \subset U_2$ if and only if the natural map*

$$i^*\colon H^1(U_1 \cup U_2, x_0; G') \to H^1(C_*^{\{U_1, U_2\}}(U_1 \cup U_2, x_0); G')$$

is one-to-one for any group G'.

Proof. First, suppose that the map i^* is one-to-one for any group G'. Then we have the commutative diagram (34) (with G replaced by G'). Moreover, the path-connectedness of $U_1 \cap U_2$ implies $H^0(U_1 \cup U_2, x_0; G') = 0$.

Suppose that X is a path-connected space and $x_0 \in X$. Take an element $\omega \in \pi_1(X, x_0)$ and consider a sequence of 1-simplices $\Delta_1^1, \dots, \Delta_k^1$ which form a loop belonging to the homotopy class of ω. For $z^1 \in Z^1(X, x_0; G')$, we set $z^1(\omega) = z^1(\Delta_1^1) + \dots + z^1(\Delta_k^1)$. It is easy to show that $z^1(\omega)$ depends only on the homotopy class of the loop; moreover, it depends only on the cohomology class of the cycle z^1, i.e., we have a map $H^1(X, x_0; G') \to \operatorname{Hom}(\pi_1(X, x_0), G')$. This map is one-to-one. Applying Theorem 4.16 and replacing H^1 by Hom, we see that $\pi_1(U_1 \cup U_2, x_0)$ is the required amalgam.

Now, suppose that $\pi_1(U_1 \cup U_2, x_0)$ is such an amalgam. Replacing Hom by H^1, we obtain the exact commutative diagram

$$\begin{array}{ccccc} 0 & \longrightarrow & H^1(U_1 \cup U_2, x_0; G') & \xrightarrow{i_1^*} & H^1(U_1, x_0; G') \\ & & {\scriptstyle i_2^*}\big\downarrow & & \big\downarrow{\scriptstyle j_1^*} \\ & & H^1(U_2, x_0; G) & \xrightarrow{j_2^*} & H^1(U_1 \cap U_2, x_0) \end{array}$$

for any group G'. Comparing it with the diagram (35) in which X_i is replaced by U_i, we see that the map i^* is one-to-one. □

2. The Poincaré and Lefschetz Isomorphisms for Topological Manifolds

The Poincaré isomorphism does hold for topological manifolds, but this cannot be proved by the same method as for smooth manifolds because not all topological manifolds are triangulable. The proofs of the Poincaré and Lefschetz isomorphisms given below follow largely [**113**] and [**141**].

2.1. Fundamental Classes. Before constructing fundamental classes for topological manifolds, we prove theorems about the vanishing of the k-dimensional homology groups of a topological manifold M^n for $k > n$ and of the n-dimensional homology groups of a noncompact topological manifold M^n without boundary. The latter is needed to construct fundamental classes.

Theorem 4.18. *If M^n is a topological manifold, then $H_k(M^n) = 0$ for $k > n$.*

Proof. Take a singular cycle $z_k \in Z_k(M^n)$. The union of the images of singular simplices in z_k is a compact set; therefore, it is contained in the union of finitely many open sets $U_1, \dots, U_m$ each of which is homeomorphic to an open subset of $\mathbb{R}^n$. We put $V_j = U_1 \cup U_2 \cup \dots \cup U_j$. Let us show that $H_k(V_j) = 0$. For $j = 1$, this follows from Theorem 4.8. For $j > 1$, consider the Mayer–Vietoris exact sequence for $V_{j-1} \cup U_j = V_j$:

$$\cdots \longrightarrow H_k(V_{j-1}) \oplus H_k(U_j) \longrightarrow H_k(V_j) \longrightarrow H_{k-1}(V_{j-1} \cap U_j) \longrightarrow \cdots .$$

The sets U_j and $V_{j-1} \cap U_j$ are homeomorphic to open subsets of $\mathbb{R}^n$; hence $H_k(U_j) = 0$ and $H_{k-1}(V_{j-1} \cap U_j) = 0$. Moreover, $H_k(V_{j-1}) = 0$ by the induction hypothesis. Therefore, $H_k(V_j) = 0$. This means that z^k is the boundary of a singular chain in $V_j \subset M^n$. □

For connected noncompact manifolds, Theorem 4.18 can be strengthened; namely, we can guarantee that $H_k(M^n) = 0$ for $k \geq n$. To prove this, we need the following auxiliary lemma.

Lemma 4.2. *Suppose that $U \subset \mathbb{R}^n$ $(n \geq 2)$ is an open set, $\alpha \in H_n(\mathbb{R}^n, U)$, and for each point $x \in \mathbb{R}^n \setminus U$, we have $(i_x)_*(\alpha) = 0$, where $i_x \colon (\mathbb{R}^n, U) \to (\mathbb{R}^n, \mathbb{R}^n \setminus \{x\})$ is the natural embedding. Then $\alpha = 0$.*

Proof. Consider the exact sequence of the pair $(\mathbb{R}^n, U)$:

$$\cdots \longrightarrow H_n(\mathbb{R}^n) \longrightarrow H_n(\mathbb{R}^n, U) \xrightarrow{\partial_*} H_{n-1}(U) \longrightarrow H_n(\mathbb{R}^n) \longrightarrow \cdots .$$

Here ∂_* is an isomorphism; therefore, it suffices to prove that $\partial_*\alpha = 0$. Let $z \in Z_{n-1}(U)$ be a cycle representing the homology class $\partial_*\alpha$. The support of the chain z is compact; hence we can choose an open set V so that $\overline{V} \subset U$, $\overline{V}$ is compact, and the support of z is contained in V. For this V, $H_{n-1}(V)$ contains an element β such that $i_*\beta = \partial_*\alpha$, where $i\colon V \to U$ is the natural embedding.

Using the compactness of $\overline{V}$, we choose an open cube Q containing V in $\mathbb{R}^n$ and put $K = Q \setminus (Q \cap V)$. We have $\overline{K} \cap \overline{V} = \varnothing$; therefore, for each point $x \in K$, we can choose a closed cube P so that $x \in P$ and $P \cap V = \varnothing$. The compact set $\overline{K}$ is covered by finitely many such cubes $P_1, \dots, P_m$.

By assumption, we have $(i_x)_*\alpha = 0$, so the commutative diagram

$$\begin{array}{ccccc} H_{n-1}(V) & \xrightarrow{\;i_*\;} & H_{n-1}(U) & \xleftarrow{\;\partial_*\;} & H_n(\mathbb{R}^n, U) \\ \downarrow & & \downarrow & & \downarrow (i_x)_* \\ H_{n-1}(\mathbb{R}^n \setminus \{x\}) & \xrightarrow[\cong]{} & H_{n-1}(\mathbb{R}^n \setminus \{x\}) & \xleftarrow[\cong]{\;\partial_*\;} & H_n(\mathbb{R}^n, \mathbb{R}^n \setminus \{x\}) \end{array}$$

shows that the image of β under the inclusion homomorphism $H_{n-1}(V) \to H_{n-1}(\mathbb{R}^n \setminus \{x\})$ is the zero element.

Now, let us show that the image of β under the inclusion homomorphism $H_{n-1}(V) \to H_{n-1}\big(Q \setminus \bigcup_{i=1}^m P_i\big)$ is zero as well. Consider the sets $Q_k = Q \setminus \bigcup_{i=1}^k P_i$ $(k = 0, 1, \dots, m)$. We prove the required assertion by induction on k. For $k = 0$, it is obvious since $Q_0 = Q$ and $H_{n-1}(Q) = 0$. To make the induction step, consider the Mayer–Vietoris exact sequence for Q and $\mathbb{R}^n \setminus P_{k+1}$ (it is defined because both sets are open):

$$H_n(Q \cup (\mathbb{R}^n \setminus P_{k+1})) \longrightarrow H_{n-1}(Q_{k+1}) \xrightarrow{j_*} H_{n-1}(Q) \oplus H_{n-1}(\mathbb{R}^n \setminus P_{k+1}).$$

Since the set $Q \cup (\mathbb{R}^n \setminus P_{k+1})$ is open, its n-dimensional homology group is trivial. Therefore, j_* is a monomorphism. It remains to note that the images of β under the homomorphisms induced by the inclusions $V \subset Q_k$ and $V \subset \mathbb{R}^n \setminus P_{k+1}$ are zero. For the former homomorphism, this is so by the induction hypothesis, and for the latter because $H_{n-1}(\mathbb{R}^n \setminus P_{k+1}) \cong H_{n-1}(\mathbb{R}^n \setminus \{x\})$.

Since the set Q_m is contained in U, we can represent the homomorphism $H_{n-1}(V) \to H_{n-1}(U)$ as the composition of homomorphisms $H_{n-1}(V) \to H_{n-1}(Q_m) \to H_{n-1}(U)$. The image of β under the first homomorphism is the zero element. Therefore, $i_*\beta = 0$, i.e., $\partial_*\alpha = 0$. □

The exact sequence of the pair $(\mathbb{R}^n, \mathbb{R}^n \setminus \{0\})$ implies $H_k(\mathbb{R}^n, \mathbb{R}^n \setminus \{0\}) \cong H_k(\mathbb{R}^n \setminus \{0\}) \cong H_{k-1}(S^{n-1})$ for all $k > 1$. Hence $H_n(\mathbb{R}^n, \mathbb{R}^n \setminus \{0\}) \cong \mathbb{Z}$ and $H_n(\mathbb{R}^n, \mathbb{R}^n \setminus \{0\}; \mathbb{Z}_2) \cong \mathbb{Z}_2$ for $n \geq 2$.

If U is an open chart homeomorphic to $\mathbb{R}^n$ and containing a point $x \in M^n$, then $H_n(M^n, M^n \setminus \{x\}) \cong H_n(U, U \setminus \{x\}) \cong \mathbb{Z}$ (or $\mathbb{Z}_2$, if the coefficient group is $\mathbb{Z}_2$). To prove this, it suffices to apply the excision theorem to the sets $M^n \setminus U \subset M^n \setminus \{x\} \subset M^n$. (The inclusion $\overline{M^n \setminus U} \subset M^n \setminus \{x\}$ follows from $\overline{M^n \setminus U} = M^n \setminus U$ and $x \in U$.)

The groups $H_n(M^n, M^n \setminus \{x\})$ play an important role in constructing fundamental classes.

Theorem 4.19. *If M^n is a connected noncompact manifold without boundary, then $H_n(M^n) = 0$.*

Proof. We show that the homomorphism $p_* : H_n(M^n) \to H_n(M^n, M^n \setminus \{x\})$ is zero for any $x \in M^n$. Consider a homology class $[z_n]$ represented by a cycle $z_n \in Z_n(M^n)$ with compact support C. First, suppose that $x \in M^n \setminus C$. Then z_n is the image of some cycle $\tilde{z}_n$ under the homomorphism $i_\#\colon Z_n(M^n \setminus \{x\}) \to Z_n(M^n)$. Therefore, $p_*[z_n] = p_* i_* [\tilde{z}_n] = 0$ because $p_* i_* = 0$. Now suppose that $x \in C$. By assumption, the manifold M^n is not compact. In particular, $M^n \neq C$. Therefore, we can choose a point $y \in M^n \setminus C$. Consider a homeomorphism $h\colon M^n \to M^n$ which is isotopic to the identity and takes x to y (such a homeomorphism exists for any connected topological manifold without boundary; the proof is similar to that for smooth manifolds (see the lemma on the homogeneity of manifolds in Part I, p. 223). The homeomorphism h induces a homeomorphism $M^n \setminus \{x\} \to M^n \setminus \{y\}$; hence we have the commutative diagram

$$\begin{array}{ccc} H_n(M^n) & \xrightarrow{(p_x)_*} & H_n(M^n, M^n \setminus \{x\}) \\ {\scriptstyle h_* = \mathrm{id}} \downarrow & & \downarrow {\scriptstyle h_*} \\ H_n(M^n) & \xrightarrow{(p_y)_*} & H_n(M^n, M^n \setminus \{y\}). \end{array}$$

The equality $(p_x)_* = 0$ follows from $(p_y)_* = 0$.

As in the proof of Theorem 4.18, we consider each singular cycle $z_n \in Z_n(M^n)$ separately, which allows us to replace M^n by a finite union of open coordinate neighborhoods $U_1, \ldots, U_m$ homeomorphic to $\mathbb{R}^n$. We set $V_k = \bigcup_{i=1}^k U_i$. Let us show that the cycle z_n is the boundary of some cycle in V_m, i.e., $H_n(V_m) = 0$.

We prove this by induction on k. For $k = 1$, the required assertion is obvious because U_1 is contractible. To make the induction step, consider the Mayer–Vietoris sequence for the open sets V_k and U_{k+1}:

$$\begin{aligned} H_n(V_k) \oplus H_n(U_{k+1}) \longrightarrow H_n(V_{k+1}) \longrightarrow H_{n-1}(V_k \cap U_{k+1}) \\ \longrightarrow H_{n-1}(V_k) \oplus H_{n-1}(U_{k+1}). \end{aligned}$$

By the induction hypothesis, we have $H_n(V_k) = 0$. Moreover, the contractibility of U_{k+1} implies $H_n(U_{k+1}) = 0$ and $H_{n-1}(U_{k+1}) = 0$. Therefore, $H_n(V_{k+1}) = 0$ if and only if the homomorphism $i_*\colon H_{n-1}(V_k \cap U_{k+1}) \to H_{n-1}(V_k)$ is a monomorphism.

Suppose that $\beta \in H_{n-1}(V_k \cap U_{k+1})$ and $i_*\beta = 0$. Consider the diagram

$$\begin{array}{ccccc}
H_n(V_{k+1}, V_k \cap U_{k+1}) & \xleftarrow{i'_*} & H_n(U_{k+1}, V_k \cap U_{k+1}) & & \\
\uparrow{\scriptstyle i''_*} & & \downarrow{\scriptstyle \partial'_*} & & \\
H_n(V_k, V_k \cap U_{k+1}) & \xrightarrow{\partial''_*} & H_{n-1}(V_k \cap U_{k+1}) & \xrightarrow{i_*} & H_{n-1}(V_k) \\
 & & \downarrow & & \\
 & & H_{n-1}(U_{k+1}) = 0. & &
\end{array}$$

The second row and the second column in this diagram are segments of exact sequences of pairs. The equality $i_*\beta = 0$ implies $\beta = \partial''_*\beta''$ for some $\beta'' \in H_n(V_k, V_k \cap U_{k+1})$. Moreover, $\beta = \partial'_*\beta'$ for some $\beta' \in H_n(U_{k+1}, V_k \cap U_{k+1})$ because ∂'_* is an epimorphism. Consider the element $\bar{\beta} = i'_*\beta' - i''_*\beta''$. Let us write a part of the exact sequence of a pair:

$$H_n(V_{k+1}) \xrightarrow{p_*} H_n(V_{k+1}, V_k \cap U_{k+1}) \xrightarrow{\partial_*} H_{n-1}(V_k \cap U_{k+1}).$$

Clearly, $\partial_* i'_* = \partial'_*$ and $\partial_* i''_* = \partial''_*$. Hence $\partial_*\bar{\beta} = \partial_* i'_*\beta' - \partial_* i''_*\beta'' = \partial'_*\beta' - \partial''_*\beta'' = \beta - \beta = 0$, and therefore $\partial_*\bar{\beta} = p_*\alpha$ for some $\alpha \in H_n(V_{k+1})$.

As at the beginning of the proof, the noncompactness of M^n implies that the homomorphism $(p_x)_*\colon H_n(V_{k+1}) \to H_n(M^n, M^n \setminus \{x\})$ is zero for each point $x \in M^n$. In particular, $p_*\alpha = 0$.

Let $x \in U_{k+1} \setminus (V_k \cap U_{k+1})$. Then $V_k \cap U_{k+1} \subset M^n \setminus \{x\}$; hence the homomorphism $(p_x)_*$ can be represented as the composition of homomorphisms

$$H_n(V_{k+1}) \xrightarrow{p_*} H_n(V_{k+1}, V_k \cap U_{k+1}) \xrightarrow{l_*} H_n(M^n, M^n \setminus \{x\}).$$

Therefore, $0 = (p_x)_*\alpha = l_*p_*\alpha = l_*\bar{\beta} = l_*i'_*\beta' - l_*i''_*\beta''$. Since x does not belong to V_k, we can replace $l_*i''_*$ by the composition of homomorphisms

$$H_n(V_k, V_k \cap U_{k+1}) \to H_n(V_k \setminus \{x\}, V_k \cap U_{k+1}) \to H_n(M^n \setminus \{x\}, M^n \setminus \{x\}) = 0.$$

Thus, $l_*i''_* = 0$. This implies $l_*i'_*\beta' = 0$. Under the identification of $H_n(M^n, M^n \setminus \{x\})$ with $H_n(U_{k+1}, U_{k+1} \setminus \{x\})$, the homomorphism $l_*i'_*$ transforms into

$$(i_x)_*\colon H_n(U_{k+1}, V_k \cap U_{k+1}) \to H_n(U_{k+1}, U_{k+1} \setminus \{x\}).$$

Applying Lemma 4.2, we obtain $\beta' = 0$; hence $\beta = 0$. □

Corollary 1. *If M^n is a connected topological manifold, then $p_*\colon H_n(M^n) \to H_n(M^n, M^n \setminus \{x\})$ is a monomorphism.*

Proof. Suppose $\alpha \in H_n(M^n)$ and $p_*\alpha = 0$. Then the group $H_n(M^n \setminus \{x\})$ has an element β for which $i_*\beta = \alpha$. But the manifold $M^n \setminus \{x\}$ is not compact; therefore, $H_n(M^n \setminus \{x\}) = 0$. □

Corollary 2. *If M^n is a connected topological manifold, then either $H_n(M^n) = 0$ or $H_n(M^n) \cong \mathbb{Z}$ and, for any $x \in M^n$, the homomorphism $p_*\colon H_n(M^n) \to H_n(M^n, M^n \setminus \{x\})$ is a monomorphism.*

Proof. By virtue of Corollary 1, the homomorphism $p_*\colon H_n(M^n) \to H_n(M^n, M^n \setminus \{x\}) \cong \mathbb{Z}$ is a monomorphism. Therefore, either $H_n(M^n) = 0$ or $H_n(M^n) \cong \mathbb{Z}$ and the image of p_* is the group $m\mathbb{Z} \subset \mathbb{Z}$. It remains to show that $m = \pm 1$.

Suppose that $m > 1$. The proof of Theorem 4.19 (and Corollary 1) is valid not only for the coefficient group $\mathbb{Z}$ but also for an arbitrary Abelian coefficient group G. Let $G = \mathbb{Z}_m$. Consider the commutative diagram

$$\begin{array}{ccc} H_n(M^n) \otimes \mathbb{Z}_m & \xrightarrow{p_*\otimes \mathrm{id}} & H_n(M^n, M^n \setminus \{x\}) \otimes \mathbb{Z}_m \\ \varphi_1 \Big\downarrow \text{mono} & & \varphi_2 \Big\downarrow \text{mono} \\ H_n(M^n; \mathbb{Z}_m) & \xrightarrow[\text{mono}]{p_*} & H_n(M^n, M^n \setminus \{x\}; \mathbb{Z}_m). \end{array}$$

Here φ_1 and φ_2 are the monomorphisms from the universal coefficient formulas. The homomorphism $p_* \otimes \mathrm{id}$ is a monomorphism as well. On the other hand, if α is a generator of the group $H_n(M^n)$, then $(p_* \otimes \mathrm{id})(\alpha \otimes 1) = p_*\alpha \otimes 1 = m \otimes 1 = 0$. □

For every point $x \in M^n$, consider the groups $T_x = H_n(M^n, M^n \setminus \{x\}) \cong \mathbb{Z}$ and $T_x(\mathbb{Z}_2) = H_n(M^n, M^n \setminus \{x\}; \mathbb{Z}_2) \cong \mathbb{Z}_2$. Let us introduce topologies on the sets $\mathcal{T} = \bigcup_{x \in M^n} T_x$ and $\mathcal{T}(\mathbb{Z}_2) = \bigcup_{x \in M^n} T_x(\mathbb{Z}_2)$ so that the natural projections of $\mathcal{T}$ and $\mathcal{T}(\mathbb{Z}_2)$ onto M^n be coverings. Note that such topologizations are equivalent to defining local systems of coefficient groups isomorphic to $\mathbb{Z}$ and $\mathbb{Z}_2$ on M^n. For coefficients in $\mathbb{Z}_2$, the local system is trivial because there exists precisely one isomorphism $\mathbb{Z}_2 \to \mathbb{Z}_2$. The space $\mathcal{T}(\mathbb{Z}_2)$ consists of two copies of the manifold M^n; one of them corresponds to the zero element of the group $\mathbb{Z}_2$, and the other corresponds to the nonzero element. We want to construct a local system of coefficient groups isomorphic to $\mathbb{Z}$ that generalizes the system $\operatorname{Or} M^n$ for smooth manifolds.

We say that a set V is *admissible* if V is the interior of a closed disk contained in some chart on M^n. To construct a topology on $\mathcal{T}$, we use the observation that if a set V is admissible and $x \in V$, then the inclusion

$M^n \setminus V \hookrightarrow M^n \setminus \{x\}$ is a homotopy equivalence. It follows that the homomorphism

$$i_{V,x}\colon H_n(M^n, M^n \setminus V) \to H_n(M^n, M^n \setminus \{x\}) = T_x$$

induced by this inclusion is an isomorphism (the proof is precisely the same as for simplicial homology; see Theorem 1.14 on p. 16). We declare the sets $\{i_{V,x}\alpha \mid x \in V\}$ to be open for all $\alpha \in H_n(M^n, M^n \setminus V)$. These open sets form a base for the topology of $\mathcal{T}$. Note that the admissible sets V form a base for the topology of M^n.

An *orientation* of a topological manifold M^n is, by definition, a continuous section s of the covering $\mathcal{T} \to M^n$ such that every element $s(x) \in T_x$ is a generator of the group $T_x \cong \mathbb{Z}$. Clearly, any connected topological manifold without boundary has either none or two orientations. A topological manifold admitting an orientation is said to be *orientable*, and a topological manifold with a given orientation is *oriented.*

An orientation of a topological manifold M^n can also be defined for the coefficient group $\mathbb{Z}_2$ as a continuous section s of the covering $\mathcal{T}(\mathbb{Z}_2) \to M^n$ such that every element $s(x) \in T_x(\mathbb{Z}_2)$ is the generator of the group $T_x(\mathbb{Z}_2) \cong \mathbb{Z}_2$. Clearly, such a section always exists and is unique because $\mathbb{Z}_2$ has only one generator.

A homology class $\alpha \in H_n(M^n)$ is said to be *fundamental* if for any $x \in M^n$, the map $p_*\colon H_n(M^n) \to H_n(M^n, M^n \setminus \{x\}) = T_x$ takes α to a generator of T_x. We shall see that there is a close relation between fundamental classes and orientations; namely, each orientation of a closed topological manifold determines a fundamental class. In particular, for a closed connected orientable manifold, we have $H_n(M^n) \cong \mathbb{Z}$. But before proceeding to orientable manifolds, we consider the nonorientable case.

Theorem 4.20. *If M^n is a connected nonorientable topological manifold without boundary, then $H_n(M^n) = 0$.*

Proof. Let $\alpha \in H_n(M^n)$. For each point $x \in M^n$, consider the image $\alpha_x \in T_x$ of α under the homomorphism $p_*\colon H_n(M^n) \to H_n(M^n, M^n \setminus \{x\}) = T_x$. The nonorientability of M^n implies the existence of a path γ such that traversing it replaces α_x by $-\alpha_x$. Therefore, $\alpha_x = 0$. According to Corollary 1 of Theorem 4.19, $\alpha = 0$. □

We use the notation α_x introduced in the proof of Theorem 4.20 in the formulation and the proof of the following theorem.

Theorem 4.21. *Let M^n be a closed connected orientable manifold endowed with an orientation $s\colon M^n \to \mathcal{T}$. Then there exists a unique fundamental class α such that $\alpha_x = s(x)$ for all $x \in M^n$.*

Proof. The most important and difficult part of the theorem is the existence of a fundamental class. We start with this part, postponing the proof of uniqueness.

We recall that if V is an admissible[6] set, then the map $i_{V,x} : H_n(M^n, M^n \setminus V) \to T_x$ is an isomorphism. This enables us to construct the required homology class α_V in $H_n(M^n, M^n \setminus V)$. Our goal is to glue such classes together by using the relative Mayer–Vietoris sequence and obtain a homology class in $H_n(M^n, \varnothing) = H_n(M^n)$. In dealing with Mayer–Vietoris sequences, open sets are more convenient to handle because they automatically satisfy the excision axiom. Thus, instead of the admissible set V, we take its closure $\overline{V}$; the map $i_{\overline{V},x}$ is an isomorphism as well, and we can construct the class $\alpha_{\overline{V}} \in H_n(M^n, M^n \setminus \overline{V})$.

Since the manifold M^n is compact, it can be covered by finitely many closures of admissible sets $\overline{V}_1, \dots, \overline{V}_m$. Suppose that we have already constructed an element

$$\alpha_{1,\dots,k} \in H_n(M^n, M^n \setminus (\overline{V}_1 \cup \dots \cup \overline{V}_k))$$

such that $i_x \alpha_{1,\dots,k} = s(x)$ for all $x \in \overline{V}_1 \cup \dots \cup \overline{V}_k$; here i_x is the homomorphism of homology groups induced by the map of pairs

$$(M^n, M^n \setminus (\overline{V}_1 \cup \dots \cup \overline{V}_k)) \to (M^n, M^n \setminus \{x\}).$$

Consider the relative Mayer–Vietoris sequence

$$\begin{aligned} &H_n(M^n, M^n \setminus (\overline{V}_1 \cup \dots \cup \overline{V}_{k+1})) \\ &\quad\longrightarrow H_n(M^n, M^n \setminus (\overline{V}_1 \cup \dots \cup \overline{V}_k)) \oplus H_n(M^n, M^n \setminus \overline{V}_{k+1}) \\ &\qquad\longrightarrow H_n(M^n, M^n \setminus ((\overline{V}_1 \cup \dots \cup \overline{V}_k) \cap \overline{V}_{k+1})). \end{aligned}$$

Take an element of the form $(\alpha_{1,\dots,k}, -\alpha_{k+1})$ in the direct sum. Let us prove that it is the image of some element

$$\alpha_{1,\dots,k+1} \in H_n(M^n, M^n \setminus (\overline{V}_1 \cup \dots \cup \overline{V}_{k+1})),$$

or, equivalently, that the image of $(\alpha_{1,\dots,k}, -\alpha_{k+1})$ in

$$H_n(M^n, M^n \setminus ((\overline{V}_1 \cup \dots \cup \overline{V}_k) \cap \overline{V}_{k+1}))$$

vanishes. We denote this image by β. By assumption, for any $x \in C = (\overline{V}_1 \cup \dots \cup \overline{V}_k) \cap \overline{V}_{k+1}$, the homomorphism $H_n(M^n, M^n \setminus C) \to T_x$ takes β to zero.

Lemma. *Let $U \subset M^n$ be an open set, and let $\beta \in H_n(M^n, U)$ be an element such that the homomorphism $H_n(M^n, U) \to T_x$ takes β to zero for any $x \in U$. Then $\beta = 0$.*

[6] Admissible sets were defined on p. 231.

Proof. The proof of the lemma follows a scheme which has already been used many times in this book. It consists in proving the required assertion for simple domains and extending it by using the Mayer–Vietoris sequence.

First, suppose that the set $M^n \setminus U$ is contained in a coordinate neighborhood $W \approx \mathbb{R}^n$. Consider the commutative diagram

$$\begin{array}{ccc} H_n(W, W \cap U) & \xrightarrow{i_*} & H_n(W, W \setminus \{x\}) \\ \downarrow \cong & & \downarrow \cong \\ H_n(M^n, U) & \xrightarrow{i_*} & H_n(M^n, M^n \setminus \{x\}). \end{array}$$

Here the vertical arrows are excision isomorphisms. The equality $p_*\beta = 0$ is equivalent to $i_*\tilde{\beta} = 0$, where $\tilde{\beta}$ is the preimage of β under the excision isomorphism. Lemma 4.2 implies $\tilde{\beta} = 0$ and, therefore, $\beta = 0$.

Now we consider the general case. Let V_i be an open coordinate neighborhood in M^n homeomorphic to $\mathbb{R}^n$. The compact set $M^n \setminus U$ can be covered by finitely many open sets V_i. Let $U_i = (M^n \setminus U) \cap V_i$. Then $M^n \setminus U = \bigcup (M^n \setminus U_i)$, i.e., $U = \bigcap U_i$. Thus, it is sufficient to show that if the assertion of the lemma holds for open sets U' and U'', then it holds for $U' \cap U''$. Consider the relative Mayer–Vietoris sequence

$$H_{n+1}(M^n, U' \cup U'') \longrightarrow H_n(M^n, U' \cap U'') \longrightarrow H_n(M^n, U') \oplus H_n(M^n, U'').$$

It is easy to see that $H_{n+1}(M^n, U' \cup U'') = 0$. Indeed, consider the exact sequence of a pair

$$H_{n+1}(M^n) \longrightarrow H_{n+1}(M^n, U' \cup U'') \longrightarrow H_n(U' \cup U'').$$

Here $H_{n+1}(M^n) = 0$ and $H_n(U' \cup U'') = 0$ if $U' \cup U'' \neq M^n$ (if $U' \cup U'' = M^n$, then the equality $H_{n+1}(M^n, U' \cup U'') = 0$ is obvious). Thus, we have a monomorphism

$$H_n(M^n, U' \cap U'') \to H_n(M^n, U') \oplus H_n(M^n, U'').$$

Suppose that it takes β to (β', β''). Let β_x, β'_x, and β''_x be the images of β, β', and β'' under the map to T_x. Then $\beta'_x = \beta_x$ and $\beta''_x = -\beta_x$. Therefore, if $\beta_x = 0$, then $\beta'_x = 0$ and $\beta''_x = 0$. By assumption, $\beta_x = 0$ for all $x \in M^n \setminus (U' \cap U'')$. Therefore, $\beta'_x = 0$ for all $x \in M^n \setminus U'$ and $\beta''_x = 0$ for all $x \in M^n \setminus U''$. It follows that $\beta' = 0$ and $\beta'' = 0$. Since the map under consideration is a monomorphism, we have $\beta = 0$. □

The element $\alpha_{1,\dots,k+1}$ that is mapped to $(\alpha_{1,\dots,k}, -\alpha_{k+1})$ has the required property, i.e., $i_x\alpha_{1,\dots,k+1} = s(x)$ for all $x \in \overline{V}_1 \cup \dots \cup \overline{V}_{k+1}$.

The uniqueness of the fundamental class is very easy to prove. If α and α' are two fundamental classes corresponding to the same orientation s, then, for each $x \in M^n$, the homomorphism $H_n(M^n) \to H_n(M^n, M^n \setminus \{x\})$

takes the class $\alpha - \alpha'$ to $s(x) - s(x) = 0$. But this homomorphism is a monomorphism (see Corollary 1 of Theorem 4.19). □

Any manifold admits a unique orientation with respect to the coefficient group $\mathbb{Z}_2$. The argument from the proof of Theorem 4.21 applies to the group $\mathbb{Z}_2$ without any changes; using it, we can construct the fundamental class in $\mathbb{Z}_2$-homology for any closed connected topological manifold.

2.2. The Thom Isomorphism. The Thom isomorphism relates the k-dimensional cohomology of a closed manifold M^n to the $(n+k)$-dimensional cohomology of the pair $(M^n \times M^n, M^n \times M^n \setminus d(M^n))$, where $d\colon M^n \to M^n \times M^n$ is the diagonal map. The construction of this isomorphism is based on Lemmas 4.3 and 4.4 below. We start by introducing some notation.

Let M^n be a closed oriented topological manifold with an orientation $s\colon M^n \to \mathcal{T}$; by $p_1, p_2\colon M^n \times M^n \to M^n$ we denote the projections onto the first and second factors. For every point $x \in M^n$, consider the map $l_x\colon M^n \to M^n \times M^n$ defined by $l_x(y) = (x, y)$. This map induces a map of pairs

$$(M^n, M^n \setminus \{x\}) \to (M^n \times M^n, M^n \times M^n \setminus d(M^n)),$$

which we denote by the same symbol l_x.

Lemma 4.3. *Suppose that $V \subset M^n$ is an open neighborhood contained in a disk (in local coordinates), which, in turn, is contained in a disk D^n of larger radius centered at the same point $x_0 \in V$. Then there exists a homeomorphism $\theta\colon p_1^{-1}(V) = V \times M^n \to p_1^{-1}(V)$ with the following properties:*

(i) *$\theta(x, y) = (x, y')$ for all $x \in V$ and $y \in M^n$ (here y' is some point of the manifold M^n);*

(ii) *$\theta(x, x) = (x, x_0)$ for all $x \in V$;*

(iii) *$(p_2\theta l_x)_*(s(x)) = s(x_0)$ for all $x \in V$ (here we mean maps of pairs because $s(x)$ is the homology class of a pair).*

The statement of (iii) *uses* (ii): $(p_2\theta l_x)(x) = p_2\theta(x, x) = p_2(x, x_0) = x_0$.

Proof. Consider the self-homeomorphism of $V \times D^n$ defined by the condition that its restriction to each line segment with endpoints (x, x) and (x, y), where $x \in V$ and $y \in \partial D^n$, is a linear map onto the segment with endpoints (x, x_0) and (x, y). Figure 4 shows the restriction of this self-homeomorphism to $\{x\} \times D^n$ for $n = 2$. On $V \times \partial D^n$, this map is the identity; therefore, it can be extended over $V \times M^n$ by the identity map outside $V \times D^n$.

The composition $p_2\theta l_x$ coincides with the map shown in Figure 4; clearly, it is orientation preserving. □

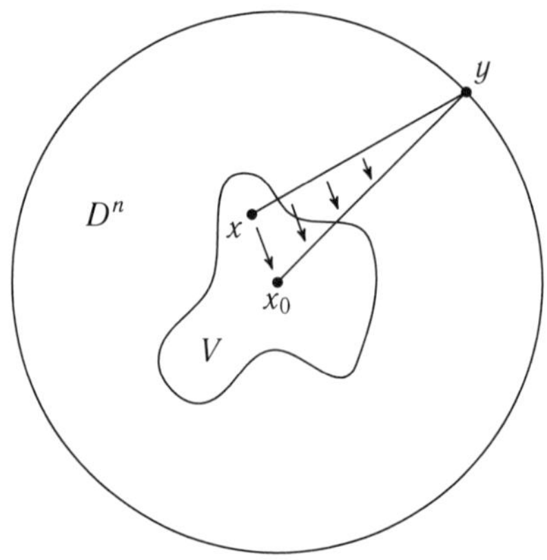

Figure 4. The construction of the homeomorphism

For any open set $V \subset M^n$, the pair

$$(p^{-1}(V), p^{-1}(V) \setminus d(M^n)) = (V \times M^n, V \times M^n \setminus d(M^n))$$

is defined; we denote it by $V^\times$. In what follows, an important role is played by the pair $M^{n\times} = (M^n \times M^n, M^n \times M^n \setminus d(M^n))$. Note that the map l_x can be regarded as a map of pairs $(M^n, M^n \setminus \{x\}) \to M^{n\times}$.

Lemma 4.4. (a) *For $i < n$, $H_i(M^{n\times}) = 0$.*

(b) *The isomorphism $H_0(M^n) \cong H_n(M^{n\times})$ holds; it is induced by the map taking the 0-chain corresponding to a point $x \in M^n$ to the relative n-cycle representing the class $(l_x)_*(s(x))$.*

Proof. We apply Lemma 4.3. Suppose that a set V satisfies the conditions of the lemma and θ is the corresponding homeomorphism. Then θ takes $d(V)$ to $V \times \{x_0\}$; therefore, it induces a homeomorphism of pairs

$$V^\times = (V \times M^n, V \times M^n \setminus d(M^n)) \to V \times (M^n, M^n \setminus \{x_0\}).$$

Thus, $H_*(V^\times) \cong H_*(V \times M^n, V \times (M^n \setminus \{x_0\}))$. Applying the Künneth theorem for relative homology, we see that the group $H_k(V^\times)$ is isomorphic to the direct sum of the groups

$$\bigoplus_{p+q=k} H_p(V) \otimes H_q(M^n, M^n \setminus \{x_0\}) = H_0(V) \otimes H_k(M^n, M^n \setminus \{x_0\})$$

and

$$\bigoplus_{p+q=k-1} \operatorname{Tor}(H_p(V), H_q(M^n, M^n \setminus \{x_0\})) = \operatorname{Tor}(H_0(V), H_{k-1}(M^n, M^n \setminus \{x_0\})).$$

But $H_i(M^n, M^n \setminus \{x_0\})) = 0$ for $i < n$. Therefore, $H_k(V^\times) = 0$ for $i < n$, and $H_k(V^\times) \cong H_0(V) \otimes H_k(M^n, M^n \setminus \{x_0\})$.

Consider the diagram

$$\begin{array}{ccccc} & & & & H_0(V)\otimes H_n(M^n, M^n\setminus\{x_0\}) \\ & & & & \uparrow \cong \\ H_n(V, V\setminus\{x\}) & \xrightarrow{(l_x)_*} & H_n(V^\times) & \xrightarrow{\theta_*} & H_n(V\times M^n, V\times(M^n\setminus\{x_0\})) \\ & & & & \downarrow (p_2)_* \\ & & & & H_n(M^n, M^n\setminus\{x_0\}). \end{array}$$

We know that $(p_2\theta l_x)_*(s(x)) = s(x_0)$. Hence the map $H_0(V)\to H_0(V)\otimes H_n(M^n, M^n\setminus\{x_0\})$ induced by the map $[x]\to[x]\otimes s(x_0)$ is an isomorphism. Returning to $H_n(V^\times)$, we obtain the map $[x]\to(l_x)_*(s(x))$ instead of this isomorphism.

Thus, the assertion of the lemma is true for the set V. Note that we have not used the connectedness of V; the isomorphism $H_n(V^\times)\cong H_0(V)$ holds even for disconnected open sets V. It remains to show that if the assertion is true for open sets U_1, U_2, and $U_1\cap U_2$, then it is also true for $U_1\cup U_2$. We take two segments of the Mayer Victoris sequences and map them to one another by the homomorphism described in the statement of the lemma:

$$\begin{array}{ccccc} H_0(U_1\cap U_2) & \longrightarrow & H_0(U_1)\oplus H_0(U_2) & \longrightarrow & H_0(U_1\cup U_2) \\ \downarrow & & \downarrow & & \downarrow \\ H_n((U_1\cap U_2)^\times) & \longrightarrow & H_{n-1}(U_1^\times)\oplus H_{n-1}(U_2^\times) & \longrightarrow & H_{n-1}((U_1\cup U_2)^\times). \end{array}$$

To this diagram we can add

$$\begin{array}{ccccc} \longrightarrow & 0 & \longrightarrow & 0 \\ & \downarrow & & \downarrow \\ \longrightarrow & H_n((U_1\cap U_2)^\times) & \longrightarrow & H_n(U_1^\times)\oplus H_n(U_2^\times) \end{array}$$

on the right without violating commutativity. The homology groups in the lower row are trivial. Applying the five lemma, we obtain the required result. □

By Lemma 4.4 and the universal coefficient theorem, $H^i(M^{n\times}) = 0$ for $i < n$ and $H^n(M^{n\times})\cong \operatorname{Hom}(H_n(M^{n\times}),\mathbb{Z})\cong\operatorname{Hom}(H_0(M^n),\mathbb{Z})$; both isomorphisms are canonical. The class $U\in H^n(M^{n\times})$ corresponding to the homomorphism $H_0(M^n)\to\mathbb{Z}$ induced by the augmentation is called the *Thom class* of the topological manifold M^n. The Thom class is uniquely determined by the condition that $\langle U, (l_x)_*(s(x))\rangle = 1$ for all $x\in M^n$, i.e., $\langle l_x^*U, s(x)\rangle = 1$. This means, in particular, that l_x^*U is a generator of the group $H^n(M^n, M^n\setminus\{x\})$.

For $X = M^n \times M^n$ and $A = M^n \times M^n \setminus d(M^n)$, we have $U \in H^n(X, A)$. For a cohomology class $\alpha^k \in H^k(M^n)$, the class $p_1^*(\alpha^k)$ belongs to $H^k(X)$; therefore, we can consider the relative cup product $H^k(X) \otimes H^n(X, A) \to H^{n+k}(X, A)$. We set $\Phi^*(\alpha^k) = p_1^*(\alpha^k) \smile U \in H^{n+k}(M^{n\times})$.

Theorem 4.22 (Thom isomorphism). *For any k, the map $\Phi^* \colon H^k(M^n) \to H^{n+k}(M^{n\times})$ is an isomorphism.*

Proof. For every open set $V \subset M^n$, let $U_V \in H^n(V^\times)$ be the restriction of the cohomology class U to V. Consider the cap product $H^n(V^\times) \otimes H_k(V^\times) \to H_{k-n}(V \times M^n)$. For $\beta_k \in H_k(V^\times)$, we set

$$\varphi_*(\beta_k) = (p_1)_*(U_V \frown \beta_k) \in H_{k-n}(V).$$

Let V be an open set satisfying the conditions of Lemma 4.3. Then the homeomorphism θ from this lemma induces an isomorphism

$$\theta^* \colon H^n(V \times M^n, V \times (M^n \setminus \{x\})) \to H^n(V^\times).$$

Using Theorem 4.15 on p. 217, we can identify the element $\theta^{*-1}(U_V)$ with $1 \otimes \gamma^n \in H^0(V) \otimes H^n(M^n, M^n \setminus \{x\})$, where γ^n is the generator of the group $H^n(M^n, M^n \setminus \{x\})$ such that $\langle \gamma^n, s(x) \rangle = 1$ (in the notation of Theorem 4.15, $\gamma^n = e^n$: we identify the groups $H^n(M^n, M^n \setminus \{x\})$ and $H^n(\mathbb{R}^n, \mathbb{R}^n \setminus \{0\})$ by means of the excision isomorphism).

The homomorphism θ has the property $p_1 = p_1 \theta$ (on the set V). Therefore, the naturality of the cap product implies

$$\begin{aligned}(p_1)_*(U_V \frown \beta_k) &= (p_1)_* \theta_*(\theta^*(\theta^{*-1}(U_V)) \frown \beta_k) \\ &= (p_1)_*(\theta^{*-1}(U_V) \frown \theta_*(\beta_k)).\end{aligned}$$

The element $\theta_*(\beta_k)$ belongs to $H_k(V \times M^n, V \times (M^n \setminus \{x\}))$. According to the Künneth theorem, we can identify it with $b_{k-n} \otimes s(x)$, where $b_{k-n} \in H_{k-n}(V)$.

The homomorphism $(p_1)_*$ takes the cap product of $1 \otimes \gamma^n \in H^0(V) \otimes H^n(M^n, M^n \setminus \{x\})$ and $b_{k-n} \otimes s(x) \in H_{k-n}(V) \otimes H_n(M^n, M^n \setminus \{x\})$ to b_{k-n}. This means that $\varphi_*(\beta_k) = b_{k-n}$. The map $\beta_k \mapsto b_{k-n}$ is the composition of isomorphisms

$$\begin{aligned}H_k(V^\times) &\xrightarrow{\theta} H_k(V \times M^n, V \times (M^n \setminus \{x\})) \\ &\cong H_{k-n}(V) \otimes H^n(M^n, M^n \setminus \{x\}) \cong H_{k-n}(V).\end{aligned}$$

Thus, $\varphi_* \colon H_k(V^\times) \to H_{k-n}(V)$ is an isomorphism.

Using the standard argument involving the Mayer–Vietoris sequence, we obtain an isomorphism $\varphi_* \colon H_k(M^{n\times}) \to H_{k-n}(M^n)$ for each $k \geq n$.

We pass to the level of chains and construct the dual map of cochains for φ_*. Then, from the map of cochains we pass to the map $\varphi^* \colon H^k(M^n) \to$

$H_{n+k}(M^{n\times})$ of cohomology groups. Let us show that $\varphi^* = \Phi^*$, i.e., $\varphi^*(\alpha^k) = p_1^*(\alpha^k) \smile U$. Indeed,

$$\begin{aligned}\langle \varphi^*(\alpha^k), \beta_{k+n}\rangle &= \langle \alpha^k, \varphi_*(\beta_{k+n})\rangle = \langle \alpha^k, (p_1)_*(U \frown \beta_{k+n})\rangle \\ &= \langle p_1^*(\alpha^k), (U \frown \beta_{k+n})\rangle = \langle p_1^*(\alpha^k) \smile U, \beta_{k+n}\rangle;\end{aligned}$$

the last equality follows from $\langle \alpha^q, \beta^p \frown \gamma_{p+q}\rangle = \langle \alpha^q \smile \beta^p, \gamma_{p+q}\rangle$ (see Theorem 2.6 on p. 71).

It remains to verify that φ^* is an isomorphism. The isomorphism φ_* induces the commutative diagram

$$\begin{array}{ccccccccc} 0 & \longrightarrow & \operatorname{Ext}(H_{k-1}(M^n), \mathbb{Z}) & \longrightarrow & H^k(M^n) & \longrightarrow & \operatorname{Hom}(H_k(M^n), \mathbb{Z}) & \longrightarrow & 0 \\ & & \downarrow{\scriptstyle \varphi_{*1}} & & \downarrow{\scriptstyle \varphi^*} & & \downarrow{\scriptstyle \varphi_{*2}} & & \\ 0 & \longrightarrow & \operatorname{Ext}(H_{k+n-1}(M^{n\times}), \mathbb{Z}) & \longrightarrow & H^{k+n}(M^{n\times}) & \longrightarrow & \operatorname{Hom}(H_{k+n}(M^{n\times}), \mathbb{Z}) & \longrightarrow & 0. \end{array}$$

Here φ_{*1} and φ_{*2} are isomorphisms because φ_* is an isomorphism. According to the five lemma, φ^* is an isomorphism as well. □

2.3. The Poincaré Isomorphism. To prove the Poincaré isomorphism theorem for topological manifolds, we need the following two properties of closed topological manifolds.

Theorem 4.23. (a) *Any closed topological manifold M^n can be embedded in $\mathbb{R}^k$ for some k.*

(b) *If $f\colon M^n \to \mathbb{R}^k$ is an embedding of a closed topological manifold M^n, then there exists an open set $U \supset f(M^n)$ such that $f(M^n)$ is its retract.*

Proof. (a) Let $U_1, \ldots, U_m$ be open sets homeomorphic to $\mathbb{R}^n$ and covering M^n. Take homeomorphisms $h_i\colon U_i \to \mathbb{R}^n \approx S^n \setminus \{y_0\}$ and consider the map $f_i\colon M^n \to S^n$ defined by

$$f_i(x) = \begin{cases} h_i(x) & \text{if } x \in U_i, \\ y_0 & \text{if } x \notin U_i. \end{cases}$$

The map f_i is continuous because $\lim_{x\to\partial U_i} f(x) = y_0$. The maps $f_1, \ldots, f_m$ can be treated as a single map $f\colon M^n \to \prod_{i=1}^m S_i^n \subset \mathbb{R}^{m(n+1)}$. Clearly, if $x \neq x'$, then $f(x) \neq f(x')$. Since M^n is compact and $\mathbb{R}^{m(n+1)}$ is Hausdorff, it follows that f is a homeomorphism from M^n to $f(M^n) \subset \mathbb{R}^{m(n+1)}$.

(b) For brevity, we denote the image of the manifold M^n in $\mathbb{R}^k$ by the same symbol M^n. Choose a sufficiently large simplex Δ^k in $\mathbb{R}^k$ so that the manifold M^n is contained strictly inside Δ^k. Let us construct an (infinite) triangulation of the space $\Delta^k \setminus M^n$ as follows. First, we take the barycentric subdivision of Δ^k and choose the simplices intersecting M^n in this subdivision. Then, we take the barycentric subdivisions of the chosen simplices and leave the remaining simplices intact. Among the new simplices we chose those intersecting M^n, and so on.

Since M^n is compact, it follows that the distance from any point $x \in \Delta^k \setminus M^n$ to M^n is positive; therefore, each point x has a neighborhood intersecting only finitely many simplices constructed above. This means that what we have constructed is indeed a triangulation K of the space $\Delta^k \setminus M^n$.

Now, let us construct a family of subsets $N_0 \subset N_1 \subset \cdots$ of Δ^k and continuous maps $r_i\colon N_i \to M^n$ such that $M^n \subset N_i$ and the restriction of each r_i to M^n is the identity map. For N_0 we take the union of M^n and the vertices of the simplicial complex K. The map r_0 takes each point $x \in N_0$ to a point of M^n nearest[7] to x. In particular, $r_0(x) = x$ for all $x \in M^n$.

The sets N_i and maps r_i are constructed by induction on i. Passing from N_i to N_{i+1}, we add some $(i+1)$-simplices from the complex K; these are the simplices $\Delta^{i+1}_\alpha \subset K$ such that their boundaries are contained in N_i and the maps $\partial\Delta^{i+1}_\alpha \xrightarrow{r_i} M^n$ can be extended to continuous maps $\Delta^{i+1}_\alpha \to M^n$. The map r_{i+1} on Δ^{i+1}_α is defined as follows. Consider all continuous extensions $\varphi\colon \Delta^{i+1}_\alpha \to M^n$ of r_i. Let $\delta(\varphi)$ be the diameter of the set $\varphi(\Delta^{i+1}_\alpha)$. Clearly, $\delta(\varphi)$ is not smaller than the diameter of $r_i(\partial\Delta^{i+1}_\alpha)$. Let a be the greatest lower bound of the set of all numbers $\delta(\varphi)$. If $a = 0$, then r_i takes $\partial\Delta^{i+1}_\alpha$ to one point $x_\alpha \in M^n$. In this case, we set $r_{i+1}(\Delta^{i+1}_\alpha) = x_\alpha$. If $a > 0$, then for the map r_{i+1} on Δ^{i+1}_α we take a map φ for which $a \le \delta(\varphi) < 2a$.

We set $N = \bigcup_{i=0}^\infty N_i$ and define $r\colon N \to M^n$ to coincide with r_i on each N_i. It is seen from the construction that this map is continuous at all points of the set $N \setminus M^n$. It remains to show that for each point $x \in M^n$, the map r is continuous at x and, moreover, N contains an open neighborhood of x.

Lemma 4.5. *For any $\varepsilon > 0$, there exists a $\delta > 0$ such that if $x \in \Delta^k \setminus M^n$ and $d(x, M^n) < \delta$, then the diameter of any k-simplex from K containing x is less than ε.*

Proof. Take a positive integer m such that the diameter of any simplex from the mth barycentric subdivision of Δ^k is less than ε. Let K_m be the subcomplex in K consisting of all simplices from the mth barycentric subdivision of Δ^k that do not intersect M^n. If $K_m = \varnothing$, then for δ we take any positive number, e.g., ε. If $K_m \ne \varnothing$, then we set $\delta = d(|K_m|, M^n)$. This number is positive because the sets $|K_m|$ and M^n are compact and disjoint.

Suppose that $d(x, M^n) < \delta$. Then $x \notin |K_m|$; therefore, the k-simplex in the mth barycentric subdivision of Δ^k that contains x must intersect M^n. By construction, such a simplex undergoes the barycentric subdivision.

[7]Since $M^n \subset \mathbb{R}^k$ is compact, it follows that, for any $x \in \mathbb{R}^k$, there exists at least one point $y \in M^n$ for which $\|x - y\| = d(x, M^n)$. Such points y are said to be nearest to x in M^n.

Thus, the k-simplex from K containing x is obtained not earlier than at the $(m+1)$th step. The diameter of any such simplex is less than ε. □

For each point $x \in M^n \subset \mathbb{R}^k$ and every $\varepsilon > 0$, we construct a system of open neighborhoods $V_0 \supset V_1 \supset \cdots \supset V_{k+1}$ in $\mathbb{R}^k$ so that $V_{k+1} \subset N$ and $r(V_{k+1}) \subset D^k_{x,\varepsilon}$, where $D^k_{x,\varepsilon} = \{y \in \mathbb{R}^k \mid \|x-y\| < \varepsilon\}$. For V_0 we take the open subset of $D^k_{x,\varepsilon}$ containing x and such that $V_0 \cap M^n$ is an admissible subset homeomorphic to $\mathbb{R}^n$. We assume that $V_0 \subset \Delta^k$. For $i \geq 1$, the neighborhood V_i is constructed from V_{i-1} as follows. First, we choose a number $\varepsilon(i) > 0$ so that $D^k_{x,5\varepsilon(i)} \subset V_{i-1}$. Then, using Lemma 4.5, we choose $\delta(i) < \varepsilon(i)$ such that if $y \in \Delta^k \setminus M^n$ and $\|x-y\| < \delta(i)$, then the diameter of any k-simplex from K containing y is less than $\varepsilon(i)$. Finally, we choose an open set $V_i \subset \Delta^k$ in $\mathbb{R}^k$ such that it is contained in $D^k_{x,\delta(i)}$ (and contains x) and $V_i \cap M^n$ is an admissible set homeomorphic to $\mathbb{R}^n$.

Suppose that $y \in V_{k+1} \setminus M^n$ and $\Delta^k(y)$ is the k-simplex from K containing y. By construction, all vertices of $\Delta^k(y)$ belong to $D^k_{x,2\delta(k+1)} \subset V_k$. If v is a vertex of $\Delta^k(y)$, then $\|v - r(v)\| \leq \|v - x\|$; therefore, the images of all vertices of $\Delta^k(y)$ under the map r belong to $D^k_{x,4\delta(k+1)} \subset V_k$.

Since the set $V_k \cap M^n$ is homeomorphic to $\mathbb{R}^n$, we can extend the map r, which is defined on the vertex set of the simplex $\Delta^k(y)$, to the edges Δ^1_α. Hence each Δ^1_α is contained in N. There is an extension of r to Δ^1_α for which the diameter of the image of Δ^1_α is less than $2\delta(k+1)$; therefore, the diameter of $r(\Delta^1_\alpha)$ is less than $4\delta(k+1) < \delta(k)$. Since the endvertices of the edge Δ^1_α belong to V_k, it follows that $r(\Delta^1_\alpha) \subset V_{k-1} \cap M^n$.

Applying the same argument to the 2-skeleton, 3-skeleton, and so on, we obtain $\Delta^k(y) \subset N$ and $r(\Delta^k(y)) \subset V_0 = D^k_{x,\varepsilon}$. In particular, $r(y) \in D^k_{x,\varepsilon}$. This means that the map r is continuous at the point y. Moreover, $V_{k+1} \subset N$. □

Lemma 4.6. *For any closed topological manifold M^n, there exists an open neighborhood W of $d(M^n)$ in $M^n \times M^n$ for which the maps $p_1|_W$ and $p_2|_W$ are homotopic.*

Proof. According to Theorem 4.23, we can assume that M^n is embedded in $\mathbb{R}^k$ for some k, and, moreover, M^n is a retract of its neighborhood U in $\mathbb{R}^k$. Since M^n is compact, we can choose $\varepsilon > 0$ such that if the distance between points $x, y \in M^n$ in $\mathbb{R}^k$ be less than ε, then the line segment joining x and y is contained in U.

The maps p_1 and p_2 coincide on the set $d(M^n)$. Therefore, using the compactness of M^n, we can choose a neighborhood W of $d(M^n)$ in $M^n \times M^n$ so that for each $w \in W$, the distance between $p_1(w)$ and $p_2(w)$ be less than ε. The segment with endpoints $p_1(w)$ and $p_2(w)$ is contained in U; therefore, the maps $p_1|_W$ and $p_2|_W$ are homotopic in U. We construct a homotopy in M^n from this homotopy in U by using a retraction $r: U \to M^n$. $\square$

Consider the map $T: M^n \times M^n \to M^n \times M^n$ defined by $T(x,y) = (y,x)$. This map induces a map of pairs $T: M^{n\times} \to M^{n\times}$.

Lemma 4.7. *If $\alpha \in H^*(M^{n\times})$, then $T^*\alpha = (-1)^n\alpha$.*

Proof. Let V be an admissible open subset of M^n homeomorphic to $\mathbb{R}^n$. Consider the commutative diagram

$$\begin{array}{ccc} H^*(M^{n\times}) & \xrightarrow{\quad T^* \quad} & H^*(M^{n\times}) \\ \downarrow{\scriptstyle i^*} & & \downarrow{\scriptstyle i^*} \\ H^n(V \times V, V \times V \setminus d(V)) & \xrightarrow{T^*} & H^n(V \times V, V \times V \setminus d(V)). \end{array}$$

Here i^* is the homomorphism induced by the inclusion of pairs $(V \times V, V \times V \setminus d(V)) \subset M^{n\times}$. It is easy to see that $H^n(V \times V, V \times V \setminus d(V)) \cong \mathbb{Z}$. Indeed, the exact sequence of a pair implies that $H^n(V \times V, V \times V \setminus d(V)) \cong H^{n-1}(V \times V \setminus d(V))$. Moreover, $V \times V \setminus d(V) \approx \mathbb{R}^n \times (\mathbb{R}^n \setminus \{0\}) \sim \mathbb{R}^n \setminus \{0\} \sim S^{n-1}$. Therefore, $H^n(V \times V, V \times V \setminus d(V)) \cong H^{n-1}(S^{n-1}) \cong \mathbb{Z}$. This group is generated by $\beta = i^*(U)$, where i^* is the homomorphism induced by the inclusion $V \subset M^n$.

Before proving the equality $T^*\beta = (-1)^n\beta$, we show that the pair $(V \times V, V \times V \setminus d(V))$ is homotopy equivalent to $(\mathbb{R}^n \times \mathbb{R}^n, \mathbb{R}^n \times (\mathbb{R}^n \setminus \{0\})) \sim (\mathbb{R}^n, \mathbb{R}^n \setminus \{0\})$. The homotopy equivalence is constructed as follows. We identify V with $\mathbb{R}^n$ and, in each layer $\{y_0\} \times \mathbb{R}^n$, move the point (y_0, y_0), which belongs to $d(V)$, to the point $(y_0, 0)$ in such a way that the layer remains invariant. Moreover, the homotopy ("shift") can be constructed so that $(y_0, 0)$ moves to $(y_0, -y_0)$.

To prove the equality $T^*\beta = (-1)^n\beta$, it suffices to track the action of T on one layer, say, $y_0 = 0$. The map T takes $(0, x_0)$ to $(x_0, 0)$. The homotopy shifts it to $(x_0, -x_0)$. Projecting this point onto the initial layer, we obtain $(0, -x_0)$. Therefore, the homomorphism $T^*: H^n(\mathbb{R}^n, \mathbb{R}^n \setminus \{0\}) \to H^n(\mathbb{R}^n, \mathbb{R}^n \setminus \{0\})$ is induced by the map $x_0 \mapsto -x_0$. The degree of this map equals $(-1)^n$.

Let W be the open neighborhood of $d(M^n)$ mentioned in Lemma 4.6. Replacing W by $TW \cap W$ if necessary, we can assume that the neighborhood W is invariant with respect to T. We set $U = M^n \times M^n \setminus W$. Then $\overline{U} \subset M^n \times M^n \setminus d(M^n)$; therefore, we can apply the excision theorem and obtain an isomorphism $H^*(W, W \setminus d(M^n)) \cong H^*(M^{n\times})$. This allows us to represent the Thom isomorphism Φ^* as the composition

$$H^*(M^n) \xrightarrow{(p_1|_W)^*} H^*(W) \xrightarrow{\smile U'} H^*(W, W \setminus d(M^n)) \cong H^*(M^{n\times}),$$

where U' is the restriction of the Thom class to W. Let $\alpha \in H^*(M^{n\times})$. Using the equality $p_1T = p_2$, we obtain

$$T^*((p_1|_W)^*(\alpha) \smile U') = (p_2|_W)^*(\alpha) \smile T^*(U') = (-1)^n(p_2|_W)^*(\alpha) \smile U'.$$

But $(p_1|_W) \sim (p_2|_W)$; hence $T^*\Phi^*(\alpha) = (-1)^n\Phi^*(\alpha)$, and therefore $T^*\alpha = (-1)^n\alpha$ because Φ^* is an isomorphism. □

We are almost ready to prove the Poincaré isomorphism theorem. It only remains to verify that the homology groups of a closed topological manifold are finitely generated.

Theorem 4.24. *If M^n is a closed topological manifold, then the group $H_*(M^n)$ is finitely generated.*

Proof. We assume the manifold M^n to be embedded in $\mathbb{R}^k$. Consider an open set $U \supset M^n$ in $\mathbb{R}^k$ for which M^n is a retract. Let Δ^k be a simplex containing U in $\mathbb{R}^k$. The distance d between the compact sets M^n and $\Delta^k \setminus U$ is positive; hence there exists an m such that the diameter of any simplex in the mth barycentric subdivision of Δ^k is less than d.

Consider the simplicial complex K consisting of the simplices from the mth barycentric subdivision of Δ^k that intersect M^n. Clearly, $M^n \subset |K| \subset U$. Let $r\colon |K| \to M^n$ be the restriction of a retraction $U \to M^n$. The composition $M^n \xrightarrow{i} |K| \xrightarrow{r} M^n$ is the identity map; therefore, $r_*\colon H_*(|K|) \to H_*(M^n)$ is an epimorphism. Since the group $H_*(|K|)$ is finitely generated, it follows that so is the group $H_*(M^n)$. □

Theorem 4.25 (the Poincaré isomorphism). *Let M^n be a closed orientable topological manifold with an orientation s, and let $[M^n] \in H_n(M^n)$ be its fundamental class corresponding to s. Then the map $D\colon H^k(M^n) \to H_{n-k}(M^n)$ defined by $D(a^k) = a^k \frown [M^n]$ is an isomorphism for any k.*

Proof. Let $\alpha' \in C_n(M^n)$ be a cycle representing the fundamental class $[M^n]$. Consider the map $D_\#\colon C^k(M^n) \to C_{n-k}(M^n)$ defined by $D_\# a' = a' \frown \alpha'$. The equality $\partial\alpha' = 0$ implies $\partial D_\# = (-1)^{n-k}D_\#\delta$ because $\partial(D_\#\alpha') = \partial(a' \frown \alpha') = (-1)^{n-k}\delta a' \frown \alpha' + a' \frown \partial\alpha' = (-1)^{n-k}\delta a' \frown \alpha' = (-1)^{n-k}D_\#(\delta a')$. Hence the map $D_\#$ takes cocycles to cycles and, therefore, induces a homomorphism of (co)homology groups, which coincides with D.

First, we shall prove the Poincaré isomorphism theorem for the case of coefficients in $\mathbb{Z}_p$, and then reduce the proof for the integral (co)homology to this case. So, instead of (co)homology with coefficients in $\mathbb{Z}$, we shall consider (co)homology with coefficients in an arbitrary commutative ring R with identity.

For homology classes $\alpha, \beta \in H_*(M^n; R)$, let $\alpha \times \beta$ denote the image of $\alpha \otimes \beta$ in $H_*(M^n \times M^n; R)$ under the monomorphism from the Künneth theorem; for cohomology classes $a, b \in H^*(M^n; R)$, $a \times b$ denotes the element $p_1^*(a) \smile p_2^*(b) \in H^*(M^n \times M^n; R)$. As above, $U \in H^n(M^{n\times})$ is the Thom class. The notation $\tilde{U} = i^*(U)$ is used for the element of the group $H^n(M^n \times M^n)$ that corresponds to U under the homomorphism $i^*\colon H^n(M^{n\times}) \to H^n(M^n \times M^n)$. Under a ring homomorphism $\mathbb{Z} \to R$ taking 1 to the identity element of R, the element $\tilde{U}$ transforms into an element of the ring $H^n(M^n \times M^n; R)$, which we denote by the same symbol $\tilde{U}$. Finally, for an arbitrary point $x \in M^n$, $[x]$ denotes the generator of $H_0(M^n)$ determined by x.

The element $[x] \times [M^n]$ is represented in $H_n(M^n \times M^n)$ by the cycle $(l_x)_*([M^n])$; therefore, the homomorphism $i_*\colon H_n(M^n \times M^n) \to H_n(M^{n\times})$ takes $[x] \times [M^n]$ to $(l_x)_*(s(x))$. Thus,

$$\langle \tilde{U}, [x] \times [M^n] \rangle = \langle i_*(U), [x] \times [M^n] \rangle = \langle U, (l_x)_*(s(x)) \rangle = 1.$$

Since the classes $[x] \times [M^n]$ and $[M^n] \times [x]$ are obtained from each other by means of T_*, it follows that $[M^n] \times [x] = (-1)^n [x] \times [M^n]$; hence

$$\langle \tilde{U}, [M^n] \times [x] \rangle = (-1)^n \langle \tilde{U}, [x] \times [M^n] \rangle = 1. \tag{36}$$

Let us prove one more formula: If $a^p \in H^p(M^n; R)$ and $b^q \in H^q(M^n; R)$, then

$$\tilde{U} \smile (a^p \times b^q) = (-1)^{pq} \tilde{U} \smile (b^q \times a^p); \tag{37}$$

it is assumed that $\tilde{U} \in H^n(M^n \times M^n; R)$. Indeed, according to Lemma 4.7, we have

$$\begin{aligned}(-1)^n U \smile (a^p \times b^q) &= T^*(U \smile (a^p \times b^q)) \\ &= T^*(U) \smile T^*(a^p \times b^q) = (-1)^{n+pq} U \smile (b^q \times a^p).\end{aligned}$$

This equality for U implies equality (37) for $\tilde{U}$.

The main purpose of our calculations is to express

$$\langle \tilde{U}, D(a^k) \times \beta_k \rangle = \langle \tilde{U}, (a^k \frown [M^n]) \times \beta_k \rangle$$

in terms of $\langle a^k, \beta_k \rangle$. To do this, we need, in addition to (36) and (37), expressions for $(a^k \times 1) \frown ([M^n] \times \beta_k)$ and $(1 \times a^k) \frown ([M^n] \times \beta_k)$. By definition, we have $a^k \times 1 = p_1^*(a^k) \smile p_2^*(1) = p_1^*(a^k) \smile 1$ and $1 \times a^k = 1 \smile p_2^*(a^k)$. Therefore,

$$(a^k \times 1) \frown ([M^n] \times \beta_k) = p_1^*(a_k) \frown ([M^n] \times \beta_k)$$

and

$$(1 \times a^k) \frown ([M^n] \times \beta_k) = p_2^*(a_k) \frown ([M^n] \times \beta_k).$$

Let $[m_0, \dots, m_n]$ and $[b_0, \dots, b_k]$ be simplices contained in representatives of the homology classes $[M^n]$ and β_k. Then $[m_0, \dots, m_n] \times [b_0, \dots, b_k]$ is the sum of simplices $[v_0, \dots, v_{n+k}]$, where $v_i = (m_{M(i)}, b_{B(i)})$ and either (i) $v_{i+1} = (m_{M(i)+1}, b_{B(i)})$ or (ii) $v_{i+1} = (m_{M(i)}, b_{B(i)+1})$. By definition,

$$\begin{aligned} p_1^*(a^k) \frown [v_0, \dots, v_{n+k}] &= \langle p_1^*(a^k), [v_n, \dots, v_{n+k}]\rangle [v_0, \dots, v_n] \\ &= \langle a^k, p_1[v_n, \dots, v_{n+k}]\rangle [v_0, \dots, v_n] \\ &= \langle a^k, [m_{M(n)}, \dots, m_{M(n+k)}]\rangle [v_0, \dots, v_n] \end{aligned}$$

and

$$p_2^*(a^k) \frown [v_0, \dots, v_{n+k}] = \langle a^k, [b_{B(n)}, \dots, b_{B(n+k)}]\rangle [v_0, \dots, v_n].$$

Nonzero expressions are obtained only if the simplices $[m_{M(n)}, \dots, m_{M(n+k)}]$ and $[b_{B(n)}, \dots, b_{B(n+k)}]$ are nondegenerate. In case (ii), the simplex $[v_0, \dots, v_{n+k}]$ is uniquely determined by this condition; therefore, $p_2^*(a_k) \frown ([M^n] \times \beta_k) = \langle a^k, \beta_k\rangle [M^n] \times [b_0]$. We can take $[x]$ instead of $[b_0]$ because all points of the connected manifold M^n determine the same zero-dimensional homology class. In case (i), only the simplex $[v_n, \dots, v_{n+k}]$ is determined uniquely. The only conditions on the simplex $[v_0, \dots, v_{n+k}]$ are $v_0 = (m_0, b_0)$ and $v_n = (m_{n-k}, b_k)$. The sum of all such simplices is equal to $[m_0, \dots, m_{n-k}] \times [b_0, \dots, b_k]$. Therefore, $p_1^*(a_k) \frown ([M^n] \times \beta_k) = (a_k \frown ([M^n]) \times \beta_k$.

We obtain

$$\begin{aligned} \langle \tilde{U}, (a^k \frown [M^n]) \times \beta_k\rangle &= \langle \tilde{U}, (a^k \times 1) \frown ([M^n] \times \beta_k)\rangle \\ &= \langle \tilde{U} \smile (a^k \times 1), [M^n] \times \beta_k\rangle \overset{(37)}{=} \langle \tilde{U} \smile (1 \times a^k), [M^n] \times \beta_k\rangle \\ &= \langle \tilde{U}, (1 \times a^k) \frown ([M^n] \times \beta_k)\rangle = \langle \tilde{U}, \langle a^k, \beta_k\rangle [M^n] \times [x]\rangle \\ &\overset{(36)}{=} (-1)^n \langle a^k, \beta_k\rangle, \end{aligned}$$

i.e.,

$$\langle \tilde{U}, D(a^k) \times \beta_k\rangle = (-1)^n \langle a^k, \beta_k\rangle. \tag{38}$$

Relation (38) leads us to conclude that if $a^k \neq 0$, then $D(a^k) \neq 0$. Hence, the maps $H^k(M^n; R) \xrightarrow{D} H_{n-k}(M^n; R)$ and $H^{n-k}(M^n; R) \xrightarrow{D} H_k(M^n; R)$ are monomorphisms. If R is a field, then $H^k(M^n; R)$ is the linear space over R dual to $H_k(M^n; R)$. These spaces are finite-dimensional by Theorem 4.24; hence they are isomorphic to the same linear space V_k. Thus, we have two monomorphisms, $V_k \to V_{n-k}$ and $V_{n-k} \to V_k$, of finite-dimensional linear spaces; clearly, they must be isomorphisms. Thus, we have proved the Poincaré isomorphism theorem for (co)homology with coefficients in $\mathbb{Z}_p$, where p is a prime.

Let us prove it for the integral (co)homology. To this end, we construct a chain complex with chain groups $C_{n-k} = C^{k+1}(M^n) \oplus C_{n-k}(M^n)$, i.e., $C_k = C^{n-k+1}(M^n) \oplus C_k(M^n)$, and define the boundary homomorphism $\bar{\partial}\colon C_k \to C_{k-1}$ by

$$\bar{\partial}(a^{n-k+1}, \beta_k) = ((-1)^k \delta a^{n-k+1}, \partial\beta_k + D_{\#}(a^{n-k+1})).$$

It is easy to see that $\bar{\partial}\bar{\partial} = 0$. Indeed,

$$\begin{aligned}\bar{\partial}\bar{\partial}&(a^{n-k+1}, \beta_k)\\ &= (-\delta\delta a^{n-k+1}, \partial(\partial\beta_k + D_{\#}(a^{n-k+1})) + (-1)^k D_{\#}(\delta a^{n-k+1})) = 0,\end{aligned}$$

because $\partial D_{\#}(a^{n-k+1}) = (-1)^{k+1} D_{\#}\delta(a^{n-k+1})$.

We have a short exact sequence

$$0 \longrightarrow C_k(M^n) \longrightarrow C^{n-k+1}(M^n) \oplus C_k(M^n) \longrightarrow C^{n-k+1}(M^n) \longrightarrow 0.$$

Such short exact sequences of groups form an exact sequence of chain complexes. The map $C_* \to C^*(M^n)$ is chain only up to sign, but this is sufficient for constructing an exact homology sequence for chain complexes. Let us show that the connecting homomorphism $H^{n-k+1}(M^n) \to H_{k-1}(M^n)$ in this exact sequence coincides with D. Suppose that $a \in C^{n-k+1}(M^n)$ and $\delta a = 0$. Then $(a, 0) \in C_k$ belongs to the preimage of a. Let $\bar{\partial}(a, 0) = (0, D_{\#}(a))$. The preimage of this element in $C_k(M^n)$ is equal to $D_{\#}(a)$. The map $a \mapsto D_{\#}(a)$ is the required connecting homomorphism (at the level of chains).

Applying the same construction to the coefficient group $\mathbb{Z}_p$, we obtain the exact sequence

$$\begin{aligned}\cdots \longrightarrow H^{n-k}(M^n; \mathbb{Z}_p) &\xrightarrow{D} H_k(M^n; \mathbb{Z}_p)\\ &\longrightarrow H_k(C_*; \mathbb{Z}_p) \longrightarrow H^{n-k+1}(M^n; \mathbb{Z}_p) \xrightarrow{D} \cdots,\end{aligned}$$

in which the homomorphisms D are isomorphisms. Therefore, $H_*(C_*; \mathbb{Z}_p) = 0$. Since all groups $H_*(C_*)$ are finitely generated, the universal coefficient theorem shows that $H_*(C_*) = 0$. Thus, D is an isomorphism for integer coefficients too. □

Any closed connected topological manifold is orientable with respect to the coefficient group $\mathbb{Z}_2$; therefore, a similar (but somewhat simpler) argument shows that the map $D\colon H^k(M^n; \mathbb{Z}_2) \to H_{n-k}(M^n; \mathbb{Z}_2)$ defined by $D(a^k) = a^k \frown [M^n]_2$ is an isomorphism.

2.4. The Lefschetz Isomorphism. For topological manifolds, the theorem on the Lefschetz isomorphism is derived from the Poincaré isomorphism theorem in the same way as for smooth manifolds. However, the reduction requires some effort, for it uses the collar theorem, whose proof for topological manifolds is more complicated than in the smooth case. Moreover, we have to construct the fundamental class of an oriented topological manifold with boundary and prove its properties.

The collar theorem for topological manifolds was proved by Brown [**20**]; we follow the exposition of [**24**].

Theorem 4.26 (on a collar). *Any compact topological manifold M^n with boundary $B = \partial M^n$ has an open subset U which contains B and is homeomorphic to $B \times [0,1)$. Moreover, the homeomorphism $h\colon U \to B \times [0,1)$ can be chosen so that $h(x) = (x,0)$ for all $x \in B$.*

Proof. For each point $x \in B$, the boundary B contains an open subset U_x for which there exists an embedding $h_x\colon \overline{U}_x \times [0,1] \to M^n$ such that $h_x(y,0) = y$ for all $y \in \overline{U}_x$. Since B is compact, the cover $\{U_x\}$ has a finite subcover $\{U_{x_1}, \dots, U_{x_m}\}$. For short, we denote its elements by $U_1, \dots, U_m$ and the corresponding embeddings by $h_1, \dots, h_m$.

Being Hausdorff and compact, the topological space B is normal. Therefore, the cover $\{U_i\}$ has an open refinement $\{V_i\}$ such that $\overline{V}_i \subset U_i$.

The main idea of the proof of the collar theorem is to consider the topological space M^+ obtained by attaching $B \times [-1,0]$ to M^n via the map $\varphi\colon B \times \{0\} \to M^n$ that takes each point $(x,0)$ to $x \in B \subset M^n$. Then, the required homeomorphism from M^n on M^+ is constructed by induction on m involving the "local collars" $h_i\colon \overline{U}_i \times [0,1] \to M^n$.

Consider the embeddings $H_i\colon \overline{U}_i \times [-1,1] \to M^+$ defined by

$$H_i(x,t) = \begin{cases} h_i(x,t) & \text{if } t \in [0,1], \\ (x,t) & \text{if } t \in [-1,0]. \end{cases}$$

These embeddings are well defined because $h_i(x,0) = (x,0)$ on $\overline{U}_i \times \{0\}$. We use them to construct functions $f_i\colon B \to [-1,0]$ and embeddings $g_i\colon M^n \to M^+$ $(i = 0, 1, \dots, m)$ with the following properties:

(a) $f_i(x) = -1$ for $x \in \bigcup_{j=1}^{i} \overline{V}_j$ (in particular, $f_m(x) = -1$ for all $x \in B$);

(b) $g_i(x) = (x, f_i(x))$ for $x \in B$;

(c) $g_i(M^n) = M^n \cup \{(x,t) \mid t \geq f_i(x)\}$.

If we constructed such maps, we would obtain the required collar because g_m must be a homeomorphism from M^n to M^+.

We construct them by induction on i. We set $g_0 = \mathrm{id}_{M^n}$ and $f_0 \equiv 0$. Suppose that f_{i-1} and g_{i-1} are already constructed. Let us construct f_i and g_i. Consider the set $H_i^{-1}(g_{i-1}(M^n)) \subset \overline{U}_i \times [-1,1]$. Since $g_{i-1}(M^n)$ contains M^n, we have $H_i^{-1}(g_{i-1}(M^n)) \supset \overline{U}_i \times [0,1]$; this is shown schematically in Figure 5. Let us construct an embedding $\varphi_i\colon H_i^{-1}(g_{i-1}(M^n)) \to \overline{U}_i \times [-1,1]$ by blowing up the fibers over $x \in B$ until the fiber over each point $x \in \overline{V}_i$

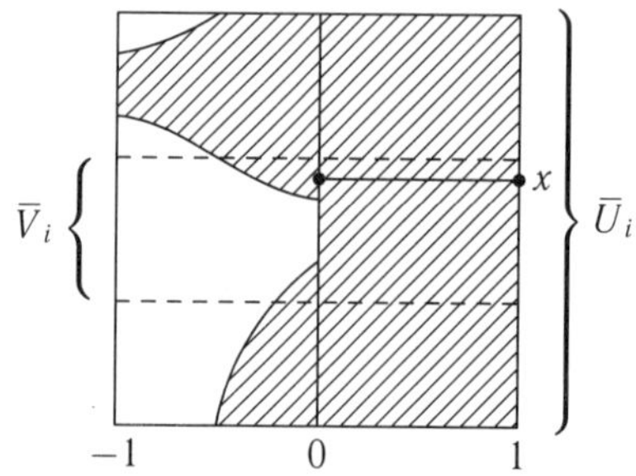

Figure 5. The set $H_i^{-1}(g_{i-1}(M^n))$

coincides with the entire interval $[-1,1]$; formally, this can be written as $\varphi_i(H_i^{-1}(g_{i-1}(\overline{V}_i))) = \overline{U}_i \times [-1,0]$. Moreover, we want the map φ_i to be the identity on $(\overline{U}_i \setminus U_i) \times [-1,1]$ and on $\overline{U}_i \times \{0\}$. To construct such an embedding, we apply the Urysohn lemma (see Part I, Theorem 3.6). The closed subsets $\overline{U}_i \setminus U_i$ and $\overline{V}_i$ of the normal space B are disjoint. Hence there exists a function $\lambda_i \colon \overline{U}_i \to [0,1]$ taking the value 0 on $\overline{U}_i \setminus U_i$ and 1 on $\overline{V}_i$. Let L_x be the affine map from the interval $[f_{i-1}(x),1]$ onto $[(1-\lambda_i(x))f_{i-1}(x)+\lambda_i(x)(-1),1]$. For $x \in \overline{V}_i$, we obtain the interval $[-1,1]$, and for $x \in \overline{U}_i \setminus U_i$, we obtain $[f_{i-1}(x),1]$; thus, in the latter case, L_x is the identity map. For $(x,t) \in H_i^{-1}(g_{i-1}(M^n))$, we set $\varphi_i(x,t) = (x, L_x(t))$; the map φ_i thus defined is continuous. Next, we define the map $\Phi_i \colon g_i(M^n) \to M^+$ by

$$\Phi_i(x) = \begin{cases} H_i \varphi_i H_i^{-1}(x) & \text{if } x \in g_{i-1}(M^n) \cap H_i(\overline{U}_i \times [-1,1]), \\ x & \text{for all other } x \in g_i(M^n). \end{cases}$$

Finally, we set $g_i = \Phi_i g_{i-1}$. The map Φ_i (and hence g_i) is well defined because φ_i is the identity map on $(\overline{U}_i \setminus U_i) \times [-1,1]$ and $\overline{U}_i \times \{0\}$. The map Φ_i (and hence g_i) is an embedding because each map L_x is one-to-one and, therefore, φ_i is an embedding. Moreover, we have

$$g_{i-1}(M^n) \cap H_i(\overline{U}_i \times [-1,1]) = H_i(\overline{U}_i \times [0,1]) \cup \{(x,t) \mid t \geq f_{i-1}(x), x \in \overline{U}_i\}$$

because (c) holds for g_{i-1}. The function $f_i(x)$ is determined by (b). Conditions (a) and (c) hold by construction. □

Let M^n be a compact topological manifold with boundary ∂M^n. Consider the closed topological manifold $\tilde{M}^n$ obtained from two copies of M^n by identifying the respective points of their boundaries.

For all $x \in M^n \setminus \partial M^n$, the groups T_x are defined (see p. 231). They determine a covering $\mathcal{T} \to M^n \setminus \partial M^n$. The manifold M^n is said to be *orientable* if this covering has a section s such that $s(x)$ generates the group T_x for every $x \in M^n \setminus \partial M^n$. The section s is then called an *orientation*. It is easy to verify that a manifold M^n is orientable if and only if $\tilde{M}^n$

is orientable. Moreover, if M^n is orientable, then so is ∂M^n (i.e., each connected component of ∂M^n is orientable).

Let α_x be a generator of the group $T_x = H_n(M^n, M^n \setminus \{x\})$. A homology class $\alpha \in H_n(M^n, \partial M^n)$ is called *fundamental* if the map $p_*\colon H_n(M^n, \partial M^n) \to H_n(M^n, M^n \setminus \{x\}) = T_x$ takes α to α_x for all $x \in M^n \setminus \partial M^n$.

Theorem 4.27. *Let M^n be a compact connected orientable topological manifold with boundary ∂M^n, and let s be an orientation of M^n. Then there exists a unique fundamental class $\alpha \in H_n(M^n, \partial M^n)$ such that $\alpha_x = s(x)$ for all $x \in M^n \setminus \partial M^n$. Also, the connecting homomorphism $\partial_*\colon H_n(M^n, \partial M^n) \to H_{n-1}(\partial M^n)$ takes α to a fundamental class of the closed manifold ∂M^n, i.e., $\partial_*(\alpha)$ determines a fundamental class on each connected component of the manifold ∂M^n.*

Proof. The orientation s of M^n determines an orientation $\tilde{s}$ of $\tilde{M}^n$. Since the manifold $\tilde{M}^n$ is closed and orientable, there exists a unique fundamental class $\tilde{\alpha} \in H_n(\tilde{M}^n)$ such that $(p_x)_*(\tilde{\alpha}) = \tilde{s}(x)$ for all $x \in \tilde{M}^n$. We shall construct the class α from $\tilde{\alpha}$. For this purpose, we need an isomorphism

$$H_n(\tilde{M}^n, \tilde{M}^n \setminus (M^n \setminus \partial M^n)) \cong H_n(M^n, \partial M^n).$$

We might have obtained such an isomorphism by excising $(M^n)' \setminus \partial M^n$, where $(M^n)'$ is the second copy of M^n. But we cannot do this because the condition

$$\overline{(M^n)' \setminus \partial M^n} \subset \operatorname{int}(M^n)'$$

does not hold. Instead, we construct the required isomorphism using the collar theorem; namely, we cut off $(M^n)' \setminus \partial M^n$ from the union of $(M^n)'$ and a collar of M^n rather than from $(M^n)'$. To do this, some obvious homotopy equivalences are needed.

We define the class α as the image of $\tilde{\alpha}$ under the composition of homomorphisms

$$H_n(\tilde{M}^n) \longrightarrow H_n(\tilde{M}^n, \tilde{M}^n \setminus (M^n \setminus \partial M^n)) \xrightarrow{\cong} H_n(M^n, \partial M^n).$$

Obviously, $\alpha_x = s(x)$.

Let $x \in \partial M^n$. Take a closed ball D^{n-1} centered at x in some chart of the manifold ∂M^n and consider the subset

$$E = \{(x,t) \mid x \in D^{n-1}, t \in [0, 1/2]\}$$

of the open set $U = \partial M^n \times [0,1)$ mentioned in the statement of the collar theorem. We put $M^n_E = M^n \setminus \operatorname{int} E$. We have the following commutative diagram:

$$\begin{array}{ccccc}
H_n(M^n,\partial M^n) & \xrightarrow{\;i_*\;} & H_n(M^n,M^n_E) & \xleftarrow[\cong]{\;k_*\;} & H_n(E,\partial E) \\
\Big\downarrow{\scriptstyle \partial_*} & & \Big\downarrow{\scriptstyle \partial_*} & & \cong\Big\downarrow{\scriptstyle \partial_*} \\
H_{n-1}(\partial M^n) & \xrightarrow{\;i'_*\;} & H_{n-1}(M^n_E) & \xleftarrow{\;k'_*\;} & H_{n-1}(\partial E) \\
\Big\downarrow{\scriptstyle (p_x)_*} & & \Big\downarrow & & \cong\Big\downarrow \\
H_{n-1}(\partial M^n,\partial M^n\setminus\{x\}) & \xrightarrow{\;\cong\;} & H_{n-1}(M^n_E,M^n_E\setminus\{x\}) & \xleftarrow{\;\cong\;} & H_{n-1}(\partial E,\partial E\setminus\{x\}).
\end{array}$$

The horizontal isomorphisms in this diagram arise from the excision theorem (in some cases, excision should be performed with care, by using obvious homotopy equivalences).

Take $y \in \operatorname{int} E$. We represent the homomorphism $(p_y)_*\colon H_n(M^n,\partial M^n) \to H_n(M^n, M^n\setminus\{y\})$ as the composition

$$H_n(M^n,\partial M^n) \xrightarrow{i_*} H_n(M^n,M^n_E) \longrightarrow H_n(M^n,M^n\setminus\{y\}).$$

The element $(p_y)_*(\alpha) = s(y)$ generates the group $H_n(M^n,M^n\setminus\{y\}) \cong \mathbb{Z}$; hence $i_*(\alpha)$ generates the group $H_n(M^n,M^n_E) \cong \mathbb{Z}$. Therefore, the group $H_n(E,\partial E) \cong \mathbb{Z}$ has a generator α' for which $k_*(\alpha') = i_*(\alpha)$.

The commutativity of the diagram implies $k'_*\partial_*(\alpha') = i'_*\partial_*(\alpha)$. Therefore, the images of $\partial_*(\alpha)$ and $\partial_*(\alpha')$ in the group $H_{n-1}(M^n_E, M^n_E\setminus\{x\}) \cong \mathbb{Z}$ coincide. Moreover, the image of $\partial_*(\alpha')$ is a generator; hence so is the image of $\partial_*(\alpha)$. Thus, the element $(p_x)_*\partial_*(\alpha)$ generates the group $H_{n-1}(\partial M^n,\partial M^n\setminus\{x\}) \cong \mathbb{Z}$. This means that $\partial_*(\alpha)$ is a fundamental class of the manifold ∂M^n.

Now it is easy to prove the remaining part of the theorem, which concerns the uniqueness of the fundamental class α. Suppose that α_1 and α_2 are two fundamental classes in $H_n(M^n,\partial M^n)$ corresponding to the same orientation. Then $\partial_*(\alpha_1)$ and $\partial_*(\alpha_2)$ are two fundamental classes in $H_{n-1}(\partial M^n)$ corresponding to the same orientation. Therefore, $\alpha_1-\alpha_2 \in \operatorname{Ker}\partial_*$.

The collar theorem implies that the spaces M^n and $M^n\setminus\partial M^n$ are homotopy equivalent to the same space $M^n\setminus U$, where $U = \partial M^n\times[0,1)$. Hence $H_n(M^n) \cong H_n(M^n\setminus\partial M^n) = 0$. Therefore, the left term in the exact sequence

$$H_n(M^n) \longrightarrow H_n(M^n,\partial M^n) \xrightarrow{\partial_*} H_{n-1}(\partial M^n)$$

is trivial, and $\operatorname{Ker}\partial_* = 0$. □

The same argument as the one used for smooth manifolds (in the triangulable case) proves the following theorem.

Theorem 4.28 (the Lefschetz isomorphism). *Let M^n be a compact orientable topological manifold with boundary ∂M^n, and $[M^n] \in H_n(M^n,\partial M^n)$*

the fundamental class of M^n. *Then the homomorphisms* $D: H_k(M^n, \partial M^n) \to H_{n-k}(M^n)$ *and* $D: H_k(M^n) \to H_{n-k}(M^n, \partial M^n)$ *that take each class to its cap product with* $[M^n]$ *are isomorphisms.*

In the nonorientable case, the Lefschetz isomorphism theorem is valid for homology and cohomology with coefficients in $\mathbb{Z}_2$.

Problem 118 ([**113**]). Let M^n be a compact orientable topological manifold with boundary ∂M^n. Suppose that ∂M^n is partitioned into disjoint sets V_1 and V_2, each consisting of several connected components of the manifold ∂M^n. Prove that the homomorphism $H^k(M^n, V_1) \to H_{n-k}(M^n, V_2)$ taking each class to its cap product with $[M^n]$ is an isomorphism.

2.5. A Generalization of Helly's Theorem. Results obtained in Section 2.1 (namely, Theorems 4.18 and 4.19) make it possible to carry over the generalized Helly theorem (Theorem 4.10 on p. 209) to an arbitrary manifold.

Theorem 4.29 (Debrunner [**29**]). *Suppose that* M^n *is a topological* n-*manifold, and* $X_1, \dots, X_m$ *are open subsets of* M^n *such that the intersection of any* $r \le n+1$ *of these sets is nonempty and has trivial reduced homology in all dimensions up to* $n-r$.

(a) *If* M^n *is not a homology* n-*sphere, then the space* $X_1 \cap \dots \cap X_m$ *is acyclic (and, in particular, nonempty).*

(b) *If* M^n *is a homology* n-*sphere and no* $n+2$ *of the sets* $X_1, \dots, X_m$ *cover* M^n, *then the space* $X_1 \cap \dots \cap X_m$ *is acyclic.*

Proof. Suppose that the assertion of the theorem does not hold for sets $X_1, \dots, X_m$, and let m be the minimum number of such sets. The minimality of m implies that the intersection of any $r \le m-1$ of the sets $X_1, \dots, X_m$ is acyclic. Therefore, Lemma 4.1 applies (see p. 208).

First, suppose that $X_1 \cap \dots \cap X_m = \varnothing$. By assumption, this is possible only if $m \ge n+2$. Let $Y = X_1 \cup \dots \cup X_m$. According to Lemma 4.1(a), the group $\tilde{H}_q(Y) = 0$ is nontrivial precisely when $q = m-2$. On the other hand, by Theorem 4.18, we have $\tilde{H}_q(Y) = 0$ for $q > n$; therefore, $m-2 \le n$. Comparing this with the inequality $m \ge n+2$, we obtain $n = m-2$. Thus, $\tilde{H}_n(Y) \ne 0$; this is possible only when $Y = M^n$. (Indeed, if $Y \ne M^n$, then Y is a noncompact n-manifold without boundary, and Theorem 4.10 implies $H_n(Y) = 0$.) Consequently, the group $\tilde{H}_q(M^n)$ is nontrivial precisely when $q = n$, i.e., M^n is a homology sphere. This contradicts the assumption that if M^n is a homology sphere, then $Y \ne M^n$.

Now suppose that the intersection $X_1 \cap \dots \cap X_m$ is nonempty (but not acyclic). Let $p \ge 0$ be the minimum number for which $\tilde{H}_p(X_1 \cap \dots \cap X_m) \ne 0$.

We set $q = p+m-1$. According to Lemma 4.1, (b), $\tilde{H}_q(X_1 \cup \cdots \cup X_m) \neq 0$. Applying Theorem 4.10 to the n-manifold $X_1 \cup \cdots \cup X_m$, we obtain $q \leq n$. Moreover, $q - m + 1 = p \geq 0$; hence $m \leq q + 1 \leq n + 1$. By assumption, the space $X_1 \cup \cdots \cup X_m$ has trivial reduced homology in all dimensions up to $n - m$. Therefore, $p > n - m$, i.e., $q - m + 1 > n - m$. Comparing this inequality with $q+1 \leq n+1$, we obtain $q = n$. The minimality of p (and q) implies that $X_1 \cup \cdots \cup X_m$ is a homology n-sphere. Thus, $M^n = X_1 \cup \cdots \cup X_m$, which contradicts the assumption. □

Problem 119. Give an example of $n + 2$ open subsets of S^n such that the intersection of any $r \leq n + 1$ of them is a (nonempty) contractible set and the union of these sets coincides with S^n (in particular, is not acyclic).

3. Characteristic Classes: Continuation

In proving many properties of characteristic classes, it is convenient to use singular homology and cohomology. Most of these properties are related to the Thom isomorphism, but we start by proving Theorem 3.43 formulated on p. 170. In the proof, we use the notation introduced in and before the formulation of this theorem.

Proof of Theorem 3.43. It is sufficient to verify that $x_i(\xi)$ satisfies all of the conditions that must be satisfied by the characteristic classes. The relations $x_0(\xi) = 1$ and $x_i(\xi) = 0$ for $i > \dim \xi$ are included in the definition. Let us show that $x_i(g^*(\xi)) = g^*(x_i(\xi))$ for an arbitrary map $g\colon B_1 \to B$.

The fibers of the bundles $g^*(\xi)$ and ξ over the points b_1 and $g(b_1)$ are canonically identified; therefore, the fibers of the bundles $Pg^*(\xi)$ and $P\xi$ over these points are canonically identified as well. Thus, a bundle map $\tilde{g}\colon E(Pg^*\xi) \to E(P\xi)$ is defined. The diagram

$$\begin{array}{ccc} E(Pg^*\xi) & \xrightarrow{\tilde{g}} & E(P\xi) \\ \downarrow & & \downarrow \\ B_1 & \xrightarrow{g} & B \end{array}$$

is commutative; therefore, the bundle $\tilde{g}(\lambda_\xi)$ is isomorphic to the bundle $\lambda_{g^*\xi}$, and $\tilde{g}(a_\xi) = a_{g^*\xi}$ in the ring $H^*(E(Pg^*\xi))$. By definition, we have $a^n_{g^*\xi} = \sum_{i=1}^{n-1} (-1)^{i+1} x_i(g^*\xi) a^{n-i}_{g^*\xi}$ and

$$\tilde{g}^*(a^n_\xi) = \tilde{g}^*\left(\sum_{i=1}^{n-1} (-1)^{i+1} x_i(\xi) a^{n-i}_\xi\right) = \sum_{i=1}^{n-1} (-1)^{i+1} g^*(x_i(\xi)) \tilde{g}^*(a^{n-i}_\xi);$$

the last equality holds because $\tilde{g}^*(\beta a_\xi^i) = g^*(\beta)\tilde{g}^*(a_\xi^i)$ for any $\beta \in H^*(B)$. The uniqueness of the decompositions implies $x_i(g^*\xi) = g^*(x_i(\xi))$, as required.

We proceed to the last (and most difficult to prove) condition, namely, that the Whitney formula $x_k(\xi \oplus \eta) = \sum_{i+j=k} x_i(\xi)x_j(\eta)$ must hold. (The Whitney formula can be written differently, but here it is more convenient to prove it for bundles over the same base.) The linear space $V_1 \oplus V_2$ contains the subspaces $V_1 \oplus \{0\}$ and $\{0\} \oplus V_2$, which are isomorphic to V_1 and V_2. Therefore, the topological space $E(P(\xi \oplus \eta))$ contains the canonically defined subspaces $E(P\xi)$ and $E(P\eta)$. These subspaces are deformation retracts of $U = E(P(\xi \oplus \eta)) \setminus E(P\xi)$ and $V = E(P(\xi \oplus \eta)) \setminus E(P\eta)$, respectively. In the cohomology ring of $E(P(\xi \oplus \eta))$, take the elements

$$\alpha_1 = \sum_{i=0}^{m} (-1)^i x_i(\xi) y^{m-i} \quad \text{and} \quad \alpha_2 = \sum_{j=0}^{n} (-1)^j x_j(\eta) y^{n-j},$$

where $y = x_1(\lambda_{\xi\oplus\eta})$, $m = \dim \xi$, and $n = \dim \eta$. To simplify notation, we set $E = E(P(\xi \oplus \eta))$. Consider the cohomology sequence of the pair (E, V):

$$\cdots \longrightarrow H^m(E,V) \xrightarrow{p^*} H^m(E) \xrightarrow{i^*} H^m(V) \longrightarrow \cdots .$$

The map i^* implements the restriction of the cohomology classes to V. Since $E(P\xi)$ is a deformation retract of V, we can consider restriction to $E(P\xi)$ rather than to V. Let us show that the restriction of the class α_1 to $E(P\xi)$ is $\sum_{i=0}^m (-1)^i x_i(\xi) a_\xi^{m-i} = 0$. Indeed, suppose that $\lambda_\xi = f^*(\gamma^1)$. The restriction of the class $y = x_1(\lambda_{\xi\oplus\eta})$ to $E(P\xi)$ is $x_1(\lambda_\xi) = x_1(f^*\gamma^1) = f^*x_1(\gamma^1) = f^*(\alpha) = a_\xi$ (we use some of the properties already proved, namely, naturality and the relation $x_1(\gamma^1) = \alpha$). Since the sequence of a pair is exact, there exists an element $\alpha_1' \in H^m(E,V)$ for which $p^*\alpha_1' = \alpha_1$. Similarly, there exists an $\alpha_2' \in H^m(E,U)$ for which $p^*\alpha_2' = \alpha_2$. The element $\alpha_1' \smile \alpha_2'$ belongs to $H^*(E, U \cup V) = 0$; hence $\alpha_1' \smile \alpha_2' = 0$. Therefore, $\alpha_1 \smile \alpha_2 = (p^*\alpha_1') \smile (p^*\alpha_2') = 0$. As a result, we obtain

$$\left(\sum_{i=0}^{m} (-1)^i x_i(\xi) y^{m-i}\right)\left(\sum_{j=0}^{n} (-1)^j x_j(\eta) y^{n-j}\right) = 0.$$

For the Stiefel–Whitney classes, we consider cohomology with coefficients in $\mathbb{Z}_2$, in which multiplication is commutative (rather than anticommutative). For the Chern classes, x_i and y have even dimension; thus, for these classes, multiplication is commutative also. Therefore, we can rewrite the obtained equality in the form

$$\sum_{k=0}^{m+n} (-1)^k \left(\sum_{i+j=k} x_i(\xi) \smile x_j(\eta)\right) y^{m+n-k} = 0.$$

On the other hand, there exist unique elements $x_k(\xi \oplus \eta)$ for which

$$\sum_{k=0}^{m+n} (-1)^k x_k(\xi \oplus \eta) y^{m+n-k} = 0.$$

Thus, $x_k(\xi \oplus \eta) = \sum_{i+j=k} x_i(\xi) \smile x_j(\eta)$, as required. □

3.1. The Thom Isomorphism for Bundles. Every smooth manifold has a tangent bundle, whereas topological manifolds have no tangent bundles. For this reason, Milnor [**91**] introduced the notion of a microbundle, which generalizes that of a vector bundle. Each topological manifold has a tangent microbundle; tangent microbundles mimic tangent bundles of smooth manifolds in many respects. For example, the proof of the Poincaré isomorphism for topological manifolds given above essentially uses microbundles.

A *microbundle* of dimension n is a map $p\colon E \to B$ with the following properties:

(i) there exists a map $i\colon B \to E$ for which $pi = \mathrm{id}_B$;

(ii) the map p is locally trivial in the sense that for each point $b \in B$, there exist open neighborhoods $U \ni b$ and $V \ni i(b)$ and a homeomorphism $h_b\colon U \times \mathbb{R}^n \to V \cap p^{-1}(U)$ such that $ph_b(u, v) = u$ for all $(u, v) \in U \times \mathbb{R}^n$ and $h_b(u, v) = i(u)$ for all $u \in U$.

Example 57. A vector bundle $p\colon E \to B$ is a microbundle. For i the zero section $i(b) = (b, 0)$ can be taken.

Example 58. Let M be a topological manifold without boundary. Then the projection $p\colon M \times M \to M$ onto the first factor is a microbundle. For i the diagonal map $d\colon M \to M \times M$ can be taken.

The proof that the map from Example 58 is indeed a microbundle is essentially contained in the proof of Lemma 4.3 on p. 235. This microbundle is called the *tangent microbundle* of the topological manifold M.

The tangent microbundle of a topological manifold M with boundary ∂M can be defined as follows. Let us attach the direct product $\partial M \times [0, 1)$ to the boundary of M. The collar theorem implies that the result is a topological manifold M'. This manifold has no boundary. The tangent microbundle of M is the restriction to $M \times M$ of the tangent microbundle of M'.

Microbundles $p\colon E \to B$ are usually considered together with given maps $i\colon B \to E$.

Two microbundles $p\colon E \to B$ and $p'\colon E' \to B$ (over the same base B) with fixed maps $i\colon B \to E$ and $i'\colon B \to E'$ are said to be *equivalent* if the sets $i(B)$ and $i'(B)$ have open neighborhoods U and U' in E and E'

for which there exists a homeomorphism $h\colon U \to U'$ such that $hi = i'$ and $p'|_{U'}h = p_U$.

Theorem 4.30. *The tangent bundle of a closed smooth manifold M is equivalent to its tangent microbundle.*

Proof. Consider a Riemannian metric on the tangent bundle. For each tangent vector $v \in T_xM$, there exists a unique geodesic $\gamma_v(t)$ such that $\gamma_v(0) = x$ and $\frac{d}{dt}\gamma_v(0) = v$. Consider the map $\exp_x\colon T_xM \to M$ that assigns the point $\gamma_v(1)$ to each tangent vector v. Since M is compact, we can choose a number $\varepsilon > 0$ such that for each $x \in M$, the restriction of the map $\exp_x$ to the open ball in T_xM that consists of vectors of length less than ε is a diffeomorphism from this ball to some open neighborhood of x in M.

Let U be an open neighborhood of the zero section in TM consisting of vectors of length less than ε. The map $h\colon U \to M \times M$ defined by $h(x, v) = (x, \exp_x(v))$ is a diffeomorphism from U to a neighborhood U' of the diagonal $d(M)$ in $M \times M$. This map h has all the required properties. □

The Thom class $U \in H^n(M \times M, M \times M \setminus d(M))$ is characterized by the condition that for any point $x \in M$, the map of pairs

$$l_x\colon (M, M \setminus \{x\}) \to (M \times M, M \times M \setminus d(M))$$

defined by $l_x(y) = (x, y)$ takes U to a generator of the group $H^n(M, M\setminus\{x\})$, i.e., l_x^*U generates $H^n(M, M \setminus \{x\})$. By analogy, the *Thom class* of an n-dimensional microbundle $p\colon E \to B$ is defined as a cohomology class $U \in H^n(E, E \setminus i(B))$ with the property that for any point $b \in B$, the natural embedding

$$i_b\colon (p^{-1}(b), p^{-1}(b) \setminus i(b)) \to (E, E \setminus i(B))$$

takes U to a generator of the group $H^n(p^{-1}(b), p^{-1}(b) \setminus i(b))$.

Now we consider the Thom classes of real vector bundles. The following theorem is valid.

Theorem 4.31. *Let ξ be an oriented real n-dimensional vector bundle over a compact CW-complex B. Suppose that $E = E(\xi)$ is the total space of this bundle and E_0 is an open subset of E consisting of all nonzero vectors. Then*

(a) *there is a unique integral cohomology class $U_\xi \in H^n(E, E_0)$ such that for any $b \in B$, the cohomology class $i_b^*U_\xi$ is the generator of $H^n(\mathbb{R}^n, \mathbb{R}^n\setminus\{0\})$ determining a given orientation of the fiber $\mathbb{R}^n$;*

(b) $H^i(E, E_0) = 0$ *for $i < n$;*

(c) *the map* $H^i(B) \to H^{i+n}(E, E_0)$ *defined by*[8] $\alpha \mapsto p^*(\alpha) \smile U_\xi$ *is an isomorphism for each* i.

Proof. If the bundle is trivial, i.e., $E = B \times \mathbb{R}^n$, $E_0 = B \times (\mathbb{R}^n \setminus \{0\})$, and $p\colon E \to B$ is the projection to the first factor, then all the required assertions are in fact contained in Theorem 4.15 on p. 217 (in the notation of this theorem, $X = B$, $A = \varnothing$, and $e^n = U_\xi$). Thus, it suffices to show that if the required assertions hold over open sets U, V, and $U \cap V$, then they also hold over $U \cup V$.

We set $E^1 = p^{-1}(U)$, $E^2 = p^{-1}(V)$, $E^3 = p^{-1}(U \cap V)$, and $E^4 = p^{-1}(U \cup V)$. Let us write the Mayer–Vietoris sequence:

$$H^{i-1}(E^3, E_0^3) \to H^i(E^4, E_0^4) \to H^i(E^1, E_0^1) \oplus H^i(E^2, E_0^2) \to H^i(E^3, E_0^3).$$

For $i < n$, the groups on the left and right of $H^i(E^4, E_0^4)$ are trivial; therefore, $H^i(E^4, E_0^4) = 0$. This proves (b). Uniqueness implies that any classes $U_1 \in H^n(E^1, E_0^1)$ and $U_2 \in H^n(E^2, E_0^2)$ determine the same element of the group $H^n(E^3, E_0^3)$. Hence there exists a class $U_4 \in H^n(E^4, E_0^4)$ whose projections are U_1 and U_2. The equality $H^{n-1}(E^3, E_0^3) = 0$ implies the uniqueness of such a class. This proves (a).

To prove (c), consider a similar Mayer–Vietoris sequence for the bases of bundles and the map from this sequence to that written above that takes each element α to $p^*(\alpha) \smile U$. Applying the five lemma, we obtain the required result. □

For coefficients in $\mathbb{Z}_2$, an analog of Theorem 4.31 for nonorientable bundles ξ is valid.

The Gysin Sequence. Let ξ be an oriented vector bundle of dimension n with projection $p\colon E \to B$. Consider the exact cohomology sequence of the pair (E, E_0):

$$\cdots \to H^{i+n}(E, E_0) \xrightarrow{j^*} H^{i+n}(E) \xrightarrow{i^*} H^{i+n}(E_0) \xrightarrow{\delta^*} H^{i+n+1}(E, E_0) \to \cdots.$$

Since the map p is a homotopy equivalence, it follows that $p^*\colon H^{i+n}(B) \to H^{i+n}(E)$ is an isomorphism. We also have the Thom isomorphism $\varphi^*\colon H^i(B) \to H^{i+n}(E, E_0)$. Using these isomorphisms, we obtain the exact sequence

$$\cdots \to H^i(B) \xrightarrow{p^{*-1}j^*\varphi} H^{i+n}(B) \xrightarrow{i^*p^*} H^{i+n}(E_0) \xrightarrow{\varphi^{-1}\delta^*} H^{i+1}(B) \to \cdots.$$

This exact sequence is called the *Gysin sequence* [**46**].

In the Gysin sequence, the map i^*p^* is induced by the restriction of p to E_0. Moreover, the map $p^{*-1}j^*\varphi$ can be expressed in terms of the Euler class $e(\xi)$ as $p^{*-1}j^*\varphi(\alpha) = \alpha \smile e(\xi)$. Indeed, we have $p^{*-1}j^*\varphi(\alpha) =$

[8] The relative cup product of the classes $p^*(\alpha) \in H^i(E)$ and $U_\xi \in H^n(E, E_0)$ belongs to $H^{i+n}(E, E_0)$.

$p^{*-1}j^*(p^*(\alpha) \smile U_\xi) = p^{*-1}(p^*(\alpha) \smile j^*(U_\xi)) = \alpha \smile (p^{*-1}j^*(U_\xi))$. It remains to verify that $p^{*-1}j^*(U_\xi) = e(\xi)$, i.e., $j^*(U_\xi) = p^*e(\xi)$. Let $\mathbb{R}^n$ be the fiber over some point of the base. Consider the cochain in E which assigns 0 to any simplex $\Delta^n \subset \mathbb{R}^n \setminus \{0\}$ and ± 1 to any simplex $\Delta^n \subset \mathbb{R}^n$ containing the origin (the sign depends on whether the orientation of the simplex is compatible with that of the fiber). The cohomology class of such a chain is determined uniquely, and it coincides with both $j^*(U_\xi)$ and $p^*e(\xi)$; this follows directly from the definitions.

In the case of a nonorientable vector bundle, the Gysin exact sequence holds for cohomology with coefficients in $\mathbb{Z}_2$; in this case, the Euler class $e(\xi)$ is reduced modulo 2, i.e., replaced by $w_n(\xi)$.

The exact Gysin sequence holds also for any oriented[9] locally trivial S^{n-1}-bundle $p\colon E \to B$ because any such bundle can be associated with an oriented n-dimensional vector bundle.

Problem 120. Calculate the cohomology groups of the manifold $G_+(2n+2,2)$ using the bundle $V(2n+2,2) \xrightarrow{S^1} G_+(2n+2,2)$.

3.2. The Thom and Wu Formulas. In this section, we consider cohomology with coefficients in $\mathbb{Z}_2$. Suppose that ξ is a real bundle over a compact simplicial complex B, U_ξ is its Thom class, and $\varphi\colon H^i(B) \to H^{i+n}(E, E_0)$ is the Thom isomorphism defined by $\varphi(\alpha) = p^*(\alpha) \smile U_\xi$.

Theorem 4.32 (Thom [**137**]). *The Stiefel–Whitney class $w_i(\xi)$ is equal to $\varphi^{-1}(\mathrm{Sq}^i U_\xi)$, where Sq^i is the Steenrod square.*

Proof. Let us show that $\varphi^{-1}(\mathrm{Sq}^i U_\xi)$ satisfies all of the conditions determining the ith Stiefel–Whitney class. Clearly, $\varphi^{-1}(\mathrm{Sq}^0 U_\xi) = \varphi^{-1}(U_\xi) = 1$. Moreover, if $i > n = \dim \xi$, then $\mathrm{Sq}^i U_\xi = 0$.

Let us verify naturality. Suppose that $f\colon B' \to B$ is a map, ξ is a bundle over B, $f^*(\xi)$ is its pullback over B', and $g\colon E \to E'$ is the fiberwise map of bundles induced by f. Then $g^*(U_\xi) = U_{f^*(\xi)}$, and we have the following commutative diagram for the Thom isomorphisms:

$$\begin{array}{ccc} H^i(B) & \xrightarrow{\varphi} & H^{i+n}(E, E_0) \\ \downarrow f^* & & \downarrow g^* \\ H^i(B') & \xrightarrow{\varphi'} & H^{i+n}(E', E'_0). \end{array}$$

[9]This means that the orientations of all fibers S^{n-1} are compatible.

Thus, $\varphi' f^* = g^* \varphi$, i.e., $f^* \varphi^{-1} = (\varphi')^{-1} g^*$. Therefore,

$$\begin{aligned} f^*(w_i(\xi)) &= f^*\varphi^{-1}(\mathrm{Sq}^i U_\xi) = (\varphi')^{-1} g^*(\mathrm{Sq}^i U_\xi) = (\varphi')^{-1}\mathrm{Sq}^i(g^*(U_\xi)) \\ &= (\varphi')^{-1}\mathrm{Sq}^i U_{f^*(\xi)} = w_i(f^*(\xi)). \end{aligned}$$

Let us show that the Whitney formula holds. This time, it is more convenient to prove it in the form $w(\xi \times \eta) = w(\xi) \times w(\eta)$. Consider the cross product of the Thom classes $U_\xi \in H^m(E, E_0)$ and $U_\eta \in H^n(E', E_0')$. The class $U_\xi \times U_\eta$ belongs to the group $H^{n+m}(E \times E', (E \times E_0') \cup (E_0 \times E'))$. The set $(E \times E_0') \cup (E_0 \times E')$ consists precisely of the nonzero vectors in $E \times E'$. Therefore, the class $U_\xi \times U_\eta$ belongs to the same group as the Thom class $U_{\xi\times\eta}$. Let us show that these two cohomology classes coincide. To this end, it suffices to verify that the restriction of $U_\xi \times U_\eta$ to $(\mathbb{R}^{n+m}, \mathbb{R}_0^{n+m})$, where $\mathbb{R}^{n+m}$ is the fiber of the bundle $\xi \times \eta$, is a nonzero cohomology class for the fiber of the bundle $\xi \times \eta$ over each point. The class under consideration equals the cross product of the restrictions of the classes U_ξ and U_η to $(\mathbb{R}^m, \mathbb{R}_0^m)$ and $(\mathbb{R}^n, \mathbb{R}_0^n)$; according to Theorem 4.15 on p. 217, this product is nonzero.

To prove the Whitney formula, we must also verify that the Thom isomorphisms of the bundles ξ, η, and $\xi \times \eta$ are related by $\varphi_\xi(\alpha) \times \varphi_\eta(\beta) = \varphi_{\xi\times\eta}(\alpha \times \beta)$, i.e.,

$$(p_\xi^*\alpha \smile U_\xi) \times (p_\eta^*\beta \smile U_\xi) = (p_{\xi\times\eta}^*(\alpha \times \beta)) \smile U_{\xi\times\eta}.$$

In the case of coefficients in $\mathbb{Z}_2$, signs do not matter; thus,

$$(p_\xi^*\alpha \smile U_\xi) \times (p_\eta^*\beta \smile U_\xi) = (p_\xi^*\alpha \times p_\eta^*\beta) \smile (U_\xi \times U_\xi).$$

The equality $U_\xi \times U_\eta = U_{\xi\times\eta}$ has already been proved. Let p_B and $p_{B'}$ be the projections of $B \times B'$ onto the first and second factors, and let p_E and $p_{E'}$ be similar maps for $E \times E'$. Then $p_{\xi\times\eta}^*(\alpha \times \beta) = p_{\xi\times\eta}^*(p_B^*\alpha \smile p_{B'}^*\beta) = (p_B p_{\xi\times\eta})^*\alpha \smile (p_{B'} p_{\xi\times\eta})^*\beta$ and $p_\xi^*\alpha \times p_\eta^*\beta = (p_E^* p_\xi^*\alpha) \smile (p_{E'}^* p_\eta^*\beta) = (p_\xi p_E)^*\alpha \smile (p_\eta p_{E'})^*\beta$. But $p_B p_{\xi\times\eta} = p_\xi p_E$ (both maps coincide with the natural projection of $E \times E'$ onto B), and $p_{B'} p_{\xi\times\eta} = p_\xi p_{E'}$.

Thus, in the case that $\tilde{w}(\xi) = \varphi^{-1}(\mathrm{Sq}\, U_\xi)$, we have $\varphi_{\xi\times\eta}(\tilde{w}(\eta) \times \tilde{w}(\xi)) = \varphi_{\xi\times\eta}((\varphi_\xi^{-1}\,\mathrm{Sq}\, U_\xi) \times (\varphi_\eta^{-1}\,\mathrm{Sq}\, U_\eta)) = \mathrm{Sq}\, U_\xi \times \mathrm{Sq}\, U_\eta = \mathrm{Sq}(U_\xi \times U_\eta) = \mathrm{Sq}\, U_{\xi\times\eta} = \varphi_{\xi\times\eta}(\tilde{w}(\eta \times \xi))$, which means that $\tilde{w}(\eta) \times \tilde{w}(\xi) = \tilde{w}(\eta \times \xi)$.

It remains to perform calculations for the bundle γ_1^1 over $\mathbb{R}P^1 \approx S^1$. For this bundle, the pair (E, E_0) consists of the Möbius band M and the Möbius band from which the central circle is removed. It has a deformation retract of the form $(M, \partial_\varepsilon M)$, where $\partial_\varepsilon M$ is the curve at distance ε from the boundary of the Möbius band (parallel to the boundary). The excision isomorphism implies $H^*(M, \partial_\varepsilon M) \cong H^*(\mathbb{R}P^2, D^2)$, and the cohomology sequence of the pair $(\mathbb{R}P^2, D^2)$ implies $H^i(\mathbb{R}P^2, D^2) \cong H^i(\mathbb{R}P^2)$

for $i \geq 1$. Thus, the Thom class U of the bundle γ_1^1 is identified with the unique nonzero element $\alpha \in H^1(\mathbb{R}P^2)$. We have $\mathrm{Sq}^1\alpha = \alpha \smile \alpha \neq 0$; therefore, $\mathrm{Sq}^1 U \neq 0$, and $\varphi^{-1}(\mathrm{Sq}^1 U) \neq 0$. □

The Thom formula is valid for all vector bundles, but its application is hindered by the necessity of calculating the Thom class. For the Stiefel–Whitney classes of tangent bundles of manifolds, there exists a more explicit formula, which does not involve the Thom class.

Let $\mathrm{Sq}^T\colon H_*(X) \to H_*(X)$ be the operation dual to $\mathrm{Sq} = \sum_{i\geq 0}\mathrm{Sq}^i$, i.e., such that $\langle \alpha, \mathrm{Sq}^T\beta\rangle = \langle \mathrm{Sq}\,\alpha, \beta\rangle$. For any closed manifold M^n, the cohomology class $v = D^{-1}\mathrm{Sq}^T[M^n] \in H^*(M^n)$ is defined, where $[M^n]$ is the fundamental class (with coefficients in $\mathbb{Z}_2$) and $D\colon H^k(M^n) \to H_{n-k}(M^n)$ is the Poincaré isomorphism. The class v is called the *Wu class* of the manifold M^n.

Theorem 4.33 (Wu [**155**]). *The Wu class of any closed manifold M^n has the following properties:*

(i) $\langle \alpha \smile v, [M^n]\rangle = \langle \mathrm{Sq}\,\alpha, [M^n]\rangle$ *for any element* $\alpha \in H^*(M^n)$;

(ii) $w(M^n) = \mathrm{Sq}\,v$.

Proof. By definition, $Dv = \mathrm{Sq}^T[M^n]$. Therefore,

$$\langle \mathrm{Sq}\,\alpha, [M^n]\rangle = \langle \alpha, \mathrm{Sq}^T[M^n]\rangle = \langle \alpha, Dv\rangle = \langle \alpha, v \frown [M^n]\rangle = \langle \alpha \smile v, [M^n]\rangle.$$

To prove the equality $w(M^n) = \mathrm{Sq}\,v$, it suffices to verify that $\langle w(M^n), \beta\rangle = \langle \mathrm{Sq}\,v, \beta\rangle$ for any element $\beta \in H_*(M^n)$. According to the Thom formula, we have $\mathrm{Sq}\,U = \varphi(w(M^n))$. Instead of the Thom class of the tangent bundle, we take the Thom class of the tangent microbundle, which belongs to $H^n(M^n \times M^n, M^n \times M^n \setminus d(M^n))$. The Thom isomorphism has the form $\varphi(\alpha) = p_1^*\alpha \smile U$, where $p_1\colon M^n \times M^n \to M^n$ is the projection onto the first factor. Hence $p_1^*\alpha = \alpha \times 1$, where $1 \in H^0(M^n)$ is the identity element. Therefore, $\mathrm{Sq}\,U = p_1^*(w(M^n)) \smile U = (w(M^n) \times 1) \smile U$.

Let us use the class $\tilde{U} \in H^n(M^n \times M^n)$ defined on p. 244 and its properties (36) and (37) (see p. 244). Similarly to U, this class satisfies the relation $\mathrm{Sq}\,\tilde{U} = (w(M^n) \times 1) \smile \tilde{U}$. Clearly,

$$\begin{aligned}\langle \mathrm{Sq}\,v, \beta\rangle &= \langle v, \mathrm{Sq}^T\beta\rangle \overset{(37)}{=} \langle \tilde{U}, \mathrm{Sq}^T\beta \times Dv\rangle = \langle \tilde{U}, \mathrm{Sq}^T\beta \times \mathrm{Sq}^T[M^n]\rangle \\ &= \langle \tilde{U}, \mathrm{Sq}^T(\beta \times [M^n])\rangle = \langle \mathrm{Sq}\,\tilde{U}, \beta \times [M^n]\rangle \\ &= \langle (w(M^n)\times 1)\smile\tilde{U}, \beta\times[M^n]\rangle \overset{(36)}{=} \langle (1\times w(M^n))\smile\tilde{U}, \beta\times[M^n]\rangle \\ &= \langle \tilde{U}, (1 \times w(M^n)) \frown (\beta \times [M^n])\rangle = \langle \tilde{U}, \beta \times (w(M^n) \frown [M^n])\rangle \\ &= \langle \tilde{U}, \beta \times Dw(M^n)\rangle \overset{(37)}{=} \langle w(M^n), \beta\rangle.\end{aligned}$$

□

It follows directly from the definition of the Wu class that if $v = v_0 + v_1 + \cdots + v_n$, where $v_i \in H^i(M^n)$, then $v_i = 0$ for $i > [n/2]$. Indeed, let $\mathrm{Sq}^T[M^n] = m_0 + m_1 + \cdots + m_n$, where $m_i \in H_i(M^n)$. If $\alpha \in H^k(M^n)$, then $\langle \alpha, m_k \rangle = \langle \mathrm{Sq}\,\alpha, [M^n] \rangle = \langle \mathrm{Sq}^{n-k}\alpha, [M^n] \rangle$. Moreover, $\mathrm{Sq}^{n-k}\alpha = 0$ if $n - k > k$, i.e., $2k < n$. Therefore, $m_k = 0$ for $2k < n$. This means that $v_{n-k} = 0$ for $2k < n$, i.e., $v_i = 0$ for $2i > n$.

Problem 121. Let $v = v_0 + v_1 + \cdots + v_{[n/2]}$ be the Wu class of a closed manifold M^n, and let $w = 1 + w_1 + \cdots + w_n$ be its Stiefel–Whitney class.

(a) Prove that if $w_1 = \cdots = w_k = 0$, then $v_1 = \cdots = v_k = 0$ and $v_{k+1} = w_{k+1}$.

(b) Prove that if $w_1 = \cdots = w_k = 0$ and $n = 2k$ or $2k + 1$, then $w_{k+1} = w_{k+2} = \cdots = w_n = 0$.

Problem 122 ([**88**]). Given a closed $(2k-1)$-connected manifold M^{4k}, prove that the diagonal elements of its intersection form are even if and only if $w_{2k}(M^{4k}) = 0$.

Problem 121 can be used to prove the following theorem.

Theorem 4.34 (Stiefel [**132**]). *Any closed orientable* 3*-manifold is parallelizable.*

Proof. To prove the parallelizability of a manifold M^3, we must construct three linearly independent sections of its tangent bundle. Since the manifold is orientable, it suffices to construct two independent sections. In constructing k independent sections of an n-dimensional real bundle, the first obstruction is involved in the extension over the $(n - k + 1)$-skeleton. If $n - k$ is odd and $k \geq 2$, then this obstruction coincides with the class w_{n-k+1}. In the case under consideration, this is the class w_2. If we construct two independent sections on the 2-skeleton, then we shall accomplish the task because the obstruction to extending them over the 3-skeleton lies in the cohomology with coefficients in the group $\pi_2(V(3,2))$, which is trivial because $V(3,2) \approx \mathbb{R}P^3$. Thus, it is sufficient prove that $w_2(M^3) = 0$. For an orientable manifold, we have $w_1 = 0$. According to Problem 121, (b), the equality $w_1 = 0$ implies $w_2 = 0$ and $w_3 = 0$. □

Remark. Stiefel proved the parallelizability of a closed orientable 3-manifold under the assumption that $w_2(M^3) = 0$; he had suggested no convincing proof of this equality.

Stiefel considered the general problem of constructing $n - 1$ linearly independent vector fields on an orientable n-manifold, that is, the parallelizability problem; he constructed the theory of characteristic classes [**132**] in order to solve this problem. On Hopf's advice, Stiefel started with the simplest nontrivial case $n = 3$. But all of the numerous examples of closed orientable 3-manifolds which he considered turned out to be parallelizable.

Hopf made a report about Stiefel's results at the First International Topological Conference in Moscow (1935). Whitney, who attended his talk, said that many of these results were contained in his paper [**149**] published just before the conference; Stiefel and Hopf were unaware of it at that time.

3.3. Obstructions to Embeddings. Suppose that M^n is a closed manifold immersed in $\mathbb{R}^{n+k}$, and ν_{M^n} is the normal bundle for this immersion. According to Whitney's theorem (Theorem 3.22 on p. 149), we have $w_i(\nu_{M^n}) = \bar{w}_i(M^n)$. Therefore, $\bar{w}_i(M^n) = 0$ for $i > k$ because $\dim \nu_{M^n} = k$. If the manifold M^n is embedded in $\mathbb{R}^{n+k}$, the additional condition $\bar{w}_k(M^n) = 0$ arises; let us prove it.

The tubular neighborhood theorem implies that if a closed manifold M^n is embedded in $\mathbb{R}^{n+k}$, then, for sufficiently small $\varepsilon > 0$, the set M_ε consisting of the points at distance at most ε from M^n is homeomorphic to the total space $E(\nu_{M^n}) = E$ of the bundle. Thus, the homeomorphism of pairs $(E, E_0) \approx (M_\varepsilon, M_\varepsilon \setminus M^n)$ arises. Using the excision isomorphism, we obtain $H^*(E, E_0) \cong H^*(M_\varepsilon, M_\varepsilon \setminus M^n) \cong H^*(\mathbb{R}^{n+k}, \mathbb{R}^{n+k} \setminus M^n)$. Consider the composition

$$H^k(\mathbb{R}^{n+k}, \mathbb{R}^{n+k} \setminus M^n) \longrightarrow H^k(\mathbb{R}^{n+k}) \longrightarrow H^k(M^n) \tag{39}$$

of restriction homomorphisms. Obviously, this composition is the zero homomorphism because $H^k(\mathbb{R}^{n+k}) = 0$. Therefore, it suffices to prove the following lemma.

Lemma. *The composition* (39) *takes the class corresponding to the Thom class $U \in H^k(E, E_0; \mathbb{Z}_2)$ to $w_k(\nu_{M^n})$.*

Proof. The zero section $s\colon M \to E$ induces an isomorphism $s^*\colon H^k(E) \to H^k(M^n)$. The composition

$$H^k(E, E_0) \longrightarrow H^k(E) \xrightarrow{s^*} H^k(M^n)$$

takes the Thom class U to $w_k(\nu_{M^n})$. Indeed, the Thom isomorphism $\varphi\colon H^k(M^n) \to H^{2k}(E, E_0)$ takes the class $s^*(U|_E)$ to $p^*s^*(U|_E) \smile U = (U|_E) \smile U = U \smile U = \mathrm{Sq}^k U$; therefore, $s^*(U|_E) = \varphi^{-1}\mathrm{Sq}^k U = w_k(\nu_{M^n})$.

Replacing the pair (E, E_0) by the homeomorphic pair $(M_\varepsilon, M_\varepsilon \setminus M^n)$, we see that the composition

$$H^k(M_\varepsilon, M_\varepsilon \setminus M^n) \longrightarrow H^k(M_\varepsilon) \longrightarrow H^k(M^n)$$

of restriction homomorphisms takes the class corresponding to the Thom class U to $w_k(\nu_{M^n})$.

The composition (39) fits into the commutative diagram

$$\begin{array}{ccc} H^k(\mathbb{R}^{n+k}, \mathbb{R}^{n+k} \setminus M^n) & \longrightarrow & H^k(\mathbb{R}^{n+k}) \\ \downarrow \cong & & \downarrow \\ H^k(M_\varepsilon, M_\varepsilon \setminus M^n) & \longrightarrow & H^k(M^n). \end{array}$$

Therefore, the composition (39) takes the class in $H^k(\mathbb{R}^{n+k}, \mathbb{R}^{n+k} \setminus M^n)$ corresponding to the Thom class U to $w_k(\nu_{M^n})$. □

Problem 123. Prove that if $n = 2^m$, then $\mathbb{R}P^n$ cannot be embedded in $\mathbb{R}^{2n-1}$.

Chapter 5

Čech Cohomology and de Rham Cohomology

1. Sheaf Cohomology

1.1. Sheaves and Presheaves. Suppose that X is a topological space, associated with each open set $U \subset X$ is an Abelian group $\mathcal{F}(U)$ so that $\mathcal{F}(\varnothing) = 0$, and associated with each inclusion of open sets $V \subset U$ is a homomorphism $r_V^U\colon \mathcal{F}(U) \to \mathcal{F}(V)$, which is called the *restriction homomorphism*. Suppose also that the correspondences $\mathcal{F}$ and r satisfy the conditions

(i) $r_U^U = \mathrm{id}_{\mathcal{F}(U)}$, and

(ii) $W \subset V \subset U$ implies $r_W^U = r_W^V r_V^U$.

Then we say that $\mathcal{F}$ is a *presheaf* of Abelian groups on X.

Let $\mathcal{F}$ and $\mathcal{G}$ be presheaves on the same space X. Suppose that for any open set $U \subset X$, a homomorphism $h_U\colon \mathcal{F}(U) \to \mathcal{G}(U)$ is defined so that the diagram

$$\begin{array}{ccc} \mathcal{F}(U) & \xrightarrow{h_U} & \mathcal{G}(U) \\ \downarrow{\scriptstyle r_V^U} & & \downarrow{\scriptstyle r_V^U} \\ \mathcal{F}(V) & \xrightarrow{h_V} & \mathcal{G}(V) \end{array}$$

is commutative whenever $V \subset U$. Then we say that a *homomorphism of presheaves*, $h\colon \mathcal{F} \to \mathcal{G}$, is given.

Presheaves of rings or R-modules (where R is a fixed ring) and homomorphisms of such presheaves are defined in a similar way.

Example 59. For any Abelian group G, we can consider the presheaf that assigns the group G to every nonempty open set $U \subset X$ and the trivial group to the empty set. This presheaf is called *constant.*

Example 60. The set $\mathcal{F}(U)$ of continuous functions on U is a presheaf; the group operation is the pointwise addition of functions.

We denote the element $r_V^U(f)$, where $f \in \mathcal{F}(U)$, by $f|_V$.

A sheaf is a presheaf satisfying certain additional conditions. These conditions are as follows. Let $\mathcal{U} = \{U\}$ be a family of open subsets of X. A family $\{f_U \in \mathcal{F}(U)\}$, where $U \in \mathcal{U}$, is said to be *consistent* if $f_U|_{U\cap U'} = f_{U'}|_{U\cap U'}$ for any $U, U' \in \mathcal{U}$. Each family $\mathcal{U}$ of open sets determines the open set $V = \bigcup_{U\in\mathcal{U}} U$. A presheaf $\mathcal{F}$ is a *sheaf* if for any family $\mathcal{U}$ of open sets, the following conditions hold:

(i) if $f \in \mathcal{F}(V)$ is such that $f|_U = 0$ for all $U \in \mathcal{U}$, then $f = 0$;

(ii) for any consistent family $\{f_U\}$, there exists an $f \in \mathcal{F}(V)$ such that $f|_U = f_U$ for all $U \in \mathcal{U}$.

Example 61. Let X be the union of two disjoint open sets U_1 and U_2. Then no (nonzero) constant presheaf on X is a sheaf.

Proof. Let $\mathcal{U} = \{U_1, U_2\}$. The family consisting of $0 \in \mathcal{F}(U_1)$ and nonzero $f \in \mathcal{F}(U_2)$ is consistent, but there exists no $f' \in \mathcal{F}(X)$ for which $f'|_{U_1} = 0$ and $f'|_{U_2} = f \neq 0$. □

Example 62. Suppose that $X = \mathbb{R}$ and $\mathcal{F}$ is the presheaf of bounded functions, i.e., $\mathcal{F}(U)$ consists of all functions bounded on U. This presheaf is not a sheaf.

Proof. Let $\mathcal{U} = \{U_k\}_{k\in\mathbb{N}}$, where $U_k = (-k, k)$. Then the family $\{f_k\}$, where $f_k(x) = x$, is consistent, but there exists no bounded function f on $\mathbb{R} = \bigcup_{k=1}^{\infty} U_k$ coinciding with f_k on each set U_k. Indeed, such a function must be $f(x) = x$, which is not bounded. □

With any presheaf we can associate a sheaf by passing to the direct limit. Below, we recall the definition of direct limit.

A set J is said to be *directed* if a relation $>$ between some of its elements is defined and the following conditions hold:

(i) $\alpha > \alpha$ for any $\alpha \in J$;

(ii) if $\alpha > \beta$ and $\beta > \gamma$, then $\alpha > \gamma$;

(iii) for any $\alpha, \beta \in J$, there exists a $\delta \in J$ for which $\alpha > \delta$ and $\beta > \delta$.

Example 63. Let J be a family of open sets that contain a fixed point $x \in X$, and let $V < U$ in J if and only if $V \subset U$. Then J is a directed set.

Let J be a directed set. A set of Abelian groups $\{G_\alpha\}$, where $\alpha \in J$, is said to be *directed* if for any pair $\alpha > \beta$, a homomorphism $f_{\alpha\beta}\colon G_\alpha \to G_\beta$ is defined and for any $\alpha < \beta < \gamma$, $f_{\alpha\alpha} = \mathrm{id}$ and $f_{\beta\gamma} \circ f_{\alpha\beta} = f_{\alpha\gamma}$.

Example 64. Suppose that $\mathcal{F}$ is a presheaf on a space X, J is the directed set from Example 63, $G_U = \mathcal{F}(U)$, and $f_{UV} = r^U_V$. Then $\{G_U\}$, where $U \ni x$, is a directed set of Abelian groups.

The *direct limit* $\varinjlim_{\alpha\in J} G_\alpha$ of a directed set of Abelian groups is defined as follows. Consider the disjoint union of all groups G_α and take its quotient by the following equivalence relation: $g_\alpha \sim g_\beta$ if $f_{\alpha\delta}(g_\alpha) = f_{\beta\delta}(g_\beta)$ for some δ (here $g_\alpha \in G_\alpha$ and $g_\beta \in G_\beta$). The group operation is defined by $\{g_\alpha\} + \{g_\beta\} = \{f_{\alpha\delta}(g_\alpha) + f_{\beta\delta}(g_\beta)\}$, where $\alpha > \delta$ and $\beta > \delta$.

For the directed set of Abelian groups from Example 64, the direct limit consists of germs at the point x, which are defined as follows. We declare elements $f \in \mathcal{F}(U)$ and $g \in \mathcal{F}(V)$, where U and V are open neighborhoods of x, to be equivalent if $f|_W = g|_W$ for some open set $W \subset U \cap V$. The equivalence class containing $f \in \mathcal{F}(U)$ is called the *germ* of f at x and denoted by f_x. The addition of germs is defined in a natural way.

We set $\mathcal{F}_x = \varinjlim_{U\ni x} \mathcal{F}(U)$ (thus, $\mathcal{F}_x$ is the group of germs at the point x) and $\tilde{\mathcal{F}} = \bigcup_{x\subset X} \mathcal{F}_x$. The set $\tilde{\mathcal{F}}$ is endowed with a topology with respect to which the natural projection $p\colon \tilde{\mathcal{F}} \to X$ taking $\mathcal{F}_x$ to x is a local homeomorphism. Let $U \subset X$ be an open set, and let $f \in \mathcal{F}(U)$. Consider the set of all germs f_x for $x \in U$. Such sets form a base for a topology on $\tilde{\mathcal{F}}$. Clearly, if $\tilde{\mathcal{F}}$ is endowed with this topology, then p is a local homeomorphism.

A *section* of a presheaf $\tilde{\mathcal{F}}$ over an open set $U \subset X$ is defined as a continuous map $s\colon U \to \tilde{\mathcal{F}}$ for which $p \circ s = \mathrm{id}_U$. The set of sections over U is denoted by $\Gamma(U, \tilde{\mathcal{F}})$; this set is a group under the operation defined by $(s_1 + s_2)(x) = s_1(x) + s_2(x)$ for each point $x \in U$. For a family of groups $\Gamma(U, \tilde{\mathcal{F}})$ with different U, the natural restriction homomorphisms are defined; they satisfy the axioms for a presheaf. Moreover, setting $\overline{\mathcal{F}}(U) = \Gamma(U, \tilde{\mathcal{F}})$, we obtain a presheaf $\overline{\mathcal{F}}$. This presheaf is a sheaf; it is called the sheaf *generated* by, or *associated* with, $\mathcal{F}$.

We can associate with each element $f \in \mathcal{F}(U)$ a section $s \in \Gamma(U, \tilde{\mathcal{F}})$ by setting $s(x) = f_x$. Thereby, we obtain a homomorphism of presheaves $\tau\colon \mathcal{F} \to \overline{\mathcal{F}}$.

Theorem 5.1. *If $\mathcal{F}$ is a sheaf, then τ is an isomorphism of sheaves, i.e., the map $\tau_U\colon \mathcal{F}(U) \to \overline{\mathcal{F}}(U)$ is an isomorphism for any open set $U \subset X$.*

Proof. First, we verify that τ_U is a monomorphism. Suppose that $f, g \in \mathcal{F}(U)$ and $f_x = g_x$ for all $x \in U$. It follows from the definition of a germ that each point $x \in U$ has a neighborhood $V \ni x$ such that $f|_V = g|_V$, i.e., $(f-g)|_V = 0$. Therefore, the first sheaf axiom implies $f = g$.

Now let us show that τ_U is an epimorphism. Take $s \in \Gamma(U, \tilde{\mathcal{F}})$. Each point $x \in U$ has a neighborhood $V \ni x$ such that there exists an $f \in \mathcal{F}(V)$ for which $f_x = s(x)$. This means that the sections s and $\tau_V(f)$ coincide at x. Two sections coinciding at a point x must coincide in some neighborhood W of this point. Therefore, $s|_W = \tau_W(f|_W)$. To each point $x \in U$ we have assigned an open set $W_x \ni x$ (contained in U) and an element $f^x \in \mathcal{F}(W_x)$ so that $f^x_y = s(y)$ for all $y \in W_x$. The family of elements $\{f^x \in \mathcal{F}(W_x)\}$ is consistent, i.e., $f^x|_{W_x \cap W_z} = f^z|_{W_x \cap W_z}$. Indeed, we have $f^x_y = s(y) = f^z_y$ for all $y \in W_x \cap W_z$. Therefore, $\tau_{W_x \cap W_z}(f^x|_{W_x\cap W_z} - f^z|_{W_x \cap W_z}) = 0$; recall that τ is already proved to be a monomorphism. Using the second sheaf axiom, we obtain an element $f \in \mathcal{F}(U)$ such that $f_x = s(x)$ for all $x \in U$. □

It is often convenient to calculate direct limits over some subsets J_0 of a directed set J rather than over the entire set J. A subset $J_0 \subset J$ is *cofinal* in a directed set J if for any $\alpha \in J$, J_0 contains an element δ such that $\alpha > \delta$.

Theorem 5.2. *Suppose that $\{G_\alpha\}$, where $\alpha \in J$, is a directed set of Abelian groups and $J_0 \subset J$ is its cofinal subset. Then J_0 is a directed set and $\varinjlim_{\alpha \in J} G_\alpha \cong \varinjlim_{\alpha \in J_0} G_\alpha$.*

Proof. For any $\alpha, \beta \in J_0$, we can choose $\delta' \in J$ so that $\alpha > \delta'$ and $\beta > \delta'$. For this δ', we can choose $\delta \in J_0$ so that $\delta' > \delta$. Therefore, J_0 is a directed set.

The identity maps $G_\alpha \to G_\alpha$, where $\alpha \in J_0$, commute with the maps $f_{\alpha\delta}$; therefore, they determine a homomorphism $\varinjlim_{\alpha \in J} G_\alpha \to \varinjlim_{\alpha \in J_0} G_\alpha$. This homomorphism is an epimorphism because for any element $g_\alpha \in G_\alpha$ with $\alpha \in J$, we can choose $\delta \in J_0$ so that $\alpha > \delta$. For these α and δ, the map $f_{\alpha\delta}\colon G_\alpha \to G_\delta$ is defined, and $g_\alpha \sim g_\delta = f_{\alpha\delta}(g_\delta)$. Now, let us check that this homomorphism is a monomorphism. Suppose that $g_\alpha \sim 0_\beta$ for some $\beta \in J$, i.e., there exists a $\delta \in J$ such that $f_{\alpha\delta}(g_\alpha) = f_{\beta\delta}(0_\beta) = 0_\delta$. Choose $\gamma \in J_0$ so that $\delta > \gamma$. We have $f_{\alpha\gamma}(g_\alpha) = f_{\delta\gamma} \circ f_{\alpha\delta}(g_\alpha) = f_{\delta\gamma}(0_\gamma) = 0_\gamma$, i.e., $g_\alpha \sim 0_\gamma$ for some $\gamma \in J_0$. □

For more information about sheaves, see [**12, 66**].

1.2. Čech Cohomology. Suppose that $\mathcal{U} = \{U_\alpha\}$ is an open cover of a space X and $\mathcal{F}$ is a presheaf on X. We assume that all sets U_α are pairwise distinct. We write $U_{\alpha_0 \ldots \alpha_k}$ to denote $U_{\alpha_0} \cap \cdots \cap U_{\alpha_k}$. A *cochain* $c^k \in C^k(\mathcal{U}; \mathcal{F})$ assigns to each ordered set $U_{\alpha_0}, \ldots, U_{\alpha_k}$ an element $c^k(U_{\alpha_0}, \ldots, U_{\alpha_k}) \in$

$\mathcal{F}(U_{\alpha_0\ldots\alpha_k})$. The *coboundary homomorphism* is defined by

$$(\delta c^k)(U_{\alpha_0},\ldots,U_{\alpha_{k+1}})=\sum_{i=0}^{k+1}(-1)^i c^k(U_{\alpha_0},\ldots,\widehat{U}_{\alpha_i},\ldots,U_{\alpha_{k+1}})|_{U_{\alpha_0\ldots\alpha_{k+1}}}.$$

A simple standard calculation shows that $\delta\delta=0$. Thus, we can define the cohomology group $\check{H}^k(\mathcal{U};\mathcal{F})$, which is called the *Čech cohomology group* of the cover $\mathcal{U}$ with coefficients in the presheaf $\mathcal{F}$.

Example 65. If $\mathcal{F}$ is the constant presheaf corresponding to an Abelian group G, then $\check{H}^k(\mathcal{U};\mathcal{F})\cong H^k(N(\mathcal{U});G)$, where $N(\mathcal{U})$ is the nerve of the cover $\mathcal{U}$.

Passing to the direct limit, we obtain the Čech cohomology of the space X, which does not depend on a particular cover $\mathcal{U}$ of this space. To this end, we construct a directed family of cohomology groups as follows. Suppose $\mathcal{U}<\mathcal{V}$, i.e., $\mathcal{U}=\{U_\alpha\mid\alpha\in\mathcal{A}\}$ is a refinement of a cover $\mathcal{V}=\{V_\beta\mid\beta\in\mathcal{B}\}$. By the definition of a refinement, there exists a map $\lambda\colon\mathcal{A}\to\mathcal{B}$ such that $U_\alpha\subset V_{\lambda(\alpha)}$ for all $\alpha\in\mathcal{A}$. To each cochain $c^k\in C^k(\mathcal{V};\mathcal{F})$ we assign a cochain $\lambda^\# c^k\in C^k(\mathcal{U};\mathcal{F})$ such that

$$(\lambda^\# c^k)(U_{\alpha_0},\ldots,U_{\alpha_k})=c^k(V_{\lambda(\alpha_0)},\ldots,V_{\lambda(\alpha_k)}).$$

It is easy to verify that $\lambda^\#$ is a cochain map, i.e., $\lambda^\#\delta=\delta\lambda^\#$. The cochain map $\lambda^\#$ induces a homomorphism $\lambda^*\colon H^k(\mathcal{V};\mathcal{F})\to\check{H}^k(\mathcal{U};\mathcal{F})$ of cohomology groups.

The homomorphism λ^* does not depend on the choice of λ. Before proving this assertion in the general form, we explain why it is true for a constant sheaf $\mathcal{F}$.

Example 66. For a constant sheaf $\mathcal{F}$, the homomorphism

$$\lambda^*\colon\check{H}^k(\mathcal{V};\mathcal{F})\to\check{H}^k(\mathcal{U};\mathcal{F})$$

is induced by the simplicial map $\tilde{\lambda}\colon N(\mathcal{U})\to N(\mathcal{V})$ taking α to $\lambda(\alpha)$. If $\mu\colon\mathcal{A}\to\mathcal{B}$ is another map for which $U_\alpha\subset V_{\mu(\alpha)}$, then the maps $\tilde{\lambda}$ and $\tilde{\mu}$ are homotopic.

Proof. If a point x belongs to a simplex $[\alpha_0,\ldots,\alpha_n]$, then the points $\tilde{\lambda}(x)$ and $\tilde{\mu}(x)$ belong to the simplex with vertices $\lambda(\alpha_0),\ldots,\lambda(\alpha_n),\mu(\alpha_0),\ldots,$ $\mu(\alpha_n)$; this simplex exists because

$$\bigcap(V_{\lambda(\alpha_i)}\cap V_{\mu(\alpha_i)})\supset\bigcap U_{\alpha_i}\neq\varnothing.$$

Joining such points $\tilde{\lambda}(x)$ and $\tilde{\mu}(x)$ by segments, we obtain the required homotopy. □

Let us prove a general assertion.

Theorem 5.3. *If $\lambda, \mu\colon \mathcal{A} \to \mathcal{B}$ are maps for which $U_\alpha \subset V_{\lambda(\alpha)}$ and $U_\alpha \subset V_{\mu(\alpha)}$, then there exists a cochain homotopy between $\lambda^\#$ and $\mu^\#$.*

Proof. Consider the map $D\colon C^k(\mathcal{V};\mathcal{F}) \to C^{k-1}(\mathcal{U};\mathcal{F})$ defined by

$$(Dc^k)(U_{\alpha_0},\dots,U_{\alpha_{k-1}}) = \sum_{i=0}^{k-1}(-1)^i c^k(V_{\lambda(\alpha_0)},\dots,V_{\lambda(\alpha_i)},V_{\mu(\alpha_i)},\dots,V_{\mu(\alpha_{k-1})}).$$

Let us show first that $\delta D + D\delta = \mu^\# - \lambda^\#$. In the expression $(\delta(Dc^k) + D(\delta c^k))(U_0,\dots,U_k)$, the restrictions of the cochains

$$(-1)^i(-1)^j c^k(V_{\lambda(0)},\dots,_{\lambda(j)},V_{\mu(j)},\dots,\hat{U}_i,\dots,V_{\mu(k)})$$

and

$$(-1)^j(-1)^{i+1}c^k(V_{\lambda(0)},\dots,V_{\lambda(j)},V_{\mu(j)},\dots,\widehat{V}_{\mu(i)},\dots,V_{\mu(k)})$$

to the set $V_{\lambda(0)\dots\lambda(j)\mu(j)\dots\widehat{\mu(i)}\dots\mu(k)}$ cancel each other (we assume that $i > j$; the case of $i < j$ is treated similarly). The only remaining terms are $c^k(\widehat{V}_{\lambda(0)},V_{\mu(0)},\dots,V_{\mu(k)})$ and $(-1)^k(-1)^{k+1}c^k(V_{\lambda(0)},\dots,V_{\lambda(k)},\widehat{V}_{\mu(k)})$. □

We have constructed a directed set of Abelian groups; so we can consider the direct limits $\varinjlim \check{H}^k(\mathcal{U};\mathcal{F}) = \check{H}^k(X;\mathcal{F})$, which are called the *Čech cohomology groups* of the space X with coefficients in the presheaf $\mathcal{F}$. We denote the Čech cohomology groups of X with coefficients in the constant presheaf corresponding to an Abelian group G by $\check{H}^k(X;G)$.

Theorem 5.4. *If K is a finite simplicial complex, then $\check{H}^k(|K|;G) \cong H^k(K;G)$.*

Proof. Let $\mathcal{U}_0$ be the cover of the space $|K|$ by the open sets $\operatorname{st} v_i$, where each v_i is a vertex of K. The nerve $N(\mathcal{U}_0)$ is identified with K; therefore, $\check{H}^k(\mathcal{U}_0;G) \cong H^k(N(\mathcal{U}_0);G) \cong H^k(K;G)$.

Suppose that K' is the barycentric subdivision of the complex K, $h_1\colon K' \to K$ is a simplicial approximation of the identity map (h_1 takes the barycenter of each face to one of the vertices of this face), and $\mathcal{U}_1$ is the cover of $|K|$ by the stars of vertices of the simplicial complex K'. Identifying $N(\mathcal{U}_1)$ with K', we can regard h_1 as a map $N(\mathcal{U}_1) \to N(\mathcal{U}_0)$. This map induces an isomorphism of cohomology groups.

Similarly, for the cover $\mathcal{U}_m$ of $|K|$ by the stars of the vertices of the mth barycentric subdivision of K, we construct a map $N(\mathcal{U}_m) \to N(\mathcal{U}_{m-1})$, which induces an isomorphism of the cohomology groups.

We have constructed the directed set $\mathcal{U}_0 > \mathcal{U}_1 > \cdots$ and the directed set of Abelian groups $\check{H}^k(\mathcal{U}_m;G)$, in which all homomorphisms $f_{\alpha\beta}$ are isomorphisms. We have $\varinjlim \check{H}^k(\mathcal{U}_m;G) \cong H^k(K;G)$.

To apply Theorem 5.2, we must show that the covers $\mathcal{U}_0 > \mathcal{U}_1 > \cdots$ form a cofinal subset in the set of all covers of $|K|$. Take an arbitrary open cover $\mathcal{U}$ of $|K|$; let δ be its Lebesgue number (we assume that the simplicial complex K is embedded in Euclidean space). Choose m so that the diameter of any simplex in the mth barycentric subdivision of K is less than δ. We have $\mathcal{U}_m < \mathcal{U}$. □

Theorem 5.5. *Let X be a normal topological space, and let Y be its compact subspace. Then $\check{H}^k(Y;G) \cong \varinjlim \check{H}^k(\mathcal{U};G)$, where $\mathcal{U}$ is the cover of Y by open sets in X (not in Y, as in the definition of Čech cohomology).*

Proof. Let $\mathcal{U}$ be a cover of Y by open sets in X. Take its finite subcover $\{U_1, \ldots, U_n\}$ and consider the cover of Y by the sets $U_i \cap Y$, which are open in Y. Any subspace of a normal space is Hausdorff, and any compact Hausdorff space is normal. Therefore, the space Y is normal, and the cover $\{U_i \cap Y\}$ has an open refinement $\{V_i\}$ such that $\overline{V}_i \subset U_i \cap Y$. Here $\overline{V}_i$ is the closure of V_i in Y; it is also the closure of V_i in X because Y is closed in X.

Let us construct open sets $W_1, \ldots, W_n$ in X such that $\overline{V}_i \subset W_i$, $\overline{W}_i \subset U_i$, and the nerves of the covers $\{\overline{V}_1, \ldots, \overline{V}_n\}$ and $\{\overline{W}_1, \ldots, \overline{W}_n\}$ coincide, i.e., $\overline{V}_{i_1} \cap \cdots \cap \overline{V}_{i_p} = \varnothing$ if and only if $\overline{W}_{i_1} \cap \cdots \cap \overline{W}_{i_p} = \varnothing$. Let C_1 be the union of all sets of the form $\overline{V}_{i_1} \cap \cdots \cap \overline{V}_{i_p}$ disjoint from $\overline{V}_1$. The sets $\overline{V}_1$, C_1, and $X \setminus U_1$ are closed in X, and the sets $\overline{V}_1$ and $C_1 \cup (X \setminus U_1)$ are disjoint. Therefore, we can choose an open set W_1 in X so that $\overline{V}_i \subset W_i$ and $\overline{W}_1 \cap (C_1 \cup (X \setminus U_1)) = \varnothing$; the latter condition means that $\overline{W}_1 \cap C_1 = \varnothing$ and $\overline{W}_1 \subset U_1$. Let us show that the nerves of the covers $\{\overline{W}_1, \overline{V}_2, \ldots, \overline{V}_n\}$ and $\{\overline{V}_1, \ldots, \overline{V}_n\}$ coincide. Clearly, if $\overline{W}_1 \cap \overline{V}_{i_1} \cap \cdots \cap \overline{V}_{i_p} = \varnothing$, then $\overline{V}_1 \cap \overline{V}_{i_1} \cap \cdots \cap \overline{V}_{i_p} = \varnothing$ because $\overline{V}_1 \subset W_1 \subset \overline{W}_1$. Now suppose that $\overline{V}_1 \cap \overline{V}_{i_1} \cap \cdots \cap \overline{V}_{i_p} = \varnothing$. Then $\overline{V}_{i_1} \cap \cdots \cap \overline{V}_{i_p} \subset C_1$. But $\overline{W}_1 \cap C_1 = \varnothing$; hence $\overline{W}_1 \cap \overline{V}_{i_1} \cap \cdots \cap \overline{V}_{i_p} = \varnothing$. Applying the same procedure to $\overline{W}_1, \overline{V}_2, \ldots, \overline{V}_n$, we construct $\overline{W}_2$, and so on.

Let us show that the nerve of the cover of Y by the sets $W_1, \ldots, W_n$, which are open in X, coincides with that of the cover by the sets $W_1 \cap Y$, $\ldots, W_n \cap Y$, which are open in Y. Clearly, if $W_{i_1} \cap \cdots \cap W_{i_p} = \varnothing$, then $W_{i_1} \cap \cdots \cap W_{i_p} \cap Y = \varnothing$. Suppose that $W_{i_1} \cap \cdots \cap W_{i_p} \cap Y = \varnothing$. Then $\overline{V}_{i_1} \cap \cdots \cap \overline{V}_{i_p} = \varnothing$, or, equivalently, $\overline{W}_{i_1} \cap \cdots \cap \overline{W}_{i_p} = \varnothing$. Therefore, $W_{i_1} \cap \cdots \cap W_{i_p} = \varnothing$.

Thus, we have shown that the cover $\{W_1, \ldots, W_n\}$ of the space Y by open sets in X, whose nerve coincides with that of $\{W_1 \cap Y, \ldots, W_n \cap Y\}$, is cofinal in the family of covers of Y by open sets in X. According to Theorem 5.2, we have $\varinjlim \check{H}^k(\mathcal{U};G) \cong \varinjlim \check{H}^k(\mathcal{W};G)$. By construction, $\check{H}^k(\mathcal{W};G) \cong \check{H}^k(\mathcal{W} \cap Y;G)$; here $\mathcal{W} \cap Y = \{W_1 \cap Y, \ldots, W_n \cap Y\}$. Moreover, the cover

$\mathcal{W} \cap Y$ is cofinal in the set of all open covers of Y. This implies the required isomorphism. □

In calculating Čech cohomology, the following assertion is often useful.

Theorem 5.6. *Let X be a compact triangulable*[1] *space, and let Y_n be a subcomplex of some triangulation of X. Suppose that $|Y_1| \supset |Y_2| \supset \cdots$ and $\bigcap_{n=1}^{\infty} |Y_n| = Y$. Then $\check{H}^k(Y; G) \cong \varinjlim H^k(Y_n; G)$.*

Proof. Let X_n be a triangulation of X for which some[2] subdivision Y_n is a full subcomplex (i.e., any simplex in X_n spanned by some vertices of Y_n is a simplex in Y_n). We can choose this triangulation to be so fine that (i) the diameter of any simplex in X_n is less than $1/n$ (we assume that X is embedded in Euclidean space) and (ii) the star of each vertex of X_n is contained in the star of some vertex of X_{n-1}. We endow Y_n with the triangulation induced by X_n.

Consider two covers $\mathcal{U}_n$ and $\mathcal{U}'_n$ of Y_n; $\mathcal{U}_n$ is the cover by the stars in Y_n of all vertices of the complex Y_n, and $\mathcal{U}'_n$ is the cover by the stars in X_n of all vertices of the same complex Y_n. The first cover is open in Y_n, and the second is open in X_n.

Condition (ii) allows us to construct a simplicial approximation $h_n \colon Y_n \to Y_{n-1}$ of the inclusion $|Y_n| \subset |Y_{n-1}|$. The nerve of the cover $\mathcal{U}_n$ can be identified with Y_n; thus, we obtain a simplicial map $h_n \colon N(\mathcal{U}_n) \to N(\mathcal{U}_{n-1})$ and the induced homomorphism (which is not necessarily an isomorphism) $(h_n)_* \colon \check{H}^k(\mathcal{U}_{n-1}) \to \check{H}^k(\mathcal{U}_n)$; we omit the coefficient group G in the notation. Thus, $\varinjlim H^k(Y_n) \cong \varinjlim \check{H}^k(\mathcal{U}_n)$.

For the covers $\mathcal{U}'_n$, condition (ii) implies $\mathcal{U}'_n < \mathcal{U}'_{n-1}$; thus, a homomorphism $(h'_n)_* \colon \check{H}^k(\mathcal{U}'_{n-1}) \to \check{H}^k(\mathcal{U}'_n)$ arises. Let us show that the nerves of the covers $\mathcal{U}_n$ and $\mathcal{U}'_n$ coincide, and therefore the homomorphisms $(h_n)_*$ and $(h'_n)_*$ coincide as well (after the nerves are identified).

Clearly, $\operatorname{st}(v, Y_n) \subset \operatorname{st}(v, X_n)$; therefore, if $\bigcap \operatorname{st}(v_i, Y_n) \neq \varnothing$, then also $\bigcap \operatorname{st}(v_i, X_n) \neq \varnothing$. Suppose that $\bigcap \operatorname{st}(v_i, X_n) \neq \varnothing$. This means that the vertices v_i of a simplex in X_n span a simplex in Y_n. Moreover, the v_i are vertices of the complex Y_n, and the simplicial subcomplex Y_n is full by construction. Therefore, the vertices v_i span a simplex in Y_n, i.e., $\bigcap \operatorname{st}(v_i, Y_n) \neq \varnothing$.

Thus, $\varinjlim \check{H}^k(\mathcal{U}_n) \cong \varinjlim \check{H}^k(\mathcal{U}'_n)$.

To apply Theorem 5.2, we must show that the covers $\mathcal{U}'_1 > \mathcal{U}'_2 > \cdots$ form a cofinal subset in the family of covers of Y by open sets in X. Let $\mathcal{U}$ be a cover of Y by open sets in X. First, we show that $\mathcal{U}$ covers some of the sets $|Y_n|$. Suppose that each $|Y_n|$ has a point y_n not covered by $\mathcal{U}$.

[1] This means that there exists a homeomorphism $X \to |K|$, where K is a simplicial complex.
[2] According to Problem 30 from Part I, it suffices to take the barycentric subdivision.

Since X is compact, the sequence $\{y_n\}$ contains a convergent subsequence $\{y_{n_k}\}$. Its limit point y belongs to $\bigcap_{n=1}^{\infty} |Y_n| = Y$. This point is covered by some open set U from the cover $\mathcal{U}$. We have $y_{n_k} \in U$ for sufficiently large k, which contradicts the assumption.

The same argument as the one used to prove the Lebesgue theorem about open covers shows that there exists a positive number δ such that any B of diameter less than δ intersecting $|Y_n|$ is contained in an open set from $\mathcal{U}$. Suppose that $1/m < \delta/2$ and $m \leq n$. Let us show that the cover $\mathcal{U}'_m$ is a refinement of $\mathcal{U}$. The diameter of the star of any vertex of Y_m is less than $2/m < \delta$. Moreover, this star intersects $|Y_n|$ because $|Y_m| \subset |Y_n|$. Therefore, it is contained in an open set from $\mathcal{U}$. □

Using the Čech cohomology, we can state and prove the following, more general, version of the Alexander duality theorem.

Theorem 5.7 (Alexander–Pontryagin duality). *If $A \subsetneq S^n$ is a closed set, then $\check{\tilde{H}}^k(A) \cong \tilde{H}_{n-k-1}(S^n \setminus A)$ for $0 \leq k \leq n-1$. (Here $\check{\tilde{H}}^*$ is the reduced Čech cohomology and $\tilde{H}_*$ is the reduced singular homology.)*

Proof. Let K be a triangulation of the sphere S^n so fine that it contains n-simplices disjoint from A, and let $K^{(m)}$ be the mth barycentric subdivision of K. Consider the subcomplex M_m of $K^{(m)}$ consisting of the simplices intersecting A. The simplicial complex M_m is not necessarily a manifold; it may have singular points (or simplices). But all of its singular points are outside A; therefore, we can turn M_m into a manifold containing A by removing small neighborhoods of singular points (see Figure 1).

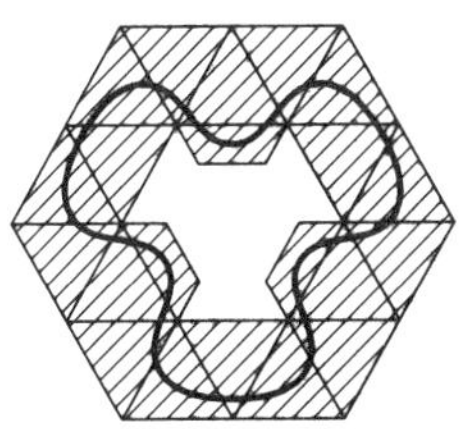

Figure 1. The correction of the complex M_m

In this way, we construct a sequence of triangulated manifolds $M_1 \supset M_2 \supset \cdots$ such that $\bigcap_{i=1}^{\infty} M_i = A$. For each M_i, the Alexander duality $\tilde{H}^k(M_i) \cong \tilde{H}_{n-k-1}(S^n \setminus M_i)$ holds; its proof is similar to that in the smooth (triangulable) case. This isomorphism commutes with the homomorphism induced by the inclusion $M_{i+1} \subset M_i$; hence

$$\varinjlim \tilde{H}^k(M_i) \cong \varinjlim \tilde{H}_{n-k-1}(S^n \setminus M_i).$$

According to Theorem 5.6, the direct limit on the left-hand side is the Čech cohomology group $\check{H}^k(A)$. Let us show that the group on the right-hand side is the singular homology group $\tilde{H}_{n-k-1}(S^n \setminus A)$.

The homomorphisms $\tilde{H}_{n-k-1}(S^n \setminus M_i) \to \tilde{H}_{n-k-1}(S^n \setminus A)$ induced by the inclusions determines a homomorphism

$$\varinjlim \tilde{H}_{n-k-1}(S^n \setminus M_i) \to \tilde{H}_{n-k-1}(S^n \setminus A).$$

Let us show that this is an isomorphism. The support of any singular chain in $C_{n-k-1}(S^n \setminus A)$ is compact. The increasing open sets $S^n \setminus M_i$ cover it; therefore, it is contained in one of these sets. This implies surjectivity. Injectivity follows from the compactness of the support of any singular $(n-k)$-chain whose boundary is the difference of two given singular $(n-k-1)$-chains. □

1.3. Bundles with Structure Groups. Suppose that G is a topological group, B is a topological space, and the group G acts effectively on a topological space F. A locally trivial bundle $p\colon E \to B$ with fiber F is called a *bundle with structure group G* if the homeomorphisms $h_i\colon p^{-1}(U_i) \to U_i \times F$ have the property that, for any pair of indices i, j, there exists a map $g_{ij}\colon U_i \cap U_j \to G$ such that $h_i h_j^{-1}(u, f) = (u, g_{ij}(u)f)$ for all $u \in U_i \cap U_j$. Since the action of G on F is effective, each map g_{ij} is uniquely determined by the homeomorphisms h_i and h_j. Such a map is called a *transition function.* The transition functions determine how the "pillars" $U_i \times F$ and $U_j \times F$ are attached to each other (see Figure 2).

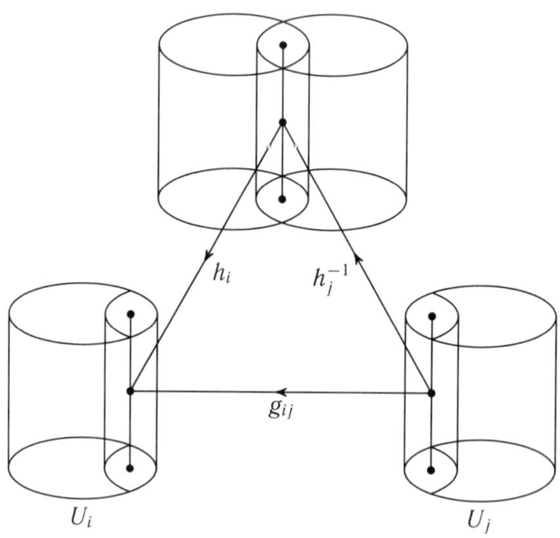

Figure 2. Transition functions

This definition of a bundle with a structure group has the essential drawback of being dependent on the cover $\{U_i\}$. To overcome it, consider all homeomorphisms $h_U\colon p^{-1}(U) \to U \times F$ that can be added to the homeomorphisms h_i, namely, those with the property that, for any U_i, there exists a map $g_{U,i}\colon U \cap U_i \to G$ such that $h_U h_i^{-1}(u, f) = (u, g_{U,i}(u)f)$ for all $u \in U \cap U_i$. Such homeomorphisms are called *admissible charts.* We assume that two covers $\{U_i\}$ and $\{U'_j\}$ with homeomorphisms $\{h_i\}$ and $\{h'_j\}$ specify the same bundle with structure group G if they determine the same set of admissible charts.

Now we introduce the notion of an isomorphism between two bundles with structure group G over the same base B. Let $p\colon E \to B$ and $p'\colon E' \to B$ be bundles with structure group G. A homeomorphism $\varphi\colon E \to E'$ is called an *isomorphism* of bundles with structure group G if, for each point $b \in B$, the following conditions hold:

(i) each fiber $p^{-1}(b)$ is mapped to $(p')^{-1}(b)$;

(ii) there exists a neighborhood $U \ni b$, a map $g_U\colon U \to G$, an admissible chart $h_U\colon p^{-1}(U) \to U \times F$ for the bundle E, and an admissible chart $h'_U\colon (p')^{-1}(U) \to U \times F$ for the bundle E' such that $h'_U \varphi h_U^{-1}(u, f) = (u, g_U(u)f)$ for all $u \in U$.

For example, an n-dimensional vector bundle is the same thing as an $\mathbb{R}^n$-bundle with structure group $\mathrm{GL}(n, \mathbb{R})$. An isomorphism of vector bundles is the same as an isomorphism of bundles with this structure group. If a vector bundle admits a Riemannian metric (e.g., if its base is compact), then we can assume that the structure group is $\mathrm{O}(n)$. Such a bundle is orientable if and only if for the structure group the group $\mathrm{SO}(n)$ can be taken.

Example 67. Suppose that the manifold $\mathbb{C}P^n$ is covered by charts U_i, $i = 0, \ldots, n$, which are specified by the equations $z_i = 0$ in homogeneous coordinates $(z_0 : \cdots : z_n)$. Then the canonical bundle γ_n^1 over $\mathbb{C}P^n$ is determined by the transition functions $g_{ij} = z_i/z_j$.

Proof. The fiber of the canonical bundle over a point $(z_0 : \cdots : z_n)$ is the complex line $(\lambda z_0, \ldots, \lambda z_n)$. The homeomorphism $h_i\colon p^{-1}(U_i) \to U_i \times \mathbb{C}$ can be defined by

$$h_i(\lambda z_0, \ldots, \lambda z_n) = ((z_0 : \cdots : z_n), \lambda_i),$$

where $\lambda_i = \lambda z_i$, i.e., $(\lambda z_0, \ldots, \lambda z_n) = \lambda_i\left(\frac{z_0}{z_i}, \ldots, \frac{z_n}{z_i}\right)$. By definition, we have $\lambda_i = g_{ij}\lambda_j$, i.e., $g_{ij} = \lambda_i/\lambda_j = z_i/z_j$. □

Example 68. Suppose that an n-dimensional vector bundle is given by transition functions g_{ij} (i.e., at each point, g_{ij} is a matrix of order n). Then the dual bundle is given by the transition functions $(g_{ij}^T)^{-1}$.

Proof. Each map $g_{ij}\colon V_1 \to V_2$ induces a dual map $V_2^* \to V_1^*$, which is determined by the matrix g_{ij}^T. We are interested in the inverse map $V_1^* \to V_2^*$. □

Clearly, if ξ is a one-dimensional bundle with transition functions g_{ij}, then the dual bundle ξ^* is determined by the transition functions $1/g_{ij}$, and if a one-dimensional bundle η is determined by transition functions h_{ij}, then the bundle $\xi \otimes \eta$ is determined by the transition functions $g_{ij}h_{ij}$. In particular, the bundle $\xi \otimes \xi^*$ is trivial.

Example 69. Consider the map

$$p\colon \mathbb{C}P^{n+1} \setminus (0 : \cdots : 0 : 1) \to \mathbb{C}P^n$$

given by $p(z_0 : \cdots : z_{n+1}) = (z_0 : \cdots : z_n)$. This map is the vector bundle dual to the canonical bundle γ_n^1.

Proof. The homeomorphism $h_i\colon p^{-1}(U_i) \to U_i \times \mathbb{C}$ can be defined by $h_i(z_0 : \cdots : z_{n+1}) = ((z_0 : \cdots : z_n), \lambda_i)$, where $\lambda_i = z_{n+1}/z_i$. Thus, $g_{ij} = \lambda_i/\lambda_j = z_j/z_i$. □

1.4. Noncommutative Čech Cohomology. We can define a sheaf of non-Abelian groups on a space X in precisely the same way as a sheaf of Abelian groups. In the non-Abelian case, the cohomology groups $\check{H}^k(X; \mathcal{F})$ for $k > 1$ cannot be defined, but there is a cohomology set $\check{H}^1(X; \mathcal{F})$ with a distinguished element. Moreover, if $\mathcal{F}$ is a sheaf of Abelian groups, then this set coincides with the Čech cohomology group, and the distinguished element is the zero element of the cohomology group.

Recall that if $\mathcal{F}$ is a sheaf of Abelian groups on a topological space X and $\mathcal{U} = \{U_\alpha\}$ is an open cover of X, then to every ordered set U_i, U_j any cochain $c^1 \in C^1(\mathcal{U}; \mathcal{F})$ assigns an element $c^1(U_i, U_j) \in \mathcal{F}(U_i \cap U_j) = \Gamma(U_i \cap U_j; \mathcal{F})$. A cochain c^1 is a cocycle if for any ordered triple U_i, U_j, U_k, we have $c^1(U_i, U_j) + c^1(U_j, U_k) = c^1(U_i, U_k)$ on $U_i \cap U_j \cap U_k$. A cochain c^1 is a coboundary if there exists a cochain $c^0 \in C^0(\mathcal{U}; \mathcal{F})$ such that $c^1(U_i, U_j) = c^0(U_j) - c^0(U_i)$ on $U_i \cap U_j$.

For sheaves of non-Abelian groups, the corresponding definitions are as follows. A (one-dimensional) cocycle f assigns an element $f_{ij} \in \mathcal{F}(U_i \cap U_j) = \Gamma(U_i \cap U_j; \mathcal{F})$ to each ordered pair U_i, U_j so that $f_{ij}f_{jk} = f_{ik}$ on $U_i \cap U_j \cap U_k$. Two cocycles f and f' are equivalent (differ by a coboundary) if for each set U_i, there exists an element $g_i \in \Gamma(U_i; \mathcal{F})$ such that $f'_{ij} = g_i^{-1} f_{ij} g_j$ on $U_i \cap U_j$. The cohomology set $\check{H}^1(\mathcal{U}; \mathcal{F})$ is the set of classes of equivalent cocycles. The distinguished element of this set is the class of the cocycle 1 that takes all ordered sets U_i, U_j to the identity element of the group $\Gamma(U_i \cap U_j; \mathcal{F})$.

As in the commutative case, the cohomology set $\check{H}^1(X;\mathcal{F})$ is constructed as a direct limit. Suppose that a cover $\mathcal{U}$ is a refinement of a cover $\mathcal{V}$ and $U_i \subset V_{\lambda(i)}$. Then every cocycle $f \in Z^1(\mathcal{V};\mathcal{F})$ can be associated with a cocycle $\lambda^{\#} f \in Z^1(\mathcal{U};\mathcal{F})$ for which $(\lambda^{\#} f)(U_i \cap U_j) = f(V_{\lambda(i)} \cap V_{\lambda(j)})$. The map $\lambda^{\#}$ induces a map $\lambda^*\colon \check{H}^1(\mathcal{V};\mathcal{F}) \to \check{H}^1(\mathcal{U};\mathcal{F})$. Let us show that if $U_i \subset V_{\mu(i)}$, then $\mu^* = \lambda^*$. The equivalence of cocycles $\lambda^{\#} f$ and $\mu^{\#} f$ is established by means of the 0-cochain $g_i = f_{\lambda(i),\mu(i)}$. Indeed, by the definition of a cocycle, we have $f_{\mu(i),\lambda(i)} f_{\lambda(i),\lambda(j)} f_{\lambda(j),\mu(j)} = f_{\mu(i),\mu(j)}$, i.e., $g_i^{-1}(\lambda^{\#} f)_{ij} g_j = (\mu^{\#} f)_{ij}$.

Let $\mathcal{G}$ be the sheaf on B for which $\Gamma(U,\mathcal{G})$ is the group of all continuous maps from U to G; in other words, $\mathcal{G}$ is the sheaf of germs of continuous functions taking values in G. Given an open cover $\mathcal{U} = \{U_i\}$ of B and a cocycle $g = \{g_{ij}\} \in Z^1(\mathcal{U};\mathcal{G})$, we can construct a bundle E_g over B with structure group G and fiber F as follows. We take the disjoint union of the sets $U_i \times F$ and for each $u \in U_i \cap U_j$, identify the points $(u,f) \in U_i \times F$ and $(u, g_{ij}(u) f) \in U_j \times F$. We can do this due to the equality in the definition of a cocycle. For the space E_g thus obtained, the projection $p_g\colon E_g \to B$ is induced by the natural projection $U_i \times F \to U_i$.

Given two cocycles $g \in Z^1(\mathcal{U};\mathcal{G})$ and $g' \in Z^1(\mathcal{U};\mathcal{G})$, the bundles $p_g\colon E_g \to B$ and $p_{g'}\colon E_{g'} \to B$ are isomorphic (as bundles with structure group G) if and only if the cocycles g and g' correspond to the same element of the cohomology set $\check{H}^1(B;\mathcal{G})$.

Clearly, any bundle over B with structure group G and fiber F can be represented as E_g for some cocycle g. We have proved the following theorem.

Theorem 5.8. *Suppose that a topological group G acts effectively and continuously on a space F. Then there is a natural one-to-one correspondence between the classes of isomorphic bundles over B with structure group G and fiber F and the elements of the cohomology set $\check{H}^1(B;\mathcal{G})$. Moreover, the trivial bundle $B \times F$ corresponds to the distinguished element.*

2. De Rham Cohomology

Recall that a differential k-form ω on a manifold M^n is a polylinear skew-symmetric function of k vector fields $\xi_1,\dots,\xi_k$ on M^n. We denote the linear space of k-forms on a manifold M^n by[3] $\Omega^k(M^n)$. For two differential forms $\omega_1 \in \Omega^p(M^n)$ and $\omega_2 \in \Omega^q(M^n)$, their exterior product $\omega_1 \wedge \omega_2 \in \Omega^{p+q}(M^n)$ is defined; it has the property $\omega_1 \wedge \omega_2 = (-1)^{pq}\omega_2 \wedge \omega_1$. Exterior multiplication makes the linear space $\Omega^*(M^n) = \bigoplus_{k\geq 0} \Omega^k(M^n)$ into an algebra.

Any smooth map $f\colon M^m \to N^n$ induces a linear map $f^*\colon \Omega^k(N^n) \to \Omega^k(M^m)$ by the rule $(f^*\omega)(\xi_1,\dots,\xi_k) = \omega(f_*\xi_1,\dots,f_*\xi_k)$, where $f_* = df$

[3]In Part I on p. 205, the notation $\Lambda^k M^n$ was used.

is the differential of f. In local coordinates, this map looks as follows. Let $x_1, \dots, x_n$ and $y_1, \dots, y_m$ be local coordinates on N^n and M^m. Then

$$f^*\Big(\sum a_{i_1 \dots i_k}\, dx_{i_1} \wedge \dots \wedge dx_{i_k}\Big) = \sum a_{i_1 \dots i_k} \frac{\partial f_{i_1}}{\partial y_{j_1}} \cdots \frac{\partial f_{i_k}}{\partial y_{j_k}}\, dy_{j_1} \wedge \dots \wedge dy_{j_k}.$$

The map $f^* \colon \Omega^*(N^n) \to \Omega^*(M^m)$ is a homomorphism of algebras.

Let us construct a polylinear map $d\colon \Omega^k(M^n) \to \Omega^{k+1}(M^n)$ such that (i) $d \circ d = 0$ and (ii) d is polylinear with respect to smooth functions, i.e., $d\omega(\varphi\xi_0, \xi_1, \dots, \xi_p) = \varphi\, d\omega(\xi_0, \dots, \xi_p)$ for any $\varphi \in C^\infty(M^n)$. For this purpose, we need the notion of the commutator of vector fields.

Let ξ and η be vector fields on M^n (we consider them as differential operators), and let $\varphi \in C^\infty(M^n)$. We set $[\xi, \eta](\varphi) = \xi(\eta(\varphi)) - \eta(\xi(\varphi))$. Using the relation $\frac{\partial^2 \varphi}{\partial x_i \partial x_j} = \frac{\partial^2 \varphi}{\partial x_j \partial x_i}$, we obtain the following expression in local coordinates:

$$[\xi, \eta](\varphi) = \sum_{i,j} \Big(\xi_j \frac{\partial \eta_i}{\partial x_j} - \eta_j \frac{\partial \xi_i}{\partial x_j}\Big) \frac{\partial \varphi}{\partial x_i}.$$

Thus, in a fixed local coordinate system, $[\xi, \eta]$ is a vector field. The operator $[\xi, \eta]$, which acts on smooth functions, does not depend on the choice of a local coordinate system; so we obtain a vector field $[\xi, \eta]$, which is called the *commutator* of the vector fields ξ and η.

It follows directly from the definition that $[\xi, \eta] = -[\eta, \xi]$. Moreover, if $\varphi \in C^\infty(M^n)$, then $[\varphi\xi, \eta] = \varphi[\xi, \eta] - \eta(\varphi)\xi$; this follows from the relation

$$\frac{\partial(\varphi\xi_i)}{\partial x_j} = \varphi \frac{\partial \xi_i}{\partial x_j} + \xi_i \frac{\partial \varphi}{\partial x_j}.$$

Now we can define the map d. Let $\omega \in \Omega^k(M^n)$. We set

$$d\omega(\xi_0, \dots, \xi_k) = \sum_{i=0}^{k} (-1)^i \xi_i(\omega(\xi_0, \dots, \hat{\xi}_i, \dots, \xi_k)) + \sum_{0 \le i < j \le k} (-1)^{i+j} \omega([\xi_i, \xi_j], \dots, \hat{\xi}_i, \dots, \hat{\xi}_j, \dots, \xi_k).$$

For $\varphi \in C^\infty(M^n) = \Omega^0(M^n)$, we put $d\varphi(\xi) = \xi(\varphi)$. By definition, the form $d\omega$ is antisymmetric and polylinear. The equality $d\omega(\varphi\xi_0, \dots, \xi_k) = \varphi\, d\omega(\xi_0, \dots, \xi_k)$ follows from

$$\begin{aligned} &\xi_i(\omega(\varphi\xi_0, \dots, \xi_k)) + \omega([\varphi\xi_0, \xi_i], \dots, \hat{\xi}_i, \dots, \xi_k) \\ &\quad = \varphi\xi_i(\omega(\varphi\xi_0, \dots, \hat{\xi}_i, \dots, \xi_k)) + \xi_i(\varphi)\omega(\varphi\xi_0, \dots, \hat{\xi}_i, \dots, \xi_k) \\ &\qquad + \varphi\omega([\xi_0, \xi_i], \dots, \hat{\xi}_i, \dots, \xi_k) - \xi_i(\varphi)\omega(\varphi\xi_0, \dots, \hat{\xi}_i, \dots, \xi_k) \\ &\quad = \varphi(\xi_i(\omega(\xi_0, \dots, \hat{\xi}_i, \dots, \xi_k)) + \omega([\xi_0, \xi_i], \dots, \hat{\xi}_i, \dots, \xi_k)). \end{aligned}$$

Let us introduce local coordinates $x_1, \dots, x_n$ and calculate the differential of the form $\omega = \varphi(x)\, dx_{i_1} \wedge \dots \wedge dx_{i_k}$, where $i_1 < \dots < i_k$. We set $\xi_0 = \frac{\partial}{\partial x_{j_0}}, \dots, \xi_k = \frac{\partial}{\partial x_{j_k}}$ for $j_0 < \dots < j_k$. The vector fields ξ_i commute, i.e., $[\xi_i, \xi_j] = 0$. Therefore,

$$d\omega(\xi_0, \dots, \xi_k) = \sum_{p=0}^{k} (-1)^p \frac{\partial}{\partial x_{j_p}} (\omega(\xi_0, \dots, \hat{\xi}_p, \dots, \xi_k)).$$

Here $\omega(\xi_0, \dots, \hat{\xi}_p, \dots, \xi_k) = \varphi(x)$ if $i_1 = j_0, \dots, i_p = j_{p-1}, i_{p+1} = j_p, \dots, i_k = j_k$; in all the other cases, we obtain 0. We have $j_{p-1} < j_p < j_{p+1}$. Hence

$$\begin{aligned}(-1)^p &\frac{\partial}{\partial x_{j_p}} (\omega(\xi_0, \dots, \hat{\xi}_p, \dots, \xi_k)) \\ &= (-1)^p \frac{\partial \varphi}{\partial x_{j_p}} dx_{i_1} \wedge \dots \wedge dx_{j_p} \wedge \dots \wedge dx_{i_k} (\xi_0, \dots, \xi_p, \dots, \xi_k) \\ &= \frac{\partial \varphi}{\partial x_{j_p}} dx_{j_p} \wedge dx_{i_1} \wedge \dots \wedge \widehat{dx_{j_p}} \wedge \dots \wedge dx_{i_k} (\xi_0, \dots, \xi_k).\end{aligned}$$

As a result, we obtain

$$d(\varphi(x)\, dx_{i_1} \wedge \dots \wedge dx_{i_k}) = \sum_{s=0}^{n} \frac{\partial \varphi(x)}{\partial x_s} dx_s \wedge dx_{i_1} \wedge \dots \wedge dx_{i_k}.$$

Using this expression, we can easily verify that $d \circ d = 0$; moreover, if ω_1 is a k-form, then $d(\omega_1 \wedge \omega_2) = d\omega_1 \wedge \omega_2 + (-1)^k \omega_1 \wedge d\omega_2$. Note also that the operation d is local in the sense that, at each point x, the form $d\omega$ is completely determined by the restriction of ω to an arbitrarily small neighborhood of x. The commutator of two vector fields is also a local operation in this sense.

A differential form ω is said to be *closed* if $d\omega = 0$. The differential of a form ω is *exact* if $\omega = d\omega'$ for some form ω'. The coset space of the set of closed k-forms on the manifold M^n modulo the exact k-forms is called the kth *de Rham cohomology space* and denoted by $H^k_{\mathrm{DR}}(M^n)$.

Example 70. $H^0_{\mathrm{DR}}(\mathbb{R}) \cong \mathbb{R}$ and $H^1_{\mathrm{DR}}(\mathbb{R}) = 0$.

Proof. For $\mathbb{R}$, there is only one nontrivial differential $d\colon \Omega^0(\mathbb{R}) \to \Omega^1(\mathbb{R})$, which is defined by $\varphi \mapsto \frac{d\varphi}{dx}\, dx$. Clearly, $\operatorname{Ker} d$ consists of constant functions. Therefore, $H^0_{\mathrm{DR}}(\mathbb{R}) = \operatorname{Ker} d \cong \mathbb{R}$. It is easy to show that $\operatorname{Im} d = \Omega^1(\mathbb{R})$. Indeed, any 1-form on $\mathbb{R}$ can be represented as $\xi\, dx$, where $\xi \in C^\infty(\mathbb{R})$. We set $\varphi(x) = \int_0^x \xi(t)\, dt$ for $x \geq 0$ and $\varphi(x) = \int_x^0 \xi(t)\, dt$ for $x \leq 0$. We have $d\varphi = \xi\, dx$. Thus, $H^1_{\mathrm{DR}}(\mathbb{R}) = \Omega^1(\mathbb{R}) / \operatorname{Im} d = \Omega^1(\mathbb{R})/\Omega^1(\mathbb{R}) = 0$. □

Example 71. If M^n is a connected manifold, then $H^0_{\mathrm{DR}}(M^n) \cong \mathbb{R}$.

Proof. The kernel of the homomorphism $d\colon \Omega^0(M^n) \to \Omega^1(M^n)$ consists of the functions satisfying the condition $\frac{\partial\varphi}{\partial x_1} = \cdots = \frac{\partial\varphi}{\partial x_n} = 0$ in each coordinate neighborhood. Such functions are constant on each connected component of the manifold. □

It is easy to see that the map $f^*\colon \Omega^k(N^m) \to \Omega^k(M^n)$ induced by a smooth map $f\colon M^n \to N^m$ is a chain map, that is, $df^*\omega = f^*\,d\omega$. Indeed, at vector fields $\xi_0, \ldots, \xi_k$, both forms take the same value

$$\sum (-1)^i \xi_i \omega(f_*\xi_0, \ldots, \hat{\xi}_i, \ldots, f_*\xi_k) + \sum_{i<j} (-1)^{i+j} \omega(f_*[\xi_i, \xi_j], \ldots, \hat{\xi}_i, \ldots, \hat{\xi}_j, \ldots, f_*\xi_k),$$

because $f_*[\xi_i, \xi_j] = [f_*\xi_i, f_*\xi_j]$.

Theorem 5.9. *If $M^n = U_1 \cup U_2$, where U_1 and U_2 are open sets, then the Mayer–Vietoris exact sequence*

$$\to H^k_{\mathrm{DR}}(M^n) \longrightarrow H^k_{\mathrm{DR}}(U_1) \oplus H^k_{\mathrm{DR}}(U_2) \longrightarrow H^k_{\mathrm{DR}}(U_1 \cap U_2) \longrightarrow H^{k+1}_{\mathrm{DR}}(M^n) \to$$

holds.

Proof. It is sufficient to construct an exact sequence of groups of cochains

$$0 \longrightarrow \Omega^k(M^n) \xrightarrow{i^*} \Omega^k(U_1) \oplus \Omega^k(U_2) \xrightarrow{j^*} \Omega^k(U_1 \cap U_2) \longrightarrow 0.$$

Let $\omega \in \Omega^k(M^n)$. We set $i^*(\omega) = (i_1^*\omega, i_2^*\omega)$, where $i_1^*\omega$ and $i_2^*\omega$ are the restrictions of ω to U_1 and U_2, respectively. Take $\omega_1 \in \Omega^k(U_1)$ and $\omega_2 \in \Omega^k(U_2)$. We set $j^*(\omega_1, \omega_2) = j_1^*\omega_1 - j_2^*\omega_2$, where $j_i^*\omega_i$ is the restriction of ω_i to U_i.

The injectivity of the homomorphism i^* is obvious because any form on M^n is uniquely determined by its restrictions to U_1 and U_2.

Let us show that j^* is an epimorphism. Let $\{\lambda_1, \lambda_2\}$ be a smooth partition of unity subordinate to the cover $\{U_1, U_2\}$. Then the form $\omega \in \Omega^k(U_1 \cap U_2)$ can be represented as $i_1^*\lambda_2\omega + i_2^*\lambda_1\omega$; here $\lambda_2\omega$ is a well-defined form on U_1 because $\lambda_2 = 0$ on $U_1 \setminus (U_1 \cap U_2)$.

Finally, let us verify that $\operatorname{Ker} j^* = \operatorname{Im} i^*$. The equality $j^*i^* = 0$ implies $\operatorname{Ker} j^* \supset \operatorname{Im} i^*$. Suppose that $(\omega_1, \omega_2) \in \operatorname{Ker} j^*$, i.e., $j_1^*\omega_1 = j_2^*\omega_2$. This means that $\omega_1(x) = \omega_2(x)$ for all $x \in U_1 \cap U_2$. Hence there exists a form ω on M^n such that it is equal to ω_1 on U_1 and to ω_2 on U_2. Clearly, $i^*\omega = (\omega_1, \omega_2)$. □

For a noncompact manifold M^n, it is often useful to consider, in addition to the algebra $\Omega^*(M^n)$ of differential forms, the algebra $\Omega^*_c(M^n)$ of compactly supported differential forms (that is, forms identically vanishing outside some compact sets). The definition of the differential d for compactly supported forms is the same as for general forms, and $d \circ d = 0$. Therefore, for compactly supported forms, cohomology can be constructed; it is denoted by $H_c^*(M^n)$. If the manifold M^n is compact, then its cohomology H_c^* coincides with H_{DR}^*. But if M^n is noncompact, then this cohomology is different.

Example 72. If M^n is a connected noncompact manifold, then $H_c^0(M^n) = 0$.

Proof. The kernel of the homomorphism $d\colon \Omega_c^0(M^n) \to \Omega_c^1(M^n)$ consists of compactly supported constant functions. But since M^n is noncompact (and connected), there exists only one such function (identical zero). □

Example 73. The isomorphism $H_c^1(\mathbb{R}) \cong \mathbb{R}$ holds.

Proof. A form $\psi(x)\,dx$ (with compact support) belongs to the image of the homomorphism $d\colon \Omega_c^0(\mathbb{R}) \to \Omega_c^1(\mathbb{R})$ if and only if the function $\varphi(x) = \int_{-\infty}^{x} \psi(t)\,dt$ has compact support, i.e., $\int_{-\infty}^{\infty} \psi(t)\,dt = 0$.

Consider the map $\Omega_c^1(\mathbb{R}) \to \mathbb{R}$ defined by $\psi(x)\,dx \mapsto \int_{-\infty}^{\infty} \psi(t)\,dt$, which clearly is an epimorphism. Its kernel coincides with $\operatorname{Im} d$. Therefore, $\Omega_c^1(\mathbb{R})/\operatorname{Im} d \cong \mathbb{R}$. □

There is a fundamental difference between cohomology with compact supports and the ordinary de Rham cohomology; it manifests itself even in the case of compact manifolds. Namely, in the Mayer–Vietoris exact sequence for compactly supported cohomology, the maps act in the reverse direction, that is, they behave as in the case of homology rather than cohomology. This is so because the operation of restriction to an open set is not defined for compactly supported forms. Indeed, if the compact support of a form contains an open set, then the restriction of this form to this set is not compactly supported. In return, for compactly supported forms, the operation of extension to a set containing a given open set U is defined. Since any point outside U has a neighborhood disjoint from the compact support $K \subset U$, we can extend any form with support contained in U by setting it to zero outside U.

Theorem 5.10. *If $M^n = U_1 \cup U_2$, where U_1 and U_2 are open sets, then the following Mayer–Vietoris exact sequence holds:*

$$\cdots \longrightarrow H_c^k(M^n) \longrightarrow H_c^k(U_1 \cap U_2) \\ \longrightarrow H_c^k(U_1) \oplus H^k(U_2) \longrightarrow H_c^{k+1}(M^n) \longrightarrow \cdots .$$

Proof. Consider the exact sequence

$$0 \longrightarrow \Omega_c^k(U_1 \cap U_2) \xrightarrow{j_c^*} \Omega_c^k(U_1) \oplus \Omega_c^k(U_2) \xrightarrow{i_c^*} \Omega_c^k(M^n) \longrightarrow 0$$

defined as follows. Let $\omega \in \Omega_c^k(U_1 \cap U_2)$. We set $j^*\omega = (j_1^*\omega, -j_2^*\omega)$, where $j_p^*\omega$ is an extension of ω to U_p. For two forms $\omega_1 \in \Omega_c^k(U_1)$ and $\omega_2 \in \Omega_c^k(U_2)$, we set $i^*(\omega_1, \omega_2) = i_1^*\omega_1 + i_2^*\omega_2$, where $i_p^*\omega_p$ is an extension of ω_p from U_p to the entire manifold.

The injectivity of j_c^* and exactness at the middle term are obvious. Let us verify the surjectivity of i_c^*. Take $\omega \in \Omega_c^k(M^n)$ and let $\{\lambda_1, \lambda_2\}$ be a smooth partition of unity subordinate to the cover $\{U_1, U_2\}$. The support of the form $\lambda_p\omega$ is the intersection of the compact set $\operatorname{supp}\omega$ and the closed set $\operatorname{supp}\lambda_p$; therefore, it is compact (any closed subset of a compact space is compact). Hence $\lambda_p\omega \in \Omega_c(U_p)$. Clearly, $i_c^*(\lambda_1\omega, \lambda_2\omega) = \omega$. □

2.1. The Stokes Theorem. Homotopy Invariance.

Integration of Forms. Let us choose coordinates in $\mathbb{R}^n$ and write a form $\omega \in \Omega^n(\mathbb{R}^n)$ as $\omega = \varphi(x_1, \dots, x_n)\, dx_1 \wedge \dots \wedge dx_n$. We shall use the abbreviated notation $\omega = \varphi(x)\, dx$. Passing to another coordinate system $y(x)$, we obtain $dy_1 \wedge \dots \wedge dy_n = \det\left(\frac{\partial y_i}{\partial x_j}\right) dx_1 \wedge \dots \wedge dx_n$, or, in the abbreviated notation, $dy = \det\left(\frac{\partial y_i}{\partial x_j}\right) dx$. In the new coordinates, we have $\omega = \psi(y)\, dy$. Let us determine the relation between the functions $\psi(y)$ and $\varphi(x)$. Clearly, $\varphi(x)\, dx = \omega = \psi(y)\, dy = \psi(y) \det\left(\frac{\partial y_i}{\partial x_j}\right) dx$; therefore, $\varphi(x) = \psi(y(x)) \det\left(\frac{\partial y_i}{\partial x_j}\right)$.

The formula for a change of variables in an integral implies

$$\int_{\mathbb{R}^n} \psi(y)\, dy = \int_{\mathbb{R}^n} \psi(y(x)) \left|\det\left(\frac{\partial y_i}{\partial x_j}\right)\right| dx = \pm \int_{\mathbb{R}^n} \varphi(x)\, dx,$$

where the sign $\pm$ coincides with that of the determinant $\det\left(\frac{\partial y_i}{\partial x_j}\right)$. Thus, we can define the integral of a form as follows. Suppose that $\mathbb{R}^n$ is oriented. Let $\omega \in \Omega_c^n(\mathbb{R}^n)$ be a compactly supported form. We set $\int_{\mathbb{R}^n} \omega = \int_{\mathbb{R}^n} \varphi(x)\, dx$, where x is a coordinate system whose orientation is compatible with that of $\mathbb{R}^n$.

We define the *integral* of a form $\omega \in \Omega_c^n(M^n)$, where M^n is an oriented manifold, as follows. Take an arbitrary locally finite atlas $\{U_\alpha\}$ with orientation-preserving charts $f_\alpha \colon U_\alpha \to \mathbb{R}^n$. Let $\{\lambda_\alpha\}$ be a smooth partition of unity subordinate to it. We set

$$\int_{M^n} \omega = \sum_\alpha \int_{\mathbb{R}^n} \lambda_\alpha (f_\alpha^{-1})^* \omega = \sum_\alpha \int_{U_\alpha} \lambda_\alpha \omega.$$

The support of the form $\lambda_\alpha\omega$ is compact because it is closed in the compact set $\operatorname{supp}\lambda_\alpha$. If $\{V_\beta\}$ is another orientation atlas and $\{\mu_\beta\}$ is a partition of unity subordinate to it, then the equality $\sum_\beta \mu_\beta = 1$ implies $\sum_\alpha \int_{U_\alpha} \lambda_\alpha \omega = \sum_{\alpha,\beta} \int_{U_\alpha} \lambda_\alpha \mu_\beta \omega$. Since the support of $\lambda_\alpha \mu_\beta \omega$ is contained in $U_\alpha \cap V_\beta$, we have $\int_{U_\alpha} \lambda_\alpha \mu_\beta \omega = \int_{V_\beta} \lambda_\alpha \mu_\beta \omega$. Hence $\sum_\alpha \int_{U_\alpha} \lambda_\alpha \omega = \sum_{\alpha,\beta} \int_{V_\beta} \lambda_\alpha \mu_\beta \omega = \sum_\beta \int_{V_\beta} \mu_\beta \omega$ because $\sum_\alpha \lambda_\alpha = 1$. This means that the integral of the form ω over the manifold M^n is well defined.

The Stokes Theorem. The *Stokes theorem* says that $\int_{M^n} d\omega = \pm \int_{\partial M^n} \omega$ for any form $\omega \in \Omega_c^n(M^n)$. Formally, in the second integral we should write $i^*\omega$ instead of ω, where $i\colon \partial M^n \to M^n$ is the natural embedding; in other words, $i^*\omega$ is the restriction of ω to ∂M^n. The sign $\pm$ before the second integral can be removed by imposing the requirement that the orientation ∂M^n must be induced by that of M^n.

Let us prove the Stokes theorem.

Case 1. $M^n = \mathbb{R}^n$.

Let $\omega = \varphi\, dx_1 \wedge \cdots \wedge dx_{n-1} \in \Omega_c^{n-1}(\mathbb{R}^n)$. Then $d\omega = \frac{\partial \varphi}{\partial x_n} dx_n \wedge dx_1 \wedge \cdots \wedge dx_{n-1} = (-1)^{n-1} \frac{\partial \varphi}{\partial x_n} dx_1 \wedge \cdots \wedge dx_n$. Therefore,

$$\int_{\mathbb{R}^n} d\omega = (-1)^{n-1} \int \left(\int_{-\infty}^{\infty} \frac{\partial \varphi}{\partial x_n} dx_n \right) dx_1 \cdots dx_{n-1}.$$

But $\int_{-\infty}^{\infty} \frac{\partial \varphi}{\partial x_n} dx_n = \varphi(x_1, \ldots, x_{n-1}, \infty) - \varphi(x_1, \ldots, x_{n-1}, -\infty) = 0$ because the function φ is compactly supported.

Case 2. $M^n = \mathbb{R}_+^n = \{(x_1, \ldots, x_n) \in \mathbb{R}^n \mid x_n \geq 0\}$.

Let $\omega = \sum_i \varphi_i\, dx_1 \wedge \cdots \wedge \widehat{dx_i} \wedge \cdots \wedge dx_n \in \Omega_c^{n-1}(\mathbb{R}^n)$. Then $d\omega = \sum_i (-1)^{n-1} \frac{\partial \varphi_i}{\partial x_n} dx_1 \wedge \cdots \wedge dx_n$. Therefore,

$$\int_{\mathbb{R}_+^n} d\omega = \sum_{1 \leq i < n} (-1)^{i-1} \left(\int_{-\infty}^{\infty} \frac{\partial \varphi_i}{\partial x_i} dx_i \right) dx_1 \cdots \widehat{dx_i} \cdots dx_n$$
$$+ (-1)^{n-1} \left(\int_0^{\infty} \frac{\partial \varphi_n}{\partial x_n} dx_n \right) dx_1 \cdots dx_{n-1}.$$

If $i \neq n$, then $\int_{-\infty}^{\infty} \frac{\partial \varphi_i}{\partial x_i} dx_i = 0$, as in Case 1. For $i = n$, we obtain the integral

$$\int_0^{\infty} \frac{\partial \varphi_n}{\partial x_n} dx_n = \varphi_n(x_1, \ldots, x_{n-1}, \infty) - \varphi_n(x_1, \ldots, x_{n-1}, 0)$$
$$= -\varphi_n(x_1, \ldots, x_{n-1}, 0).$$

Hence

$$\int_{\mathbb{R}_+^n} d\omega = (-1)^n \int_{\mathbb{R}^{n-1}} \varphi_n(x_1, \ldots, x_{n-1}, 0)\, dx_1 \cdots dx_{n-1} = (-1)^n \int_{\partial \mathbb{R}_+^n} \omega;$$

the last equality holds because the form dx_n vanishes on $\partial\mathbb{R}^n_+$, and therefore the restriction of ω to $\partial\mathbb{R}^n_+$ contains only one term, $\varphi_n\, dx_1 \wedge \dots \wedge dx_{n-1}$.

There is no sign $(-1)^n$ if the orientation of $\partial\mathbb{R}^n_+$ is determined by the form $(-1)^n\, dx_1 \wedge \dots \wedge dx_{n-1}$. This condition can be stated differently as follows. Suppose that a basis $e_1, \dots, e_{n-1}$ determines the positive orientation of the boundary and ε is the outward normal to the boundary. Then the basis $\varepsilon, e_1, \dots, e_{n-1}$ determines the positive orientation of the manifold. Indeed, we have $\varepsilon = -e_n$, and hence the basis $\varepsilon, e_1, \dots, e_{n-1}$ has the same orientation as $(-1)^n e_1, \dots, e_{n-1}, e_n$.

In what follows, we assume that the orientation of $\partial\mathbb{R}^n_+$ satisfies this condition and omit the sign from the Stokes formula. Note that on p. 96, we were led to the same condition on the orientation of the boundary by different considerations.

Case 3. M^n is an arbitrary oriented manifold.

The passage from $\mathbb{R}^n$ and $\mathbb{R}^n_+$ to an arbitrary manifold is performed by means of partitions of unity. Let $\{U_\alpha\}$ be an orientation atlas for M^n in which all sets U_α are diffeomorphic to $\mathbb{R}^n$ or $\mathbb{R}^n_+$, and let $\{\lambda_\alpha\}$ be a smooth partition of unity subordinate to it. We represent $\omega \in \Omega^{n-1}_c(M^n)$ as $\omega = \sum \lambda_\alpha \omega$. It is sufficient to prove the Stokes theorem for each form $\lambda_\alpha\omega$ separately. The support of such a form is a closed subset of the compact set $\operatorname{supp}\lambda_\alpha \subset U_\alpha$; therefore, it is compact.

Homotopy Invariance of the de Rham Cohomology.

Theorem 5.11. *Let $i_0, i_1\colon \mathbb{R}^n \to \mathbb{R}^n \times \mathbb{R} = \mathbb{R}^{n+1}$ be the embeddings defined by $x \mapsto (x, 0)$ and $x \mapsto (x, 1)$. Then, for any $k \geq 0$, there exists a linear map $D\colon \Omega^k(\mathbb{R}^{n+1}) \to \Omega^{k-1}(\mathbb{R}^n)$ such that $dD + Dd = i_1^* - i_0^*$.*

Proof. For $\varphi \in \Omega^0(\mathbb{R}^{n+1}) = C^\infty(\mathbb{R}^{n+1})$, we set $D\varphi = 0$. The basis forms in $\Omega^k(\mathbb{R}^{n+1})$, where $k \geq 1$, are of two types. Namely, let t be a coordinate on the line $\mathbb{R}$, and let $x_1, \dots, x_n$ be coordinates on $\mathbb{R}^n$. We denote the forms not containing dt by $\alpha = a\, dx_{i_1} \wedge \dots \wedge dx_{i_k}$ and the forms containing dt by $\beta = b\, dt \wedge dx_{j_1} \wedge \dots \wedge dx_{j_{k-1}}$. Any form is a sum of such forms; so it is sufficient to prove the required assertion only for forms of types α and β.

We set $D\alpha = 0$ and $D\beta = \left(\int_0^1 b\, dt\right) dx_{j_1} \wedge \dots \wedge dx_{j_{k-1}}$.

Let us prove the equality $dD + Dd = i_1^* - i_0^*$ for (i) the function φ, (ii) the form α, and (iii) the form β.

(i) $dD\varphi = 0$ and $Dd\varphi = \int_0^1 \frac{\partial\varphi}{\partial t}\, dt = \varphi(1) - \varphi(0) = (i_1^* - i_0^*)\varphi$.

(ii) $dD\alpha = 0$ and $Dd\alpha = \left(\int_0^1 \frac{\partial a}{\partial t}\right) dx_{i_1} \wedge \dots \wedge dx_{i_k} = (i_1^* - i_0^*)\alpha$.

(iii) Note that $i_0^*\beta = i_1^*\beta = 0$ because the restrictions of dt to $i_0(\mathbb{R}^n)$ and to $i_1(\mathbb{R}^n)$ vanish identically (the t-coordinate of any tangent vector is zero). We have

$$dD\beta = \sum_i \left(\int_0^1 \frac{\partial b}{\partial x_i} \, dt \right) dx_i \wedge dx_{j_1} \wedge \cdots \wedge dx_{j_{k-1}}$$

and

$$\begin{aligned} Dd\beta &= D\left(\sum_i \frac{\partial b}{\partial x_i} \, dx_i \wedge dt \wedge dx_{j_1} \wedge \cdots \wedge dx_{j_{k-1}} \right) \\ &= -\sum_i \left(\int_0^1 \frac{\partial b}{\partial x_i} \, dt \right) dx_i \wedge dx_{j_1} \wedge \cdots \wedge dx_{j_{k-1}}. \end{aligned}$$

Therefore $dD\beta + Dd\beta = 0$. □

Corollary (the Poincaré lemma). *In $\mathbb{R}^n$, any closed differential k-form (with $k > 0$) is exact; i.e., $H^k_{\mathrm{DR}}(\mathbb{R}^n) = 0$ for $k > 0$.*

Proof. Consider the map $H\colon \mathbb{R}^n \times \mathbb{R} \to \mathbb{R}^n$ defined by $(x,t) \mapsto tx$. We have $Hi_1 = \mathrm{id}$ and $Hi_0 = *$ ($*$ denotes the constant map to 0). Suppose that $\omega \in \Omega^k(\mathbb{R}^n)$ and $d\omega = 0$. Then $\omega = (i_1^* - i_0^*)H^*\omega = dD(H^*\omega) + Dd(H^*\omega)$. But $D\,d(H^*\omega) = DH^*\,d\omega$; therefore, $\omega = dD(H^*\omega)$, i.e., the form ω is exact. □

Theorem 5.11 remains valid for maps $i_0, i_1\colon M^n \times \mathbb{R} \to M^n$; no substantial changes in the proof are required. This implies that homotopic maps induce the same map of the de Rham cohomology.

Example 74. $H^k_c(\mathbb{R}^n) \cong H^k_{\mathrm{DR}}(S^n) \cong \begin{cases} \mathbb{R} & \text{if } k = n, \\ 0 & \text{if } k \neq n. \end{cases}$

Proof. Using the diffeomorphism $\mathbb{R}^n \approx S^n \setminus \{x_0\}$, we can construct an exact sequence

$$0 \to \Omega^k_c(\mathbb{R}^n) \to \Omega^k(S^n) \to \Omega^k(S^n)/\Omega^k_c(\mathbb{R}^n) \to 0.$$

Here $\Omega^k(S^n)/\Omega^k_c(\mathbb{R}^n)$ is the space of germs of k-forms at the point x_0. The cohomology of this space is easy to calculate by using the Poincaré lemma. Indeed, any sufficiently small neighborhood of x_0 can be assumed to be contained in $\mathbb{R}^n$; therefore, for a k-form with $k \geq 0$, closedness in a small neighborhood of x_0 implies exactness.

Using a short exact sequence of forms, we can construct the exact sequence of cohomology. As a result, we obtain $H^k_c(\mathbb{R}^n) \cong H^k_{\mathrm{DR}}(S^n)$ for

$k \geq 2$. Moreover, we obtain an exact sequence

$$
\begin{array}{ccccccccc}
0 & \longrightarrow & H_c^0(\mathbb{R}^n) & \longrightarrow & H^0_{\mathrm{DR}}(S^n) & \longrightarrow & \mathbb{R} & & & & \\
 & & \| & & \| & & \downarrow & & & & \\
 & & 0 & & \mathbb{R} & & H_c^1(\mathbb{R}^n) & \longrightarrow & H^1_{\mathrm{DR}}(S^n) & \longrightarrow & 0.
\end{array}
$$

Therefore, $H_c^1(\mathbb{R}^n) \cong H^1_{\mathrm{DR}}(S^n)$. In particular, $H^1_{\mathrm{DR}}(S^1) \cong \mathbb{R}$.

It remains to calculate the de Rham cohomology of the sphere S^n for $n \geq 2$. Let us represent S^n as the union of open sets $U_1 = S^n \setminus \{x_1\}$ and $U_2 = S^n \setminus \{x_2\}$ diffeomorphic to $\mathbb{R}^n$. We have $U_1 \cap U_2 \approx S^{n-1} \times \mathbb{R}$; hence $H^k_{\mathrm{DR}}(U_1 \cap U_2) \cong H^k_{\mathrm{DR}}(S^{n-1})$. Therefore, for $k > 0$, the Mayer–Vietoris sequence implies $H^k_{DR}(S^{n-1}) \cong H^{k+1}_{\mathrm{DR}}(S^n)$. For $k = 0$ and $n \geq 2$, we obtain an exact sequence

$$
\begin{array}{ccccccc}
H^0_{\mathrm{DR}}(S^n) & \longrightarrow & H^0_{\mathrm{DR}}(\mathbb{R}^n) \oplus H^0_{\mathrm{DR}}(\mathbb{R}^n) & \longrightarrow & H^0_{\mathrm{DR}}(S^{n-1}) & \longrightarrow & H^1_{\mathrm{DR}}(S^n) \\
\| & & \| & & \| & & \downarrow \\
\mathbb{R} & & \mathbb{R} \oplus \mathbb{R} & & \mathbb{R} & & 0.
\end{array}
$$

Exactness at the term $\mathbb{R} \oplus \mathbb{R}$ implies that Ker $\cong \mathbb{R}$ and Im $\cong \mathbb{R}$ in this term; hence the kernel of the map $H^0_{\mathrm{DR}}(S^{n-1}) \to H^1_{\mathrm{DR}}(S^n)$ is $\mathbb{R}$. Therefore, $H^1_{\mathrm{DR}}(S^n) = 0$. □

Any n-form $\omega \in \Omega^n(S^n)$ is closed. If this form is exact, then it belongs to the kernel of the map $\Omega^n(S^n) \to \mathbb{R}$ defined by $\omega \mapsto \int_{S^n} \omega$. Indeed, $\int_{S^n} d\alpha = 0$. The equality $H^n_{\mathrm{DR}}(S^n) = \mathbb{R}$ shows that the space of exact n-forms coincides with the kernel of this map. Therefore, the cohomology class of any n-form ω for which $\int_{S^n} \omega \neq 0$ is a nonzero element of the group $H^n_{\mathrm{DR}}(S^n) = \mathbb{R}$.

To obtain a nonzero cohomology class in $H^n_c(\mathbb{R}^n)$, we take the form $\varphi(x_1, \dots, x_n)\, dx_1 \wedge \dots \wedge dx_n$, where φ is a compactly supported function such that $\int_{\mathbb{R}^n} \varphi(x_1, \dots, x_n)\, dx_1 \dots dx_n \neq 0$.

2.2. The Poincaré Isomorphism for de Rham Cohomology. Suppose that M^n is a compact oriented manifold, $\alpha^k \in \Omega^k(M^n)$, and $\beta^{n-k} \in \Omega^{n-k}(M^n)$. Then $\alpha^k \wedge \beta^{n-k} \in \Omega^n(M^n)$; thus, we can consider the integral $\int_{M^n} \alpha^k \wedge \beta^{n-k}$. For a noncompact manifold m^n, this integral may be divergent. To prevent this, we assume that $\beta^{n-k} \in \Omega_c^{n-k}(M^n)$.

If M^n is a manifold without boundary, then the map

$$\Omega^k(M^n) \times \Omega_c^{n-k}(M^n) \to \mathbb{R}$$

thus constructed carries over to the de Rham cohomology. Indeed, suppose that $d\alpha^k = 0$ and $d\beta^{n-k} = 0$. Then $d(\omega^{k-1} \wedge \beta^{n-k}) = d\omega^{k-1} \wedge \beta^{n-k}$ and

$d(\alpha^k \wedge \omega^{n-k-1}) = \pm\alpha^k \wedge d\omega^{n-k-1}$; therefore, the forms $d\omega^{k-1} \wedge \beta^{n-k}$ and $\alpha^k \wedge d\omega^{n-k-1}$ are exact. But the integral of any exact (compactly supported) form over a manifold without boundary vanishes. This means that the map carries over to the coset space of the set of closed forms modulo the exact forms. As a result, we obtain a bilinear map $H^k_{\mathrm{DR}}(M^n) \times H^{n-k}_c(M^n) \to \mathbb{R}$, i.e., a linear map $D\colon H^k_{\mathrm{DR}}(M^n) \to (H^{n-k}_c(M^n))^*$.

Theorem 5.12 (the Poincaré isomorphism). *For any oriented manifold M^n without boundary, the map*

$$D\colon H^k_{\mathrm{DR}}(M^n) \to (H^{n-k}_c(M^n))^*$$

is an isomorphism.

Proof (see [**42**]). First, let us show that the required assertion is true for $M^n = \mathbb{R}^n$. We have already calculated the groups $H^k(\mathbb{R}^n)$ and $H^k_c(\mathbb{R}^n)$ (see the Poincaré lemma and Example 74). The only nontrivial groups among these are $H^0(\mathbb{R}^n)$ and $H^n_c(\mathbb{R}^n)$, and both of them are isomorphic to $\mathbb{R}$. Therefore, it suffices to verify that the map

$$D\colon H^0_{\mathrm{DR}}(M^n) \to (H^n_c(M^n))^*$$

is nonzero.

Let $1 \in H^0(\mathbb{R}^n)$ be the cohomology class of the function identically equal to 1 on $\mathbb{R}^n$, and let $\varphi(x_1, \dots, x_n)\, dx_1 \wedge \dots \wedge dx_n \in \Omega^n_c(\mathbb{R}^n)$ be a form whose cohomology class is nonzero. Then

$$\langle D(1), \varphi(x_1, \dots, x_n)\rangle = \int_{\mathbb{R}^n} \varphi(x_1, \dots, x_n)\, dx_1 \wedge \dots \wedge dx_n \neq 0,$$

and therefore $D(1) \neq 0$.

We have proved the required assertion for $M^n = \mathbb{R}^n$, and thereby for all open subsets of any manifold M^n diffeomorphic to $\mathbb{R}^n$. To transfer it to open subsets of M^n (and, thereby, to the entire manifold M^n), we apply the standard technique based on maps of Mayer–Vietoris sequences.

Consider the Mayer–Vietoris sequence for compactly supported cohomology:

$$\begin{aligned} \dots \longleftarrow H^{n-k}_c(M^n) \longleftarrow H^{n-k}_c(U_1) \oplus H^{n-k}_c(U_2) \\ \longleftarrow H^{n-k}_c(U_1 \cap U_2) \longleftarrow H^{n-k-1}_c(M^n) \longleftarrow \cdots . \end{aligned}$$

Let us dualize it, i.e., take the sequence of dual spaces and dual maps. The Mayer–Vietoris sequence for the de Rham cohomology can be mapped to the dualized sequence by means of D:

$$\begin{array}{ccc} H^k(M^n) & \longrightarrow & H^k(U_1)\oplus H^k(U_2) \\ \downarrow D_{M^n} & & \downarrow D_{U_1}\oplus D_{U_1} \\ H_c^{n-k}(M^n)^* & \to & H_c^{n-k}(U_1)^*\oplus H_c^{n-k}(U_2)^* \end{array}$$

$$\begin{array}{ccccc} \longrightarrow & H^k(U_1\cap U_2) & \longrightarrow & H^{k+1}(M^n) \\ & \downarrow D_{U_1\cap U_2} & & \downarrow D_{M^n} \\ \longrightarrow & H_c^{n-k}(U_1\cap U_2)^* & \longrightarrow & H_c^{n-k-1}(M^n)^*. \end{array}$$

Let us show that this diagram is commutative up to sign. The commutativity of the left square and of the central square, in which only the restriction and extension operations are used, is obvious. Consider the right square.

The upper map in the right square is as follows. Suppose that $\alpha^k\in\Omega^k(U_1\cap U_2)$ and $d\alpha^k=0$. Let us represent the form α^k as $\alpha^k=\alpha_1^k-\alpha_2^k$, where $\alpha_i^k\in\Omega^k(U_i)$. The upper map takes α^k to $\alpha^{k+1}\in\Omega^{k+1}(M^n)$, which coincides with $d\alpha_i^k$ on U_i. The map $H_c^{n-k-1}(M^n)\to H_c^{n-k}(U_1\cap U_2)$ dual to the lower map in the square is as follows. Suppose that $\beta^{n-k-1}\in\Omega_c^{n-k-1}(M^n)$ and $d\beta^{n-k-1}=0$. Let us represent the form β^{n-k-1} as $\beta^{n-k-1}=\beta_1^{n-k-1}+\beta_2^{n-k-1}$, where $\beta_i^{n-k-1}\in\Omega_c(U_i)$. The form β^{n-k-1} is taken to the form $\beta^{n-k}\in\Omega_c^{n-k}(U_1\cap U_2)$, which coincides with $d\beta_1^{n-k-1}=-d\beta_2^{n-k-1}$ (on $U_1\cap U_2$). We must prove that $\int_{U_1\cap U_2}\alpha^k\wedge\beta^{n-k}=\pm\int_{M^n}\alpha^{k+1}\wedge\beta^{n-k-1}$. Clearly,

$$\begin{aligned}\int_{M^n}\alpha^{k+1}\wedge\beta^{n-k-1}&=\int_{U_1}\alpha^{k+1}\wedge\beta_1^{n-k-1}+\int_{U_2}\alpha^{k+1}\wedge\beta_2^{n-k-1}\\&=\int_{U_1}d\alpha_1^k\wedge\beta_1^{n-k-1}+\int_{U_2}d\alpha_2^k\wedge\beta_2^{n-k-1}.\end{aligned}$$

The form $\alpha_i^k\wedge\beta_i^{n-k-1}$ has compact support in U_i; consequently, the integral $\int_{U_i}d(\alpha_i^k\wedge\beta_i^{n-k-1})$ vanishes. Therefore,

$$(-1)^{k+1}\int_{M^n}\alpha^{k+1}\wedge\beta^{n-k-1}=\int_{U_1}\alpha_1^k\wedge d\beta_1^{n-k-1}+\int_{U_2}\alpha_2^k\wedge d\beta_2^{n-k-1}.$$

The support of the form $d\beta_1^{n-k-1}=-d\beta_2^{n-k-1}$ is contained in the intersection of the supports of the forms β_1^{n-k-1} and β_2^{n-k-1}; hence it is contained in $U_1 \cap U_2$. Thus, this form can be regarded as an extension of β^{n-k}, and

$$\begin{aligned}(-1)^{k+1} \int_{M^n} \alpha^{k+1} \wedge \beta^{n-k-1} &= \int_{U_1 \cap U_2} \alpha_1^k \wedge \beta^{n-k} - \int_{U_1 \cap U_2} \alpha_2^k \wedge \beta^{n-k} \\ &= \int_{U_1 \cap U_2} (\alpha_1^k - \alpha_2^k) \wedge \beta^{n-k} \\ &= \int_{U_1 \cap U_2} \alpha^k \wedge \beta^{n-k},\end{aligned}$$

as required.

According to the five lemma, if the Poincaré isomorphism holds for the manifolds U_1, U_2, and $U_1 \cap U_2$, then it holds also for the manifold $U_1 \cup U_2 = M^n$.

The next (and most important) step is proving the Poincaré isomorphism for any open set $U \subset \mathbb{R}^n$. Let $\mathcal{U}$ be the base for the topology of $\mathbb{R}^n$ which consists of all open rectangles given by $a_i < x_i < b_i$ for $i = 1, \dots, n$. This base has two properties important for our purposes; namely, (i) the Poincaré isomorphism theorem is valid for any set from $\mathcal{U}$, and (ii) the intersection of any two sets from $\mathcal{U}$ belongs to $\mathcal{U}$.

Lemma 5.1. *Let $\mathcal{U}$ be a base for the topology of M^n with properties* (i) *and* (ii). *Then, for any finite union of sets from $\mathcal{U}$, the Poincaré isomorphism theorem is valid.*

Proof. Take $U_1, \dots, U_k \in \mathcal{U}$. We shall prove that the Poincaré isomorphism holds for $U_1 \cup \dots \cup U_k$ by induction on k. For $k = 1$, this follows from (i). For $k > 1$, we represent the union under consideration in the form $U' \cup U_k$, where $U' = U_1 \cup \dots \cup U_{k-1}$. The Poincaré isomorphism theorem is valid for U' and U_k by the induction hypothesis. It remains to show that it is valid for $U' \cap U_k = (U_1 \cap U_k) \cup \dots \cup (U_{k-1} \cap U_k)$. According to (ii), the theorem is true for each set $U_i \cap U_k$; therefore, it is also true for $U' \cap U_k$. □

Lemma 5.2. *Let $\mathcal{U}$ be a base for the topology of M^n with property* (i). *Then, for any* (*not necessarily finite*!) disjoint *union of elements of $\mathcal{U}$, the Poincaré isomorphism theorem is valid.*

Proof. Suppose that $\{U_\alpha\}_{\alpha \in A} \subset \mathcal{U}$ and $U_\alpha \cap U_\beta = \varnothing$ for any $\alpha, \beta \in A$. The natural embeddings $\Omega^k(\bigcup_{\alpha \in A} U_\alpha) \to \prod_{\alpha \in A} \Omega^k(U_\alpha)$ and $\bigoplus_{\alpha \in A} \Omega_c^{n-k}(U_\alpha) \to \Omega_c^{n-k}(\bigcup_{\alpha \in A} U_\alpha)$ are isomorphisms. The second map acts on the direct sum rather than product because a compactly supported form can differ from zero only on finitely many sets U_α. Dualizing the second map, we obtain an isomorphism $\Omega_c^{n-k}(\bigcup_{\alpha \in A} U_\alpha)^* \to \prod_{\alpha \in A} \Omega_c^{n-k}(U_\alpha)^*$ (this time, to the direct product). The situation is the same as for chains and cochains: each chain contains only finitely many simplices, but a cochain may assign nonzero values to arbitrarily many simplices.

Passing from forms to cohomology, we obtain isomorphisms of cohomology groups, which form the commutative diagram

$$\begin{array}{ccc} H^k_{\mathrm{DR}}\Big(\bigcup\limits_{\alpha\in A} U_\alpha\Big) & \xrightarrow{\cong} & \prod\limits_{\alpha\in A} H^k_{\mathrm{DR}}(U_\alpha) \\ \Big\downarrow D & & \cong\Big\downarrow \prod_{\alpha\in A} D_\alpha \\ H^{n-k}_c\Big(\bigcup\limits_{\alpha\in A} U_\alpha\Big)^* & \xrightarrow{\cong} & \prod\limits_{\alpha\in A} H^{n-k}_c(U_\alpha). \end{array}$$

Hence D is an isomorphism. □

Let $U \subset \mathbb{R}^n$ be an arbitrary open set. It can be represented as $U = \bigcup_{i=1}^{\infty} V_i$, where $\overline{V}_i \subset U$; for V_i we take open balls of rational radii centered at points with rational coordinates. We set $K_1 = \overline{V}_1$. The boundary of K_1 is a compact subset of U; hence it can be covered by finitely many sets $V_{i_1}, \dots, V_{i_m}$. We put $K_2 = \overline{V}_2 \cup \overline{V}_{i_1} \cup \dots \cup \overline{V}_{i_m}$. In a similar way, we construct K_3, and so on. By construction, we have $\bigcup_{i=1}^{\infty} K_i = U$ and $K_i \subset \operatorname{int} K_{i+1}$. The inclusion $K_i \subset \operatorname{int} K_{i+1}$ implies the existence of a compact set L_i for which $K_i \subset \operatorname{int} L_i \subset L_i \subset \operatorname{int} K_{i+1}$ (see Figure 3).

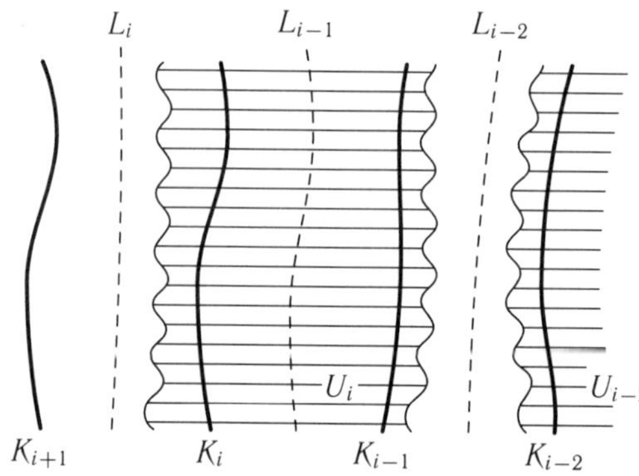

Figure 3. The sets K_i and L_i

We cover $\overline{K_i \setminus K_{i-1}}$ with the base sets from $\mathcal{U}$ that are contained entirely in $\operatorname{int}(L_i \setminus L_{i-1})$. Since $\overline{K_i \setminus K_{i-1}}$ is compact, it follows that this cover has a finite subcover. Let U_i be the union of all sets from this subcover. According to Lemma 5.1, the Poincaré isomorphism theorem is valid for U_i. Moreover, by construction, we have $\bigcup_{i=1}^{\infty} U_i = U$ and $U_i \cap U_j = \varnothing$ for $|i-j| > 1$. We set $W_1 = \bigcup_{i=0}^{\infty} U_{2i+1}$ and $W_2 = \bigcup_{i=1}^{\infty} U_{2i}$. By Lemma 5.2, the Poincaré isomorphism theorem is valid for W_1 and W_2. We have $U = W_1 \cup W_2$. Thus, it remains to prove the theorem for $W_1 \cap W_2$. The set $W_1 \cap W_2$ is the disjoint union of the sets $U_i \cap U_{i+1}$, $i = 1, 2, \dots$. Each of these sets is a finite union of elements of $\mathcal{U}$. By Lemma 5.1, the Poincaré isomorphism theorem is valid for $U_i \cap U_{i+1}$, and by Lemma 5.2, it is valid for $W_1 \cap W_2$.

The proof of the Poincaré isomorphism theorem for an arbitrary manifold M^n follows approximately the same scheme as that for an open set $U \subset \mathbb{R}^n$. The only essential difference is related to the choice of a base $\mathcal{U}$ for the topology of M^n. For $\mathcal{U}$ we take the family of all open sets homeomorphic to $\mathbb{R}^n$ (they cover M^n) and their finite intersections. This base has property (ii) by definition; (i) holds because any set from $\mathcal{U}$ can be considered as an open subset of $\mathbb{R}^n$, and for such sets, the Poincaré isomorphism is already proved. Instead of balls with rational radii and centers, we take a suitable countable base for the topology of M^n. Otherwise the proof is the same. □

3. The de Rham Theorem

The de Rham theorem [**109**] asserts that, for any closed manifold M^n, the de Rham cohomology $H^*_{\mathrm{DR}}(M^n)$ is isomorphic to the simplicial (or singular) cohomology $H^*(M^n; \mathbb{R})$. They are isomorphic as algebras, that is, $\smile$-products of cochains correspond to $\wedge$-products of forms. There exist many different proofs of the de Rham theorem, but none of them is simple. We give a proof which goes back to Whitney [**154**] (see also [**121**]). Other proofs of the de Rham theorem can be found in [**28, 51, 114, 146, 147**]. We also prove an analog of the de Rham theorem for simplicial complexes.

3.1. Proof of the de Rham Theorem. A triangulation $f\colon |K| \to M^n$ is said to be *smooth* if for any n-simplex Δ^n in K, there exists an open set $U \supset \Delta^n$ in $\mathbb{R}^n$ and an extension of the map $f|_{\Delta^n}$ to U such that this extension is a (smooth) embedding of the manifold U into M^n. We assume that M^n is endowed with a fixed smooth triangulation (its existence was proved in Part I). Let $C^k(M^n) = C^k(K; \mathbb{R})$ be the group of simplicial k-cochains. Consider the map $\rho\colon \Omega^k(M^n) \to C^k(M^n)$ defined by $\langle \rho(\omega^k), \Delta^k \rangle = \int_{\Delta^k} \omega^k$. According to the Stokes theorem, we have $\rho\, d\omega = (\delta\rho)\omega$; therefore, ρ induces a homomorphism $\rho^*\colon H^k_{\mathrm{DR}}(M^n) \to H^k(M^n; \mathbb{R})$ of linear spaces.

Step 1. The map ρ^* is an epimorphism.

We shall show that this map is an epimorphism even at the level of cochains; i.e., for each cochain $c^k \in C^k(M^n)$, there is a form $\omega^k \in \Omega^k(M^n)$ such that $\rho(\omega^k) = c^k$. To construct such a form for a given cochain, we need a special partition of unity subordinate to the cover $\{\operatorname{st} v_i\}$, where $v_1, \dots, v_r$ are the vertices of the complex K. Namely, let x_i be the barycentric coordinate corresponding to the vertex v_i (it is assumed that $x_i = 0$ for any point outside the simplices for which v_i is a vertex). Suppose that F_i is the set of all points for which $x_i \geq \frac{1}{n+1}$ and G_i is the set of all points for which $x_i \leq \frac{1}{n+2}$ (see Figure 4). Let λ_i be a nonnegative smooth function which is positive on F_i and vanishes on G_i. The sets F_i cover M^n. Indeed, the sum of the $n+1$ barycentric coordinates of any point $x \in M^n$ equals

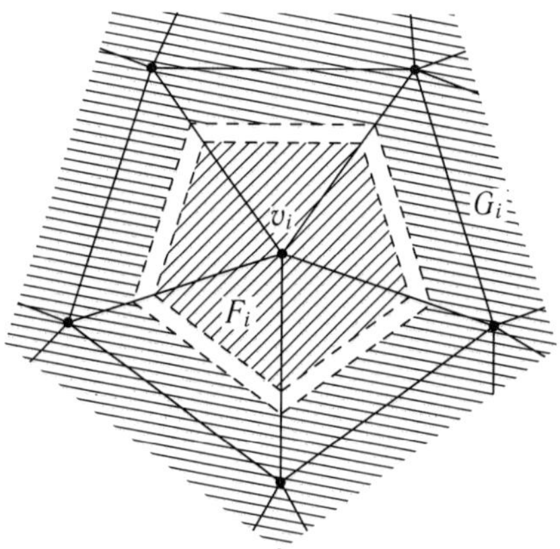

Figure 4. The sets F_i and G_i

1; therefore, one of these coordinates is at least $\frac{1}{n+1}$. Thus, the function $\lambda(x) = \lambda_1(x) + \cdots + \lambda_n(x)$ is positive. Hence the functions $\{\mu_i\}$, where $\mu_i(x) = \lambda_i(x)/\lambda(x)$, form a smooth partition of unity subordinate to the cover $\{M^n \setminus G_i\}$.

Consider the simplex $\Delta_s^k = [v_{i0}, \dots, v_{ik}]$ and the cochain c_s^k dual to Δ_s^k, i.e., such that $\langle c_s^k, \Delta_r^k \rangle = \delta_{sr}$. To this cochain c_s^k we assign the form

$$\varphi_{i0\dots i_k} = k! \sum_{j=0}^{k} (-1)^j \mu_{i_j}\, d\mu_{i0} \wedge \cdots \wedge \widehat{d\mu_{i_j}} \wedge \cdots \wedge d\mu_{i_k}.$$

As a result, we obtain a linear map $\varphi\colon C^k(M^n) \to \Omega^k(M^n)$ (any linear map is determined by its values at the elements of a basis). Below, we prove several properties of this map, which imply that ρ^* is an epimorphism.

Property 1. $d\varphi(c_s^k) = \varphi(\delta c_s^k)$.

Clearly, $d\varphi(c_s^k) = (k+1)!\, d\mu_{i0} \wedge \cdots \wedge d\mu_{i_k}$. Let us calculate $\varphi(\delta c_s^k)$. The cochain δc_s^k is the sum of the cochains dual to simplices of the form $[v_p, v_{i0}, \dots, v_{i_k}]$. Therefore, the cochain $\frac{1}{(k+1)!}\varphi(\delta c_s^k)$ is equal to

$$\sum{}' \left(\mu_p\, d\mu_{i0} \wedge \cdots \wedge d\mu_{i_k} - \sum_{j=0}^{k} (-1)^j \mu_{i_j}\, d\mu_p \wedge d\mu_{i0} \wedge \cdots \wedge \widehat{d\mu_{i_j}} \wedge \cdots \wedge d\mu_{i_k} \right),$$

where $\sum'$ denotes summation over all $(k+1)$-simplices $[v_p, v_{i_0}, \dots, v_{i_k}]$. Note that if $p \notin \{i_0, \dots, i_k\}$ but no simplex is spanned by $v_p, v_{i_0}, \dots, v_{i_k}$, then $\mu_p\, d\mu_{i_0} \wedge \dots \wedge d\mu_{i_k} = 0$. Indeed, if $x \notin \operatorname{st} v_p$, then $\mu_p(x) = 0$. If $x \in \operatorname{st} v_p$, then $x_p \neq 0$. In the latter case, $x_{i_j} = 0$ for some j (otherwise, $x_p \neq 0$, $x_{i_0} \neq 0, \dots, x_{i_k} \neq 0$, and x belongs to the interior of the simplex $[v_p, v_{i_0}, \dots, v_{i_k}]$). Consider the open set U consisting of all points $y \in M^n$ for which $y_{i_j} < \frac{1}{n+2}$. The point x belongs to U, and μ_{i_j} vanishes on U because $U \subset G_{i_j}$. Therefore, $d\mu_{i_j} = 0$. Thus,

$$\sum{}' \mu_p\, d\mu_{i_0} \wedge \dots \wedge d\mu_{i_k} = \sum_{p \notin \{i_0, \dots, i_k\}} \mu_p\, d\mu_{i_0} \wedge \dots \wedge d\mu_{i_k}.$$

Combining this equality with $\sum \mu_i(x) = 1$ (which implies $\sum d\mu_i = 0$), we obtain

$$\begin{aligned}
&\sum{}' \sum_{j=0}^{k} (-1)^j \mu_{i_j}\, d\mu_p \wedge d\mu_{i_0} \wedge \dots \wedge \widehat{d\mu_{i_j}} \wedge \dots \wedge d\mu_{i_k} \\
&\quad = \sum_{j=0}^{k} (-1)^j \sum{}' \mu_{i_j}\, d\mu_p \wedge d\mu_{i_0} \wedge \dots \wedge \widehat{d\mu_{i_j}} \wedge \dots \wedge d\mu_{i_k} \\
&\quad = \sum_{j=0}^{k} (-1)^j \sum_{p \notin \{i_0, \dots, i_k\}} \mu_{i_j}\, d\mu_p \wedge d\mu_{i_0} \wedge \dots \wedge \widehat{d\mu_{i_j}} \wedge \dots \wedge d\mu_{i_k} \\
&\quad = \sum_{j=0}^{k} (-1)^j \sum_{p \neq i_j} \mu_{i_j}\, d\mu_p \wedge d\mu_{i_0} \wedge \dots \wedge \widehat{d\mu_{i_j}} \wedge \dots \wedge d\mu_{i_k} \\
&\quad = \sum_{j=0}^{k} (-1)^j \mu_{i_j} \Big(\sum_{p \neq i_j} d\mu_p \Big) d\mu_{i_0} \wedge \dots \wedge \widehat{d\mu_{i_j}} \wedge \dots \wedge d\mu_{i_k} \\
&\quad = \sum_{j=0}^{k} (-1)^j \mu_{i_j} (-d\mu_{i_j})\, d\mu_{i_0} \wedge \dots \wedge \widehat{d\mu_{i_j}} \wedge \dots \wedge d\mu_{i_k} \\
&\quad = -\sum_{j=0}^{k} \mu_{i_j}\, d\mu_{i_0} \wedge \dots \wedge d\mu_{i_k}
\end{aligned}$$

Thus,

$$\begin{aligned}
\varphi(\delta c_s^k) &= (k+1)! \sum_p \mu_p\, d\mu_{i_0} \wedge \dots \wedge d\mu_{i_k} \\
&= (k+1)!\, d\mu_{i_0} \wedge \dots \wedge d\mu_{i_k} = d\varphi(c_s^k).
\end{aligned}$$

Property 2. If $\langle c^0, v_i \rangle = 1$ for each vertex v_i, then $\varphi(c^0)$ is identically equal to 1.

Clearly, $c^0 = \sum_i c_i^0$, where c_i^0 is the cochain dual to the simplex $[v_i]$. Therefore, $\varphi(c^0) = \sum_i \mu_i = 1$.

Property 3. If c_s^k is the cochain dual to a simplex $\Delta_s^k = [v_{i_0}, \ldots, v_{i_k}]$, then the form $\varphi(c_s^k)$ vanishes identically in a neighborhood of the set $M^n \setminus \operatorname{st} \Delta_s^k$.

By definition,

$$\varphi(c_s^k) = k! \sum_{j=0}^{k} (-1)^j \mu_{i_j} \, d\mu_{i_0} \wedge \cdots \wedge \widehat{d\mu_{i_j}} \wedge \cdots \wedge d\mu_{i_k}.$$

For each i, the function μ_i and the form $d\mu_i$ vanish identically on G_i; therefore, the form $\varphi(c_s^k)$ vanishes on the set $G_{i_0} \cup \cdots \cup G_{i_k}$. This set contains a neighborhood of $M^n \setminus \operatorname{st} \Delta_s^k$.

Property 4. $\rho \circ \varphi = \mathrm{id}$.

We prove this property by induction on k. First, consider the case $k = 0$. Let c_i^0 be the cochain dual to the simplex $[v_i]$. Then $\varphi(c_i^0) = \mu_i$ and $\langle \rho\varphi(c_i^0), [v_j] \rangle = \int_{[v_j]} \mu_i = \mu_i(v_j)$. If $i \neq j$, then $v_j \notin \operatorname{st} v_i$; outside $\operatorname{st} v_i$, the function μ_i vanishes identically. Moreover, $\mu_i(v_i) = \sum_j \mu_i(v_j) = 1$. Therefore, the cochain $\rho\varphi(c_i^0)$ is dual to $[v_i]$, i.e., $\rho\varphi(c_i^0) = c_i^0$.

Now, suppose that $k \geq 1$ and Property 4 holds for all cochains of dimension $k-1$. Let c_s^k be the cochain dual to a simplex $\Delta_s^k = [v_{i_0}, \ldots, v_{i_k}]$. According to Property 3, the form $\varphi(c_s^k)$ vanishes outside $\operatorname{st} \Delta_s^k$; therefore, it vanishes identically on any k-simplex different from Δ_s^k. It remains to show that $\langle \rho\varphi(c_s^k), \Delta_s^k \rangle = 1$. Let $c_{s,j}^{k-1}$ be the cochain dual to the simplex $\Delta_{s,j}^{k-1} = [v_{i_0}, \ldots, \widehat{v}_{i_j}, \ldots, v_{i_k}]$. Then $\delta c_{s,0}^{k-1} = c_s^k + c_t^k + \cdots + c_\tau^k$, where $c_t^k, \ldots, c_\tau^k$ are the cochains dual to k-simplices containing $\Delta_{s,0}^{k-1}$ as a face (and different from Δ_s^k). But we just proved that $\langle \rho\varphi(c_t^k), \Delta_s^k \rangle = 0$ for $t \neq s$. Therefore, $\langle \rho\varphi(\delta c_{s,0}^{k-1}), \Delta_s^k \rangle = \langle \rho\varphi(c_s^k), \Delta_s^k \rangle$. On the other hand, Property 1 and the Stokes formula imply

$$\begin{aligned}\langle \rho\varphi(\delta c_{s,0}^{k-1}), \Delta_s^k \rangle &= \langle d\rho\varphi(c_{s,0}^{k-1}), \Delta_s^k \rangle \\ &= \int_{\partial \Delta_s^k} \rho\varphi(c_{s,0}^{k-1}) = \sum_j \int_{\Delta_{s,j}^{k-1}} \rho\varphi(c_{s,j}^{k-1}) = \int_{\Delta_{s,0}^{k-1}} \rho\varphi(c_{s,0}^{k-1}) = 1\end{aligned}$$

(the last equality follows from Property 4 for cochains of dimension $k-1$).

This completes the proof of the surjectivity of ρ^*. Indeed, Property 1 implies that α induces the homomorphism $\varphi^*\colon H^k(M^n; \mathbb{R}) \to H^k_{\mathrm{DR}}(M^n)$, and Property 4 implies $\rho^* \circ \varphi^* = \mathrm{id}$. Therefore, ρ^* is an epimorphism.

Step 2. The map ρ^* is a monomorphism.

Suppose that $m \geq 1$ and Δ^m is a simplex in $\mathbb{R}^n$. Then the following assertions are valid.

(A_k), $k \geq 0$. Let ω^k be a closed form defined in a neighborhood of $\partial\Delta^m$; if $m = k+1$, then we additionally assume that $\int_{\partial\Delta^m}\omega^k = 0$. Then there exists a closed form α^k defined in a neighborhood of Δ^m and such that $\alpha^k = \omega^k$ in a neighborhood of $\partial\Delta^m$.

(B_k), $k \geq 1$. Let ω^k be a closed form defined in a neighborhood of the simplex Δ^m, and let α^{k-1} be a form defined in a neighborhood of its boundary $\partial\Delta^m$. Suppose that $\omega^k = d\alpha^{k-1}$ in a neighborhood of $\partial\Delta^m$. If $m = k$, then we additionally assume that $\int_{\partial\Delta^k}\alpha^{k-1} = \int_{\Delta^k}\omega^k$. Then there exists a form $\beta^{k-1} = \alpha^{k-1}$ defined in a neighborhood of $\partial\Delta^m$ and such that $\beta^{k-1} = \alpha^{k-1}$ in a neighborhood of $\partial\Delta^m$ and $d\beta^{k-1} = \omega^k$ in a neighborhood of Δ^m.

The additional assumptions are needed because the Stokes theorem must hold for the required form α, and this imposes certain constraints on the initial form ω.

We start by proving (A_0); then, we prove that $(A_{k-1}) \Rightarrow (B_k)$ and $(B_k) \Rightarrow (A_k)$.

(A_0) The closedness of the 0-form (function) ω^0 means that this form is constant on each connected component. If $m \geq 2$, then the set $\partial\Delta^m$ is connected; we can assume that a neighborhood of $\partial\Delta^m$ is connected as well. In this case, the form ω^0 is equal to a constant c, and we set $\alpha^0 = c$. If $m = 1$, then $\partial\Delta^m$ is disconnected; but then $m = k+1$, and the condition $\int_{\partial\Delta^1}\omega^0 = 0$ must hold. Let $\Delta^1 = [v_0, v_1]$. Then this condition has the form $\omega^0(v_1) = \omega^0(v_0)$; therefore, the form ω^0 takes the same value c on both connected components, and we again set $\alpha^0 = c$.

$(A_{k-1}) \Rightarrow (B_k)$ Let ω^k be a closed form defined in a neighborhood of Δ^m. According to the Poincaré lemma, we have $\omega^k = d\beta_1^{k-1}$ for some form β_1^{k-1} (we assume that the neighborhood of the simplex Δ^m is diffeomorphic to the open n-disk). In a neighborhood of $\partial\Delta^m$, consider the form $\gamma^{k-1} = \alpha^{k-1} - \beta_1^{k-1}$. We have $d\gamma^{k-1} = \omega^k - \omega^k = 0$. Moreover, if $m = k$, then $\int_{\partial\Delta^k}\gamma^{k-1} = \int_{\partial\Delta^k}\alpha^{k-1} - \int_{\partial\Delta^k}\beta_1^{k-1} = \int_{\Delta^k}\omega^k - \int_{\Delta^k} d\beta_1^{k-1} = 0$. Applying (A_{k-1}) to the form γ^{k-1}, we see that there exists a closed form γ_1^{k-1} defined in a neighborhood of Δ^m such that $\gamma_1^{k-1} = \gamma^{k-1} = \alpha^{k-1} - \beta_1^{k-1}$ in a neighborhood of $\partial\Delta^m$. The form $\beta^{k-1} = \beta_1^{k-1} + \gamma_1^{k-1}$ has the required properties.

$(B_k) \Rightarrow (A_k)$ Let ω^k be a closed form defined in a neighborhood of $\partial\Delta^m$, where $\Delta^m = [v_0, \dots, v_m]$. Consider the set $(\partial\Delta^m) \setminus [v_1, \dots, v_m]$, which is homeomorphic to the open $(m-1)$-disk. We can assume that its neighborhood is homeomorphic to the open n-disk. Applying the Poincaré lemma, we obtain a form β^{k-1} defined in this neighborhood and satisfying the equality $d\beta^{k-1} = \omega^k$. In particular, $d\beta^{k-1} = \omega^k$ in a neighborhood of $\partial\Delta^{m-1}$, where $\Delta^{m-1} = [v_1, \dots, v_m]$. If $m > 1$, then we can apply assertion (B_k) to the forms ω^k and β^{k-1} and the simplex Δ^{m-1}. We must only verify that $\int_{\Delta^{m-1}} \omega^k = \int_{\partial\Delta^{m-1}} \beta^{k-1}$ for $m - 1 = k$. Let $c = \partial\Delta^m - \Delta^{m-1}$. Then $\partial c = -\partial\Delta^{m-1}$ and any simplex from the chain c is contained in the neighborhood in which the form β^{k-1} is defined. Therefore, $\int_{\Delta^{m-1}} \omega^k - \int_{\partial\Delta^{m-1}} \beta^{k-1} = \int_{\Delta^{m-1}} \omega^k + \int_{\partial c} \beta^{k-1} = \int_{\Delta^{m-1}} \omega^k + \int_c \omega^k = \int_{\partial\Delta^m} \omega^k = 0$. Applying (B_{k-1}), we obtain a form β_1^{k-1} defined in a neighborhood of the simplex Δ^{m-1} and such that $\beta_1^{k-1} = \beta^{k-1}$ in a neighborhood of $\partial\Delta^{m-1}$. The forms β_1^{k-1} and β^{k-1} coincide on the intersection of their domains; hence we can sew them together so as to obtain a form α_1^{k-1} defined in a neighborhood of $\partial\Delta^{m-1}$ and satisfying the equality $d\alpha_1^{k-1} = \omega_k$. Let us extend it to a form α^{k-1} defined in a neighborhood of Δ^m. Consider a function λ that takes the value 1 in a small neighborhood of $\partial\Delta^m$ and vanishes outside a slightly larger neighborhood of $\partial\Delta^m$. The form $\alpha^{k-1} = \lambda\alpha_1^{k-1}$ has the required properties. It remains to consider the case $m = 1$. In this case, the form ω^k is defined in neighborhoods of the vertices v_0 and v_1; we can assume them to be disjoint (and homeomorphic to the open n-disk). According to the Poincaré lemma, there exists a form α_1^{k-1} defined in these neighborhoods and satisfying the equality $d\alpha_1^{k-1} = \omega^k$. A form α^{k-1} defined in a neighborhood of the segment $[v_0, v_1]$ is constructed from α_1^{k-1} in precisely the same way as above.

The injectivity of ρ^* is implied by the following lemma, which is proved by using (B_k).

Lemma. *Suppose that ω^k is a closed form on M^n and $\rho(\omega^k) = \delta c^{k-1}$ for some cochain $c^{k-1} \in C^{k-1}(M^n; \mathbb{R})$. Then there exists a form α^{k-1} on M^n for which $d\alpha^{k-1} = \omega^k$ and $\rho(\alpha^{k-1}) = c^{k-1}$.*

Proof. Recall that we have fixed a triangulation $f\colon |K| \to M^n$. We construct the α^{k-1} by induction on the dimension of the skeleton of the complex K. To be more precise, we shall construct a sequence of forms $\alpha_0^{k-1}, \dots, \alpha_n^{k-1}$ such that

(i) each form α_m^{k-1} is defined in a neighborhood of the m-skeleton of K, and $d\alpha_m^{k-1} = \omega^k$ in this neighborhood;

(ii) the form α_{m+1}^{k-1} coincides with α_m^{k-1} in a neighborhood of the m-skeleton;

(iii) $\rho(\alpha_{m-1}^{k-1}) = c^{k-1}$.

The last equality is understood as follows. The map ρ is defined only for forms defined on the entire complex K, but in reality, only the restrictions of k-forms to the k-skeleton is used. Therefore, we can assume that ρ is defined for the forms α_m^{k-1} with $m \geq k-1$; moreover, $\rho(\alpha_{m-1}^{k-1}) = \rho(\alpha_n^{k-1})$, and for α^{k-1} we can take the form α_n^{k-1}.

The form α_0^{k-1} is constructed as follows. Take disjoint neighborhoods of the vertices $v_1, \dots, v_r$ of K. In each of these neighborhoods, we apply the Poincaré lemma and construct a form with differential ω^k. If $k \geq 1$, then these forms determine the required form α_0^{k-1}. For $k = 1$, we must ensure that $\rho(\alpha_0^0) = c^0$, i.e., $\alpha_0^0(v_i) = \langle c^0, v_i \rangle$. This is easy to achieve by adding a constant C_i to the function α_0^0 in a neighborhood of each point v_i.

We can construct the form α_m^{k-1} from α_{m-1}^{k-1} for each simplex Δ^m separately. Indeed, the common domain of the forms thus constructed is contained in a neighborhood of the $(m-1)$-skeleton, and these forms coincide with α_{m-1}^{k-1} on this domain. The construction of α_m^{k-1} is based on applying (B_k) to the forms ω_k and α_{m-1}^{k-1}. To apply (B_k), we must check that if $k = m$, then $\int_{\partial\Delta^k} \alpha_{m-1}^{k-1} = \int_{\Delta^k} \omega^k$. But $\int_{\Delta^k} \omega^k = \rho(\omega^k) = \langle \delta c^{k-1}, \Delta^k \rangle = \langle c^{k-1}, \partial\Delta^k \rangle = \langle \rho(\alpha_{k-1}^{k-1}), \partial\Delta^k \rangle = \int_{\partial\Delta^k} \alpha_{m-1}^{k-1}$. If $m = k-1$, then it is also required that $\int_{\Delta^{k-1}} \alpha_{k-1}^{k-1} = \langle c^{k-1}, \Delta^{k-1} \rangle$. Let us try to achieve this by adding a closed form β^{k-1} to the form already constructed. The form to be added must be defined in a neighborhood of the $(k-1)$-skeleton and vanish in a neighborhood of the $(k-2)$-skeleton, and the integral $\int_{\Delta^{k-1}} \beta^{k-1}$ must take a given value for each simplex Δ^{k-1}; in other words, $\rho(\beta^{k-1})$ must coincide with a given cochain d^{k-1}. Let us show that the form $\beta^{k-1} = \varphi(d^{k-1})$, where $\varphi\colon C^k(M^n) \to \Omega^k(M^n)$ is the map defined on p. 290, has the required properties. Recall that by Property 3, the form $\varphi(c_s^k)$ vanishes identically in a neighborhood of $M^n \setminus \operatorname{st} \Delta_s^k$. In particular, it vanishes in a neighborhood of the $(k-1)$-skeleton. Therefore, the form β^{k-1} vanishes in a neighborhood of the $(k-2)$-skeleton, and $d\beta^{k-1} = \varphi(\delta d^{k-1})$ vanishes in a neighborhood of the $(k-1)$-skeleton. □

Finally, let us show that $\wedge$-products of forms correspond to $\smile$-products of cohomology classes under the de Rham isomorphism. Suppose that forms ω_1^p and ω_2^q correspond to cohomology classes α^p and β^q. Consider the form $(p_1^*\omega_1^p) \wedge (p_2^*\omega_2^q)$ on the manifold $M^n \times M^n$; here $p_1, p_2\colon M^n \times M^n \to M^n$ are the natural projections onto the first and second factor. The Fubini theorem implies that this form corresponds to the cohomology class $\alpha^p \otimes \beta^q$ because

$$\int_{\Delta_1^p \times \Delta_2^q} (p_1^*\omega_1^p) \wedge (p_2^*\omega_2^q) = \delta_{pr}\delta_{qs}\left(\int_{\Delta_1^r} \omega_1^p\right)\left(\int_{\Delta_2^q} \omega_2^q\right).$$

Let $d\colon M^n \to M^n \times M^n$ be the diagonal map. Then $d^*(\alpha^p \otimes \beta^q) = \alpha^p \smile \beta^q$ and $d^*((p_1^*\omega_1^p) \wedge (p_2^*\omega_2^q)) = \omega_1^p \wedge \omega_2^q$.

3.2. The Simplicial de Rham Theorem. Differential forms can be considered not only on smooth manifolds but also on simplicial complexes. The idea of constructing such forms goes back to Whitney [**154**] and Thom [**138**]. But the most important role in the development of this theory was played by Sullivan's work [**135**], in which piecewise polynomial differential forms were applied to solving some problems of homotopy topology. Moreover, Sullivan developed the theory not only over the field $\mathbb{R}$ but also over the field $\mathbb{Q}$ (and, in general, over an arbitrary field of characteristic zero).

Let $x_0, \ldots, x_n$ be the barycentric coordinates on a simplex Δ^n; it is assumed that $\sum x_i = 1$. We refer to any expression of the form

$$\sum P_{i_1,\ldots,i_k}(x_0, \ldots, x_n)\, dx_{i_1} \wedge \cdots \wedge dx_{i_k},$$

where $P_{i_1,\ldots,i_k}$ is a polynomial with rational coefficients, as a *polynomial* k-form on Δ^n. Polynomials with real coefficients can also be considered. Again, it is assumed that $\sum x_i = 1$ and, accordingly, $\sum dx_i = 0$.

In addition to polynomial differential forms, we consider *smooth* differential k-forms on the simplex Δ^n. They are expressed as

$$\sum f_{i_1,\ldots,i_k}(x_0, \ldots, x_n)\, dx_{i_1} \wedge \cdots \wedge dx_{i_k},$$

where $f_{i_1,\ldots,i_k}$ is a smooth function on an open subset of $\mathbb{R}^n$ containing Δ^n.

A *polynomial* differential form on a simplicial complex K is obtained by sewing together polynomial differential forms defined on the simplices of K. These forms must be compatible in the sense that if two simplices Δ_1 and Δ_2 have a common face Δ_{12} and forms ω_1 and ω_2 are defined on these simplices, then the restrictions of ω_1 and ω_2 to Δ_{12} must coincide. The restriction of a form to a face $x_i = 0$ is defined by setting $x_i = 0$ and $dx_i = 0$.

Smooth differential forms on a simplicial complex K are defined similarly. Note that smooth forms on a triangulation of a manifold are not the same as smooth forms on this manifold. For example, smooth functions (0-forms) on a triangulation of the circle are piecewise smooth (they may be nondifferentiable at the vertices of the triangulation).

We denote the linear spaces of polynomial and smooth k-forms on a simplicial complex K by $A^k(K)$ and $A^k_{C^\infty}(K)$, respectively.

On the linear spaces $A^k(K)$ and $A^k_{C^\infty}(K)$, the operator d acts and exterior product is defined. Importantly, for polynomial forms over $\mathbb{Q}$, the form $d(P_{i_1,\ldots,i_k}\, dx_{i_1} \wedge \cdots \wedge dx_{i_k})$ can be expressed in terms of the partial derivatives of the polynomial $P_{i_1,\ldots,i_k}$, which are also polynomials with rational coefficients.

Polynomial forms have properties similar to those of differential forms on manifolds.

Lemma 1. *Suppose that K is a finite simplicial complex, CK is a cone over K, $k \geq 1$, and $\omega^k \in A^k(CK)$ is a closed form. Then $\omega^k = d\alpha^{k-1}$ for some form $\alpha^{k-1} \in A^{k-1}(CK)$.*

Proof. Let us represent each point of the cone CK as $\lambda x + (1-\lambda)a$, where $x \in K$, a is the vertex of the cone, and $0 \leq \lambda \leq 1$. Consider the map $\mu\colon CK \times I \to CK$,

$$\mu(\lambda x + (1-\lambda)a) = \lambda(1-t)x + (\lambda + t(1-\lambda))a,$$

and the form $\mu^*(\omega^k)$ on $CK \times I$. Let Δ be a simplex in K. The restriction of $\mu^*(\omega^k)$ to $CK \times I$ can be represented as $\sum_{p\geq 0}(t^p\alpha_p(\Delta) + t^p\beta_p(\Delta) \wedge dt)$, where $\alpha_p(\Delta)$ and $\beta_p(\Delta)$ are forms on Δ not containing t and dt. These forms determine forms α_p and β_p on the entire complex K.

For $t = 0$, μ is the identity map; thus, for $t = 0$ and $dt = 0$, the form $\mu^*(\omega^k)$ coincides with ω^k. This means that $\alpha_0 = \omega^k$. For $t = 1$, the map μ is constant; thus, for $t = 1$ and $dt = 0$, the form $\mu^*(\omega^k)$ is zero (any constant map induces the zero map of k-forms for $k \geq 1$; the condition $k \geq 1$ is essential). This means that $\sum_{p\geq 0} \alpha_p = 0$.

Now it is time to use the closedness of the form ω^k, which implies $d(\mu^*(\omega^k)) = \mu^*(d\omega^k) = 0$, i.e., $d\big(\sum_{p\geq 0} t^p\alpha_p + t^p\beta_p \wedge dt\big) = 0$. Here $d(t^p\alpha_p) = t^p d\alpha_p + (-1)^{\deg \alpha_p} pt^{p-1}\alpha_p \wedge dt$ and $d(t^p\beta_p \wedge dt) = t^p\, d\beta_p \wedge dt$. Thus, $d\alpha_p = 0$ and $(-1)^{\deg \alpha_p} p\alpha_p + d\beta_{p-1} = 0$ for all $p \geq 0$.

The above equalities imply

$$d\Big(\sum_{p\geq 0} (-1)^{\deg \alpha_p} \frac{\beta_p}{p+1}\Big) = \sum_{p\geq 0} (-1)^{\deg \alpha_p} \frac{d\beta_p}{p+1} = -\sum_{p\geq 1} \alpha_p = \alpha_0 = \omega^k. \quad \square$$

Lemma 2 (on extension)**.** *For any form $\omega \in A^k(\partial\Delta^n)$, there exists a form $\Omega \in A^k(\Delta^n)$ such that its restriction to $\partial\Delta^n$ coincides with ω.*

Proof. Consider the projection of the simplex $\Delta^n = [v_0, \ldots, v_n]$ minus the vertex v_0, from the point v_0 onto the opposite face $[v_1, \ldots, v_n]$. In barycentric coordinates, this map is written as

$$(x_0, \ldots, x_n) \mapsto \Big(\frac{x_1}{1-x_0}, \ldots, \frac{x_n}{1-x_0}\Big).$$

We denote it by π_0. Let ω_0 be the restriction of the form ω to the face $[v_1, \ldots, v_n]$. Consider the form $\pi_0^*\omega_0$ on $\Delta^n \setminus \{x_0\}$. If

$$\omega_0 = \sum P_{i_1\ldots i_k}(y_1, \ldots, y_n)\, dy_{i_1} \wedge \cdots \wedge y_{i_k},$$

where $y_1, \ldots, y_n$ are the barycentric coordinates on $[v_1, \ldots, v_n]$, then

$$\pi_0^*\omega_0 = \sum P_{i_1\ldots i_k}\Big(\frac{x_1}{1-x_0}, \ldots, \frac{x_n}{1-x_0}\Big)\, d\Big(\frac{x_{i_1}}{1-x_0}\Big) \wedge \cdots \wedge d\Big(\frac{x_{i_k}}{1-x_0}\Big).$$

But $d(\frac{x_i}{1-x_0}) = \frac{(1-x_0)\,dx_i + x_i\,dx_0}{(1-x_0)^2}$. Hence there exists a positive integer N such that the form $(1-x_0)^N \pi_0^* \omega_0 = \tilde{\omega}_0$ is polynomial. The form $\tilde{\omega}_0$ is well defined on the entire simplex Δ^n; indeed, $x_0 = 1$ at the vertex v_0, so $\tilde{\omega}_0$ vanishes at v_0.

The face $[v_1, \ldots, v_n]$ is determined by the equation $x_0 = 0$; therefore, the form $\tilde{\omega}_0$ coincides with ω on this face. Thus, the form $\omega - \tilde{\omega}_0|_{\partial\Delta^n}$ vanishes on the face $[v_1, \ldots, v_n]$.

By construction, if the form ω vanishes on a face $x_i = 0$, then the form $\omega - \tilde{\omega}_0|_{\partial\Delta^n}$ also vanishes on this face. In other words, given a form ω on $\partial\Delta^n$ that vanishes on several faces, we can construct a form $\tilde{\omega}_1$ on Δ^n that vanishes on the same faces and, in addition, coincides with ω on some other face. Now we can construct the required form by induction because the form $\omega = 0$ can be extended to Δ^n in an obvious way. □

Remark. A similar assertion is valid for smooth functions on Δ^n. Its proof coincides with the proof of the extension lemma with the only difference that the function $(1-x_0)^N$ should be replaced by a smooth function taking the value 1 in a neighborhood of the face $[v_1, \ldots, v_n]$ and vanishing in a neighborhood of the vertex v_0.

Lemma 3. *Let ω^k be a closed polynomial form on a simplex Δ^n vanishing on $\partial\Delta^n$; for $k = n$, we assume in addition that $\int_{\Delta^n} \omega^k = 0$. Then $\omega^k = d\alpha^{k-1}$ for some form $\alpha^{k-1} \in A^{k-1}(\Delta^n)$ vanishing on $\partial\Delta^n$ ($\omega^k = 0$ for $k = 0$).*

Proof. For $n = 0$, only the zero form exists. Suppose that $n = 1$ and $\Delta^1 = [0, 1]$. If $k = 0$, then there exists a polynomial $P(x)$ such that $P'(x) = 0$ for all x and $P(0) = P(1) = 0$. Therefore, $P(x) = 0$ for all x. If $k = 1$, then the 1-form $P(x)\,dx$ is defined, and, by assumption, $\int_0^1 P(x)\,dx = 0$. Let $Q(x) = \int_0^x P(t)\,dt = 0$. Then $d(Q(x)) = P(x)\,dx$ and $Q(0) = Q(1) = 0$.

Let us apply induction on n. The implication $(A_{n-1}) \Rightarrow (A_n)$ (where (A_n) is the assertion of the lemma for Δ^n) follows from $(A_{n-1}) \Rightarrow (B_n) \Rightarrow (A_n)$, where (B_n) is the following assertion: *Suppose that $k \geq 1$ and ω^k is a closed polynomial form on $\partial\Delta^n$; for $k = n-1$, we assume in addition that $\int_{\partial\Delta^n} \omega^{n-1} = 0$. Then $\omega^k = d\alpha^{k-1}$ for some polynomial form α^{k-1} on $\partial\Delta^n$.*

First, we prove the implication $(A_{n-1}) \Rightarrow (B_n)$. Let ω^k be a closed polynomial form on $\partial\Delta^n$. Consider the face $[v_1, \ldots, v_n]$ of the simplex $\Delta^n = [v_0, \ldots, v_n]$. We set $K = \partial[v_1, \ldots, v_n]$ and denote the union of all faces of Δ^n except $[v_1, \ldots, v_n]$ by CK. The Poincaré lemma (Lemma 1) implies that $\omega^k|_{CK} = d\beta^{k-1}$ for some form $\beta^{k-1} \in A^k(CK)$. Applying the extension lemma, we extend the form β^{k-1} to a form $\tilde{\beta}^{k-1}$ on $\partial\Delta^n$. The form $\omega^k - \tilde{\beta}^{k-1}$ is closed, and it may differ from zero only on the face $[v_1, \ldots, v_n]$; on

$K = \partial[v_1, \dots, v_n]$, this form vanishes. If $k = n-1$, then

$$0 = \int_{\partial\Delta^n} \omega^k = \int_{\partial\Delta^n} (\omega^k - \tilde{\beta}^{k-1}) = \int_{[v_1,\dots,v_n]} (\omega^k - \tilde{\beta}^{k-1}).$$

Thus, we can apply (A_{n-1}) to the restriction of the form $\omega^k - \tilde{\beta}^{k-1}$ to the simplex $[v_1, \dots, v_n]$ and obtain $(\omega^k - \tilde{\beta}^{k-1})|_{[v_1,\dots,v_n]} = \partial\gamma^{k-1}$ for some form γ^{k-1} on $[v_1, \dots, v_n]$ vanishing on $\partial[v_1, \dots, v_n]$. The form γ^{k-1} can be extended to a form $\tilde{\gamma}^{k-1}$ on $\partial\Delta^n$ vanishing on CK. Moreover, $\omega^k = d(\tilde{\beta}^{k-1} + \tilde{\gamma}^{k-1})$.

Now, let us prove the implication $(B_n) \Rightarrow (A_n)$. Suppose that ω^k is a closed polynomial form on Δ^n vanishing on $\partial\Delta^n$; for $k = n-1$, we assume in addition that $\int_{\partial\Delta^n} \omega^k = 0$. Obviously, if $k = 0$, then $\omega^k = 0$. Suppose that $k \geq 1$. In this case, we can apply the Poincaré lemma and construct a form $\beta^{k-1} \in A^{k-1}(\Delta^n)$ for which $d\beta^{k-1} = \omega^k$. The form β^{k-1} may be nonzero on $\partial\Delta^n$. We make it vanish as follows. Suppose that $k = 1$ and $n \geq 2$. Then the function β^{k-1} is constant on $\partial\Delta^n$. Therefore, we can assume that β^{k-1} vanishes on $\partial\Delta^n$. Now, suppose that $k \geq 2$. If $k = n$, then

$$\int_{\partial\Delta^n} \beta^{k-1} = \int_{\Delta^n} d\beta^{k-1} = \int_{\Delta^n} d\omega^k = 0.$$

Thus, we can apply the assertion (B_n) and obtain $\beta^{k-1}|_{\partial\Delta^n} = d\tilde{\gamma}^{k-2}$, where $\tilde{\gamma}^{k-2}$ is a polylinear form on $\partial\Delta^n$. Take an extension γ^{k-2} of $\tilde{\gamma}^{k-2}$ over Δ^n and let $\alpha^{k-1} = \beta^{k-1} - d\gamma^{k-2}$. We have $d\alpha^{k-1} = d\beta^{k-1} = \omega^k$ and $\alpha^{k-1}|_{\partial\Delta^n} = \beta^{k-1}|_{\partial\Delta^n} - d\tilde{\gamma}^{k-2} = 0$. □

We are ready to formulate and prove the *simplicial de Rham theorem.* Let $\rho\colon A^*(K) \to C^*(K; \mathbb{Q})$ be the map defined by $\langle \rho(\omega), \Delta^n \rangle = \int_{\Delta^n} \omega^n$. According to the Stokes theorem, $\rho\, d\omega = \delta\rho\omega$; hence ρ induces the map ρ^* of cohomology algebras.

Theorem 5.13. *The map ρ^* is an algebra isomorphism.*

Proof. We start by proving that ρ^* is an isomorphism of vector spaces; then, we prove that ρ^* preserves multiplication.

First, let us show that ρ is an epimorphism. Consider the form $\omega_i^n = dx_1 \wedge \cdots \wedge dx_n$ on the simplex Δ_i^n. This form is closed because there are no nonzero $(n+1)$-forms on an n-simplex. Let us verify that $\omega_i^n|_{\partial\Delta_i^n} = 0$. This is obvious for the faces given by the equations $x_1 = 0, \dots, x_n = 0$; for the face given by $x_0 = 0$, this follows from the relation $dx_1 + \cdots + dx_n = -dx_0 = 0$. Thus, we can take a form on the n-skeleton which coincides with ω_i^n on Δ_i^n and vanishes on the other n-simplices. Applying the extension lemma, we extend this form to the entire simplicial complex K. As a result, we obtain a form $\Omega_i^n \in A^n(K)$ such that $\langle \rho(\Omega_i^n), \Delta_j^n \rangle = c\delta_{ij}$, where $c \neq 0$ is a rational number. It follows that ρ is an epimorphism.

We have obtained the short exact sequence

$$0 \longrightarrow \operatorname{Ker} \rho \longrightarrow A^*(K) \xrightarrow{\rho} C^*(K; \mathbb{Q}) \longrightarrow 0.$$

Here $\operatorname{Ker} \rho$ consists of all forms $\omega^n \in A^*(K)$ such that $\int_{\Delta^n} \omega^n = 0$ for any simplex Δ^n in K. To prove that ρ^* is an isomorphism, it suffices to verify that $H^*(\operatorname{Ker} \rho) = 0$; in other words, we must show that if $\omega^n \in A^*(K)$, $d\omega^n = 0$, and $\int_{\Delta^n} \omega^n = 0$ for all simplices Δ^n in K, then $\omega^n = d\alpha^{n-1}$ for some form $\alpha^{n-1} \in \operatorname{Ker} \rho$ (this means that $\int_{\Delta^{n-1}} \alpha^{n-1} = 0$ for any simplex Δ^{n-1} in K (if $n > 0$) or $\omega^n = 0$ (if $n = 0$)). According to Lemma 3, there exists a form $\beta_i^{n-1} \in A^{n-1}(\Delta_i^n)$ for which $\omega^n|_{\Delta_i^n} = d\beta_i^{n-1}$ and $\beta_i^{n-1}|_{\partial \Delta_i^n} = 0$. Sewing all such forms β_i^{n-1} together, we obtain a form $\tilde{\beta}^{n-1}$ on the n-skeleton of K which vanishes on the $(n-1)$-skeleton. Using the extension lemma, we extend the form $\tilde{\beta}^{n-1}$ to a form β^{n-1} on the entire K. The closed form $\omega^n - d\beta^{n-1}$ vanishes on the n-skeleton, i.e., on the boundary of any $(n+1)$-simplex. Again applying Lemma 3, we construct a form $\gamma_j^{n-1} \in A^{n-1}(\Delta_j^{n+1})$ for which $(\omega^n - d\beta^{n-1})|_{\Delta_j^{n+1}} = d\gamma_j^{n-1}$ and $\gamma_j^{n-1}|_{\partial \Delta_j^{n+1}} = 0$. Then, we sew together the forms γ_j^{n-1} so as to obtain a form $\tilde{\gamma}^{n-1}$ on the $(n+1)$-skeleton, which we extend to a form γ^{n-1} on the entire K. The closed form $\omega^n - d(\beta^{n-1} + \gamma^{n-1})$ vanishes on the $(n+1)$-skeleton. Thus, we can repeat the construction. At the end, we obtain $\alpha^{n-1} = \beta^{n-1} + \gamma^{n-1} + \cdots$. This sum is well defined because the form obtained at the kth step vanishes on the $(n+k-2)$-skeleton. The form α^{n-1} has the required properties because the forms $\beta^{n-1}, \gamma^{n-1}, \ldots$ vanish on the $(n-1)$-skeleton.

Now let us prove that ρ^* preserves multiplication. Take $\alpha, \beta \in Z^*(K; \mathbb{Q})$. Recall that the class $[\alpha] \smile [\beta] \in H^*(K; \mathbb{Q})$ is the image of the class $[\alpha \otimes \beta]$ under the map induced by the diagonal map $d\colon |K| \to |K \times K|$. Take $\omega_1^n, \omega_2^m \in A^*(K)$. Let L be the triangulation of $|K \times K|$ whose vertices are the products of vertices of K. We define the form $\omega_1^n \times \omega_2^m \in A^{n+m}(L)$ as follows. Any simplex Δ in L is contained in the product $\Delta_1 \times \Delta_2$ of two simplices from K. Consider the canonical projections $p_i\colon \Delta_1 \times \Delta_2 \to \Delta_i$ $(i = 1, 2)$. Sewing together the restrictions of the forms $(p_1^* \omega_1^n) \wedge (p_2^* \omega_2^m)$ (which are defined on $\Delta_1 \times \Delta_2$) to Δ for all simplices Δ, we obtain a form on L, which we denote by $\omega_1^n \times \omega_2^m$.

The map $\rho\colon A^*(L) \to C^*(L; \mathbb{Q})$ takes this form $\omega_1^n \times \omega_2^m$ to a cochain $\rho(\omega_1^n \times \omega_2^m)$ for which $\langle \rho(\omega_1^n \times \omega_2^m), \Delta_1^p \times \Delta_2^q \rangle = \int_{\Delta_1^p \times \Delta_2^q} (p_1^* \omega_1^n) \wedge (p_2^* \omega_2^m) = \delta_{np}\delta_{mq} \big(\int_{\Delta_1^n} \omega_1^n\big)\big(\int_{\Delta_2^m} \omega_2^m\big)$; the second equality follows from the Fubini theorem. Thus, $\rho(\omega_1^n \times \omega_2^m) = \rho(\omega_1^n) \otimes \rho(\omega_2^m)$. Moreover, it follows directly from the definition that $d^*(\omega_1^n \times \omega_2^m) = \omega_1^n \wedge \omega_2^m$. □

Chapter 6

Miscellany

1. The Alexander Polynomial

A natural way to construct an invariant of a link $L \subset S^3$ is to consider the space $S^3 \setminus L$. The homology groups of this space form an invariant that is too rough because these groups are the same for all links with the same number of components. There is a finer invariant related to homology, namely, the Alexander polynomial. It can be defined in several ways. One of them is to paste up the link L with a surface F (i.e., take a self-avoiding surface F with boundary L) and construct the so-called Seifert form by using this surface. This approach involves certain difficulties, which arise in the proof of the independence of the invariant thus constructed of the choice of F. An approach avoiding these difficulties is as follows. We construct an infinite cyclic covering X_∞ over $S^3 \setminus L$. There is a natural automorphism $t\colon X_\infty \to X_\infty$, which depends only on L. This automorphism defines the structure of a module over the ring $\mathbb{Z}[t^{-1}, t]$ on $H_1(X_\infty)$. It turns out that this module structure is closely related to the matrix determining the Seifert form (see Theorem 6.2).

Such is the program. Let us implement it.

1.1. The Seifert Form. In this section, for a surface F embedded in the sphere S^3, we construct a bilinear form on $H_1(F)$. We use this form in what follows to define the Alexander polynomial.

Let F be a sphere with g handles from which $n \geq 1$ open disks are removed (we assume that F is compact). Then $\chi(F) = 2 - 2g - n$. Clearly, the surface F is homotopy equivalent to a wedge of circles. The Euler characteristic of the wedge of k circles is equal to $k - 1$; therefore, the surface F is homotopy equivalent to the wedge of $2g + n - 1$ circles. Thus,

$H_1(F) \cong \mathbb{Z}^{2g+n-1}$. For generators of the group $H_1(X)$ we can take the homology classes of the cycles shown in two equivalent ways in Figures 1 and 2. Figure 1 shows a sphere with g handles with n disks removed, which is standardly embedded in 3-space. The parallels and meridians on the handles represent the generators of the 1-dimensional homology groups; the remaining generating cycles are homologous to the boundaries of $n-1$ removed disks. This surface is homeomorphic to a disk with $2g$ pairs of bands attached to the upper part and $n-1$ bands attached to the lower part, as shown in Figure 2. Indeed, both surfaces are orientable, their Euler characteristics coincide, and their boundaries have equally many components. It is easy to explicitly construct a homeomorphism between the surfaces and see that the system of generating cycles on one surface is mapped to a system of generating cycles on the other surface. It is seen from Figure 2 that each of the cycles shown in this figure corresponds to a circle from a wedge of circles homotopy equivalent to the surface under consideration.

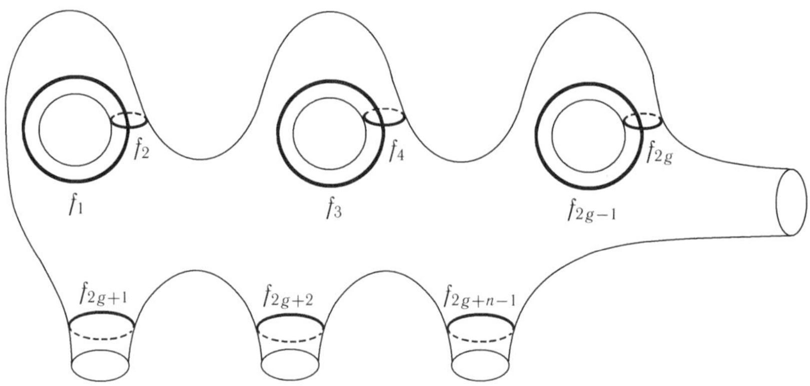

Figure 1. A homology basis

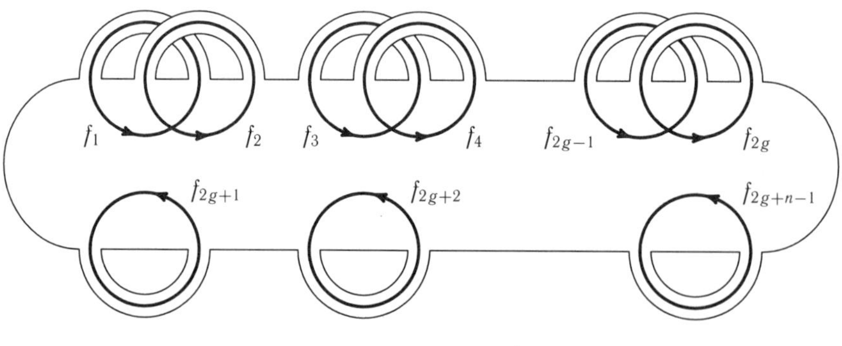

Figure 2. A homology basis (an equivalent presentation)

Suppose that the surface F is embedded in S^3. Then, according to the Alexander duality theorem, we have $H^1(F;\mathbb{Z}) \cong H_1(S^3 \setminus F)$. The space F is homotopy equivalent to the wedge of $2g+n-1$ circles; therefore, $H^1(F;\mathbb{Z}) \cong \mathbb{Z}^{2g+n-1} \cong H_1(F)$. The isomorphism $H_1(S^3 \setminus F) \cong H_1(F)$ can also be proved directly. Let us do this in order to obtain an explicit description of this isomorphism. The most important point in the proof is that the bilinear form $\beta\colon H_1(S^3 \setminus F) \times H_1(F) \to \mathbb{Z}$ (which assigns to any two closed oriented curves in $S^3 \setminus F$ and in F the linking number of these curves) is nondegenerate.

If $\varepsilon > 0$ is sufficiently small, then the closed ε-neighborhood V of the surface F in S^3 is a 3-disk to which $2g+n-1$ handles $D^2 \times I$ are attached. Moreover, the inclusion $F \subset V$ is a homotopy equivalence. The boundary ∂V is a sphere with $2g+n-1$ handles. Corresponding to each handle are two generators, e_i and f_i', in the group $H_1(\partial V)$; e_i is the boundary of a small disk transversally intersecting f_i in one point and disjoint from f_j for $j \neq i$, and f_i' is the element of $H_1(\partial V)$ mapped to f_i under the homomorphism of homology groups induced by the inclusion $\partial V \subset V$ (this homomorphism takes e_i to 0). We choose the orientation of e_i so that $\operatorname{lk}(e_i, f_i) = 1$. Then $\operatorname{lk}(e_i, f_j) = \delta_{ij}$. Let V' be the closure of the complement $S^3 \setminus V$. Then the inclusion $V' \subset S^3 \setminus F$ is a homotopy equivalence because so is $\partial V \subset V \setminus F$. In the Mayer–Vietoris sequence

$$H_2(S^3) \longrightarrow H_1(\partial V) \longrightarrow H_1(V) \oplus H_1(V') \longrightarrow H_1(S^3),$$

the first and last terms are zero; therefore, $H_1(\partial V) \cong H_1(V) \oplus H_1(V')$. In particular, $H_1(V)$ and $H_1(V')$ are free Abelian groups. We know the ranks of the groups $H_1(\partial V)$ and $H_1(V)$; thus, $H_1(V') \cong \mathbb{Z}^{2g+n-1}$. This group is generated by the images of the elements e_i under the homomorphism induced by the inclusion $\partial V \subset V'$. Thus, the groups $H_1(V) \cong H_1(F)$ and $H_1(V') \cong H_1(S^3 \setminus F)$ have bases $f_1, \dots, f_{2g+n-1}$ and $e_1, \dots, e_{2g+n-1}$ such that $\operatorname{lk}(e_i, f_j) = \delta_{ij}$.

The bilinear form $\beta\colon H_1(S^3 \setminus F) \times H_1(F) \to \mathbb{Z}$ determines the bilinear form $\alpha\colon H_1(F) \times H_1(F) \to \mathbb{Z}$ defined as follows. Since the surface F is orientable, we can choose a positively oriented basis v_1, v_2 at each of its points. Let n be the unit normal vector to F for which the basis v_1, v_2, n is positively oriented. Note that it is not always possible to choose the vectors v_1 and v_2 so that they depend continuously on the corresponding point of F, but the vector n depends only on the orientation determined by v_1 and v_2, rather than on these vectors themselves. Therefore, the vector n continuously depends on the point. Choose a sufficiently small number $\varepsilon > 0$. For each curve γ on the surface F, consider the curve γ^+ obtained by translating each point of γ in the direction of the chosen normal by a distance of ε; the curve obtained by applying a similar translation in the opposite direction is denoted by γ^-. We set $\alpha(x,y) = \beta(x^-, y)$. The bilinear form α is called the *Seifert form*.

For the basis of $H_1(F)$ we again take $f_1, \dots, f_{2g+n-1}$. In this basis, the Seifert form is determined by the *Seifert matrix* $A = (a_{ij})$, where $a_{ij} = \alpha(f_i, f_j) = \mathrm{lk}(f_i^-, f_j) = \mathrm{lk}(f_i, f_j^+)$. Let $e_1, \dots, e_{2g+n-1}$ be the basis of the group $H_1(S^3 \setminus F)$ dual to $f_1, \dots, f_{2g+n-1}$ with respect to the form β. Then $f_i^- = \sum_j a_{ij} e_j$ and $f_j^+ = \sum_i a_{ij} e_i$.

1.2. Infinite Cyclic Coverings. Consider a compact orientable surface F with boundary $\partial F = L$ embedded in S^3. Let us construct a space X_∞ that is an infinite covering of $S^3 \setminus L$. To this end, we cut the sphere S^3 along F. To be more precise, we take the space $S^3 \setminus F$, add L to it, and for each $a \in F \setminus L$, add points a^+ and a^-, the respective limits of points converging to a from each of the two sides of F. As a result, we obtain a 3-manifold M^3 with boundary $F^+ \cup F^-$; corresponding to each point $a \in F \setminus L$ are two points, $a^+ \in F^+$ and $a^- \in F^-$ (since the surface F is orientable, we can choose compatible signs $+$ and $-$ for all points). We take countably many copies M_i^3 $(i \in \mathbb{Z})$ of this manifold and identify F_{i-1}^+ with F_i^- (by identifying the points a_{i-1}^+ and a_i^- that correspond to the same point a). This construction is schematically shown in Figure 3. The space X_∞ thus obtained covers $S^3 \setminus L$. The maps $M_i^3 \to M_{i+1}^3$ taking each point of M_i^3 to the corresponding point of the manifold $M_{i+1}^3 \approx M_i^3$ induce an automorphism $t\colon X_\infty \to X_\infty$ of this covering. The automorphism group of the covering is generated by t and is isomorphic to $\mathbb{Z}$; it acts transitively on the covering.

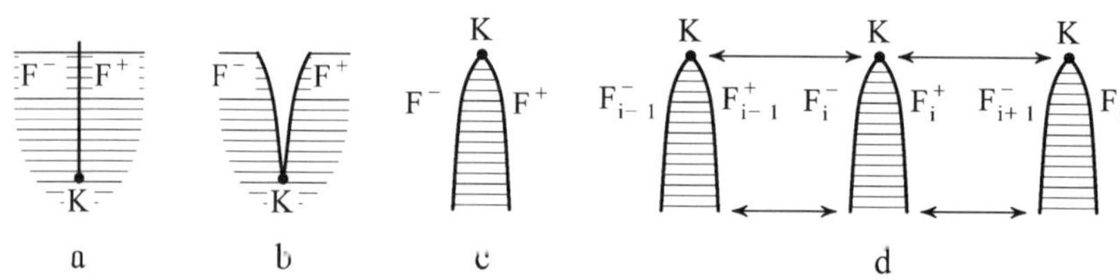

Figure 3. The space X_∞

We assume that the surface F is oriented and the points a^+ and a^- correspond to the positive and negative directions of the normal.

We have constructed the covering $X_\infty \to S^3 \setminus L$ by using the surface F. It turns out that this construction does not depend on the choice of F; it depends only on the oriented link L (the boundary of F). Indeed, any covering over $S^3 \setminus L$ is uniquely determined by the image of the fundamental group of the covering space in $\pi_1(S^3 \setminus L)$. Let us show that this image consists of those loops γ for which the linking number with L is zero (we assume that if L has connected components $L_1, \dots, L_n$, then $\mathrm{lk}(\gamma, L) = \sum \mathrm{lk}(\gamma, L_i)$).

Clearly, the image consists of the loops γ whose liftings to X_∞ are closed. In other words, going from any point of M_0^3 along the lifting of γ, we must return to M_0^3. The transitions between M_i^3 and $M_{i\pm1}^3$ correspond to the intersection points of γ with F; the sign is the same as in the definition of the intersection number. We arrive at the condition that the intersection number of the curve γ with the surface F equals 0. According to one of the definitions of linking number (see [**108**]), this is equivalent to the vanishing of the linking number of γ with $L = \partial F$.

The automorphism of the covering $t\colon X_\infty \to X_\infty$ does not depend on the surface F either. It is determined by those loops in $S^3 \setminus L$ for which the linking number with L equals 1 (the starting point of the lifting of γ is mapped to the ending point of this lifting).

For each oriented link $L \subset S^3$, there exists a connected oriented surface F embedded in S^3 so that its boundary is L (see, e.g., [**105**]). This surface is called a *Seifert surface* of the link L. Using the surface F, we can construct a covering $X_\infty \to S^3 \setminus L$ and an automorphism $t\colon X_\infty \to X_\infty$ of this covering which depends only on L. Thus, with the link L we can associate the group $H_1(X_\infty)$ on which the automorphism t_* is defined. This means that $H_1(X_\infty)$ is a module over the ring $\mathbb{Z}[t^{-1}, t]$, whose elements are polynomials in t^{-1} and t with integer coefficients. To apply this algebraic structure, we need a matrix describing the specification of a module by generators and relations (a presentation matrix). Below, we give a definition of such a matrix and prove some of its properties.

Let M be a module over a commutative ring R with identity; it is assumed that $1m = m$ for all $m \in M$. The module M is said to be *free* if there exist elements $m_1, \dots, m_n \in M$ such that any element $m \in M$ has a unique representation in the form $m = r_1 m_1 + \dots + r_n m_n$, where $r_i \in R$; the elements $m_1, \dots, m_n$ form a *basis* of the module M. A module M is *finitely generated* if there exists an exact sequence

$$F \xrightarrow{\alpha} E \xrightarrow{\varphi} M \longrightarrow 0, \tag{40}$$

where E and F are free modules (over the same ring R). Take bases $e_1, \dots, e_m$ and $f_1, \dots, f_n$ in E and F. Let $A = (a_{ij})$ be the matrix of the map α in these bases (i.e., $\alpha(f_i) = \sum_{j=1}^m a_{ji} e_j$). The matrix A is called a *presentation matrix* for the module M. Each of the m rows of A corresponds to a generator of M, and each of the n columns corresponds to a relation between generators. For a given module M, the exact sequence (40) is not unique; moreover, we can take different bases in E and F.

Theorem 6.1. *Let A and A' be two presentation matrices for the same module M. Then they can be obtained from each other by the following transformations (and their inverses):*

(1) *a permutation of rows or columns;*

(2) *the replacement of a matrix B by the matrix* $\left(\begin{smallmatrix} B & 0 \\ 0 & 1 \end{smallmatrix}\right)$;

(3) *the addition of the zero column;*

(4) *the addition to a row (column) of another row (column) multiplied by a number.*

Proof. The exact sequences corresponding to the matrices A and A' give the commutative diagram

$$\begin{array}{ccccccc} F & \xrightarrow{\alpha} & E & \xrightarrow{\varphi} & M & \longrightarrow & 0 \\ \downarrow{\scriptstyle f} & & \downarrow{\scriptstyle g} & & \downarrow{\scriptstyle \mathrm{id}} & & \\ F' & \xrightarrow{\alpha'} & E' & \xrightarrow{\varphi'} & M & \longrightarrow & 0\,. \end{array}$$

Indeed, the homomorphisms g and f can be specified on the bases and then extended by linearity. The element $g(e_j) \in E'$ is defined as follows. The map φ' is an epimorphism; hence there exists an element $e'_j \in E'$ for which $\varphi'(e'_j) = \varphi(e_j)$. We set $g(e_j) = e'_j$; then $\varphi' g(e_j) = \varphi'(e'_j) = \varphi(e_j)$. The element $f(f_i) \in F'$ is defined as follows. The equality $\varphi' g\alpha(f_i) = \varphi\alpha(f_i) = 0$ implies the existence of an element $f'_i \in F'$ for which $\alpha'(f'_i) = g\alpha(f_i)$. We set $f(f_i) = f'_i$; then $\alpha' f(f_i) = \alpha'(f'_i) = g\alpha(f_i)$.

Let X and Y be the matrices of the maps f and g in the bases in which the matrices A and A' are written. The equality $g\alpha = \alpha' f$ means that $XA = A'Y$. The maps f' and g' with matrices X' and Y' for which $X'A' = AY'$ are constructed in the same way as f and g.

The equalities $\varphi' g = \varphi$ and $\varphi g' = \varphi'$ imply $\varphi g' g = \varphi' g = \varphi$. Therefore, $\mathrm{Im}(g'g - \mathrm{id}_E) \subset \mathrm{Ker}\,\varphi = \mathrm{Im}\,\alpha$. Using this property and the freeness of the module E, we can construct a homomorphism $h\colon E \to F$ such that $\alpha h = g'g - \mathrm{id}_E$. For matrices, this equality takes the form $AZ = X'X - I$, where Z is the matrix of h.

We shall write $P \sim Q$ if the matrix Q can be obtained from the matrix P by applying transformations (1)–(4) and their inverses. Clearly,

$$A \overset{(2)}{\sim} \begin{pmatrix} A & 0 \\ 0 & I \end{pmatrix} \overset{(4)}{\sim} \begin{pmatrix} A & X' \\ 0 & I \end{pmatrix} \overset{(3),\,(4)}{\sim} \begin{pmatrix} A & X' & X'A' \\ 0 & I & A' \end{pmatrix}.$$

Taking into account the relation $X'A' = AY'$, we obtain

$$\begin{pmatrix} A & X' & X'A' \\ 0 & I & A' \end{pmatrix} \overset{(4)}{\sim} \begin{pmatrix} A & X' & 0 \\ 0 & I & A' \end{pmatrix} \overset{(3),\,(4)}{\sim} \begin{pmatrix} A & X' & 0 & XX' \\ 0 & I & A' & X \end{pmatrix}.$$

Next, using the equality $XX' = AZ + I$, we obtain

$$\begin{pmatrix} A & X' & 0 & XX' \\ 0 & I & A' & X \end{pmatrix} \overset{(4)}{\sim} \begin{pmatrix} A & X' & 0 & I \\ 0 & I & A' & X \end{pmatrix}.$$

Therefore,

$$A \sim \begin{pmatrix} A & X' & 0 & I \\ 0 & I & A' & X \end{pmatrix} \overset{(1)}{\sim} \begin{pmatrix} A' & X & 0 & I \\ 0 & I & A & X' \end{pmatrix} \sim A',$$

as required. □

Let A be a presentation matrix of a module M over a ring R. If A has m rows and n columns, then the ideal of R generated by all minors of order $m-r+1$ of the matrix A is called the rth *elementary ideal* of the module M and denoted by $\mathcal{E}_r$. Using standard properties of determinants and applying Theorem 6.1, we see that the ideal $\mathcal{E}_r$ does not depend on the choice of A. We assume that $\mathcal{E}_r = R$ for $r > m$ and $\mathcal{E}_r = 0$ for $r \leq 0$. It is easy to show that $\mathcal{E}_{r-1} \subset \mathcal{E}_r$.

Note that if $m = n$, then $\mathcal{E}_1$ is the ideal in $\mathbb{Z}[t^{-1}, t]$ generated by the polynomial $\det A$.

Problem 124. Let G be a finite Abelian group regarded as a $\mathbb{Z}$-module. Prove that for this module, the ideal $\mathcal{E}_1$ is generated by the number $|G|$ equal to the order of G.

1.3. Fundamental Theorem. The fundamental theorem, which relates the Seifert matrix A to the structure of the module $H_1(X_\infty)$ over the ring $\mathbb{Z}[t^{-1}, t]$, is as follows.

Theorem 6.2. *The matrix $tA - A^T$ is a presentation matrix of the module $H_1(X_\infty)$ over the ring $\mathbb{Z}[t^{-1}, t]$.*

Proof. We use the notation of the preceding section. Let us represent the space X_∞ as the union of its subspaces $M' = \bigcup M^3_{2i+1}$ and $M'' = \bigcup M^3_{2i}$. The intersection of these subspaces is a countable set of copies of the Seifert surface. In the Mayer–Vietoris sequence

$$\cdots \longrightarrow H_i(M' \cap M'') \longrightarrow H_i(M') \oplus H_i(M'') \longrightarrow H_i(X_\infty) \longrightarrow \cdots,$$

all groups are modules over the ring $\mathbb{Z}[t^{-1}, t]$. The action of t interchanges the groups $H_i(M')$ and $H_i(M'')$; hence it is not defined on any of these groups, but it is defined on $H_i(M') \oplus H_i(M'')$. The maps in the Mayer–Vietoris sequence are module homomorphisms.

We show that the segment

$$H_1(M' \cap M'') \xrightarrow{j_*} H_1(M') \oplus H_1(M'') \xrightarrow{i_*} H_1(X_\infty)$$

of the Mayer–Vietoris sequence determines a presentation of the module $H_1(X_\infty)$ by generators and relations. First, we verify that i_* is an epimorphism. This is so if and only if $\operatorname{Ker} \partial_* = H_1(X_\infty)$, i.e., $\operatorname{Im} \partial_* = 0$. The latter equality is equivalent to the map $j_*\colon H_0(M' \cap M'') \to H_0(M') \oplus H_0(M'')$

being a monomorphism. The groups $H_0(M' \cap M'')$ and $H_0(M') \oplus H_0(M'')$ are direct sums of the groups $H_0(M^3_k \cap M^3_{k+1})$ and $H_0(M^3_k)$, and each of these groups is isomorphic to $\mathbb{Z}$ because the Seifert surface F is connected. Therefore, the modules $H_0(M' \cap M'')$ and $H_0(M') \oplus H_0(M'')$ are isomorphic to $\mathbb{Z}[t^{-1}, t]$; the element t^k corresponds to $1 \in H_0(M^3_k \cap M^3_{k+1})$ and $1 \in H_0(M^3_k)$. Under this identification, the map j_* takes t^k to $t^k + t^{k+1}$. In particular, $j_*(1) = 1 + t \neq 0$; therefore, j_* is a monomorphism.

Choose a basis $\{f_i\}$ in $H_1(F) = H_1(M^3_k \cap M^3_{k+1})$, and let $\{e_i\}$ be the basis in $H_1(S^3 \setminus F) = H_1(M^3_k)$ dual to $\{f_i\}$ with respect to the form β. The modules $H_1(M' \cap M'')$ and $H_1(M') \oplus H_1(M'')$ are free. Their bases consist of the elements $1 \otimes f_i$ and $1 \otimes e_i$. Corresponding to the elements $f_i \in H_1(M^3_k \cap M^3_{k+1})$ and $e_i \in H_1(M^3_k)$ are $t^k \otimes f_i$ and $t^k \otimes e_i$. It is easy to show that

$$j_*(1 \otimes f_i) = t \otimes f_i^+ - 1 \otimes f_i^-.$$

Indeed, consider $f_i \in H_1(M^3_0 \cap M^3_1)$. The restriction of f_i to $M^3_0 \subset M''$ is f_i^- (the curve is translated in the direction opposite to that of the normal), and the restriction of f_i to $M^3_1 \subset M'$ is f_i^+ (the curve is translated in the direction of the normal). By definition, $j_* = (j', -j'')$; thus, $j'(1 \otimes f_i) \in H_1(M^3_1) = t \otimes H_1(M^3_0)$ and $j''(1 \otimes f_i) \in H_1(M^3_0)$.

Using the expressions for f_i^+ and f_i^- in terms of the Seifert matrix $A = (a_{ij})$, we obtain

$$j_*(1 \otimes f_i) = \sum_j a_{ji} t \otimes e_j - \sum_j a_{ij} \otimes e_j.$$

Thus, the matrix of the map j_* in the chosen bases is $tA - A^T$. But the matrix of j_* coincides with the presentation matrix of $H_1(X_\infty)$. $\square$

The elementary ideal $\mathcal{E}_r$ of the module $H_1(X_\infty)$ over the ring $\mathbb{Z}[t^{-1}, t]$ is called the rth *Alexander ideal* of the oriented link L. A generator of the minimal principal ideal containing the rth Alexander ideal is said to be an rth *Alexander polynomial* of the oriented link L. The first Alexander polynomial is called simply the *Alexander polynomial*; it is denoted by $\Delta_L(t)$.

A generator of the principal ideal is determined up to multiplication by a unity (an invertible element) of the ring. In the ring $\mathbb{Z}[t^{-1}, t]$, the unities are $\pm t^n$, where $n \in \mathbb{Z}$. Therefore, the rth Alexander polynomial is determined up to multiplication by $\pm t^n$.

According to Theorem 6.2, the module $H_1(X_\infty)$ has a square presentation matrix, namely, $tA - A^T$. Therefore, the first elementary ideal of this module is principal, and hence $\Delta_L(t) \doteq \det(tA - A^T)$; here and in what follows, the symbol $\doteq$ denotes equality up to multiplication by $\pm t^n$.

1.4. Properties of the Alexander Polynomial. First, note that if K is a trivial knot, then $\Delta_K(t) \doteq 1$. Indeed, in this case, we have $X_\infty = D^2 \times \mathbb{R}$, whence $H_1(X_\infty) = 0$.

Calculating the Alexander polynomial reduces to calculating the Seifert matrix $A = (a_{ij})$, where $a_{ij} = \operatorname{lk}(f_i, f_j^+)$. Let us calculate, for example, the Seifert matrix for the trefoil. By cutting and gluing, we can easily show that the surface in Figure 4 is a torus with a hole (see [**105**] for more details). Hence, to calculate the Seifert matrix for the trefoil, we can choose generators of the homology groups as in Figure 4. It is easy to show that the curves f_1^+ and f_2^+ are arranged as shown in Figure 5. Therefore, $A = \left(\begin{smallmatrix} 1 & -1 \\ 0 & 1 \end{smallmatrix}\right)$ and $tA - A^T = \left(\begin{smallmatrix} t-1 & -t \\ 1 & t-1 \end{smallmatrix}\right)$. Thus, the Alexander polynomial for the trefoil is equal to $t^2 - t + 1$.

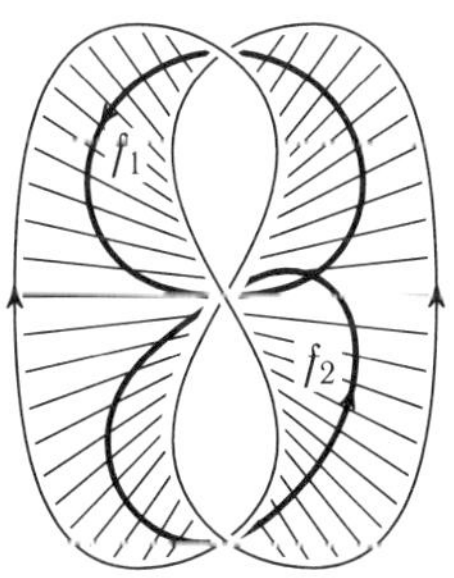

Figure 4. The trefoil

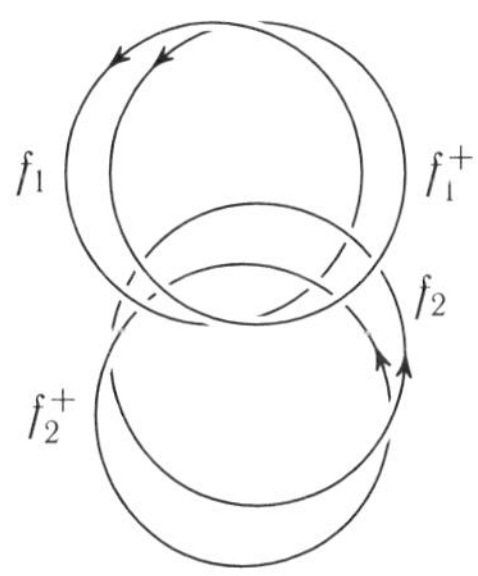

Figure 5. The curves f_i and f_i^+

Note that, in calculating the Seifert matrix, it is very convenient to use the "plane" Seifert surface. This surface for the trefoil is shown in Figure 6. Such a Seifert surface exists for any link. Indeed, look again at Figure 2. Let

Figure 6. The plane Seifert surface

us deform the Seifert surface so that the handles attached to the disk become long and narrow. Were it not for the possible twisting of the handles (see Figure 7a) it would be obvious how to arrange the Seifert surface. Since the surface is oriented, it follows that the number of twistings of each handle

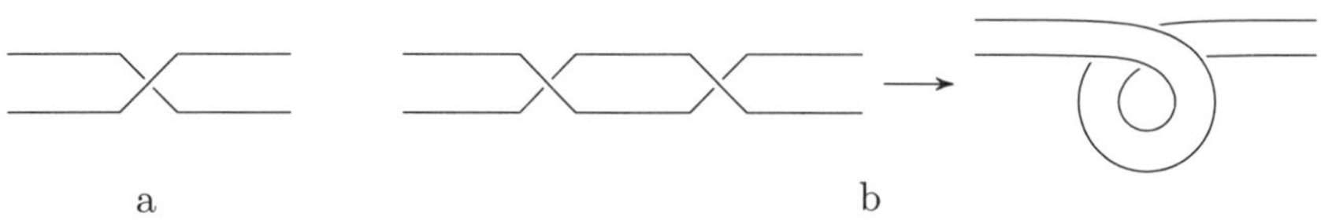

Figure 7. Twistings of a handle

is even. Two twistings in opposite directions cancel each other, and two twistings in the same direction are replaced by a loop (see Figure 7b).

Problem 125. Calculate the Alexander polynomial for the knot shown in Figure 8.

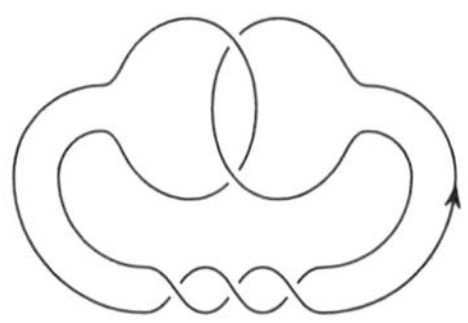

Figure 8. The knot diagram for Problem 125

Problem 126. (a) Given odd numbers p, q, and r, we can construct a knot $P(p,q,r)$ as shown in Figure 9 (the numbers p, q, and r may be negative; for each negative number, the types of crosses change). Calculate the Alexander polynomial for the knot $P(p,q,r)$.

(b) Prove that the Alexander polynomial of the knot $P(-3,5,7)$ is t; i.e., it coincides with the Alexander polynomial of the trivial knot. (This knot is called the *Seifert knot*.)

Now we prove some general properties of the Alexander polynomial. It follows from the properties of determinants that $\Delta_L(t) \doteq \Delta_L(t^{-1})$ for any oriented link L. Indeed, $\Delta_L(t) \doteq \det(tA - A^T) = (-t)^n \det(t^{-1}A - A^T) \doteq \Delta_L(t^{-1})$.

Theorem 6.3. (a) *If K is a knot, then $\Delta_K(1) \doteq 1$.*

(b) *If L is a link with at least two connected components, then $\Delta_L(1) \doteq 0$.*

Proof. Let L be an oriented knot or a link. Take the standard generating cycles $f_1, \dots, f_{2g+n-1}$ on its Seifert surface (a knot is characterized by $n = 1$). By definition, $\Delta_L(1) \doteq \det B$, where $B = A - A^T$, i.e., $b_{ij} = a_{ij} - a_{ji} = \operatorname{lk}(f_i, f_j^+) - \operatorname{lk}(f_j, f_i^+) = \operatorname{lk}(f_i^-, f_j) - \operatorname{lk}(f_i^+, f_j)$. To calculate this difference

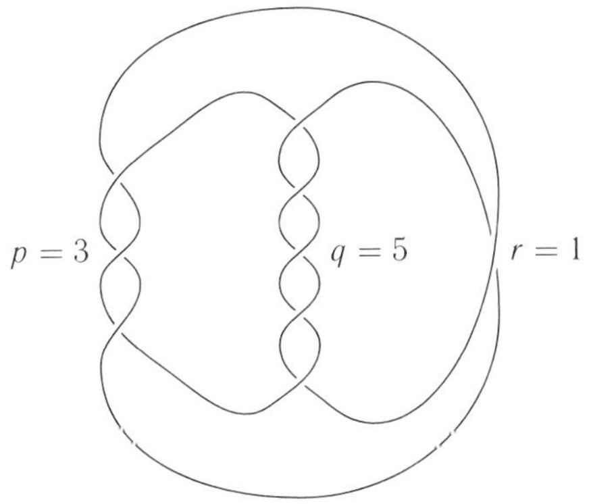

Figure 9. The knot diagram for Problem 126

of linking numbers, consider the strip F_i spanned by the curves f_i^- and f_i^+ and orient it so that its boundary is the cycle $f_i^- f_i^+$. The required difference of linking numbers is equal to the intersection number $\langle\langle F_i, f_j \rangle\rangle = \pm\langle\langle f_i, f_j \rangle\rangle$, where the last intersection number is calculated on the Seifert surface. It is easy to show that under the sign convention which we use, the sign in the last expression is minus.

When two 1-dimensional cycles are interchanged, their intersection number changes sign. Therefore, the matrix B consists of g diagonal blocks $\left(\begin{smallmatrix} 0 & 1 \\ -1 & 0 \end{smallmatrix}\right)$ and $n-1$ zeros on the diagonal. If $n = 1$, then $\det B = 1$, and if $n > 1$, then $\det B = 0$. $\square$

The Multivariable Alexander Polynomial. In constructing the Alexander polynomial $\Delta_L(t)$, we used the infinite cyclic covering $X_\infty \to S^3 \setminus L$. It is constructed by using the subgroup of $\pi_1(S^3 \setminus L)$ consisting of the loops for which the linking number with L is zero. For a knot, this is the commutator subgroup $[\pi_1, \pi_1]$. For a link, we can also consider the covering constructed modulo the subgroup $[\pi_1, \pi_1]$. This subgroup is normal; therefore, we obtain a regular covering $X \to S^3 \setminus L$ with automorphism group $\pi_1/[\pi_1, \pi_1] \cong H_1(S^3 \setminus L) \cong \mathbb{Z}^r$, where r is the number of components in the link L. The automorphism group of this covering is the free Abelian group with generators $t_1, \dots, t_r$, which correspond to small circles put on the components of L. Thus, $H_1(X; \mathbb{Z})$ is a module over the ring $\mathbb{Z}[t_1^{\pm 1}, \dots, t_r^{\pm 1}]$, and we can use this module to construct the Alexander polynomial $\Delta_L(t_1, \dots, t_r)$ in the variables $t_1, \dots, t_r$.

1.5. The Conway Polynomial.

The Alexander polynomial is determined up to multiplication by $\pm t^r$; actually, this is a whole class of polynomials rather than a single polynomial. The Conway polynomial is a uniquely determined polynomial in the class of Alexander polynomials, which is obtained by a change of variables. The Conway polynomials of different links can be summed, as opposed to the Alexander polynomials (sums of different representatives of two Alexander polynomials may belong to different Alexander polynomials). Moreover, there is a remarkable relation between the Conway polynomials of links depending on each other in a certain way, which makes it possible to calculate the Conway polynomial of any link directly from the diagram of this link.

To prove that the Conway polynomials are well defined, we need an auxiliary assertion concerning relationships among different Seifert surfaces of the same oriented link L. This assertion will also be used to prove that the Arf invariant of a link is well defined.

Let F be a Seifert surface of an oriented link $L \subset S^3$. Suppose that the cylinder $I \times D^2$ is embedded in S^3 so that its intersection with F is precisely the bases of the cylinder, i.e., $F \cap (I \times D^2) = \partial I \times D^2$; moreover, in a small neighborhood of the bases, the cylinder lies on one side of the surface F. Let us remove the bases of the cylinder from F and attach $I \times \partial D^2$ instead of them. As a result, we obtain another Seifert surface F' of the link L. We say that the Seifert surface F' is obtained from the Seifert surface F by *attaching a handle.*

Theorem 6.4. *Let F_1 and F_2 be Seifert surfaces of an oriented link L. Then there exists a sequence of Seifert surfaces $S_1, \dots, S_k$, where $S_1 = F_1$ and $S_k = F_2$, such that, for any $i = 1, \dots, n-1$, one of the surfaces S_i and S_{i+1} is obtained from the other by either attaching a handle or applying an isotopy (fixed on L).*

Proof. Applying a small deformation (fixed on L), if necessary, we can assume that the surfaces F_1 and F_2 transversally intersect each other in finitely many closed self-avoiding curves (the intersection along L is also transversal). Let M^3 be the closure of one of the domains into which the surfaces F_1 and F_2 divide S^3. Suppose that M^3 lies on one side of each of the surfaces F_1 and F_2; this means that if different parts of M^3 are adjacent to F_1 (or F_2), then all of them lie on one side of F_1 (or F_2); below, we shall show that such a manifold M^3 always exists. Let $\partial M^3 = \partial_1 M^3 \cup \partial_2 M^3$, where $\partial_i M^3 = \partial M^3 \cap F_i$. Any triangulation of the sphere S^3 for which F_1 and F_2 are subcomplexes can be restricted to M^3. Let A^3 be the union of all simplices from the second barycentric subdivision of this triangulation of M^3 that intersect $\partial_1 M^3$ or some 1-simplices in the initial triangulation. Informally, A^3 is the union of the ε-neighborhood (collar) of the boundary $\partial_1 M^3$ in M^3 and the ε-neighborhood of the 1-skeleton of the initial triangulation. Let F_1' be the surface obtained from F_1 by removing $\partial_1 M^3$ and attaching the closure of the manifold $\partial A^3 \setminus \partial_1 M^3$ instead. This operation reduces to a sequence of isotopies and attachments of handles.

Indeed, the transformation in question is as follows. We have a surface to which several edges of a graph are attached so that all of the edges are on one side of the surface. Consider the boundary of the ε-neighborhood of the graph. To the initial surface we attach the part of this boundary that lies on the same side of the surface as the graph; the fragments of the initial surface that are contained in the ε-neighborhood of the graph are removed. This procedure reduces to successive isotopies and attachments of handles.

The surface F_2 is changed as follows. Let B^3 be the closure of $M^3 \setminus A^3$, i.e., the union of the cones whose vertices are the centers of the 3-simplices from the initial triangulation and bases are the stars (in the second barycentric subdivision) of the centers of those 2-faces of these simplices which are not contained in $\partial_1 M^3$. Removing $B^3 \cap F_2$ from the surface F_2 and attaching the closure of $\partial B^3 \setminus (B^3 \cap F_2)$ instead, we obtain a surface F'. This transformation is of the same form. Indeed, if some 3-simplex has n 2-faces not contained in $\partial_1 M^3$, then this simplex corresponds to the attachment of the boundary of the ε-neighborhood of a graph consisting of n edges going from one vertex.

The surfaces F_1' and F_2' coincide inside M^3, and $F_1' \cap F_2' = (F_1 \cap F_2) \cup (\partial A^3 \setminus \partial_1 M^3)$. But, translating F_1 along the collar $\partial_1 M^3$, we might violate the necessary condition $\partial F_1 = L$. This must be fixed. Each connected component of the set $F_1 \cap F_2$ is either contained entirely in the boundary of M^3 or does not intersect it. If some component K of the link L is contained in the boundary of M^3, then we change F_1' by an isotopy that moves $\partial F_1'$ back along the collar until $\partial F_1'$ coincides with K.

These transformations of the Seifert surfaces F_1 and F_2 decrease the number of domains into which these surfaces divide the sphere S^3. Therefore, we can achieve the situation that F_1 and F_2 do not divide S^3, i.e., coincide.

It remains to prove that there always exists a manifold M^3 such that it lies on one side of both F_1 and F_2. For this purpose, we use the infinite cyclic covering $p\colon X_\infty \to S^3 \setminus L$ constructed in Section 1.2. The space X_∞ can be constructed by using either of the surfaces F_1 and F_2. Therefore, X_∞ contains the two families of surfaces $t^i \tilde{F}_1$ and $t^j \tilde{F}_2$ (these are the boundaries of the 3-manifolds constituting X_∞). We are interested in the case where F_1 and F_2 intersect not only in L. The idea is to consider the maximum integer n for which $\tilde{F}_2 \cap t^n \tilde{F}_1 \neq \varnothing$ (see Figure 10). For such an integer to exist, we assume that the set L is removed from S^3 together with its open ε-neighborhood, so that the space obtained is compact. Now we deal with compact sets, and the required number n does exist.

The surface $\tilde{F}_2$ cuts X_∞ into two parts; let Y be the part that contains the surfaces $t^r \tilde{F}_2$ for $r < 0$. The surface $t^n \tilde{F}_1$ also cuts X_∞; let Z be the part

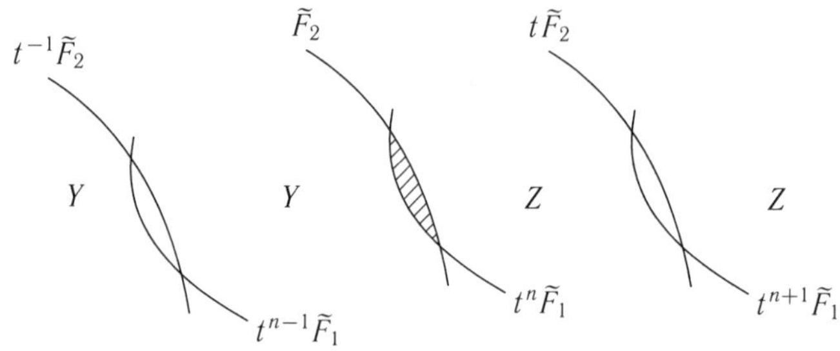

Figure 10. The choice of the manifold M^3

of X_∞ containing the $t^{n+r}\tilde{F}_1$ for $r > 0$. It is easy to see that the projection $p(Y \cap Z)$ is the required manifold. $\square$

Two square matrices are said to be *S-equivalent* if one of them can be obtained from the other by applying several operations of the following two types:

(1) $A \mapsto PAP^T$, where P is an integer matrix with determinant ± 1;

(2) $A \leftrightarrow \begin{pmatrix} A & \alpha & 0 \\ 0 & 0 & 1 \\ 0 & 0 & 0 \end{pmatrix}$ or $A \leftrightarrow \begin{pmatrix} A & 0 & 0 \\ \beta & 0 & 0 \\ 0 & 1 & 0 \end{pmatrix}$, where α is a column and β is a row.

Theorem 6.5. *Any two Seifert matrices of the same oriented link are S-equivalent.*

Proof. A basis change in $H_1(F; \mathbb{Z})$ leads to a type (1) transformation of the Seifert matrix. According to Theorem 6.4, it remains to determine what happens to the Seifert matrix when a handle is attached. We can assume that, under this operation, the basis of $H_1(F; \mathbb{Z})$ is supplemented by the generators f_{n+1} and f_{n+2} shown in Figure 11. If $i \leq n$, then $\operatorname{lk}(f_{n+2}^{\pm}, f_i) = 0$. Choosing

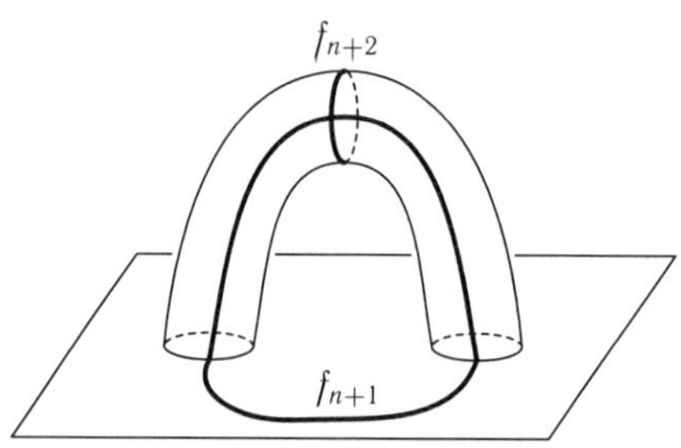

Figure 11. The additional generators

appropriate orientations of the curves f_{n+1} and f_{n+2}, we can assume that either $\operatorname{lk}(f_{n+1}^+, f_{n+2}) = 0$ and $\operatorname{lk}(f_{n+1}^-, f_{n+2}) = 1$ or $\operatorname{lk}(f_{n+1}^+, f_{n+2}) = 1$ and $\operatorname{lk}(f_{n+1}^-, f_{n+2}) = 0$. In the former case, the new Seifert matrix is $\begin{pmatrix} A & \alpha & 0 \\ * & * & 1 \\ 0 & 0 & 0 \end{pmatrix}$.

Transformations of type (1) reduce this matrix to the form $\begin{pmatrix} A & \alpha & 0 \\ 0 & 0 & 1 \\ 0 & 0 & 0 \end{pmatrix}$. Similarly, in the latter case, the matrix reduces to the form $\begin{pmatrix} A & 0 & 0 \\ \beta & 0 & 0 \\ 0 & 1 & 0 \end{pmatrix}$. □

Let us introduce the formal variable $t^{1/2}$, for which $(t^{1/2})^2 = t$, and define the *Alexander polynomial in Conway's normalization* $\Delta_L(t) \in \mathbb{Z}[t^{-1/2}, t^{1/2}]$ by $\Delta_L(t) = \det(t^{1/2}A - t^{-1/2}A^T)$, where A is the Seifert matrix of the link L. If A is a matrix of order r, then $\Delta_L(t) = t^{-r/2}\det(tA - A^T)$; therefore, the Alexander polynomial in Conway's normalization coincides with the Alexander polynomial up to multiplication by a unit of the ring $\mathbb{Z}[t^{-1/2}, t^{1/2}]$.

Theorem 6.6. *For any oriented link L, the Alexander polynomial in Conway's normalization is determined uniquely; i.e., it does not depend on the choice of the Seifert matrix A.*

Proof. First, note that

$$\begin{aligned}\det(t^{1/2}P^TAP - t^{-1/2}P^TA^TP) &= \det(P^T(t^{1/2}A - t^{-1/2}A^T)P) \\ &= (\det P)^2 \det(t^{1/2}A - t^{-1/2}A^T).\end{aligned}$$

Let $B = \begin{pmatrix} A & \alpha & 0 \\ 0 & 0 & 1 \\ 0 & 0 & 0 \end{pmatrix}$. Then

$$t^{1/2}B - t^{-1/2}B^T = \begin{pmatrix} t^{1/2}A - t^{-1/2}A^T & t^{1/2}\alpha & 0 \\ -t^{-1/2}\alpha^T & 0 & t^{1/2} \\ 0 & -t^{-1/2} & 0 \end{pmatrix}.$$

Multiplying the last row of this matrix by suitable elements of the ring $\mathbb{Z}[t^{-1}, t]$ and adding the result to the remaining rows, we kill the column $t^{1/2}\alpha$. The row $-t^{-1/2}\alpha^T$ can be killed in a similar way. Therefore,

$$\begin{aligned}\det(t^{1/2}B - t^{-1/2}B^T) &= \det(t^{1/2}A - t^{-1/2}A^T)\det\begin{pmatrix} 0 & t^{1/2} \\ -t^{-1/2} & 0 \end{pmatrix} \\ &= \det(t^{1/2}A - t^{-1/2}A^T)\end{aligned}$$

because $\det\begin{pmatrix} 0 & t^{1/2} \\ -t^{-1/2} & 0 \end{pmatrix} = 1$. For a transformation of type (2), the proof is similar. □

The proof of Theorem 6.3 implies the following properties of the Alexander polynomial in Conway's normalization.

(i) If K is a knot, then $\Delta_K(t) \in \mathbb{Z}[t^{-1}, t]$, $\Delta_K(1) = 1$, and $\Delta_K(t) = \Delta_K(t^{-1})$.

(ii) If L is a link with at least two connected components, then $\Delta_L(1) = 0$.

Theorem 6.7. *Suppose that L_+, L_-, and L_0 are oriented links whose diagrams coincide everywhere except in a small disk, in which they are as shown in Figure 12. Then their Alexander polynomials in Conway's normalization are related by*

$$\Delta_{L_+} - \Delta_{L_-} = (t^{-1/2} - t^{1/2})\Delta_{L_0}. \tag{41}$$

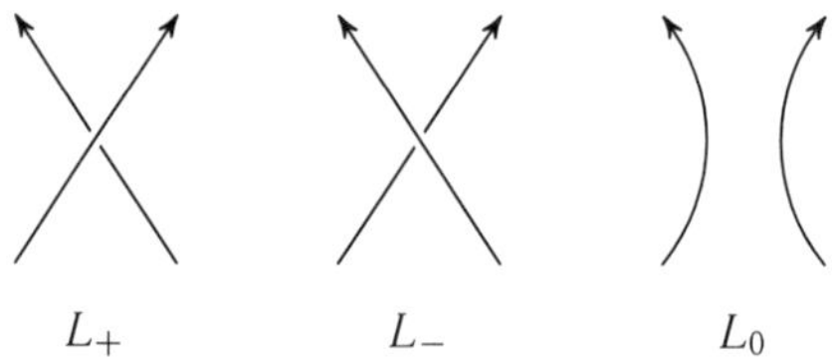

Figure 12. The three links

Proof. Let us construct a Seifert surface F_0 for the link L_0 so that, in the disk under consideration, it has the form shown in Figure 13. Attaching

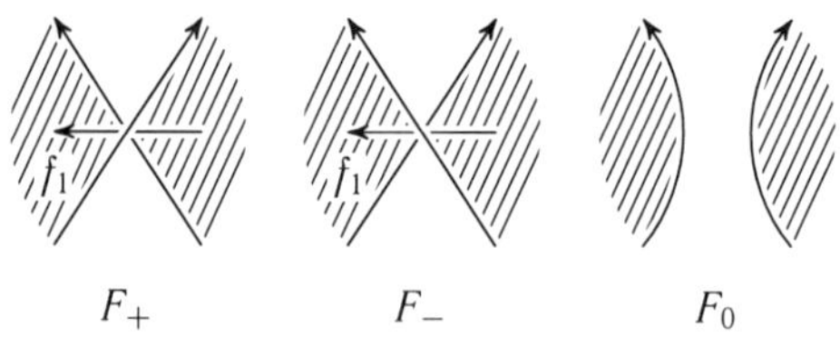

Figure 13. The three Seifert surfaces

a twisted strip to L_0, we obtain Seifert surfaces F_+ and F_- for the links L_+ and L_-. Take the basis in $H_1(F_\pm)$ that consists of the class of a closed curve f_1 going along the strip and the basis elements of $H_1(F_0)$. If A_0 is the Seifert matrix for L_0, then the Seifert matrices for L_- and L_+ are $\begin{pmatrix} n & \alpha \\ \beta & A_0 \end{pmatrix}$ and $\begin{pmatrix} n-1 & \alpha \\ \beta & A_0 \end{pmatrix}$, where n is an integer. Considering the determinants of the matrices $t^{1/2}A - t^{-1/2}A^T$ for the three Seifert matrices and taking into account the fact that $\det \begin{pmatrix} a & x \\ y & A_0 \end{pmatrix} = a \det A_0 + c$, where c does not depend on a, we obtain the required result. □

Relation (41), together with the condition that the Alexander polynomial in Conway's normalization for the trivial knot equals 1, makes it possible to calculate Δ_L for any oriented link L (see [**108**] for details). Therefore, Δ_L is a polynomial in $z = t^{-1/2} - t^{1/2}$. The polynomial $\nabla_L(z)$ such that $\nabla_L(t^{-1/2} - t^{1/2}) = \Delta_L(t)$, where Δ_L is the Alexander polynomial in Conway's normalization, is called the *Conway polynomial*. The Conway polynomial satisfies the *skein relation*

$$\nabla_{L_+}(z) - \nabla_{L_-}(z) = z\nabla_{L_0}(z). \tag{42}$$

Problem 127. Prove that if a link L lies on both sides of a plane disjoint from this link, then $\nabla_L(z) = 0$.

Let us write the Conway polynomial in the form $\nabla_L(z) = a_0(L) + a_1(L)z + a_2(L)z^2 + \cdots$.

Problem 128. Let $n(L)$ denote the number of components in the link L. Prove that $a_i(L) = 0$ for $i \equiv n(L) \pmod 2$.

Problem 129. (a) Prove that if K is a knot, then $a_0(K) = 1$.

(b) Prove that if L is a link with more than one component, then $a_0(L) = 0$.

Problem 130. Let L be a link with $n(L)$ components. Prove that $a_i(L) = 0$ for $i < n(L) - 1$.

Problem 131. Prove that if L is a link with precisely two components L_1 and L_2, then $a_1(L) = \mathrm{lk}(L_1, L_2)$.

Problem 132. Suppose that L_+ and L_- are knots and L_0 is the link with components L_1 and L_2. Prove that $a_2(L_+) - a_2(L_-) = \mathrm{lk}(L_1, L_2)$.

2. The Arf Invariant

2.1. The Arf Invariant of Quadratic Forms. Associated with each real quadratic form $q(x)$ is the symmetric bilinear form $B(x, y)$ for which $B(x, x) = q(x)$. The identity $q(x + y) - q(x) - q(y) = 2B(x, y)$ makes it possible to reconstruct B from q. For the field $\mathbb{Z}_2$, the definition of a quadratic form and the associated bilinear form changes as follows. Let V be a finite-dimensional space over $\mathbb{Z}_2$. A map $q\colon V \to \mathbb{Z}_2$ is called a *quadratic form* if there exists a bilinear form $B(x, y)$ for which $q(x+y)+q(x)+q(y) = B(x, y)$. Of course, the signs do not matter; it only matters that $2B(x, y)$ is replaced by $B(x, y)$, because division by 2 is not defined over the field $\mathbb{Z}_2$. Note that, in contrast with the real case, $q(x)$ is not equal to $B(x, x)$. The quadratic form q is said to be *nondegenerate* if the bilinear form B is nondegenerate.

It is seen from the definition that $B(x, y) = B(y, x)$ and $B(x, x) = q(2x)+2q(x) = 0$. Therefore, the same argument as in the case of real skew-symmetric matrices shows that if q is a nondegenerate quadratic form, then the space V has a basis $e_1, f_1, \ldots, e_n, f_n$ such that $B(e_i, e_j) = B(f_i, f_j) = 0$ and $B(e_i, f_j) = \delta_{ij}$; such a basis is called a *symplectic* basis.

Let q be a nondegenerate quadratic form, and let $e_1, \ldots, f_n$ be a symplectic basis. The element $c(q) = \sum_{i=1}^n q(e_i)q(f_i) \in \mathbb{Z}_2$ of the group $\mathbb{Z}_2$ is called the *Arf invariant* of q.

Theorem 6.8 (Arf [**8**]). *Any nondegenerate quadratic form q can be reduced to the form $x_1y_1 + \cdots + x_ny_n + c(q)(x_n^2 + y_n^2)$, where $c(q)$ is the Arf invariant. Moreover, the Arf invariant $c(q)$ does not depend on the choice of a symplectic basis.*

Proof. First, we consider the case $n = 1$. The equality $q(e_1) + q(f_1) + q(e_1 + f_1) = 1$ implies that either one of the elements $q(e_1)$, $q(f_1)$, and $q(e_1 + f_1)$ is equal to 1 and the two others are 0 or all of these elements are equal to 1. The elements e_1, f_1, and $e_1 + f_1$ are equivalent in the sense that any two of them can be taken for a symplectic basis because $B(e_1 + f_1, e_1 + f_1) = 0$ and $B(e_1 + f_1, e_1) = 1$. Therefore, we can assume that $q(e_1 + f_1) = 1$. Thus, for $n = 1$, we obtain two nondegenerate quadratic forms, q_0 and q_1; here $q_0(e_1) = q_0(f_1) = 0$ and $q_1(e_1) = q_1(f_1) = 1$. In coordinates, these forms can be written as $q_0 = x_1y_1$ and $q_1 = x_1y_1 + x_1^2 + y_1^2$. Clearly, $c(q_0) = 0$ and $c(q_1) = 1$.

To prove the independence of the Arf invariant on the choice of a symplectic basis, we must show that the forms q_0 and q_1 are not equivalent, that is, cannot be obtained from each other by a basis change. Indeed, the form q_0 takes the value 1 precisely once, while q_1 takes this value three times.

Now consider an arbitrary nondegenerate quadratic form q. Choose a symplectic basis $e_1, \dots, f_n$. Let φ_i be the restriction of q to the subspace spanned by the vectors e_i and f_i. Then $q = \varphi_1 \oplus \cdots \oplus \varphi_n$, where each form φ_i is equivalent to q_0 or to q_1.

Lemma 1. *The quadratic forms $\psi_0 = q_0 \oplus q_0$ and $\psi_1 = q_1 \oplus q_1$ are equivalent.*

Proof. Let e_1, f_1, e_2, f_2 be a symplectic basis for which $\psi_0(e_i) = \psi_0(f_i) = 0$ and $\psi_1(e_i) = \psi_1(f_i) = 1$. Consider the basis $e_1' = e_1 + e_2$, $f_1' = e_1 + f_2$, $e_2' = e_1 + f_1 + e_2 + f_2$, $f_2' = e_1 + f_1 + f_2$. It is easy to show that this basis is symplectic. Let us show that $\psi_1(e_i') = \psi_1(f_i') = 0$, i.e., $\psi_1(e_i') = \psi_0(e_i)$ and $\psi_1(f_i') = \psi_0(f_i)$. Indeed, we have $\psi_1(e_1 + e_2) = \psi_1(e_1) + \psi_1(e_2) + B(e_1, e_2) = 1 + 1 + 0 = 0$, $\psi_1(e_1 + f_2) = \psi_1(e_1) + \psi_1(f_2) + B(e_1, f_2) = 1 + 1 + 0 = 0$, $\psi_1(e_1 + f_1 + e_2 + f_2) = \psi_1(e_1 + e_2) + \psi_1(f_1 + f_2) + B(e_1 + e_2, f_1 + f_2) = 0 + 0 + 0 = 0$, and $\psi_1(e_1 + f_1 + f_2) = \psi_1(e_1 + f_2) + \psi_1(f_1) + B(e_1 + f_2, f_1) = 0 + 1 + 1 = 0$. □

We introduce the notation

$$\underbrace{q \oplus \cdots \oplus q}_{n \text{ times}} = nq.$$

Lemma 1 implies that the form q is equivalent to either nq_0 or $(n-1)q_0 \oplus q_1$. In coordinates, these forms are written as $x_1y_1 + \cdots + x_ny_n$ and $x_1y_1 + \cdots + x_ny_n + x_n^2 + y_n^2$, respectively. Clearly, $c(nq_0) = 0$ and $c((n-1)q_0 \oplus q_1) = 1$.

To prove that the Arf invariant does not depend on the choice of a symplectic basis, we must prove that the forms nq_0 and $(n-1)q_0 \oplus q_1$ are not equivalent.

Lemma 2. *The form nq_0 takes the value 1 precisely $2^{2n-1} - 2^{n-1}$ times, and the form $(n-1)q_0 \oplus q_1$ takes this value precisely $2^{2n-1} + 2^{n-1}$ times.*

Proof. We prove the lemma by induction on n. For $n = 1$, the assertion is easy to check. Suppose that we have already proved that the form nq_0 takes the value 1 precisely $2^{2n-1} - 2^{n-1}$ times. The form q_0 takes the value 1 precisely once and the value 0 precisely three times. Therefore, the form $(n+1)q_0$ takes the value 1 precisely $3(2^{2n-1} - 2^{n-1}) + 2^{2n-1} + 2^{n-1} = 2^{2n+1} - 2^n$ times (because nq_0 takes the value 0 precisely $2^{2n-1} + 2^{n-1}$ times). The form q_1 takes the value 1 precisely three times and the value 0 precisely once. Therefore, the form $nq_0 \oplus q_1$ takes the value 1 precisely $3(2^{2n-1} + 2^{n-1}) + 2^{2n-1} - 2^{n-1}$ times. □

This completes the proof of the theorem. □

2.2. The Arf Invariant of Oriented Links. Let $L \subset S^3$ be an oriented link, and let F be its Seifert surface. Consider the map $q\colon H_1(F;\mathbb{Z}_2) \to \mathbb{Z}_2$ defined by $q(x) = \alpha_2(x,x)$, where α_2 is the Seifert form α reduced modulo 2. For example, if the homology class of x is represented by a closed self-avoiding curve on the surface F, then $q(x)$ is the number of full turns (modulo 2) of the ε-neighborhood of this curve in the surface F. Clearly, $q(x+y) = \alpha_2(x,x) + \alpha_2(x,y) + \alpha_2(y,x) + \alpha_2(y,y)$; therefore, $q(x+y) + q(x) + q(y) = \alpha_2(x,y) + \alpha_2(y,x)$. Let us show that the right-hand side of this equality equals $f(x,y)$, where $f(x,y)$ is the intersection form reduced modulo 2. Indeed, $\alpha(x,y) + \alpha(y,x) = \operatorname{lk}(x^-,y) + \operatorname{lk}(y^-,x) = \operatorname{lk}(x^-,y) + \operatorname{lk}(y,x^+)$. It is seen from the definition of linking number that the parity of $\operatorname{lk}(x^-,y) + \operatorname{lk}(y,x^+)$ coincides with that of the number of intersection points of curves representing y and x.

Thus, q is the quadratic form corresponding to the bilinear form f. Look at Figure 2 on p. 302. It shows that the intersection form is nondegenerate if and only if the boundary of the Seifert surface is connected, i.e., when we deal with a knot. For this reason, the definition of the Arf invariant for a knot is simpler than for links with several components: the *Arf invariant* of a knot is defined as the Arf invariant of the quadratic form q. For links with several components, the definition of the Arf invariant is not so simple; moreover, the Arf invariant is not defined for some links.

To obtain a nondegenerate form from the intersection form, we must take the quotient of the space $H_1(F;\mathbb{Z}_2)$ modulo the subspace generated by the cycles $f_{2g+1}, \dots, f_{2g+n-1}$ corresponding to $n-1$ boundary components. The cycle corresponding to the remaining boundary component is homologous to the sum of these cycles. Thus, consider the quotient space $H_1(F;\mathbb{Z}_2)/i_*H_1(\partial F;\mathbb{Z}_2)$, where $i\colon \partial F \to F$ is the natural embedding.

For the form q to be well defined on this quotient space, each component L_i must satisfy the condition $q(L_i)=0$, or, equivalently, $\operatorname{lk}(L_i^-, L_i)=0$. The Seifert surface does not intersect L_i^-; therefore, the linking coefficient of L_i^- and L_i is congruent modulo 2 to that of L_i^- and $L-L_i$, which, in turn, is equal to the linking coefficient of L_i and $L-L_i$. Thus, we can assume that the link L has the property

$$\operatorname{lk}(L_i^-, L_i) \equiv 0 \pmod 2 \quad \text{for all } i. \tag{43}$$

Then the form q is well defined on the quotient space. The *Arf invariant* $\mathcal{A}(L)$ of a link L with property (43) is defined as the Arf invariant of the quadratic form q on the quotient space $H_1(F;\mathbb{Z}_2)/i_*H_1(\partial F;\mathbb{Z}_2)$.

Theorem 6.9. *The Arf invariant $\mathcal{A}(L)$ of an oriented link L is well defined, i.e., it does not depend on the choice of the surface F.*

Proof. According to Theorem 6.4, it suffices to show that the Arf invariant does not change when a handle is attached to F. The first $2g$ curves in Figure 2 constitute a symplectic basis $e_1, f_1, \dots, e_g, f_g$ for the form q in the space $H_1(F)/i_*H_1(\partial F)$. After a handle is attached, we complete it by the curves e_{g+1} and f_{g+1} chosen as shown in Figure 11 on p. 314. If e_{g+1} is the curve f_{n+2} in this figure, then $q(e_{g+1})=0$. Therefore, $\sum_{k=1}^{g} q(e_k)q(f_k) = \sum_{k=1}^{g+1} q(e_k)q(f_k)$. □

Clearly, if K is the unknot, then $\mathcal{A}(K)=0$. It is also easy to verify that if K is the trefoil, then $\mathcal{A}(K)=1$. Indeed, in calculating the Alexander polynomial for the trefoil (see p. 309), we constructed a basis f_1, f_2. By the definition of the form $f(x,y)$, we have $f(x,x)=0$. Moreover, $f(f_1,f_2)=1$ because f_1 and f_2 intersect in one point. Hence $\mathcal{A}(K)=q(f_1)q(f_2)=1$.

Theorem 6.10. *Suppose that K_+ and K_- are two knots and L is a two-component link, and their diagrams coincide everywhere except for a small disk, in which they are as shown in Figure* 14*. Then*

$$\mathcal{A}(K_+) \equiv \mathcal{A}(K_-) + \operatorname{lk}(L_1, L_2) \pmod 2,$$

where L_1 and L_2 are the components of the link L.

Proof. For K_+ and K_-, we can construct Seifert surfaces F_+ and F_- which are arranged as shown in Figure 15 over the small disk under consideration and coincide otherwise. Using F_+ and F_-, we construct a surface F_0 for the

Figure 14. The diagrams of the two knots and the link

link L. If this surface is disconnected, then, to obtain a Seifert surface for L, we attach a handle to all of the three surfaces, thus rendering F_0 connected.

Let $e_1, f_1, \ldots, e_n, f_n$ be a symplectic basis for the form q on the space $H_1(F_0; \mathbb{Z}_2)$. Let us complete it to a symplectic basis by adding the cycles e_{n+1} and $f_{n+1,\pm}$ shown in Figure 15; the cycle e_{n+1} is represented by the

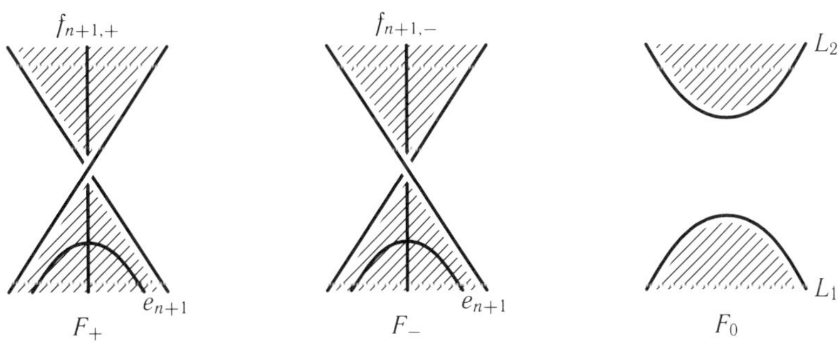

Figure 15. The three Seifert surfaces

curve L_1. The strips corresponding to the curves $f_{n+1,+}$ and $f_{n+1,-}$ differ by precisely one full turn; therefore, $q(f_{n+1,+}) - q(f_{n+1,-}) = 1$. Thus,

$$\begin{aligned}\mathcal{A}(K_+) - \mathcal{A}(K_-) &= q(e_{n+1})(q(f_{n+1,+}) - q(f_{n+1,-})) = q(e_{n+1}) \\ &= \operatorname{lk}(L_1^-, L_1) = \operatorname{lk}(L_1, L - L_1) = \operatorname{lk}(L_1, L_2). \qquad \square\end{aligned}$$

Theorem 6.11. (a) *For any knot K, $\mathcal{A}(K) \equiv a_2(K) \pmod 2$, where $a_2(K)$ is the coefficient of z^2 in the Conway polynomial $\nabla_K(z)$.*

(b) *For any knot K,*

$$\mathcal{A}(K) = \begin{cases} 0 & \textit{if } \Delta_K(-1) \equiv \pm 1 \pmod 8, \\ 1 & \textit{if } \Delta_K(-1) \equiv \pm 3 \pmod 8, \end{cases}$$

where $\Delta_K(t)$ is the Alexander polynomial in Conway's normalization.

Proof. (a) If L_+ and L_- are knots, then L_0 is a two-component link. According to Problem 132,we have $a_2(L_+) - a_2(L_-) = \operatorname{lk}(L_1, L_2)$, where L_1

and L_2 are the components of the link L. Theorem 6.10 implies $\mathcal{A}(L_+) \equiv \mathcal{A}(L_-) + \operatorname{lk}(L_1, L_2) \pmod 2$. Moreover, if K is the unknot, then $a_2(K) = 0 = \mathcal{A}(K)$. Therefore, $\mathcal{A}(K) \equiv a_2(K) \pmod 2$ for any knot K.

(b) By definition, $\Delta_K(t) = \nabla_K(t^{-1/2} - t^{1/2})$; therefore, $\Delta_K(-1) = \nabla_K(-2i)$. Problem 128 implies $\nabla_K(z) = 1 + a_2(K)z^2 + a_4(K)z^4 + \cdots$. Hence $\nabla_K(-2i) \equiv 1 - 4a_2(K) \pmod 8$. If $a_2(K)$ is even, then $\nabla_K(-2i) \equiv 1 \pmod 8$, and if $a_2(K)$ is odd, then $\nabla_K(-2i) \equiv -3 \pmod 8$. □

2.3. Knotting of Embeddings of the Graph K_7. Using properties of the Arf invariant, Conway and Gordon [**26**] proved the following theorem.

Theorem 6.12. *Any embedding of K_7 (the complete graph on 7 vertices) in $\mathbb{R}^3$ contains a nontrivial knot.*

Proof. For every embedding of K_7 in $\mathbb{R}^3$, we define $\sigma \in \mathbb{Z}_2$ by $\sigma = \sum \mathcal{A}(C)$, where the summation is over all of the $360 = \frac{6!}{2}$ cycles C of length 7 passing through all vertices of the graph K_7. Any embedding of the graph K_7 in $\mathbb{R}^3$ can be transformed into a given embedding by applying the transformations of edges shown in Figure 16. Let us prove that σ does not change under such transformations.

Generally, we must consider three types of transformations of crosses, namely, a self-intersection of an edge, an intersection of two adjacent edges,

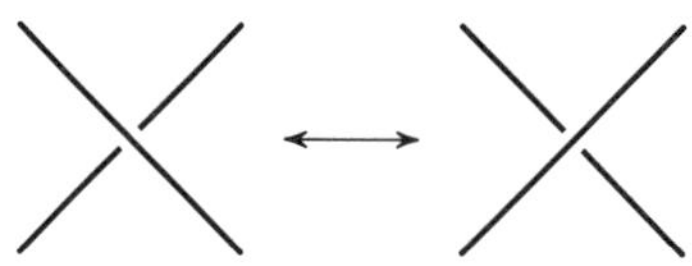

Figure 16. The transformations of edges

and an intersection of two nonadjacent edges. But it is seen from Figure 17 that in K_7 one intersection of an edge with itself can be replaced by five intersections of different edges; therefore, it suffices to consider only intersections of different edges (adjacent or nonadjacent).

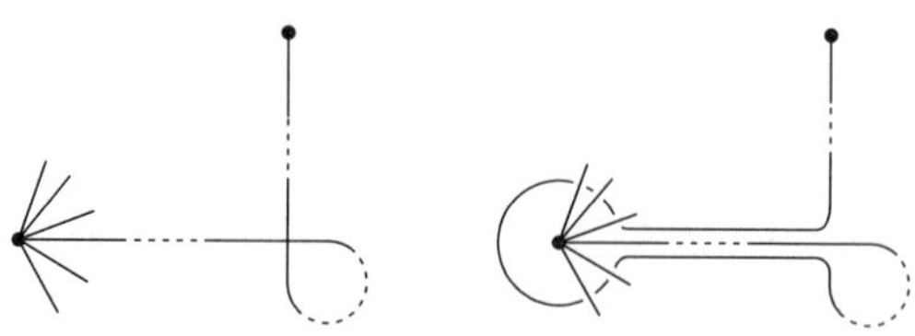

Figure 17. A self-intersection of an edge

Dragging the vertices of the intersecting edges[1] to the cross under consideration, we obtain a diagram of the form shown in Figure 18. Figure 18a refers to an intersection of adjacent edges, and Figure 18b, to an intersection of nonadjacent edges (of course, the cross itself may be of the opposite type).

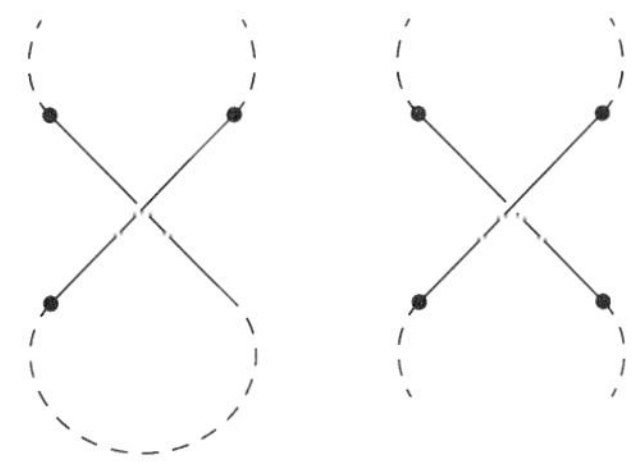

Figure 18. Simplification of a cross

Let us see what happens to $\mathcal{A}(C)$ under a change of the type of a cross. We denote the variation of $\mathcal{A}(C)$ by $\varepsilon(C)$. If the cycle C does not contain at least one of the two edges in the cross under consideration, then $\varepsilon(C) = 0$. First, suppose that the cycle C contains both of these edges. Then, according to Theorem 6.10, we have $\varepsilon(C) \equiv \operatorname{lk}(L_1, L_2) \pmod 2$, where L_1 and L_2 are the components of the link L constructed as shown in Figure 14 (p. 321).

Suppose that the intersecting edges are adjacent. The construction of the link $L = L_1 \cup L_2$ in this case is shown in Figure 19a. Note that the component

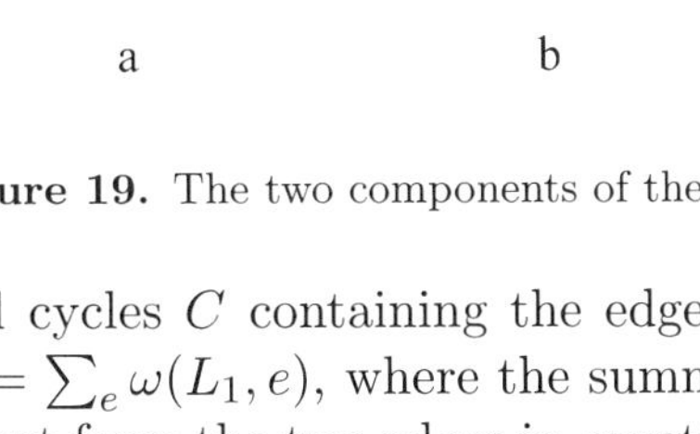

Figure 19. The two components of the link

L_1 is the same for all cycles C containing the edges under consideration. Therefore, $\operatorname{lk}(L_1, L_2) = \sum_e \omega(L_1, e)$, where the summation is over all edges e of the cycle C different from the two edges in question, and $\omega(L_1, e)$ is the number of crosses at which L_1 passes over e with signs taken into account. The change of σ equals $\sum_C \sum_e \omega(L_1, e) \pmod 2$, where the first sum is

[1]Together with the edges incident to them.

over the cycles C containing both edges under consideration. Thus, it is sufficient to prove that for each edge e of K_7 different from the two edges in question, the number of cycles C containing e is even. The following cases are possible.

1. The edge e and the two edges under consideration are either incident with the same vertex or form a cycle. Then no cycles C contain e.

2. The edge e shares a vertex with precisely one of the edges under consideration. Then precisely six cycles C contain e.

3. The edge e has no common vertices with the edges under consideration. Then precisely 12 cycles C contain e.

Now, suppose that the intersecting edges are nonadjacent. The construction of the link $L = L_1 \cup L_2$ in this case is shown in Figure 19b. We have $\mathrm{lk}(L_1, L_2) = \sum \omega(e_1, e_2)$, where the summation is over all pairs of edges $e_1 \subset L_1$, $e_2 \subset L_2$.

We number the vertices of the graph K_7 starting with the endvertices of the two edges under consideration as shown in Figure 20; the arrows indicate

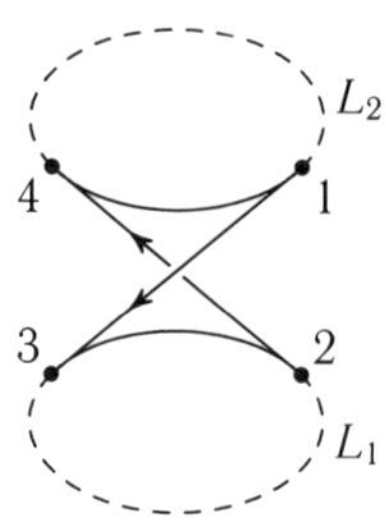

Figure 20. The numbering of graph vertices

the direction of traversing the cycle C. The curves L_1 and L_2 may contain the following edges e_1 and e_2 (up to a renumbering of vertices): (a) (23), (45); (b) (23), (56); (c) (27), (45); (d) (27), (56). It is easy to see that in these cases, the cycles C are as follows:

(a) (1234567), (1234576);

(b) (1234567), (1234657), (1234756), (1234765);

(c) (1273456), (1276345);

(d) (1273456), (1273465).

In every case, an even number of cycles C is obtained; therefore, σ does not change.

A simple verification shows that for the embedding of K_7 in $\mathbb{R}^3$ shown in Figure 21, all but one of the cycles C are unknotted. Moreover, the knotted

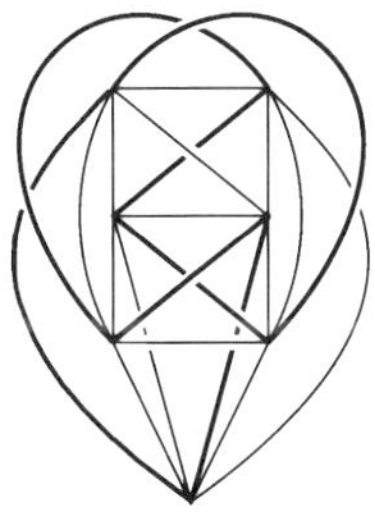

Figure 21. An example of an embedding

cycle is the trefoil, and the Arf invariant of the trefoil is 1. Thus, we have $\sigma = 1$ for any embedding, so there always exists a knotted cycle C. □

Remark. In Part I, we proved that for any embedding of the graph K_6 in $\mathbb{R}^3$, there exists a pair of linked cycles (Theorem 1.10). The proof us much simpler than that of Theorem 6.12 about knotting of cycles. It is known that for any link, there exists a positive integer N such that this link can be inscribed in any generic set of N points in $\mathbb{R}^3$ (i.e., the line segments joining some points from this set form a link equivalent to the given one); a simple proof of this fact is given in [**107**].

3. Embeddings and Immersions

3.1. The Strong Whitney Embedding Theorem. In Part I, we proved the weak Whitney theorem about embeddings and immersions, which says that, for $n \geq 2$, any n-manifold without boundary can be immersed in $\mathbb{R}^{2n}$ and embedded in $\mathbb{R}^{2n+1}$. In this section, we prove the strong Whitney embedding theorem that any closed manifold M^n can be embedded in $\mathbb{R}^{2n}$.

The idea of the proof is to first construct a *regular* (*generic*) immersion $f\colon M^n \to \mathbb{R}^{2n}$, that is, an immersion such that it has no triple points (in other words, no three different points are mapped to one point) and, at each *double point* (*self-intersection*) $a = f(x) = f(y)$, where $x \neq y$, the equality $df(T_xM^n) \oplus df(T_yM^n) = T_a\mathbb{R}^{2n}$ holds.

The so-called Whitney trick makes it possible to kill all pairs of double points. If the manifold M^n is orientable and n is even, then the points in each pair to be killed must match in a certain sense. Thus, to kill all double points, it is sometimes required to first add several double points.

Regular Immersions.

Theorem 6.13. *Suppose that M^n be a closed manifold and $f\colon M^n \to \mathbb{R}^{2n}$ is an immersion. Then for any $\varepsilon > 0$, there exists a regular immersion $g\colon M^n \to \mathbb{R}^{2n}$ such that $\|f(x) - g(x)\| < \varepsilon$ for all $x \in M^n$.*

Proof. We use the same construction as in the proof of the existence of immersions (see Part I, Theorem 5.17), but now, instead of the change $f_i(y) = f_{i-1}(y) + Ay$, we make the change $f_i(y) = f_{i-1}(y) + v$, where v is a sufficiently small constant vector. First, let us see what happens in a small neighborhood of the image of a point $x \in M^n$. We can assume that $f(x) = 0$ and the image of a neighborhood $U_{i,1} \ni x$ in $\mathbb{R}^{2n}$ is locally (i.e., in some neighborhood V) determined by the equations $y_1 = 0, \dots, y_n = 0$. For the manifold $f^{-1}(V)$, consider the composition

$$f^{-1}(V) \xrightarrow{f} V \xrightarrow{p} \mathbb{R}^n,$$

where $p(y_1, \dots, y_{2n}) = (y_1, \dots, y_n)$. Let $v' = (v_1, \dots, v_n)$ be a regular value of this composition different from the images of the double points belonging to V. The n-dimensional subspace Π^n given by $y_1 = v_1, \dots, y_n = v_n$ does not pass through the double points of the image of M^n that belong to V, and at each point $a \in \Pi^n \cap f(M^n) \cap V$, the tangent space $df(T_{f^{-1}(a)}M^n)$ is complementary to Π^n, i.e., the direct sum of this space and Π^n has dimension $2n$. This means, in particular, that all intersection points of Π^n and $f(M^n) \cap V$ are isolated.

Thus, adding the vector $(v_1, \dots, v_n, 0, \dots, 0)$ to each point from the image of $U_{i,1}$ and leaving all the other parts of the intersection $f(M^n) \cap V$ intact, we obtain an extension of the regular immersion over $U_{i,1}$.

We described what is going on in $U_{i,1}$. The global map that remains the same outside $U_{i,2}$, is constructed by using the bell-shaped function λ, as in the proof of the weak Whitney theorem. The only thing to care about is the vector v', which should be sufficiently small for the map obtained to be an immersion. □

If a closed manifold M^n is regularly immersed in $\mathbb{R}^{2n}$, then the number of double points is finite. Indeed, suppose that there are infinitely many double points. Let x_0 be a limit point of their preimages. It has a neighborhood U with self-avoiding image. Let $\{x_i\}$ be a sequence of preimages of double points from U converging to x_0. The corresponding points y_i do not belong to U. Let y_0 be a limit point for $\{y_i\}$. Choosing a neighborhood of y_0 whose image intersects the image of U in at most one point, we obtain a contradiction.

Double Points. Suppose that $f\colon M^n \to \mathbb{R}^{2n}$ is a regular immersion and $a = f(x) = f(y)$ is a double point. If the manifold M^n is orientable and n is even, then, for the point a, the *self-intersection number* is defined. Namely, choose an orientation of M^n and take positively oriented bases of $df(T_xM^n)$ and $df(T_yM^n)$. From these bases, we compose a basis of space $\mathbb{R}^{2n}$ by consecutively writing the elements of the first and second bases. If the orientation of the basis thus obtained is positive (negative), then the self-intersection number is 1 (-1). The self-intersection number does not depend on the choice of orientation of M^n because a simultaneous change of the orientations of the bases of $df(T_xM^n)$ and $df(T_yM^n)$ does not affect the orientation of the basis of $\mathbb{R}^{2n}$ composed of them. The self-intersection number does not change under transpositions of the points x and y either. Indeed, interchanging the first n vectors and the last n vectors requires n^2 transpositions.

The *self-intersection number* of a regular immersion f is defined as the sum of the self-intersection numbers of all double points. If n is odd or the manifold M^n is nonorientable, then the self-intersection numbers of double points have indefinite signs; in these cases, the self-intersection number of the immersion f is defined to be the remainder of the total number of double points on dividing by 2. We denote the self-intersection number of a regular immersion f by I_f.

Theorem 6.14. *Let M^n be a closed manifold. Then there exists a regular immersion $f\colon M^n \to \mathbb{R}^{2n}$ for which I_f takes any given value (it is assumed that $I_f \in \mathbb{Z}$ if M^n is orientable and n is even, and $I_f \in \mathbb{Z}_2$ otherwise).*

Proof. First, we construct a regular immersion $\mathbb{R}^n \to \mathbb{R}^{2n}$ with one self-intersection. For $n = 1$, we set

$$y_1 = x - \frac{2x}{1+x^2} \quad \text{and} \quad y_2 = \frac{1}{1+x^2}$$

(see Figure 22). The points $x = \pm 1$ are mapped to $(0, 1/2)$; at the other

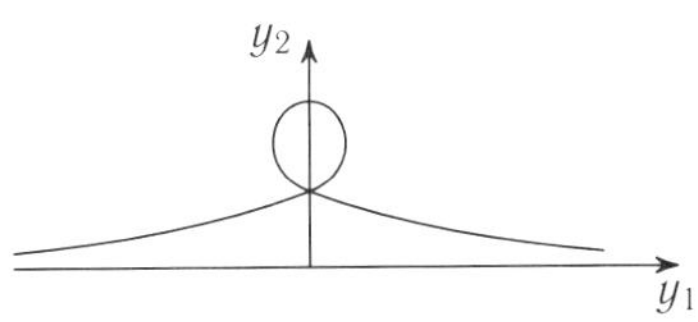

Figure 22. A self-intersection

points, no self-intersection occurs. The Jacobian matrix of this map is $\big(1 - 2\frac{1-x^2}{1+x^2}, -\frac{2x}{1+x^2}\big)$; it is nowhere vanishing. For $x = \pm 1$, we obtain the linearly independent matrices $(1, \mp 1)$.

For $n \geq 2$, consider the map defined by $y_1 = x_1 - \frac{2x_1}{X}$, $y_i = x_i$ for $i = 2, \ldots, n$, where $X = (1 + x_1^2) \cdots (1 + x_n^2)$, $y_{n+1} = \frac{1}{X}$, and $y_{n+i} = \frac{x_1 x_i}{X}$ for $i = 2, \ldots, n$. Let us show that this map is a regular immersion with a unique self-intersection. Its Jacobian matrix is

$$\begin{pmatrix} 1-2\frac{1-x_1^2}{X(1+x_1^2)} & \frac{4x_1x_2}{X(1+x_2^2)} & \dots & \frac{4x_1x_n}{X(1+x_n^2)} \\ 0 & 1 & \dots & 0 \\ \dots & \dots & \dots & \dots \\ 0 & 0 & \dots & 1 \\ \frac{-2x_1}{X(1+x_1^2)} & \frac{-2x_2}{X(1+x_2^2)} & \dots & \frac{-2x_n}{X(1+x_n^2)} \\ \frac{x_2(1-x_1^2)}{X(1+x_1^2)} & \frac{x_1(1-x_2^2)}{X(1+x_2^2)} & \dots & \frac{-2x_1x_2x_n}{X(1+x_n^2)} \\ \dots & \dots & \dots & \dots \\ \frac{x_n(1-x_1^2)}{X(1+x_1^2)} & \frac{-2x_1x_2x_n}{X(1+x_2^2)} & \dots & \frac{x_1(1-x_n^2)}{X(1+x_n^2)} \end{pmatrix}.$$

The last n elements of the first column vanish if and only if $x_1 = \dots = x_n = 0$. In this case, we have

$$1 - 2\frac{1-x_1^2}{X(1+x_1^2)} = -1 \neq 0.$$

Therefore, the Jacobian matrix has rank n at each point.

Now let us find all pairs of points $(x_1', \dots, x_n') = (x_1, \dots, x_n)$ such that $y_i' = y_i$ for all i. The equalities $y_i = x_i$, where $i = 2, \dots, n$, imply $x_i' = x_i$ for $i = 2, \dots, n$. We have $y_{n+1}' = y_{n+1}$; therefore, $X' = X$, and $(x_1')^2 = x_1^2$, i.e., $x_1' = \pm x_1$. Thus, $x_1' = -x_1 \neq 0$. The equalities $y_{n+i} = \frac{x_1x_i}{X}$, where $i = 2, \dots, n$, imply $x_2 = \dots = x_n = 0$. Therefore, $X = 1+x_1^2$. Consequently, $y_1 = x_1 - \frac{2x_1}{1+x_1^2}$ and $y_1' = -y_1$. Hence $y_1 = 0$ and $x_1^2 = 1$. Thus, only the points $(\pm 1, 0, \dots, 0)$ are mapped to the same point.

The Jacobian matrices at the points $(\pm 1, 0, \dots, 0)$ have the form

$$\begin{pmatrix} 1 & 0 & \dots & 0 \\ 0 & 1 & \dots & 0 \\ \dots & \dots & \dots & \dots \\ 0 & 0 & \dots & 1 \\ \mp 1/2 & 0 & \dots & 0 \\ 0 & \pm 1/2 & \dots & 0 \\ \dots & \dots & \dots & \dots \\ 0 & 0 & \dots & \pm 1/2 \end{pmatrix}.$$

These two matrices constitute the matrix

$$\begin{pmatrix} 1 & 0 & \dots & 0 & 1 & 0 & \dots & 0 \\ 0 & 1 & \dots & 0 & 0 & 1 & \dots & 0 \\ \hdotsfor{8} \\ 0 & 0 & \dots & 1 & 0 & 0 & \dots & 1 \\ -1/2 & 0 & \dots & 0 & 1/2 & 0 & \dots & 0 \\ 0 & 1/2 & \dots & 0 & 0 & -1/2 & \dots & 0 \\ \hdotsfor{8} \\ 0 & 0 & \dots & 1/2 & 0 & 0 & \dots & -1/2 \end{pmatrix}$$

$$\sim \begin{pmatrix} 1 & 0 & \dots & 0 & 1 & 0 & \dots & 0 \\ 0 & 1 & \dots & 0 & 0 & 1 & \dots & 0 \\ \hdotsfor{8} \\ 0 & 0 & \dots & 1 & 0 & 0 & \dots & 1 \\ 0 & 0 & \dots & 0 & 1 & 0 & \dots & 0 \\ 0 & 0 & \dots & 0 & 0 & -1 & \dots & 0 \\ \hdotsfor{8} \\ 0 & 0 & \dots & 0 & 0 & 0 & \dots & 1 \end{pmatrix}.$$

This matrix is nonsingular, and hence the immersion is regular.

It follows directly from the definition of the self-intersection number (provided that n is even and the manifold M^n is orientable) that it changes sign when the orientation of $\mathbb{R}^{2n}$ is changed. Therefore, using the regular immersion constructed above and its composition with the symmetry of $\mathbb{R}^{2n}$ about a hyperplane, we can construct two regular immersions with one self-intersection each and self-intersection number $+1$ for one immersion and -1 for the other.

Suppose that $f\colon M^n \to \mathbb{R}^{2n}$ is a regular immersion and $U \subset M^n$ is a sufficiently small neighborhood homeomorphic to $\mathbb{R}^n$ such that the restriction of f to U is an embedding. It is easy is construct a regular immersion g coinciding with f outside U and such that the sets $g(\operatorname{int} U)$ and $f(M^n \setminus \operatorname{int} U)$ are disjoint, the image $g(U)$ has a unique self-intersection, and the self-intersection number for this point has a given sign (provided that n is even and M^n is orientable). Compared to f, the immersion g has precisely one additional self-intersection point. This construction proves the required assertion. □

The Whitney Trick. To prove the strong Whitney embedding theorem, it remains to learn how to kill pairs of self-intersection points of the manifold M^n in $\mathbb{R}^{2n}$ (if M^n is orientable and n is even, then it is assumed that the self-intersection numbers of the points in these pairs have opposite signs). Note that the Whitney trick only works for $n \geq 3$. For $n = 2$, the Whitney embedding theorem is proved by a different method (see p. 334). In what follows, we assume that $n \geq 3$.

Let p and q be two self-intersection points for a regular immersion $f\colon M^n \to \mathbb{R}^{2n}$. Consider $f^{-1}(p) = \{p_1, p_2\}$ and $f^{-1}(q) = \{q_1, q_2\}$. Let $\bar{\gamma}_i$ be a path joining p_i and q_i in M^n $(i = 1, 2)$. We set $\gamma_i = f(\bar{\gamma}_i)$. We can assume that the curves γ_1 and γ_2 are self-avoiding and intersect each other only at the points p and q.

First, we show that γ_1 and γ_2 span a disk transversal to $f(M^n)$. To be more precise, let τ be a closed convex set in $\mathbb{R}^2$ with two corners (see Figure 23), and let τ' be its open ε-neighborhood. Then there exists an

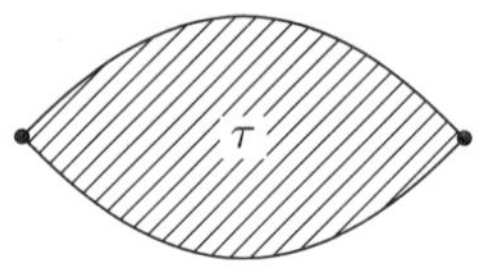

Figure 23. The domain τ

embedding $\psi\colon \tau' \to \mathbb{R}^{2n}$ with the following properties:

(1) the corners are mapped to p and q;

(2) $\psi(\tau') \cap f(M^n) = \gamma_1 \cup \gamma_2$;

(3) at any point of $\gamma_1 \cup \gamma_2$ different from p and q, the tangent space $T_x\psi(\tau')$ is not contained in $T_x f(M^n)$.

The required map ψ is constructed as follows. First, we construct a vector field on γ_1 transversal[2] to $f(M_1^n)$ in $\mathbb{R}^{2n}$; here M_1^n is a small neighborhood of the curve $\bar{\gamma}_1$ in M^n. For the vectors at p and q we take the tangent vectors to the curve γ_2 directed toward each other (see Figure 24). They are transversal to $f(M_1^n)$ because $f(M_2^n)$ is. Consider the bundle over γ_1 whose fiber over each point x is the space $\mathbb{R}^{2n} \setminus T_x M^n \sim S^{n-1}$. This bundle is trivial because its base space is an interval. The sections of this bundle at the endpoints of γ_1 are given. Since the space S^{n-1} is path-connected, these sections can be extended over the entire curve γ_1. As a result, we obtain the required vector field. In a similar way, we construct a vector field on the curve γ_2.

Outside small neighborhoods of the points p and q, we take the endpoints of the vectors $tv(x)$, where $x \in \gamma_1 \cup \gamma_2$, $v(x)$ is the vector field at x, and $|t| \leq \varepsilon$ (ε is a fixed sufficiently small positive number). In small neighborhoods of p and q, we take parts of the planes spanned by the tangent vectors to the curves γ_1 and γ_2. In constructing the vector fields at p and q, we used

[2] *Transversal* here means that, at each point of γ_1, the vector does not belong to the tangent space to $f(M_1^n)$ at this point.

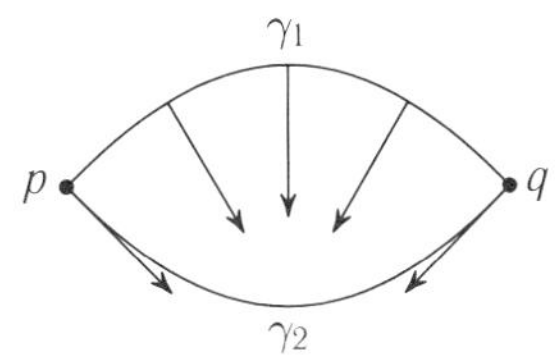

Figure 24. The vector field on γ_1

tangent vectors to γ_1 and γ_2; therefore, slightly moving the surfaces under consideration, we can glue them together so as to obtain an embedding of a small neighborhood of the boundary of the disk τ (see Figure 25). Transversality is not violated.

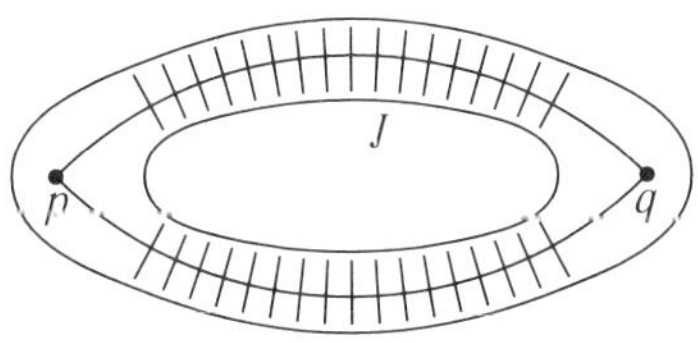

Figure 25. The embedding of a neighborhood of the boundary of τ

To the interior of the obtained strip (this is the curve J in Figure 25) we can attach the disk D^2 because any continuous map $S^1 \to \mathbb{R}^{2n}$ extends to a continuous map $D^2 \to \mathbb{R}^{2n}$. Since $2n > 5$, we can approximate the constructed map $D^2 \to \mathbb{R}^{2n}$ by a smooth embedding without changing it near $\gamma_1 \cup \gamma_2$. Then, using the inequality $2 + n < 2n$, we can remove all intersections of $f(M^n)$ with the image of D^2 that arise in the approximation. This completes the construction of the required map ψ.

The next step is to construct an orthonormal system of vector fields $w_3, \dots, w_{2n}$ on the disk $\psi(\tau)$ that are orthogonal to $\psi(\tau)$. First, we suppose that n is even, M^n is orientable, and the self-intersection numbers of the points p and q have opposite signs.

Consider the restriction to γ_1 of the tangent bundle of M^n and take the orthogonal complement of the tangent bundle of γ_1 in this restriction. As a result, we obtain an $(n-1)$-dimensional bundle. This bundle is trivial because its base is an interval. Therefore, it has linearly independent sections $w_3, \dots, w_{n+1}$. In a similar way, we construct vector fields $w_{n+2}, \dots, w_{2n}$ on the curve γ_2. At the points p and q, all of the $2n-2$ vectors $w_3, \dots, w_{2n}$ are defined, but they determine opposite orientations of the normal bundle to the disk $\psi(\tau)$ at these points. Indeed, transferring vectors tangent to the curves γ_1 and γ_2 along these curves, we obtain bases of the tangent spaces $T_p\psi(\tau)$ and $T_q\psi(\tau)$ at p and q with opposite orientations (see Figure 26).

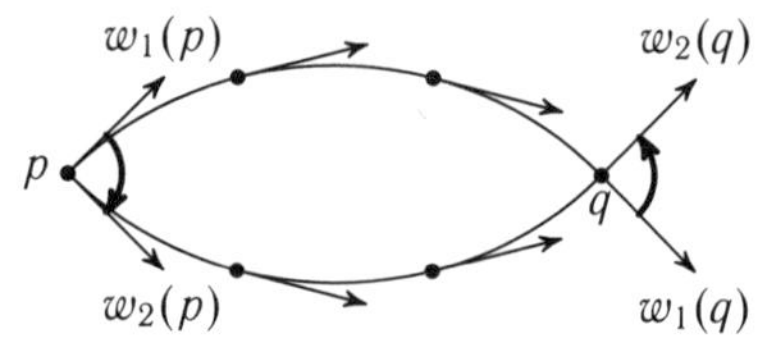

Figure 26. Change of orientation

We shall assume for convenience that the curves γ_1 and γ_2 are not only transversal but also orthogonal to TM_2^n and TM_1^n. On the curve γ_1, the vector fields $w_3, \ldots, w_{n+1}$ are defined. Consider the bundle over γ_1 whose fiber over each point x is the $(n-1)$-dimensional subspace orthogonal to the vectors $w_3(x), \ldots, w_{n+1}(x)$ and tangent to the space $T_x\psi(\tau)$. At the points p and q, the vectors $w_{n+2}, \ldots, w_{2n}$ are defined; these are orthonormal sections of this bundle. We need to extend them over the entire curve γ_1. First, we extend the section w_{n+2}. We can do this because the bundle is trivial and the fiber $\mathbb{R}^{n-1} \setminus \{0\} \sim S^{n-2}$ is connected. Then, we extend a section of the orthogonal complement of the bundle determined by the extended section, and so on. The bundle used to extend the section w_{2n-1} has fiber $\mathbb{R}^2 \setminus \{0\} \sim S^1$. When we try to extend the section w_{2n}, the fiber will be the disconnected set $\mathbb{R}^1 \setminus \{0\}$. To extend this section, we need the orientations to be compatible; but they are compatible in the case under consideration.

Now we have the vectors $w_3, \ldots, w_{2n}$ on the curve γ_1 and the vectors $w_{n+2}, \ldots, w_{2n}$ on the curve γ_2. By Theorem 3.16, $\pi_1(V(2n-2, n-1)) = 0$ for $1 < (2n-2) - (n-1)$, i.e., $n > 2$. Therefore, the vector fields $w_{n+2}, \ldots, w_{2n}$ can be extended to the entire disk $\psi(\tau)$. Consider the orthogonal complement of the vectors $w_{n+2}, \ldots, w_{2n}$ and of the tangent space $T_x\psi(\tau)$ at each point $x \in \psi(\tau)$. On the curve γ_1, the sections $w_3, \ldots, w_{n+1}$ of this bundle are given. Each point of the disk $\psi(\tau)$ can be represented as $\gamma_t(s)$, where $1 \leq s, t \leq 2$. Let us extend the sections $w_3, \ldots, w_{n+1}$, which are given at $\gamma_1(s)$, to constant sections of the trivial bundle over the curve $\gamma_t(s)$, where s is fixed and t varies. As a result, we obtain the required vector fields $w_3, \ldots, w_{n+1}$ on the disk $\psi(\tau)$.

Now, we can prove the strong Whitney embedding theorem. We assume that $\tau \subset \mathbb{R}^2 \subset \mathbb{R}^{2n}$. Let us continue the embedding $\psi\colon \tau \to \mathbb{R}^{2n}$ to a map from a small neighborhood of τ to $\mathbb{R}^{2n}$ as follows. For $x = (x_1, \ldots, x_{2n})$, we set

$$\psi(x) = \psi(\overline{x}) + \sum_{i=3}^{2n} x_i w_i(\psi(\overline{x})), \quad \text{where } \overline{x} = (x_1, x_2).$$

Since the vectors $w_3, \ldots, w_{2n}$ form a basis of the orthogonal complement of the tangent space $T_x\psi(\tau)$ at $x \in \psi(\tau)$, the Jacobian of the map ψ is nonzero at x. Therefore, ψ is a diffeomorphism from some neighborhood U of the disk $\psi(\tau)$ onto a domain in $\mathbb{R}^{2n}$. We set $N_1^n = \psi^{-1}(f(M_1^n))$ and $N_2^n = \psi^{-1}(f(M_2^n))$. Our immediate goal is to deform N_2^n in the domain U so as to kill two intersection points with N_1^n while preventing the appearance of new intersection points.

To describe the required deformation, it is convenient to assume that the preimage of the curve γ_1 under ψ lies on the x_1-axis (see Figure 27). Moreover, we assume that the curves γ_1 and γ_2 are naturally extended so

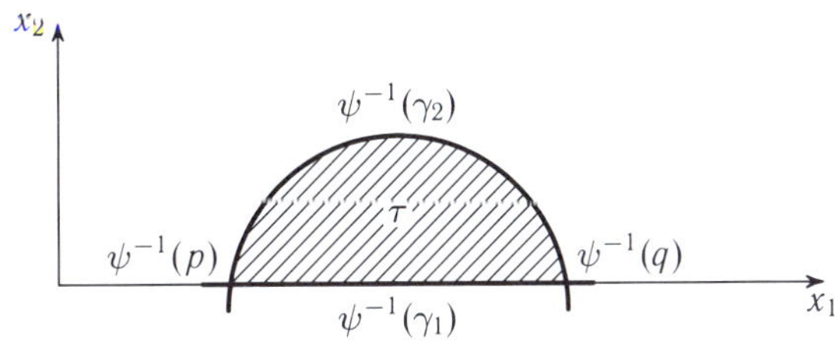

Figure 27. The location of the disk τ

as to exit the small neighborhood of the disk $\psi(\tau)$; on their extensions, the vector fields $w_3, \dots, w_{2n}$ are constructed in the same way as on the curves themselves.

At each point $x \in \psi^{-1}(\gamma_1)$, the tangent space $T_x N_1^n$ is the plane $(x_1, x_3, \dots, x_{n+1})$; therefore, N_1^n differs little from this plane. Performing a small deformation inside a small neighborhood of the curve $\psi^{-1}(\gamma_1)$, we can assume that N_1^n coincides with a part of this plane (possibly, we need to replace U with a smaller neighborhood of $\psi^{-1}(\gamma_1)$). Similarly, we can assume that N_2^n coincides with a part of the manifold obtained by applying all vectors of the space $(x_{n+2}, \dots, x_{2n})$ to each point of the curve $\psi^{-1}(\gamma_2)$ and its extension. Consider the projection

$$\pi\colon (x_1, x_2, x_3, \dots, x_{2n}) \mapsto (x_1, 0, x_3, \dots, x_{2n}).$$

Clearly, $\pi(N_1^n) = N_1^n$, and the set $\pi(N_2^n)$ is contained in the space $(x_1, x_{n+2}, \dots, x_{2n})$. Therefore, the set $N_1^n \cap \pi(N_2^n)$ lies on the x_1-axis.

Let us subject N_2^n to the following deformation, which is the identity outside the ε-neighborhood of the disk τ. First, all points of N_2^n are translated parallel to the x_2-axis; thus, the set $\pi(N_2^n)$ does not change. The curve $\psi^{-1}(\gamma_2)$ and its extension are moved to the domain $x_2 < 0$ (see Figure 28). For any other point $(x_1, x_2, \dots, x_{2n}) \in N_2^n$, the displacement vector is obtained upon multiplying the displacement vector of $(x_1, x_2, 0, \dots, 0)$ by $\lambda(x_3^2 + \dots + x_{2n}^2)$, where $\lambda(0) = 1$ and $\lambda(t) = 0$ for $t > \varepsilon^2$.

The projection π takes the intersection points of N_1^n and N_2^n to those of N_1^n and $\pi(N_2^n)$; hence if the plane (x_1, x_2) contains no intersection points of

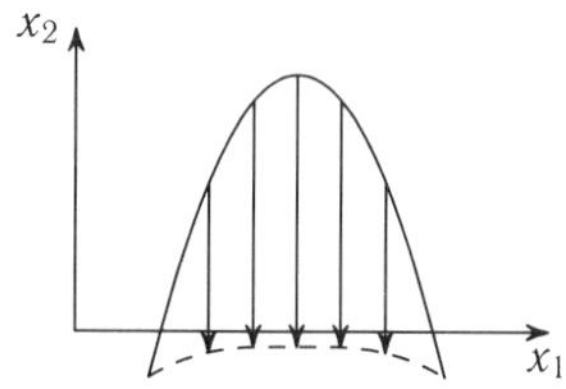

Figure 28. The Whitney trick

the manifold N_1^n and the deformed manifold N_2^n, then these manifolds do not intersect at all. Clearly, if ε is sufficiently small, then the image of the ε-neighborhood of the disk τ under the map ψ contains no other parts of the manifold $f(M^n)$. Therefore, no new self-intersections arise.

We have proved the Whitney theorem in the case where the self-intersection numbers for the points p and q are of opposite signs. In the cases where the manifold M^n is nonorientable and where it is orientable and n is odd, the main construction is the same. But to apply it, we may need to change the curves γ_1 and γ_2. It is required that under the translation of the pair of normal spaces to M_1^n and M_2^n from p to q along the curves γ_1 and γ_2, the orientation of $\mathbb{R}^{2n}$ determined by this ordered pair of spaces change. If it does not change and the manifold M^n is nonorientable, then we can extend one of the curves γ_1 and γ_2 so that the translation along the added part of the curve will change orientation (a change of the orientation of the tangent space entails a change of the orientation of the normal space). If the manifold is orientable but has odd dimension $n > 1$, we replace the curves γ_1 and γ_2 with γ_1' and γ_2', where γ_1' starts from the point p along the curve γ_1 and arrives at the point q along γ_2, and γ_2' starts along γ_2 and ends along γ_1. As a result, the spaces in the ordered pair are interchanged, which leads to a change of orientation for odd n.

Remark. A similar argument proves that for $n > 2$, any continuous map $S^n \to M^{2n}$, where M^{2n} is a simply connected manifold, is homotopic to an embedding. But it turns out that there exist maps $S^2 \to M^4$, where M^4 is a simply connected manifold, that are not homotopic to embeddings. An example is given in [**70**].

Now, let us prove that any closed 2-manifold can be embedded in $\mathbb{R}^4$. Clearly, a sphere with g handles can be embedded even in $\mathbb{R}^3$. It is also clear that if M_2^1 and M_2^2 can be embedded in $\mathbb{R}^4$, then the connected sum $M_2^1 \# M_2^1$ can be embedded in $\mathbb{R}^4$ as well. Therefore, it is sufficient to prove that $\mathbb{R}P^2$ embeds in $\mathbb{R}^4$.

An embedding of $\mathbb{R}P^2$ into $\mathbb{R}^4$ can be constructed as follows. Consider the map from the unit sphere $S^2 \subset \mathbb{R}^3$ to $\mathbb{R}^4$ defined by $(x, y, z) \mapsto$

$(xy, yz, zx, x^2 - y^2)$. Let us show that this map induces an embedding of $\mathbb{R}P^2$ into $\mathbb{R}^4$, i.e., that it is an immersion and the preimage of each point consists of precisely two antipodal points of the sphere.

First, we prove the assertion concerning the preimage. If $abc \neq 0$, then the equalities $xy = a$, $yz = b$, and $zx = c$ imply $x = \pm\sqrt{ac/b}$, $y = a/x$, and $z = c/x$. If $a = b = c = 0$, then two of the numbers x, y, and z are zero and the third is equal to ± 1. It remains to consider the case where two of the numbers a, b, and c are zero and the third is nonzero. In this case, one of the numbers x, y, and z is zero and the two others are not. Now, we employ the equality $x^2 - y^2 = d$ (or the equivalent equality $2x^2 + z^2 = 1 + d$). If $x = 0$, then $y = \pm\sqrt{-d}$ and $z = b/y$. If $y = 0$, then $x = \pm\sqrt{d}$ and $z = c/x$. If $z = 0$, then $x = \pm\frac{1}{2}\sqrt{1+d}$ and $y = a/x$.

Let us prove that the map $f\colon \mathbb{R}P^2 \to \mathbb{R}^4$ is an immersion, i.e., its Jacobian matrix has rank 2 at each point $(x, y, z) \in S^2$. The projective plane can be covered by three charts with coordinates (x, y), (x, z), and (y, z), in which $z \neq 0$, $y \neq 0$, and $x \neq 0$, respectively. In the first chart, the Jacobian matrix of f is

$$\begin{pmatrix} y & -\frac{x}{z} & z - \frac{x^2}{z} & 2x \\ x & z - \frac{y^2}{z} & -\frac{y}{z} & -2y \end{pmatrix}.$$

The matrix formed by the first and last columns is singular if and only if $x^2 + y^2 = 0$, i.e., $x = y = 0$ and $z = \pm 1$. But in this case, the matrix formed by the second and third columns is nonsingular. In the second chart, the Jacobian matrix of f has the form

$$\begin{pmatrix} -\frac{x}{y} & y - \frac{x^2}{y} & z & 4x \\ y - \frac{z^2}{y} & -\frac{z}{y} & x & 2z \end{pmatrix}.$$

We are interested only in the points with $z = 0$. At these points, the matrix formed by the two last columns is nonsingular if $x \neq 0$, and if $z = 0$ and $x = 0$, then the matrix formed by the two first columns is nonsingular. The nonsingularity of the Jacobian matrix at the point $(\pm 1 : 0 : 0)$ in the third chart is proved similarly.

3.2. The Normal Degree of an Immersion. Let $f\colon M^n \to \mathbb{R}^{n+1}$ be an immersion of a closed connected oriented manifold. Since M^n is oriented, at each point $x \in f(M^n)$, the unit normal vector $N(x)$ is determined uniquely. Therefore, the map f induces a Gaussian map $N\colon M^n \to S^n$. The degree of this map is called the *normal degree* of the immersion f. The question of what normal degree an immersion f can be, was considered, in particular, by Hopf [**55**, **56**] and Milnor [**87**]. In this section, we present some of their results.

First, note that if an immersion $f\colon M^n \to \mathbb{R}^{n+1}$ has normal degree d, then the immersion sf, where $s\colon \mathbb{R}^{n+1} \to \mathbb{R}^{n+1}$ is the symmetry with respect to a hyperplane, has degree $(-1)^n d$. Indeed, if vectors $e_1, \dots, e_n, N(x)$ form a positively oriented basis, then $se_1, \dots, se_n, sN(x)$ form a negatively oriented basis because the map s is orientation-reversing. Therefore, for the map sf, the new map $M^n \to S^n$ is the composition of the old map $N\colon M^n \to S^n$ and the map $-s|_{S^n}$. The restriction $s|_{S^n}$ has degree -1, and the antipodal map $S^n \to S^n$ has degree $(-1)^{n-1}$.

Two oriented manifolds M_1^n and M_2^n determine an oriented manifold $M_1^n \# M_2^n$, which is called their *connected sum*. It is constructed as follows. Let $D_1^n \subset M_1^n$ and $D_2^n \subset M_2^n$ be unit disks embedded in these manifolds. We remove disks of radius $1/2$ from these disks and glue together the remaining cylinders $I \times S_1^{n-1}$ and $I \times S_2^{n-1}$ so as to preserve orientation (see Figure 29). More formally, the construction is as follows. Suppose that $i_1\colon D^n \to M_1^n$ is

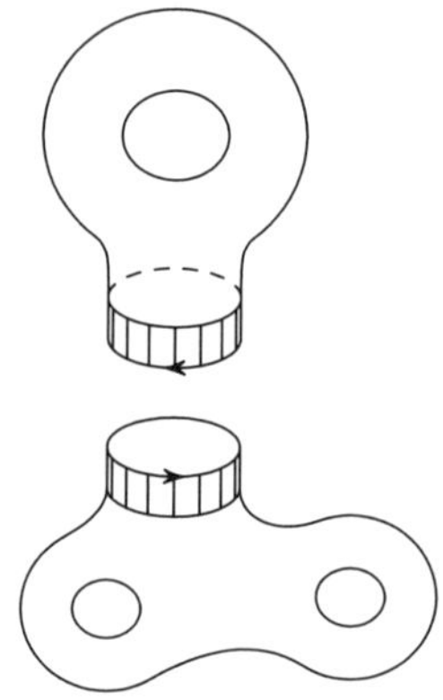

Figure 29. Connected sum

an orientation preserving embedding, and $i_2\colon D^n \to M_2^n$ is an orientation-reversing embedding. In $M_1^n \setminus i_1(0)$ and $M_2^n \setminus i_2(0)$, we identify $i_1(tu)$ with $i_2((1-t)u)$ for every unit vector u and each t, $0 < t < 1$. The map $i_1(tu) \mapsto i_2((1-t)u)$ is orientation-preserving; therefore, on the obtained manifold $M_1^n \# M_2^n$, we can choose an orientation compatible with those of the manifolds M_1^n and M_2^n. Clearly, $M^n \# S^n \approx M^n$; in particular, $S^n \# S^n \approx S^n$.

Theorem 6.15. *If M_1^n and M_2^n admit immersions into $\mathbb{R}^{n+1}$ with normal degrees d_1 and d_2, then $M_1^n \# M_2^n$ admits an immersion into $\mathbb{R}^{n+1}$ with normal degree $d_1 + d_2 - 1$.*

Proof. Take any points $x_1 \in M_1^n$ and $x_2 \in M_2^n$. We apply to the manifold M_2^n an orientation-preserving Euclidean motion of $\mathbb{R}^{n+1}$ so as to render the normal vectors at x_1 and x_2 directed toward each other (see Figure 30); this is needed for the orientations to be compatible. Then, we perform a

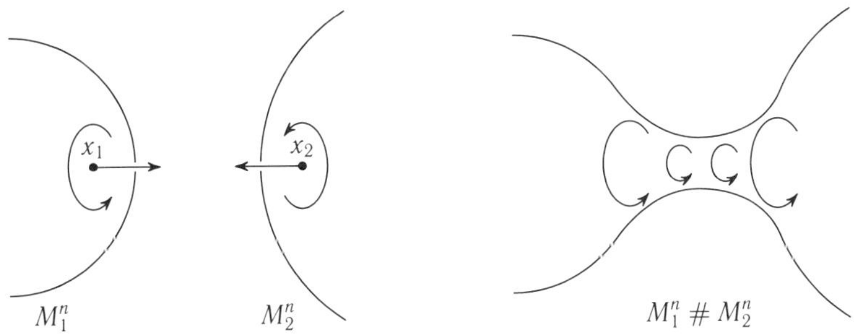

Figure 30. The immersion of a connected sum

surgery and obtain an immersion of the manifold $M_1^n \# M_2^n$. To calculate the normal degree of this immersion, it suffices to trace how many times the map $M_1^n \# M_2^n \to S^n$ takes an arbitrary given value. The most convenient value is $N(x_1)$. Slightly moving the manifold M_1^n if necessary, we can assume that the contribution of the value $N(x_1)$ taken by the map $M_1^n \to S^n$ at a neighborhood of x_1 is $+1$. This means that in this neighborhood of x_1, the manifold M_1^n looks as a sphere, and $N(x_1)$ is the outward normal to this sphere. In this situation, no new points taken to $N(x_1)$ arise under the surgery. Thus, the maps $M_1^n \to S^n$ and $M_2^n \to S^n$ take to $N(x_1)$ not only x_1 but also some other points. Consequently, the normal degree of the constructed immersion is equal to $d_1 + d_2 - 1$. □

Theorem 6.16 (Hopf). *For any odd n, the sphere S^n can be immersed into $\mathbb{R}^{n+1}$ with any odd normal degree.*

Proof. Applying the symmetry about a hyperplane, we construct an immersion of S^n into $\mathbb{R}^{n+1}$ with normal degree -1. We take two copies of the sphere thus immersed and apply Theorem 6.15. As a result, we obtain an immersion of the manifold $S^n \# S^n \approx S^n$ with normal degree $(-1) + (-1) - 1 = -3$. Then, we construct an immersion of S^n with normal degree $(-3) + (-1) - 1 = -5$, and so on.

We have constructed immersions of S^n into $\mathbb{R}^{n+1}$ with all negative odd degrees. Applying symmetries about hyperplanes, we obtain immersions with all positive odd degrees. □

Corollary. *Suppose that n is odd and a manifold M^n is immersed into $\mathbb{R}^{n+1}$ with normal degree d. Then M^n can be immersed into $\mathbb{R}^{n+1}$ with any normal degree of the same parity as d.*

Proof. We take an immersion of S^n with normal degree $2k+1$ and construct an immersion of $M^n \approx M^n \# S^n$ with normal degree $d + 2k$. $\square$

Theorem 6.17. *If a closed oriented manifold M^n can be immersed into $\mathbb{R}^{n+1}$ with normal degree zero, then M^n is parallelizable.*

Proof. It is easy to see that $TM^n = N_*(TS^n)$, i.e., the tangent bundle of the manifold M^n is a pullback of the tangent bundle of the sphere S^n by means of a map $N\colon M^n \to S^n$. Indeed, $N_*(TS^n)$ consists of the pairs (x, v) with $x \in M^n$ and $v \perp N(x)$.

If the map $N\colon M^n \to S^n$ has degree zero, then, by the Hopf theorem, it is null-homotopic and, therefore, induces the trivial bundle. $\square$

Corollary. *Suppose that a manifold M^n is not parallelizable and n is odd. Then, if M^n can be immersed into $\mathbb{R}^{n+1}$, then it can be immersed into $\mathbb{R}^{n+1}$ with any odd normal degree, but it admits no immersions into $\mathbb{R}^{n+1}$ of even normal degree.*

Proof. Apply Theorem 6.17 and the corollary of Theorem 6.16. $\square$

Problem 133. Prove that if the sphere S^n is parallelizable, then so is any closed orientable manifold M^n that can be immersed into $\mathbb{R}^{n+1}$.

Every embedding of a manifold M^n into $\mathbb{R}^{n+1}$ is an immersion. Therefore, for any embedding, its normal degree is defined. The normal degree of an embedding satisfies fairly strong conditions. To formulate them, we introduce the quantity $\Sigma(M^n) = \sum_{i\ge 0} \dim H_i(M^n; F)$, where F is the additive group of a field.

Theorem 6.18 (Milnor [**87**]). *If d is the normal degree of an embedding of M^n into $\mathbb{R}^{n+1}$, then $d \equiv \frac{1}{2}\Sigma(M^n) \pmod 2$ and $|d| \le \frac{1}{2}\Sigma(M^n)$.*

Proof. The set $\mathbb{R}^{n+1} \setminus M^n$ has two connected components, one of which is bounded; we denote it by A. We can assume that the manifold M^n is oriented so that the normal vector is directed outside of A. The normal degree d is equal to $\chi(\overline{A})$. This is proved in the same way as the main lemma in the proof of the Poincaré–Hopf theorem (Theorem 5.28 in Part I). It is seen from the definitions that $\chi(\overline{A}) \equiv \Sigma(\overline{A}) \pmod 2$ and $-\Sigma(\overline{A}) \le \chi(\overline{A}) \le \Sigma(\overline{A})$. Moreover, $\dim H_0(\overline{A}; F) = 1$; therefore, $2 - \Sigma(\overline{A}) \le \chi(\overline{A})$.

Now let us prove that $\Sigma(\overline{A}) = \frac{1}{2}\Sigma(M)$. Consider the one-point compactification S^{n+1} of $\mathbb{R}^{n+1}$. We set $B = S^{n+1} \setminus A$. According to the Alexander duality theorem, we have $\tilde{H}^k(M^n) \cong \tilde{H}_{n-k}(S^{n+1} \setminus M^n)$ for $0 \le k \le n$. If $1 \le k \le n-1$, then

$$\begin{aligned}\dim H_{n-k}(M^n) = \dim H^k(M^n) &= \dim H_{n-k}(S^{n+1} \setminus M^n)\\ &= \dim H_{n-k}(A) + \dim H_{n-k}(B).\end{aligned}$$

The equality $\dim \tilde{H}^0(M^n) = 0$ implies $\dim H_n(A) = \dim H_n(B) = 0$. Therefore, $\Sigma(M) = \Sigma(A) + \Sigma(B)$ because $\dim H_0(M^n) + \dim H_n(M^n) = 2 = \dim H_0(A) + \dim H_0(B)$. We can apply the Alexander duality theorem not only to M^n but also to $\overline{A}$. As a result, we obtain $\tilde{H}^k(\overline{A}) \cong \tilde{H}_{n-k}(S^{n+1}\backslash\overline{A}) = \tilde{h}_{n-k}(B)$ for $0 \le k \le n$, whence $\Sigma(\overline{A}) = \Sigma(B)$. Clearly, $\Sigma(\overline{A}) = \Sigma(A)$ because the spaces A and $\overline{A}$ are homotopy equivalent. Thus, $2\Sigma(M^n) = \Sigma(A)$. □

4. Complex Manifolds

4.1. Complete Intersections. Let Y be a complex manifold of dimension n, and let $X \subset Y$ be its complex submanifold of dimension $n-1$. This means that X can be covered by open subsets U_i of Y so that in each U_i, X is determined by the equation $f_i = 0$, where f_i is a holomorphic function such that $\operatorname{grad} f_i(x) \neq 0$ for all $x \in U_i \cap X$. We can assume that the sets U_i cover Y (if $U_i \cap X = \varnothing$, then we set $f_i = \text{const} \neq 0$). We associate with the submanifold $X \subset Y$ a one-dimensional complex bundle[3] $\xi(X)$ over Y with transition functions $g_{ij}(x) = f_i(x)/f_j(x)$. If $x \in X$, then $g_{ij}(x) = \lim_{y\notin X, y\to x} f_i(x)/f_j(x)$; this limit is equal to the proportionality coefficient of the vectors $\operatorname{grad} f_i(x)$ and $\operatorname{grad} f_j(x)$.

The bundle $\xi(X)$ does not depend on the choice of the functions f_i. Indeed, suppose that in U_i, the set X is determined by the equations $f_i = 0$ and $\tilde{F}_i = 0$. Then the function $h_i = f_i/\tilde{F}_i$ does not vanish on U_i, and $\tilde{g}_{ij} = \frac{h_j}{h_i} g_{ij}$.

The bundle $\xi(X)$ has a holomorphic section that vanishes precisely on X. On U_i, this section is determined by f_i. Such a section is well defined because $f_i = g_{ij} f_j$.

Example 75. Let X_d be a nonsingular hypersurface of degree d in $\mathbb{C}P^n$ given by a homogeneous equation $F(z_0, \ldots, z_n) = 0$ of degree d. Then $(\xi(X_d))^* \cong \underbrace{\gamma_n^1 \otimes \cdots \otimes \gamma_n^1}_{d}$.

Proof. In the chart U_i, the hypersurface X_d is defined by the equation $f_i(w_0, \ldots, \hat{w}_i, \ldots, w_n) = 0$, where

$$f_i\left(\frac{z_0}{z_i}, \ldots, \hat{1}, \ldots, \frac{z_n}{z_i}\right) = \frac{F(z_0, \ldots, z_n)}{z_i^d}.$$

[3]In algebraic geometry, this bundle is called the *line bundle associated with the divisor* X and denoted by $[X]$. We do not use this notation because in topology, $[X]$ usually denotes the fundamental class of the manifold X.

Therefore, the bundle $\xi(X_d)$ is determined by the transition functions $g_{ij} = \frac{f_i}{f_j} = \left(\frac{z_j}{z_i}\right)^d$, while the canonical bundle γ_n^1 over $\mathbb{C}P^n$ is determined by the transition functions $g_{ij} = \frac{z_i}{z_j}$ (see Example 67 on p. 273). $\square$

In particular, for a hypersurface $\mathbb{C}P^{n-1} \subset \mathbb{C}P^n$ specified by a linear equation, the bundle $\xi(\mathbb{C}P^n)$ is isomorphic to the bundle defined in Example 69 on p. 274.

It is easy to show that the restriction of the bundle $\xi(X)$ to X is the normal (in the Hermitian metric) bundle to X in Y. Indeed, consider the bundle $\bar{\nu}$ over X whose fibers over U_i are vectors proportional to $\operatorname{grad} f_i$. If $\lambda_i \operatorname{grad} f_i = \lambda_j \operatorname{grad} f_j$, then $\lambda_i = \frac{f_j}{f_i}\lambda_j$. Therefore, $\bar{\nu}$ is determined by the transition functions $g_{ij} = f_j/f_i$. The bundle $\bar{\nu}$ is complex conjugate to the normal bundle ν; hence ν is determined by the transition functions $g_{ij} = f_i/f_j$, which coincide with those for $\xi(X)$.

Example 76. Let X_d be a nonsingular hypersurface of degree d in $\mathbb{C}P^n$. Then

$$c(X_d) = (1+\tilde{\alpha})^{n+1}(1 - d\tilde{\alpha} + d^2\tilde{\alpha}^2 - d^3\tilde{\alpha}^3 + \cdots),$$

where $\tilde{\alpha}$ is the restriction to X_d of the canonical[4] generator of the group $H^2(\mathbb{C}P^n;\mathbb{Z})$ (the sum is finite because $\tilde{\alpha}^n = 0$).

Proof. Let $j\colon X_d \to \mathbb{C}P^n$ be the natural embedding. Then $\tilde{\alpha} = j^*\alpha$, and $j^*(\xi(X_d)) = \nu_{X_d}$ is the normal bundle to X_d in $\mathbb{C}P^n$. Therefore, $j^*(\tau_{\mathbb{C}P^n}) = \tau_{X_d} \oplus \nu_{X_d} = \tau_{X_d} \oplus j^*(\xi(X_d))$. Taking into account the equalities $c(\xi(X_d)) = -dc(\gamma_n^1) = d\alpha$ and $c(\mathbb{C}P^n) = (1+\alpha)^{n+1}$, we see that $(1+\tilde{\alpha})^{n+1} = c(X_d)(1+d\tilde{\alpha})$, and hence

$$\begin{aligned} c(X_d) &= (1+\tilde{\alpha})^{n+1}(1+d\tilde{\alpha})^{-1} \\ &= (1+\tilde{\alpha})^{n+1}(1 - d\tilde{\alpha} + d^2\tilde{\alpha}^2 - d^3\tilde{\alpha}^3 + \cdots). \end{aligned}$$

$\square$

Example 77. Let X_d be a nonsingular hypersurface of degree d in $\mathbb{C}P^n$. Then

$$\chi(X_d) = \sum_{k=0}^{n-1} (-1)^k d^{k+1} \binom{n+1}{n-1-k}.$$

Proof. In Example 76, we calculated the total Chern class of the manifold X_d. In particular, we obtained

$$c_{n-1}(X_d) = \sum_{k=0}^{n-1} (-1)^k d^k \binom{n+1}{n-1-k} \tilde{\alpha}^{n-1}.$$

To calculate $\chi(X_d) = \langle c_{n-1}(X_d), [X_d]\rangle$, it remains to find $\langle \tilde{\alpha}^{n-1}, [X_d]\rangle$.

[4]That is, taking the value $+1$ on $\mathbb{C}P^2$ with canonical complex orientation.

The number $\langle\tilde{\alpha}^{n-1}, [X_d]\rangle$ is equal to the number of the intersection points in $\mathbb{C}P^n$ of the cycle dual to α^{n-1} and X_d (with signs taken into account). The cycle dual to α^{n-1} is the complex line $\mathbb{C}P^1$. In general position, it intersects a surface of degree d at d points because the restriction to this line of a homogeneous polynomial F in $n+1$ variables is a homogeneous polynomial of degree d in two variables, which corresponds to an inhomogeneous polynomial of degree d in one variable. All intersection numbers are positive because we deal with complex manifolds. Thus, the homology class $[X_d]$ is $d[\mathbb{C}P^{n-1}]$. □

In particular, for a smooth algebraic curve X_d of degree d in $\mathbb{C}^2$, we have $\chi(X_d) = \binom{3}{1}d - \binom{3}{0}d^2 = 3d - d^2$. Therefore, X_d is a sphere with g handles, where $2 - 2g = 3d - d^2$, i.e., $g = \frac{(d-1)(d-2)}{2}$.

In a similar way, we can calculate the total Chern class and the Euler characteristic not only for a nonsingular hypersurface but also for an intersection of generic nonsingular hypersurfaces, as in [**22**]. A *complete intersection* is the complex n-manifold $M = M_1 \cap \cdots \cap M_p \subset \mathbb{C}P^{n+p}$, where $M_1, \ldots, M_p$ are the hypersurfaces defined by the homogeneous equations $f_1 = 0, \ldots, f_p = 0$ such that the vectors $\operatorname{grad} f_1, \ldots, \operatorname{grad} f_p$ are linearly independent at each point $x \in M$.

Example 78. Suppose $M = M_1 \cap \cdots \cap M_p \subset \mathbb{C}P^{n+p}$ is a complete intersection, and the hypersurfaces $M_1, \ldots, M_p$ have degrees $d_1, \ldots, d_p$, respectively. Suppose also that the hypersurfaces $M_1, \ldots, M_p$ are nonsingular. Then

$$\chi(M) = \left(\sum_{k=0}^{n} (-1)^k \binom{n+p+1}{n-k} \sigma_k\right) \prod_{i-1}^{p} d_i,$$

where $\sigma_k = \sum_{a_1+\cdots+a_p=k} d_1^{a_1} \cdots d_p^{a_p}$.

Proof. Suppose that $j\colon M \to \mathbb{C}P^{n+p}$ is the natural embedding, and $\alpha \in H^2(\mathbb{C}P^{n+p}; \mathbb{Z})$ is the canonical generator. We set $\tilde{\alpha} = j^*\alpha$. Then $c(\mathbb{C}P^{n+p}) = (1+\alpha)^{n+p+1}$ and $j^*\tau_{(\mathbb{C}P^{n+p})} \cong \tau_M \oplus \nu$, where ν is the normal bundle to M in $\mathbb{C}P^{n+p}$. Clearly, ν is the direct sum of the bundles $\xi(M_i)$; therefore, $c(\nu) = (1 + d_1\tilde{\alpha}) \cdots (1 + d_p\tilde{\alpha})$. Thus,

$$c(M) = (1+\tilde{\alpha})^{n+p+1}(1 + d_1\tilde{\alpha})^{-1} \cdots (1 + d_p\tilde{\alpha})^{-1}.$$

In particular,

$$c_n(M) = \left(\sum_{k=0}^{n} (-1)^k \binom{n+p+1}{n-k} \sigma_k\right) \tilde{\alpha}^n.$$

It remains to prove that $\langle\tilde{\alpha}^n, [M]\rangle = d_1 \cdots d_p$, i.e., that any generic p-dimensional subspace of $\mathbb{C}^p$ intersects M in $d_1 \cdots d_p$ points. This can be done in different ways. First, we can employ the fact that the cycle M_i is homologous to $d_i[\mathbb{C}P^{n+p-1}]$. Indeed, consider the intersection of p generic copies of $\mathbb{C}P^{n+p-1}$ in $\mathbb{C}P^{n+p}$ (this intersection is $\mathbb{C}P^n$). The intersection number of $[\mathbb{C}P^p]$ and $[M]$ is equal to that of $[\mathbb{C}P^p]$ and $[\mathbb{C}P^n]$ multiplied by $d_1 \cdots d_p$. Secondly, we can use the method of elimination of variables, which is based on calculating the resultant of two polynomials. Applying this method to the restrictions of homogeneous forms of degrees $d_1, \dots, d_p$ to a p-dimensional subspace, we obtain a polynomial of degree $d_1 \cdots d_p$. □

Remark. A more careful argument shows that we can dispense with the assumption that the hypersurfaces $M_1, \dots, M_p$ are nonsingular; see [**9**].

4.2. The Homology of the Hypersurface $z^{a_0} + \cdots + z^{a_n} = 1$. Let $a_0, \dots, a_n$ be positive integers, and let $a = (a_0, \dots, a_n)$. Consider the hypersurface V_a in $\mathbb{C}^{n+1}$ defined by the equation

$$z^{a_0} + \cdots + z^{a_n} = 1.$$

In [**101**], Pham calculated the homology groups of the space V_a by finding a deformation retract U_a of V_a whose homology groups are easy to calculate. In [**53**], it was shown that U_a is homotopy equivalent to the wedge of $(a_0 - 1) \cdots (a_n - 1)$ n-spheres. Thus,

$$\tilde{H}_k(V_a) = \begin{cases} 0 & \text{if } k \neq n, \\ \mathbb{Z}^{(a_0-1)\cdots(a_n-1)} & \text{if } k = n. \end{cases}$$

The subspace $U_a \subset V_a$ is given by

$$U_a = \{x \in V_a \cap \mathbb{R}^{n+1} \mid x_i^{a_i} \geq 0 \text{ for } i = 0, \dots, n\},$$

and the deformation retraction from V_a to U_a is constructed as follows. Consider the map $\mathbb{C}^{n+1} \to \mathbb{C}^{n+1}$ defined by

$$(z_0, \dots, z_n) \mapsto (z_0^{a_0}, \dots, z_n^{a_n}) = (u_0, \dots, u_n).$$

The image of V_a under this map is the hyperplane X in $\mathbb{C}^{n+1}$ determined by the equation $u_0 + \cdots + u_n = 1$. The map $p\colon V_a \to X$ is an $(a_0 \cdots a_n)$-fold covering branched over the intersections of X with the hyperplanes $u_i = 0$. The standard simplex Δ^n in $\mathbb{R}^{n+1} \subset \mathbb{C}^{n+1}$, which is given by the inequalities $t_0 \geq 0, \dots, t_n \geq 0$ and the equation $t_0 + \cdots + t_n = 1$, is a deformation retract of the hyperplane X. Indeed, a deformation retraction can be obtained from a deformation retraction of $\mathbb{C}^{n+1}$ onto $\mathbb{R}^{n+1}$, a deformation retraction of $\mathbb{R}^{n+1}$ onto the hyperplane $X \cap \mathbb{R}^{n+1}$, and a deformation retraction of the hyperplane onto the simplex Δ^n, which is shown in Figure 31 for $n = 2$.

This deformation retraction does not affect the type of branching over any moving point; therefore, the deformation retraction of X onto the simplex Δ^n can be lifted to a deformation retraction of V_a onto the preimage

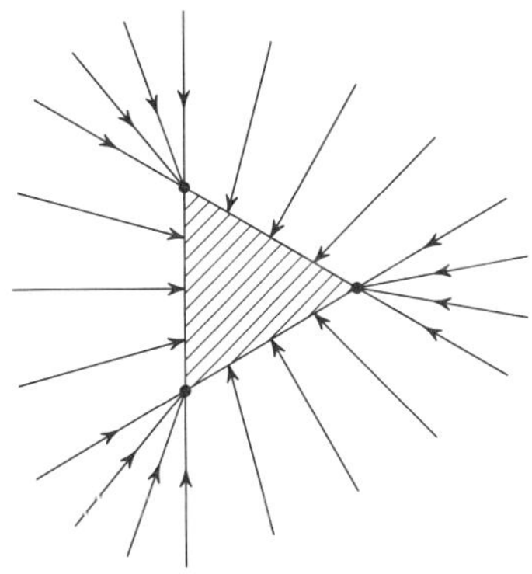

Figure 31. The deformation retraction

of Δ^n under the branched covering $p\colon V_a \to X$. Clearly, this preimage coincides with U_a.

Example 79. For the hypersurface $z_0^n + z_1^m = 1$, the space U_a coincides with the graph $K_{n,m}$ (see Part I, p. 7).

Proof. The corresponding branched covering over the 1-simplex Δ^1 for $n = 2$ and $m = 3$ is shown in Figure 32. □

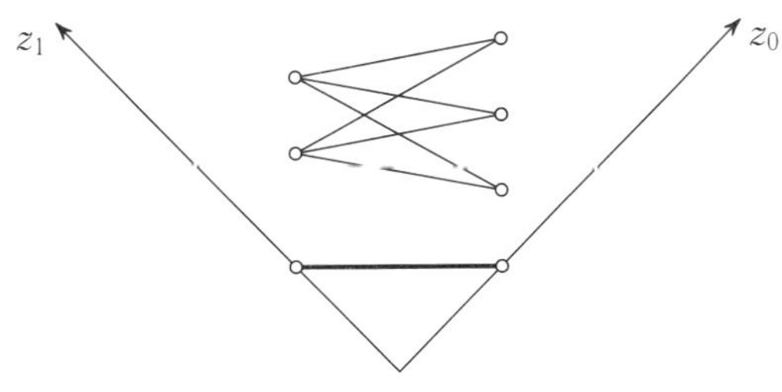

Figure 32. The branched covering

Now, let us prove that the space U_a is homotopy equivalent to the wedge of $(a_0 - 1)\cdots(a_n - 1)$ n-spheres. We identify U_a with the space of n-tuples $(u_0 t_0, \ldots, u_n t_n)$, where each u_i is an a_ith root of unity, $t_i \geq 0$, and $\sum t_i = 1$. This means that U_a coincides with the join $\mathbb{Z}_{a_0} * \mathbb{Z}_{a_1} * \cdots * \mathbb{Z}_{a_n}$, where $\mathbb{Z}_m$ is the m-point discrete space. Since $\mathbb{Z}_m = \underbrace{S^0 \vee \cdots \vee S^0}_{m-1}$, we have

$$\mathbb{Z}_{a_0} * \mathbb{Z}_{a_1} * \cdots * \mathbb{Z}_{a_n} \sim \underbrace{S^n \vee \cdots \vee S^n}_{(a_0-1)\cdots(a_n-1)},$$

because $\underbrace{S^0 * \cdots * S^0}_{n+1} = S^n$; we have also used the relation

$$(X_1 \vee X_2) * Y \sim (X_1 * Y) \vee (X_2 * Y).$$

Problem 134. Consider $\mathbb{Z}_m$ as the group of mth roots of unity. Then $U_a = \mathbb{Z}_{a_0} * \cdots * \mathbb{Z}_{a_n}$ is an n-dimensional simplicial complex, and there is a natural one-to-one correspondence between its n-simplices and the elements of the group $\mathbb{Z}_a = \mathbb{Z}_{a_0} \oplus \cdots \oplus \mathbb{Z}_{a_n}$: each n-simplex is obtained from the simplex e corresponding to the identity element of $\mathbb{Z}_a$ under the action of an element of $\mathbb{Z}_a$. Let $w_0, \ldots, w_n$ be the elements of $\mathbb{Z}_a$ corresponding to the identity elements of the groups $\mathbb{Z}_{a_0}, \ldots, \mathbb{Z}_{a_n}$.

(a) Prove that $\varepsilon = (1 - w_0) \cdots (1 - w_n)e$ is a cycle.

(b) Prove that the kernel I_a of the self-map of the group algebra $\mathbb{Z}[\mathbb{Z}_a]$ defined by $w \mapsto w(1 - w_0) \cdots (1 - w_n)$ is generated by the elements

$$1 + w_i + w_i^2 + \cdots + w_i^{a_i - 1}, \text{ where } i = 0, 1, \ldots, n.$$

(c) Prove that the elements

$$w_0^{k_0} w_1^{k_1} \cdots w_n^{k_n} \varepsilon, \text{ where } 0 \le k_i \le a_i - 2,$$

form a basis in the integral homology of the space U_a.

5. Lie Groups and H-Spaces

5.1. Some Properties of Lie Groups. A *Lie group* is a manifold G that is also a group for which the maps $G \times G \to G$ and $G \to G$ defined by $(g, h) \mapsto gh$ and $g \mapsto g^{-1}$ are smooth. To avoid delving deeply into the details of Lie group theory not related to algebraic topology, we consider only matrix Lie groups, which are submanifolds in the space of nonsingular square matrices over the field $\mathbb{R}$; the multiplication in this group is the usual multiplication of matrices, and the identity element is the identity matrix I.

To each square matrix A we assign the matrix $\exp A = I + A + \frac{A^2}{2!} + \frac{A^3}{3!} + \cdots$ (the proofs of the convergence of this series and of the basic properties of the map $\exp A$ are given, e.g., in [**104**]). The map $\exp A$ is a diffeomorphism from some neighborhood of the zero matrix to a neighborhood of the identity matrix. The tangent vector to the curve $\exp(tA)$ at $t = 0$ equals A. Therefore, the tangent space to the (matrix) Lie group at the identity is the linear space spanned by the preimage of a neighborhood of the identity under the exp map; this linear space is called the *Lie algebra* of the given Lie group. Any connected Lie group is completely determined by a neighborhood of the identity element (which we regard as a subset of the space of matrices); therefore, it is determined by its Lie algebra (which we also regard as a subset of the space of matrices).

A Lie group is commutative if and only if all matrices in its Lie algebra pairwise commute.

Problem 135. Prove that the image of a linear subspace V of the space of matrices under the map exp is a Lie group if and only if, for any two matrices $A, B \in V$, the matrix $[A, B] = AB - BA$ belongs to V.

The Lie group $T^n = \underbrace{S^1 \times \cdots \times S^1}_{n} = \mathbb{R}^n/\mathbb{Z}^n$ is called the *torus*.

Theorem 6.19. *Any connected commutative Lie group G is isomorphic to $T^n \times \mathbb{R}^m$.*

Proof. For a commutative Lie group G, the map $\exp\colon T_eG \to G$ is a group homomorphism because if $AB = BA$, then $e^A e^B = e^{A+B}$. For a connected Lie group, this homomorphism is an epimorphism. Therefore, $G \cong T_eG/K$, where $K = \operatorname{Ker}\exp$. The group K is discrete because exp is a diffeomorphism in a neighborhood of zero. Hence this group is a discrete subgroup in $T_eG \cong \mathbb{R}^{n+m}$. There is a basis in $\mathbb{R}^{n+m}$ in which the elements of K have the form $(k_1, \ldots, k_n, 0, \ldots, 0)$, where $k_i \in \mathbb{Z}$. Thus, $\mathbb{R}^{n+m}/K \cong T^n \times \mathbb{R}^m$. □

Corollary. *Any compact connected commutative Lie group is the torus.*

Theorem 6.20. *Any closed subgroup H of a Lie group G is a submanifold.*

Proof. Take a norm on the space of matrices, say $\|A\| = \max|a_{ij}|$ for $A = (a_{ij})$. Suppose that in the space T_eG, a matrix A and a sequence of matrices A_n such that $\exp A_n \in H$, $\|A_n\| \to 0$, and $A_n/\|A_n\| \to A$ are given. Let us show that $e^{tA} \in H$ for any real number t. Since $\|A_n\| \to 0$, we can choose integers m_n such that $m_n\|A_n\|$ converges to any given number t. We have $m_nA_n \to tA$ and $e^{m_nA_n} \to e^{tA}$. But $e^{m_nA_n} = (e^{A_n})^{m_n} \in H$, and the group H is closed. Therefore, $e^{tA} \in H$.

Let W be the space of all matrices of the form tA, where A is obtained by using the passage to the limit specified above and t is an arbitrary real number. Let us show that W is a linear space. To this end, it suffices to check that if $A, B \in W$, then $A + B \in W$. In a small neighborhood of the identity matrix, the map exp is invertible; therefore, for small t, we have $e^{tA}e^{tB} = e^{F(t)}$, where $F(t)$ is a smooth curve in T_eG and $F(0) = 0$. Taking into account the relation $e^{tA}e^{tB} = I + t(A + B) + \cdots$, we see that $F(t)/t \to A + B$ as $t \to 0$. Consider the sequence of matrices $C_n = F(1/n)$. Clearly, $\|C_n\| \to 0$, $e^{C_n} \in H$ (because $e^{tA}e^{tB} \in H$ for all t), and $C_n/\|C_n\| \to \lambda(A + B)$, where $\lambda = \lim\limits_{n\to\infty} \frac{1}{n}\left|F\left(\frac{1}{n}\right)\right|^{-1}$. Hence $A + B \in W$.

Let us prove that the set $\exp W$ contains some neighborhood of the identity element in H. We decompose T_eG into the direct sum $W \oplus W'$ and put $\varphi(w, w') = e^w e^{w'}$ for $w \in W$ and $w' \in W'$. On some neighborhood of the zero matrix in T_eG, the map φ acts as a diffeomorphism to a neighborhood of the identity matrix in G because $\varphi(T_eG) \subset G$ and the differential of φ at zero coincides with that of exp, since $e^w e^{w'} \approx e^{w+w'}$ for small w and w'. Suppose that there exists a sequence of points (w_n, w'_n) such that $\varphi(w_n, w'_n) \in H$, $(w_n, w'_n) \to 0$, and $w'_n \neq 0$. Then $e^{w'_n} \in H$ because $e^{w_n}e^{w'_n} \in H$ and $e^{w_n} \in H$. Choose a subsequence w'_{n_k} such that the sequence $w'_{n_k}/\|w'_{n_k}\|$ has a limit w'. On one hand, $w' \in W'$ and $w' \neq 0$. On the other hand, $w' \in W$. We have arrived at a contradiction.

Thus, the identity element of G has a neighborhood U whose intersection with H is a submanifold in U. For an arbitrary point $h \in H$, we take the neighborhood hU. □

Theorem 6.21. *For an arbitrary action of the torus T^n on a vector space V, there exists an invariant inner product in V (that is, an inner product such that $(u, v) = (tu, tv)$ for any $t \in T^n$ and $u, v \in V$).*

Proof. Take any inner product $(u, v)_1$ in V and consider the function $f(s) = (su, sv)_1$ on the torus. Let $(u, v) = \int_{T^n} f(s)\,d\mu$, where $d\mu = dx_1 \cdots dx_n$ is the canonical measure on T^n. This measure is translation-invariant; therefore, if $g(s) = f(s + t)$, then $\int_{T^n} g(s)\,d\mu = \int_{T^n} f(s)\,d\mu$. Thus, $(u, v) = \int_{T^n}(su, sv)_1\,d\mu = \int_{T^n} g(s)\,d\mu = \int_{T^n} f(s)\,d\mu = \int_{T^n}((t + s)u, (t + s)v)_1\,d\mu = (tu, tv)$. □

5.2. Cohomology of Lie Algebras. The de Rham cohomology of a compact connected Lie group G can be expressed in terms of the Lie algebra $\mathfrak{g}$ of G, which reduces calculating the real cohomology of G to a purely algebraic problem. In this section, we describe this reduction. As in Subsection 5.1, we consider only matrix Lie groups and algebras.

The Lie Algebra of a Lie Group. As mentioned on p. 344, the Lie algebra $\mathfrak{g}$ of a Lie group G is the tangent space at the identity element $e \in G$. But this definition refers only to the linear structure of $\mathfrak{g}$. To define an algebra structure, we must define the product of any two matrices $A, B \in \mathfrak{g}$ in $\mathfrak{g}$; this product is usually denoted by $[A, B]$ and called the *commutator* of A and B.

The commutator is defined as follows. Each matrix $A \in \mathfrak{g}$ determines the *left-invariant vector field* $v_A(g) = (L_g)_*A$, where $L_g(x) = gx$ is the left translation by g on the group G. Since the translation is by a constant matrix g, it follows that, in the matrix form, this left-invariant vector field can be written as $v_A(g) = gA$. A left-invariant vector field can be defined as a vector field on G invariant with respect to left translations. Then any left-invariant vector field is uniquely determined by the vector at the point e; this vector is spread over the entire group by left translations. For two left-invariant vector fields v_A and v_B, their commutator $[v_A, v_B]$ is defined. This is again a left-invariant vector field v_C. The matrix C, i.e., the vector $[v_A, v_B]$ at $e \in G$, is defined to be the commutator of the matrices A and B.

Theorem 6.22. *For matrix Lie algebras, $[A, B] = AB - BA$ is the usual commutator of matrices.*

Proof. Any matrix Lie group G is a subgroup in the group $\mathrm{GL}(n, \mathbb{R})$ of nonsingular matrices of order n. The commutator of two vector fields on G is the same as the commutator of these vector fields considered on $\mathrm{GL}(n, \mathbb{R})$. Therefore, is suffices to prove the required assertion for $\mathrm{GL}(n, \mathbb{R})$. In this case, the elements x^i_j of the matrix $\|x^i_j\|^n_1$ can be taken for local coordinates. We are interested in the commutator of vector fields with coordinates $v_A(x)^i_j = x^i_p A^p_j$ and $v_B(x)^i_j = x^i_p B^p_j$ (with summation over repeated indices). By definition, $[v_A(x), v_B(x)] = v_C(x)$, where

$$\begin{aligned} v_C(x)^k_l &= x^i_p A^p_j \frac{\partial(x^k_q B^q_l)}{\partial x^i_j} - x^i_p B^p_j \frac{\partial(x^k_q A^q_l)}{\partial x^i_j} \\ &= x^i_p A^p_j B^q_l \delta^j_q \delta^k_i - x^i_p B^p_j A^q_l \delta^j_q \delta^k_i = x^k_p (A^p_q B^q_l - B^p_q A^q_l). \end{aligned}$$

Thus, $v_C(x)^k_l = x^k_p C^p_l$, where $C = AB - BA$. □

The Maps Ad and ad. Consider the transformation $x \mapsto gxg^{-1}$ on the Lie group G. In a neighborhood of the identity, it can be represented as $e^{tX} \mapsto ge^{tX}g^{-1}$, where X is an element of the Lie algebra $\mathfrak{g}$. It is easy to see that the derivatives with respect to t of e^{tX} and $ge^{tX}g^{-1}$ at $t = 0$ are X and gXg^{-1}, respectively. Therefore, the map $X \mapsto gXg^{-1}$, where $g \in G$, is a self-map of the Lie algebra $\mathfrak{g}$. It is denoted by Ad_g.

Moreover, any fixed matrix $A \in \mathfrak{g}$ determines the map $X \mapsto [A, X] = AX - XA$ of the Lie algebra $\mathfrak{g}$. It is denoted by ad_A.

The maps Ad and ad are related as follows: $\mathrm{ad}_A X$ is the derivative of $\mathrm{Ad}_{\exp(tA)} X$ with respect to t at $t = 0$. Indeed,

$$e^{tA} X e^{-tA} = (I + tA + \cdots)X(I - tA + \cdots) = X + t(AX - XA) + \cdots .$$

Theorem 6.23. *If G is a compact connected Lie group, then* $\det \mathrm{Ad}_g = 1$ *for any $g \in G$.*

Proof. It is easy to see that the map $g \mapsto \mathrm{Ad}_g$ is a homomorphism from the group G to the automorphism group of the space $\mathfrak{g}$. Therefore, the map $g \mapsto \det \mathrm{Ad}_g$ is a homomorphism from the group G to the multiplicative group $\mathbb{R}^*$ of nonzero real numbers. Clearly, this homomorphism is continuous; hence its image is a compact connected subgroup in $\mathbb{R}^*$. But such a subgroup can contain only the element 1. □

Integration over Compact Lie Groups. A differential k-form ω on a Lie group G is said to be *left-invariant* if $L_g^*\omega = \omega$ for any $g \in G$. Similarly, a *right-invariant* k-form is a form for which $R_g^*\omega = \omega$, where $R_g(x) = xg$ is the right translation by g. A differential form is *bi-invariant* if it is both left- and right-invariant.

Theorem 6.24. *On any compact connected Lie group G of dimension n, there exists a nonzero bi-invariant n-form, and it is unique up to proportionality.*

Proof. Let ω_e be a nonzero n-form on the tangent space at the identity element $e \in G$. It uniquely determines a left-invariant form ω^L and a right-invariant form ω^R by $\omega_g^L(X_1, \ldots, X_n) = \omega_e(gX_1, \ldots, gX_n)$ and by $\omega_g^R(X_1, \ldots, X_n) = \omega_e(X_1 g, \ldots, X_n g)$ at every $g \in G$. Up to multiplication by a nonzero constant, $\omega_e(X_1, \ldots, X_n)$ is the determinant of the matrix composed of the coordinates of the vectors $X_1, \ldots, X_n$. Hence $\omega_g^R = (\det A)\omega_g^L$, where A is the determinant of the transformation taking each vector $Y = Xg$ to $gX = gYg^{-1}$; i.e., $A = \mathrm{Ad}_g$. Thus, on any Lie group of dimension n, there exists at most one (up to proportionality) nonzero bi-invariant n-form, and such an n-form exists if and only if $\det \mathrm{Ad}_g = 1$ for all $g \in G$. According to Theorem 6.23, any compact connected Lie group does satisfy this condition. $\square$

Theorem 6.24 implies that, on any compact connected Lie group G of dimension n, there exists a unique bi-invariant n-form ω for which $\int_G \omega = 1$. Using this n-form, we can define the integral $\int_G f = \int_{x\in G} f(x)\omega$ for any continuous function $f\colon G \to \mathbb{R}$. This integral has the properties $\int_G 1 = 1$ and

$$\int_{x\in G} f(xg) = \int_{x\in G} f(gx) = \int_{x\in G} f(x) \qquad \text{for any } g \in G.$$

Exercise. Prove that Theorem 6.21 is valid not only for the torus T^n but also for any compact connected Lie group.

To define *averaging*, which is an analog of the arithmetic mean of functions on the group $\mathbb{Z}_n$, for functions on a compact Lie group G, we need integration over G.

Left-Invariant Forms. Any left-invariant differential form is determined by its values on the sets of vectors of the tangent space at the identity; therefore, it is also determined by its values on the sets of left-invariant vector fields. If ω is a left-invariant differential k-form and $\xi_1, \ldots, \xi_k$ are left-invariant vector fields, then $\omega(\xi_1, \ldots, \xi_k) = \text{const}$; hence $\eta\omega(\xi_1, \ldots, \xi_k) = 0$

for any vector field η. As a consequence, the formula for the differential of a form becomes simpler, namely,

$$(44)\qquad d\omega(\xi_0,\ldots,\xi_k)=\sum_{0\le i<j\le k}(-1)^{i+j}\omega([\xi_i,\xi_j],\ldots,\hat{\xi}_i,\ldots,\hat{\xi}_j,\ldots,\xi_k).$$

For the Lie algebra $\mathfrak{g}$ of a Lie group G, we define the space of cochains $C^k(\mathfrak{g};\mathbb{R})$ as the space of skew-symmetric k-linear maps $\mathfrak{g}\times\cdots\times\mathfrak{g}\to\mathbb{R}$; the map $d\colon C^k\to C^{k+1}$ is defined by (44). The cohomology of the cochain complex thus obtained is called the *cohomology of the Lie algebra* $\mathfrak{g}$ with coefficients in $\mathbb{R}$. This is the same as the cohomology of the complex of left-invariant differential forms. Generally, it does not coincide with the cohomology of the complex of all differential forms, i.e., with the de Rham cohomology of the Lie group G. But these cohomologies do coincide for compact (and connected) Lie groups G.

Theorem 6.25. *If a Lie group G is compact and connected, then the cohomology of its Lie algebra $\mathfrak{g}$ is isomorphic to the de Rham cohomology of the manifold G.*

Proof. The proof uses the existence of the integral $\int_{g\in G} f(g)$ on a compact connected Lie group G and its properties specified on p. 348.

First, note that if forms ω and $\tilde{\omega}$ are left-invariant and $\omega-\tilde{\omega}=d\alpha$ for some form α, then $\omega-\tilde{\omega}=d\alpha^L$ for some left-invariant form α^L. Indeed, since the form $d\alpha$ is left-invariant, it follows that $d\alpha=\int_{g\in G}L_g^*(d\alpha)=d\int_{g\in G}L_g^*\alpha$. Clearly, the form $\alpha^L=\int_{g\in G}L_g^*\alpha$ is left-invariant as well.

Now we prove that if ω is an arbitrary closed form, then there exists a left-invariant form ω^L such that $\omega=\omega^L+d\alpha$ for some form α. Since the group G is path-connected, for each $g\in G$ there exists a path $g(t)$ from $g(0)=e$ to $g(1)=g$. Using this path, we obtain a family of forms $\omega_t=L_{g(t)}^*\omega$ which join $\omega_0=\omega$ to $\omega_1=L_g^*\omega$. Clearly, the form $L_g^*\omega$ is closed because so is ω. Therefore, the homotopy invariance of the de Rham cohomology (see p. 282) implies that $\omega-L_g^*\omega=d\alpha_g$ for some form α_g. The dependence of α_g on g is not necessarily continuous, but, in any case, it can be assumed to be integrable (piecewise continuous) in g. Consider the left-invariant form $\omega^L=\int_{g\in G}L_g^*\omega$. Clearly,

$$\omega-\omega^L=\int_{g\in G}(\omega-L_g^*\omega)=\int_{g\in G}d\alpha_g=d\int_{g\in G}\alpha_g=d\alpha,$$

where $\alpha=\int_{g\in G}\alpha_g$. □

Corollary. *If G_1 and G_2 are compact connected Lie groups one of which covers the other, then $H^*(G_1;\mathbb{R})\cong H^*(G_2;\mathbb{R})$.*

Example. For the torus T^n, all differentials in the complex of left-invariant differential forms are zero; therefore, $H^k(T^n;\mathbb{R}) \cong \Lambda^k(\mathbb{R}^n)$ is the space of skew-symmetric k-forms on $\mathbb{R}^n$.

Remark. For noncompact Lie groups, Theorem 6.25 is false. For example, T^n and $\mathbb{R}^n$ have isomorphic Lie algebras, but the spaces $H^*(T^n;\mathbb{R})$ and $H^*(\mathbb{R}^n;\mathbb{R})$ are not isomorphic.

Bi-Invariant Forms.

Theorem 6.26. *If G is a compact connected Lie group, then the de Rham cohomology algebra of the manifold G is isomorphic to the algebra of bi-invariant differential forms.*

Proof. According to Theorem 6.25, the de Rham cohomology algebra of the manifold G is isomorphic to the cohomology algebra of the complex of left-invariant differential forms. For the latter, we can repeat the argument from the proof of Theorem 6.25 with left translations replaced by right ones. As a result, we obtain an isomorphism between the de Rham cohomology algebra of the manifold G and the cohomology algebra of the complex of bi-invariant differential forms. It remains to prove that the differential on this complex is zero, i.e., $d\omega = 0$ for any bi-invariant form ω.

Consider the map $i\colon G \to G$ defined by $i(g) = g^{-1}$. It takes the curve ge^{tX} to $e^{-tX}g^{-1}$; therefore, the tangent vector $Y = gX$ at g is taken to $-Xg^{-1} = -g^{-1}Yg^{-1}$. Thus, if ω is a bi-invariant k-form, then

$$\begin{aligned}(i^*\omega_g)(X_1,\dots) &= \omega_{g^{-1}}(i_*X_1,\dots) = \omega_{g^{-1}}(-g^{-1}X_1g^{-1},\dots)\\ &= \omega_e(-X_1g^{-1},\dots) = \omega_g(-X_1,\dots) = (-1)^k\omega_g(X_1,\dots),\end{aligned}$$

i.e., $i^*\omega = (-1)^k\omega$. Hence

$$(-1)^{k+1}\,d\omega = i^*\,d\omega = di^*\omega = (-1)^k\,d\omega,$$

which means that $d\omega = 0$. □

5.3. Maximal Tori. Let G be a topological group. An element $g \in G$ is called a *topological generator* of G if the closure of the subgroup generated by g coincides with the entire group G.

Theorem 6.27. *The torus T^n has a topological generator.*

Proof. Consider a countable open base $U_1, U_2, \dots$ for the topology of T^n. By a *cube* in T^n we mean the image of the cube given by $|x_i - a_i| \le \varepsilon$ in $\mathbb{R}^n$ under the canonical projection. Clearly, any cube is a compact subset of the torus. We refer to the number $\min(1, 2\varepsilon)$ as the edge length of the cube. If a cube has edge length 1, then it coincides with T^n.

Take any cube C_0 in T^n. Suppose that we have constructed a sequence of cubes $C_0 \supset C_1 \cdots \supset C_{k-1}$. Let us construct a cube C_k as follows. Suppose that the edge length of C_{k-1} equals ε. Choose a number $N(k)$ so that $2N(k)\varepsilon > 1$. The $N(k)$th power of the cube C_{k-1} coincides with T^n. Therefore, we can choose a cube $C_k \subset C_{k-1}$ so that $C_k^{N(k)} \subset U_k$.

Take $g \in \bigcap_{k=0}^{\infty} C_k$. We have $g^{N(k)} \in U_k$; therefore, g is a topological generator of the torus T^n. □

Let G be a compact connected Lie group. The term *maximal torus* is used for a subgroup $T \subset G$ that is a torus which is maximal in the sense that if $T \subset T' \subset G$, where T' is a torus, then $T = T'$.

Any torus $T \subset G$ is contained in some maximal torus. Indeed, if tori $T \subset T_1 \subset T_2 \subset \cdots \subset G$ form a strictly increasing sequence, then their dimensions strictly increase as well; indeed, if a closed n-manifold is a submanifold in an n-dimensional manifold, then it is both open and closed. Therefore, any strictly increasing sequence of tori is finite.

Let $S \subset G$ be an arbitrary subset of a group G. The *normalizer* of S is the group

$$N(S) = \{g \in G \mid gS = Sg\},$$

and the *centralizer* of S is the group

$$Z(S) = \{g \in G \mid gs = sg \ \forall s \in S\}.$$

If S is a subgroup of G, then S is a normal subgroup of $N(S)$. If G is a Lie group and S is its subgroup, then $N(S)$ and $Z(S)$ are closed subgroups of G, and therefore, according to Theorem 6.20, they are Lie groups.

For any element $g \in N(S)$, the inner automorphism $x \mapsto gxg^{-1}$ maps S to S; thus, $N(S)$ acts on S.

Theorem 6.28. *Let $T \subset G$ be a maximal torus. Then*

(a) *the quotient group $N(T)/T$ is finite;*

(b) *the dimension of the space $G/N(T)$ is even, $\chi(G/N(T)) = 1$, and $\chi(G/T) = |N(T)/T|$.*

Proof. (a) Consider the map $N(T) \to \mathrm{Aut}(T)$ that assigns the automorphism $x \mapsto gxg^{-1}$ to each element $g \in N(T)$. The group $\mathrm{Aut}(T)$ is discrete because any automorphism of the torus is determined by an integer matrix with determinant ± 1 (any such automorphism leaves the integer lattice invariant). Therefore, the connected component $N(T)_e$ of the identity element is mapped to the identity automorphism, i.e., $N(T)_e \subset Z(T)$.

Let A be a tangent vector to $Z(T)$ at e. Consider the subgroup H generated by the torus T and the elements e^{At}, where $t \in \mathbb{R}$. The subgroup H is closed; therefore, it is a Lie subgroup. Moreover, the group H is connected and commutative. Therefore, it is a torus containing T. The maximality of T implies $H = T$; in particular, $e^{At} \in T$. Hence $Z(T)_e = T$. But $T \subset N(T)_e \subset Z(T)_e$, whence $Z(T)_e = N(T)_e = T$. Thus, $N(T)/T = N(T)/N(T)_e$ is a discrete compact group, and it must be finite.

(b) Consider the action of the torus T on the quotient space $G/N(T)$ by left translations. Under the action of $t \in T$, the class $gN(T)$ is mapped to $tgN(T)$. The class $N(T)$ remains fixed because $T \subset N(T)$.

Consider the action of T on G by left translations. This action can be transferred to the tangent space T_eG. Take a T-invariant inner product in T_eG. Let $(T_eN(T))^\perp$ be the orthogonal complement to $T_eN(T)$ in T_eG. We have

$$T_{N(T)}(G/N(T)) \cong T_eG/T_eN(T) \cong (T_eN(T))^\perp;$$

each element of T acts on $T_{N(T)}(G/N(T))$ as an orthogonal transformation. Using the restriction of the map exp to the space $(T_eN(T))^\perp$, we construct a disk D^m in $G/N(T)$, where $m = \dim(G/N(T))$, such that it is centered at the point $N(T)$ and the three closed sets D^m, $S^{m-1} = \partial D^m$, and $M = G/N(T) \setminus \operatorname{int} D^m$ are invariant with respect to the action of T.

Let us show that the coset $N(T)$ is the unique fixed point of the action of T on $G/N(T)$; moreover, this is so not only for the action of the entire torus T but also for the action of its topological generator t. Indeed, $tgN(T) = gN(T)$ if and only if $g^{-1}tg \in N(T)$, or, equivalently, $g^{-1}t^ng \in N(T)$ for $n = 1, 2, \dots$; the last condition, in turn, is equivalent to $g^{-1}Tg \subset N(T)$ because the set $\{t^n\}$ is everywhere dense in T. Finally, the condition $g^{-1}Tg \subset N(T)$ is equivalent to $g^{-1}Tg = T$, i.e., $g \in N(T)$. Indeed, $N(T)_e = T$, and the set $g^{-1}Tg$ is connected and contains the identity element e.

The torus T contains a path from t to e; therefore, the actions of t on $G/N(T)$ and on G are homotopic to the identity map, and the actions on T_eG during the homotopy respect the inner product in T_eG chosen above. Thus, we have a homotopy in the class of maps leaving invariant D^m, S^{m-1}, and M. Consequently, there exist fixed-point-free maps $S^{m-1} \to S^{m-1}$ and $M \to M$ homotopic to the identity. By the Lefschetz theorem, we have $\chi(M) = 0$ and $\chi(S^{m-1}) = 0$; hence m is even.

The formula for the Euler characteristic of the union of two CW-complexes implies

$$\chi(G/N(T)) = \chi(D^m) + \chi(M) - \chi(S^{m-1}) = \chi(D^m) = 1.$$

The natural projection $G/T \to G/N(T)$ is an $|N(T)/T|$-fold covering; therefore, $\chi(G/T) = |N(T)/T|\chi(G/N(T)) = |N(T)/T|$. □

Theorem 6.28 and the Lefschetz theorem readily imply the following assertion.

Theorem 6.29. *Let $T \subset G$ be a maximal torus. Then, for any element $g \in G$, there exists an $x \in G$ such that $g \in xTx^{-1}$.*

Proof. Consider the left coset space $G/T = \{xT\}$ and the map $f(xT) = gxT$ (this map is well defined because the coset gxT, when x is replaced by xt, where $t \in T$, does not change, $gxtT = gxT$). Suppose that a coset xT is a fixed point of f, i.e., $gxT = xT$. Then $g \in xTx^{-1}$. Therefore, it suffices to prove that the map $f\colon G/T \to G/T$ has a fixed point. For this purpose, we employ the Lefschetz theorem.

To calculate the Lefschetz number $\Lambda(f)$, we can take any map f_0 homotopic to f. This means that we can replace g by any other element $g_0 \in G$ because the group G is connected. Take the identity element e for g_0. Then $f_0 = \mathrm{id}_{G/T}$ and $\Lambda(f_0) = \chi(G/T)$. According to Theorem 6.28, we have $\chi(G/T) = |N(T)/T| \neq 0$. Therefore, $\Lambda(f) = \Lambda(f_0) \neq 0$, and hence the map f has a fixed point. □

Corollary 1. *All maximal tori are conjugate; i.e., if T and T' are maximal tori of a compact Lie group G, then $T' = gTg^{-1}$ for some $g \in G$.*

Proof. Let t' be a topological generator of the torus T'. Take $g \in G$ for which $t' \in gTg^{-1}$. We have $T' \subset gTg^{-1}$. Clearly, gTg^{-1} is a torus. Therefore, the maximality of T' implies $T' = gTg^{-1}$. □

Corollary 2. *For any compact connected Lie group G, the image of the map* $\exp\colon T_eG \to G$ *is the entire group G.*

Proof. Let T be a maximal torus of the Lie group G. For each $x \in G$, we can choose $g \in G$ so that x belongs to the torus gTg^{-1}. In the proof of Theorem 6.19, we showed that the map exp is an epimorphism for a torus. □

The dimension of a maximal torus of a compact connected Lie group G is called the *rank* of G.

Problem 136. Given a compact connected Lie group G and a positive integer n, prove that the image of the map $f\colon G \to G$ defined by $f(g) = g^n$ coincides with the entire group G.

5.4. Regular Elements. Let G be a compact connected Lie group. An element $g \in G$ is said to be *regular* if it is contained in precisely one maximal torus; if g is contained in more than one of the maximal tori, then it is called *singular*. Regular elements can also be defined as follows: An element $g \in G$ is regular if and only if the dimension of its normalizer $N(g)$ is equal to that of a maximal torus. To prove the equivalence of these definitions, we need two lemmas.

Lemma 1. *Suppose that S is a connected Abelian subgroup of a compact connected Lie group G and $g \in G$ commutes with all elements of S. Then there exists a torus T containing g and S.*

Proof. Consider the subgroup H generated by g and S in G. Since H is Abelian, it follows that its closure $\overline{H}$ is a compact Abelian group; hence the connected component $(\overline{H})_e$ of the identity is a torus. The compactness of $\overline{H}$ implies the finiteness of the quotient group $\overline{H}/(\overline{H})_e$. This quotient group is generated by one coset $g(\overline{H})_e$ because the connectedness of S implies $S \subset (\overline{H})_e$. Thus, $\overline{H}/(\overline{H})_e \cong \mathbb{Z}_k$ for some positive integer k. Let t be a topological generator of the torus $(\overline{H})_e$. Clearly, $g^k \in (\overline{H})_e$ and $tg^{-k} \in (\overline{H})_e$. The group $(\overline{H})_e$ is divisible; therefore, $tg^{-k} = s^k$ for some $s \in (\overline{H})_e$. Consider the element $h = gs \in \overline{H}$. Clearly, $h^k = (gs)^k = t$; hence the powers of h are everywhere dense in $(\overline{H})_e$. Applying the translation by h^r, we see that the powers of h are also everywhere dense in the coset of g^r in $(\overline{H})_e$. In this way, all cosets are obtained; therefore, h is a topological generator of the group $\overline{H}$. The element h belongs to some maximal torus T, and $\{g\} \cup S \subset H \subset \overline{H} \subset T$. □

Lemma 2. *Let $N(g)_e$ be the connected component of the identity in the normalizer of some element $g \in G$. Then $N(g)_e$ is the union of all maximal tori of G containing g.*

Proof. First, suppose that g belongs to some maximal torus T. Then $N(g) \supset N(T)$. Since T is maximal, we have $N(g)_e = T$. Therefore, $N(g)_e$ contains all maximal tori containing g.

Now suppose that $n \in N(g)_e$. Consider a maximal torus S of $N(g)_e$ containing n. Any element of S commutes with g; this allows us to use Lemma 1, which implies the existence of a maximal torus T containing S and g. This maximal torus contains n and g. □

Theorem 6.30. *An element $g \in G$ is regular of and only if the dimension of its normalizer $N(g)$ is equal to that of a maximal torus. For any singular element, the dimension of its normalizer is larger than that of a maximal torus.*

Proof. First, suppose that g is regular, i.e., belongs to precisely one maximal torus T. Then, according to Lemma 2, we have $N(g)_e = T$, whence $\dim N(g) = \dim N(g)_e = \dim T$.

Now, suppose that g belongs to two different maximal tori, T_1 and T_2. According to Lemma 2, the normalizer $N(g)$ contains both T_1 and T_2. Therefore, the tangent space $N(g)$ at e contains the sum of the tangent spaces to T_1 and T_2. These spaces are different, and hence the dimension of their sum is strictly larger than that of a maximal torus. □

Roots. Suppose that G is a compact connected Lie group, $\mathfrak{g} = T_eG$ is its Lie algebra, and $(u, v)_1$ is an inner product in $\mathfrak{g}$. Consider the inner product defined by $(u, v) = \int_{g\in G}(\mathrm{Ad}_g\, u, \mathrm{Ad}_g\, v)_1$. This new inner product is invariant with respect to Ad_g; that is, the map $\mathrm{Ad}_g\colon \mathfrak{g} \to \mathfrak{g}$ is an orthogonal transformation for any $g \in G$. In what follows, we assume that $\mathfrak{g}$ is endowed with this inner product.

Theorem 6.31. *Let $T \subset G$ be a torus. Then the space $\mathfrak{g}$ decomposes into a direct sum as $\mathfrak{g} = V_0 \oplus \sum_{r=1}^{m} V_r$, where the spaces $V_0, V_1, \dots, V_m$ are invariant with respect to the action of T by means of* Ad *and this action is trivial on V_0, each of the spaces V_r with $1 \le r \le m$ is two-dimensional, and, in some orthogonal basis, the action of each element $t \in T$ on V_r is determined by the matrix*

$$\begin{pmatrix} \cos 2\pi\bar{\theta}_r(t) & -\sin 2\pi\bar{\theta}_r(t) \\ \sin 2\pi\bar{\theta}_r(t) & \cos 2\pi\bar{\theta}_r(t) \end{pmatrix},$$

where $\bar{\theta}_r\colon T \to \mathbb{R}/\mathbb{Z}$ is a nontrivial homomorphism.

Proof. Take a topological generator t_0 of the torus T. The transformation Ad_{t_0} is orthogonal; therefore, in some orthonormal basis, its matrix is the direct sum of the identity matrix and several matrices of the form

$$\begin{pmatrix} \cos 2\pi\alpha_r & -\sin 2\pi\alpha_r \\ \sin 2\pi\alpha_r & \cos 2\pi\alpha_r \end{pmatrix}.$$

Clearly, $\mathrm{Ad}_{t_0^k} = (\mathrm{Ad}_{t_0})^k$, and the map Ad_t depends continuously on t. Hence the matrix of Ad_t in the same orthonormal basis has the required form. □

The set of numbers $\{\pm\alpha_1, \dots, \pm\alpha_m\}$ that appear in the proof of Theorem 6.31 is determined uniquely; therefore, so is the set of homomorphisms $\{\pm\bar{\theta}_1, \dots, \pm\bar{\theta}_m\}$ in the statement of this theorem. For a maximal torus T, these homomorphisms $\bar{\theta}_r\colon T \to \mathbb{R}/\mathbb{Z}$ are called *roots*. Together with the homomorphisms $\bar{\theta}_r$, the linear maps $\theta_r\colon \mathfrak{t} \to \mathbb{R}$ for which the diagram

$$\begin{array}{ccc} \mathfrak{t} & \xrightarrow{\ \theta_r\ } & \mathbb{R} \\ {\scriptstyle \exp}\downarrow & & \downarrow{\scriptstyle p} \\ T & \xrightarrow{\ \bar{\theta}_r\ } & \mathbb{R}/\mathbb{Z} \end{array}$$

is commutative are often considered. The map θ_r is the lifting of $\bar{\theta}_r \circ \exp$ for which $\theta_r(0) = 0$. Such a lifting exists and is unique because the topological space $\mathfrak{t}$ is simply connected. The linear maps $\theta_r\colon \mathfrak{t} \to \mathbb{R}$ are called *roots* too.

Theorem 6.32. *A torus T is maximal if and only if $V_0 = \mathfrak{t}$.*

Proof. Clearly, $\mathfrak{t} \subset V_0$ for any torus. Suppose that T is a maximal torus and $V_0 \ne \mathfrak{t}$, i.e., there exists an $X \in V_0$ such that $X \notin \mathfrak{t}$. On the group

H consisting of the elements e^{tX}, where $t \in \mathbb{R}$, the torus T acts trivially; therefore, the subgroup generated by T and H is connected and Abelian, and it contains T as a proper subset. This contradicts the maximality of T.

Now, suppose that $V_0 = \mathfrak{t}$, T and T' are tori, and T is contained in T'. Then T acts trivially on $\mathfrak{t}'$, and therefore $\mathfrak{t}' \subset V_0$. Hence $\mathfrak{t} \subset \mathfrak{t}' \subset V_0 = \mathfrak{t}$, which means that $\dim T = \dim T'$. Thus, $T = T'$, and the torus T is maximal. □

Corollary. *The dimension of any compact connected Lie group G is equal to $l + 2m$, where l is the rank of G and $2m$ is the number of roots.*

For each root $\bar{\theta}_r \colon T \to \mathbb{R}/\mathbb{Z}$, we denote its kernel by U_r. Clearly, U_r is a subgroup in T of dimension $l - 1$, where $l = \dim T$ (= the rank of the Lie group G).

Theorem 6.33. *If an element $t \in T$ belongs to precisely μ groups U_r, then* $\dim N(t) = l + 2\mu$.

Proof. Let $V \subset \mathfrak{g}$ be the subspace on which the given element t acts trivially. By Theorem 6.31, we have $V = V_0 \oplus \sum_{U_r \ni t} V_r$; therefore, $\dim V = l + 2\mu$. Thus, it is sufficient to prove that $V = T_e N(t)$.

The elements of $N(t)$ commute with t; hence $tnt^{-1} = n$ for any $n \in N(t)$, which means that t acts trivially on $T_e N(t)$. Therefore, $T_e N(t) \subset V$.

Now, suppose that $X \in V$. Then $tXt^{-1} = X$, and hence $te^{\lambda X}t^{-1} = e^{\lambda X}$ for any $\lambda \in \mathbb{R}$. Therefore, $e^{\lambda X} \in N(t)$ for all $\lambda \in \mathbb{R}$; thus, $X \in T_e N(t)$. □

Corollary. *An element $t \in T$ is regular if and only if it does not belong to any of the subgroups U_r.*

Theorem 6.34. *The singular elements of a Lie group G form a subset of codimension at least* 3 *in G in the sense that this set is the image of one or several smooth compact manifolds of dimension at most* $\dim G - 3$ *under a smooth map.*

Proof. First, let us show that the group U_r has a topological generator. The connected component $(U_r)_e$ of the identity in U_r is a torus of codimension 1 in the torus T. The quotient of a torus by a torus is a compact connected commutative Lie group, that is, a torus. Therefore, $T/(U_r)_e$ is a torus of dimension 1, i.e., the circle S^1. The group $U_r/(U_r)_e$ is a finite subgroup in $T/(U_r)_e \cong S^1$; hence $U_r/(U_r)_e \cong \mathbb{Z}_k$ for some positive integer k. Let t be a topological generator of the torus $(U_r)_e$, and let $g \in U_r$ be an element such that the group $\mathbb{Z}_k$ is generated by the coset $g(U_r)_e$. As in the proof of Lemma 1 on p. 354, we choose an element $s \in (U_r)_e$ for which $tg^{-k} = s^k$ and consider $h = gs$. The same argument shows that h is a topological generator of U_r.

Let h be a topological generator of U_r. According to Theorem 6.33, we have $\dim N(h) \geq l+2$. If $n \in N(h)$, then n commutes with h^m for any integer m; therefore, n commutes with all elements of U_r.

Consider the map $f\colon G \times U_r \to G$ defined by $f(g,u) = gug^{-1}$. If $n \in N(h)$ and $u \in U_r$, then $nun^{-1} = u$; hence $f(gn,u) = f(g,u)$. Thus, in fact, we have a map $f_r\colon G/N(h) \times U_r \to G$. The image of f_r consists of all elements of G conjugate to the elements of U_r. The dimension of the manifold $G/N(h) \times U_r$ is equal to

$$\dim G - \dim N(h) + \dim U_r \leq \dim G - (l+2) + (l-1) = \dim G - 3.$$

An element of the group G is regular if and only if it does not belong to the image of any of the maps $f_1, \dots, f_m$. Indeed, the normalizers of conjugate elements always have the same dimension, and therefore an element is regular if and only if none of the elements of the maximal torus T conjugate to it belongs to any of the subgroups $U_1, \dots, U_m$. □

Proof of the Equality $\pi_2(G) = 0$. Let G_{reg} be the set of regular elements of a compact connected Lie group G, and let $G_{\text{sing}} = G \setminus G_{\text{reg}}$. By Theorem 6.34, the set G_{sing} has codimension at least 3 in G. Therefore, any smooth map $S^2 \to G$ admits an arbitrarily close approximation whose image is contained in G_{reg}. This means that the equality $\pi_2(G) = 0$ for a compact connected Lie group G is implied by the following theorem (as applied to $X = S^2$).

Theorem 6.35. *Any continuous map from a simply connected space X to a compact connected Lie group G whose image is contained in G_{reg} is null-homotopic.*

Proof. Let T be a maximal torus in G with Lie algebra $\mathfrak{t}$. Consider the map $\varphi\colon G \times \mathfrak{t} \to G$ defined by $\varphi(g,X) = ge^Xg^{-1}$. If $t \in T$ and $X \in \mathfrak{t}$, then $te^Xt^{-1} = e^X$; we obtain a map $\varphi\colon (G/T) \times \mathfrak{t} \to G$, where G/T is the space of left cosets gT.

Let $\mathfrak{t}_{\text{reg}}$ be the set of those $X \in \mathfrak{t}$ for which $e^X \in G_{\text{reg}}$. Clearly, $X \in \mathfrak{t}_{\text{reg}}$ if and only if $ge^Xg^{-1} \in G_{\text{reg}}$ for any $g \in G$. Thus, we have a map $\varphi_{\text{reg}}\colon (G/T) \times \mathfrak{t}_{\text{reg}} \to G_{\text{reg}}$, and this map is surjective because so is φ. Let us show that φ_{reg} is a local homeomorphism. It suffices to verify that the Jacobian φ_{reg} vanishes nowhere.

Lemma. *A point (g, X) is regular for the map $\varphi\colon (G/T) \times \mathfrak{t} \to G$ if and only if $X \in \mathfrak{t}_{\text{reg}}$.*

Proof. Since e^X and dX commute, we have $e^{X+dX} = e^X + e^X\,dX$. Moreover, $(g+dg)^{-1} = g - g^{-1}(dg)g^{-1}$. Therefore,

$$\begin{aligned}(g+dg)&e^{X+dX}(g+dg)^{-1} - ge^X g^{-1}\\ &= (dg)e^X g^{-1} - ge^X g^{-1}(dg)g^{-1} + g(e^X\,dX)g^{-1}\\ &= g(g^{-1}\,dg - e^X(g^{-1}\,dg)e^{-X} + e^X(dX)e^{-X})e^X g^{-1}.\end{aligned}$$

Here dg is an element of the tangent space at g; hence $dg = g\,dz$, where dz is an element of the Lie algebra. Thus, $g^{-1}\,dg = dz$.

Below, we use the notation of Theorem 6.31. The map $\mathrm{id} - \mathrm{Ad}_{e^X}$ vanishes on $V_0 = \mathfrak{t}$, and it is nonsingular on V_r ($r = 1, \ldots, m$) if and only if $\theta_r(e^X) \neq 1$, i.e., $e^X \notin U_r$. Thus, the restriction of $\mathrm{id} - \mathrm{Ad}_{e^X}$ to $\sum_{r=1}^m V_r$ is nonsingular if and only if the element e^X is regular, i.e., $X \in \mathfrak{t}_{\mathrm{reg}}$.

Clearly, the restriction of Ad_{e^X} to $\mathfrak{t}$ is an automorphism, so it is always nonsingular. $\square$

Consider the commutative diagram

$$\begin{array}{ccccc} & & (G/T)\times\mathfrak{t}_{\mathrm{reg}} & \xrightarrow{\subset} & (G/T)\times\mathfrak{t}\\ & \overset{\tilde u}{\nearrow} & \downarrow{\scriptstyle\varphi_{\mathrm{reg}}} & & \downarrow{\scriptstyle\varphi}\\ X & \xrightarrow{u} & G_{\mathrm{reg}} & \xrightarrow{\subset} & G.\end{array}$$

Here $\tilde u$ is a lifting of u. Similarly to the case of coverings, it is constructed using liftings of paths starting at a base point $x_0 \in X$. The lifting does not depend on the choice of paths from x_0 to x, since X is simply connected. Using natural embeddings, we obtain the commutative diagram

$$\begin{array}{ccc} & & (G/T)\times\mathfrak{t}\\ & \overset{\tilde u}{\nearrow} & \downarrow{\scriptstyle\varphi}\\ X & \xrightarrow{u} & G.\end{array}$$

The space $\mathfrak{t}$ is contractible, and therefore the map $\tilde u$ is homotopic to a map $X \to (G/T)\times\{0\}$. Since $\varphi(g,0) = e$ is the identity element of G, it follows that φ takes $(G/T)\times\{0\}$ to the identity element e. $\square$

5.5. *H*-Spaces and Hopf Algebras. A topological space X with a base point $x_0 \in X$ is called an *H-space* if it is equipped with a continuous map $\mu\colon X\times X \to X$ such that the maps $x \mapsto \mu(x, x_0)$ and $x \mapsto \mu(x_0, x)$ are homotopic to the identity map (it is assumed that $\mu(x_0, x_0) = x_0$). The map μ is called a *multiplication.*

The Hopf theorem (Theorem 6.41) shows that the cohomology algebra of an H-space over a field of characteristic zero has a very special form.

One of the corollaries of this theorem is that $H^2(G;\mathbb{R}) = 0$ for any simply connected Lie group G.

Natural examples of H-spaces are Lie groups and topological groups. But there exist other examples.

Example 80. The spaces $\mathbb{C}P^\infty$ and $\mathbb{R}P^\infty$ are H-spaces.

Proof. Let us identify $\mathbb{C}^\infty \setminus \{0\}$ with the space of polynomials by assigning the polynomial $a_0 + a_1 z + \cdots + a_n z^n$ to each point $(a_0, a_1, \ldots, a_n, 0, \ldots)$. Multiplication of polynomials induces a multiplication in $\mathbb{C}^\infty \setminus \{0\}$ for which $(1, 0, 0, \ldots)$ is an identity element. Identifying proportional polynomials, we obtain a multiplication in the space $\mathbb{C}P^\infty$ with the identity element $(1 : 0 : 0 : \cdots)$. For $\mathbb{R}P^\infty$, the construction is similar. □

Example 81. The sphere S^7 is an H-space.

Proof. Let us represent S^7 as the set of Cayley numbers with unit norm. Multiplication of Cayley numbers induces a (nonassociative) multiplication on S^7. □

Theorem 6.36. *If X is an H-space with base point x_0, then the group $\pi_1(X, x_0)$ is Abelian.*

Proof. A homotopy between loops $\alpha\beta$ and $\beta\alpha$ can be constructed as follows. Consider the map of the inner rhombus in Figure 33 defined by $(s,t) \mapsto$

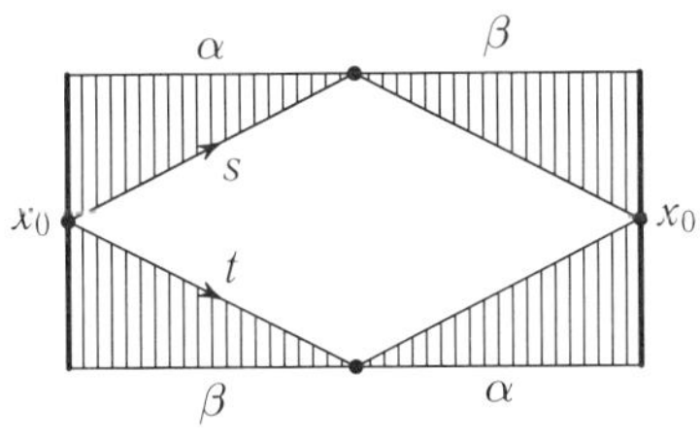

Figure 33. The construction of the homotopy

$\mu(\alpha(s), \beta(t))$. Maps of the remaining four triangles are constructed by using homotopies between the maps $s \mapsto \alpha(s)$ and $s \mapsto \mu(\alpha(s), x_0)$ and between the maps $t \mapsto \beta(t)$ and $t \mapsto \mu(x_0, \beta(t))$. □

Theorem 6.36 has the following generalization.

Theorem 6.37. *Any H-space X is simple.*

Proof. Clearly, $\pi_n(X \times Y, (x_0, y_0)) \cong \pi_n(X, x_0) \times \pi_n(Y, y_0)$, and the action of the group $\pi_1(X \times Y, (x_0, y_0))$ corresponds to a product of actions; namely, $(\alpha, \beta)(u, v) = (\alpha u, \beta v)$.

The multiplication $\mu\colon X \times X \to X$ induces the commutative diagram

$$\begin{array}{ccc} \pi_1(X\times X)\times \pi_n(X\times X) & \longrightarrow & \pi_n(X\times X) \\ \big\downarrow {\scriptstyle \mu_*\times\mu_*} & & \big\downarrow {\scriptstyle \mu_*} \\ \pi_1(X)\times\pi_n(X) & \longrightarrow & \pi_n(X), \end{array}$$

where the horizontal arrows represent the action of the group π_1 on π_n.

Let e_1 and e_n be the identity elements of $\pi_1(X)$ and $\pi_n(X)$. Then, for any $\alpha \in \pi_1(X)$ and $u \in \pi_n(X)$, we have $(\alpha, e_1)(e_n, u) = (\alpha e_n, e_1 u) = (e_n, u)$. Therefore, $\mu_*(\alpha, e_1)\mu_*(e_n, u) = \mu_*(e_n, u)$, and hence $\alpha u = u$. $\square$

Let F be the additive group of some field, and let X be a connected H-space for which all groups $H_i(X;F) \cong H^i(X;F)$ are finite-dimensional spaces over F. Consider the cohomology algebra $A = \bigoplus_{i\ge 0} A^i$, where $A^i = H^i(X;F)$. The diagonal map $d\colon X \to X\times X$ determines a multiplication $A\otimes A \to A$ in the cohomology algebra. The multiplication $\mu\colon X\times X\to X$ induces a map $\Delta = \mu^*\colon A \to A\otimes A$. This map is called a *coproduct*. The coproduct Δ has the following properties.

1. The map $\Delta\colon A \to A\otimes A$ is a homomorphism of algebras, i.e., $\Delta(1) = 1\otimes 1$, and if $\Delta(a) = \sum a_{1i}\otimes a_{2i}$ and $\Delta(b) = \sum b_{1j}\otimes b_{2j}$, then

$$\Delta(a \smile b) = \sum_{i,j} (-1)^{\dim b_{1j}\dim a_{2i}} (a_{1i}\smile b_{1j})\otimes(a_{2i}\smile b_{2j}).$$

This follows from the Künneth theorem for cohomology with coefficients in a field and from the multiplication law in the algebra $H^*(X\times X)$.

2. Let $p_1\colon A\otimes A \to A$ be the map defined by $\alpha^p\otimes 1 \mapsto \alpha^p$ and $\alpha^p\otimes\alpha^q \mapsto 0$ for $q>0$, and let $p_2\colon A\otimes A\to A$ be a similar map taking $1\otimes\alpha^p$ to α^p. Then both compositions in the diagram

$$\begin{array}{ccc} A & \xrightarrow{\ \Delta\ } & A\otimes A \\ {\scriptstyle \Delta}\big\downarrow & & \big\downarrow{\scriptstyle p_1} \\ A\otimes A & \xrightarrow{\ p_2\ } & A \end{array}$$

are the identity. In other words, we have $\Delta(\alpha^p) = \alpha^p\otimes 1 + 1\otimes\alpha^p + \sum \alpha^r\otimes\alpha^s$ for $r,s\ge 1$. This is so because both compositions in the diagram

$$\begin{array}{ccc} X & \xleftarrow{\ \mu\ } & X\times X \\ {\scriptstyle \mu}\big\uparrow & & \big\uparrow{\scriptstyle \mu(\cdot,x_0)} \\ X\times X & \xleftarrow{\ \mu(x_0,\cdot)\ } & X \end{array}$$

are homotopic to the identity map.

3. We have $\Delta(A^p) \subset \bigoplus_{i+j=p} A^i \otimes A^j$. This follows from the isomorphism $H^p(X \times X) \cong \bigoplus_{i+j=p} H^i(X) \otimes H^j(X)$ for cohomology with coefficients in a field.

A graded[5] associative algebra $A = \bigoplus_{i \geq 0} A^i$ for which a coproduct Δ with properties 1–3 is defined is called a *Hopf algebra*. If A^0 coincides with the ground field F, then the Hopf algebra is said to be *connected* (this condition corresponds to the path-connectedness of the space X).

Theorem 6.38. *Suppose that* $\operatorname{char} F = 0$ *and A is a connected Hopf algebra generated multiplicatively by an element $x \in A^k$, where $k \geq 1$ and k is even. Then A is the polynomial algebra $F[x]$.*

Proof. Clearly, $\Delta(x) = 1 \otimes x + x \otimes 1$ because there are no elements of positive dimension strictly less than k. Since Δ is an algebra homomorphism, $\Delta(x^n) = \sum_{r=0}^{n} x^r \otimes x^{n-r}$. (Note that for odd k, the corresponding equality is quite different. For example, $\Delta(x^2) = 1 \otimes x^2 + x^2 \otimes 1$ because the two terms $x \otimes x$ have opposite signs in this case.) Suppose that $x^n = 0$ for $n \geq 2$ and n is the minimum number with this property. Then $0 = \sum_{r=1}^{n-1} x^r \otimes x^{n-r}$, which is impossible. □

Remark. If k is odd, then $x^2 = 0$. Therefore, $A = \Lambda(x)$ is an exterior algebra; it consists of all elements of the form $a + bx$, where $a, b \in F$.

Example 82. The sphere S^{2n} is not an H-space.

Proof. Suppose that the sphere S^{2n} is an H-space. Then $A = H^*(S^{2n}; \mathbb{R})$ is a connected Hopf algebra. The algebra A is generated multiplicatively by an element $x \in H^{2n}(S^{2n}; \mathbb{R})$. Therefore, by Theorem 6.38, we have $x^k \neq 0$ for all k. On the other hand, $x^2 = 0$. □

For a field F with characteristic $p \neq 0$, Theorem 6.38 is false because the equality $x^n = 0$ may hold for $n = p^m$. To prove a version of Theorem 6.38 for fields of nonzero characteristic, we need the following lemma.

Lemma. *Suppose that $n = \sum n_i p^i$ and $r = \sum r_i p^i$, where $0 \leq n_i, r_i \leq p-1$. Then $\binom{n}{r} \equiv \prod_i \binom{n_i}{r_i} \pmod{p}$.*

Proof. Suppose that $1 \leq k \leq p-1$. Then $\binom{p}{k}$ is divisible by p. Therefore, $(1+x)^p \equiv 1 + x^p \pmod{p}$. Applying induction on k, we obtain the relation

[5] We assume that $ab = (-1)^{pq} ba$ for $a \in A^p$ and $b \in A^q$.

$(1+x)^{p^k} \equiv 1 + x^{p^k} \pmod{p}$. Hence

$$\begin{aligned}(1+x)^n &= (1+x)^{n_0+n_1p+\dots+n_mp^m}\\ &= (1+x)^{n_0}(1+x)^{n_1p}\cdots(1+x)^{n_mp^m}\\ &\equiv (1+x)^{n_0}(1+x^p)^{n_1}\cdots(1+x^{p^m})^{n_m}\\ &\equiv \prod_{i=0}^{m}\sum_{k=0}^{n_i}\binom{n_i}{k}x^{kp^i} \pmod{p}.\end{aligned}$$

Thus, the coefficient of $x^r = x^{r_0+r_1p+\dots+r_mp^m}$ in the expression for $(1+x)^n$ is congruent to $\binom{n_0}{r_0}\binom{n_1}{r_1}\cdots\binom{n_m}{r_m}$ modulo p. On the other hand, it equals $\binom{n}{r}$. □

This lemma implies, in particular, that $\binom{p^m}{k} \equiv 0 \pmod{p}$ for $0 < r < p^m$. Therefore, $\sum_{r=1}^{n-1} x^r \otimes x^{n-r} = 0$ for $n = p^m$. Moreover, if $\sum_{r=1}^{n-1} x^r \otimes x^{n-r} = 0$ and $x^r \neq 0$ for $1 \leq r \leq n-1$, then $n = p^m$ for some m. Indeed, suppose that $n = n_0 + n_1p + \dots + n_sp^s$, where $n_s \neq 0$ and either $n_q \neq 0$ for $q < s$ or $n_s \neq 1$. In the former case, we set $r = n_qp^q$ and obtain $\binom{n}{r} \equiv \binom{n_q}{r_q} \pmod{p}$ and $\binom{n_q}{r_q} = 1$. In the latter case, we set $r = p^s$ and obtain $r < n$, $\binom{n}{r} \equiv \binom{n_s}{1} \pmod{p}$, and $\binom{n_s}{1} = n_s$, where $2 \leq n_s \leq p-1$.

These congruences imply the following version of Theorem 6.38.

Theorem 6.39. *Suppose that* $\operatorname{char} F = p$, *where p is an odd number, and A is a connected Hopf algebra generated multiplicatively by $x \in A^k$, where $k \geq 1$. Then $A = \Lambda(x)$ if k is odd, and either $A = F[x]$ or $A = F[x]/(x^{p^m})$ if k is even.*

This theorem has an analog for $p = 2$; its formulation and proof are similar to those of the above theorem, except that the equality $x^2 = 0$ may not hold for odd k and $\Lambda(x) = F[x]/(x^2)$. Thus, for a field of characteristic 2, the analog of Theorem 6.38 is as follows.

Theorem 6.40. *Suppose that* $\operatorname{char} F = 2$ *and A is a connected Hopf algebra generated multiplicatively by an element $x \in A^k$, where $k \geq 1$. Then either $A = F[x]$ or $A = F[x]/(x^{2^m})$.*

Theorem 6.41 (Hopf [**61**]). *Suppose that A is a connected Hopf algebra over a field F of characteristic zero and the space A^n is finite-dimensional for any n. Then the algebra A is isomorphic (as a Hopf algebra) to the tensor product of an exterior algebra on odd-dimensional generators and a polynomial algebra on even-dimensional generators.*

Proof. Since the space A^n is finite-dimensional for any n, the algebra A has generators $x_1, x_2, \dots$ such that $\dim x_i \leq \dim x_{i+1}$ (it is assumed that

$x_i \in A^{\dim x_i}$). Let A_n be the subalgebra of A generated by $x_1, \ldots, x_n$. It is a Hopf subalgebra, i.e., $\Delta(A_n) \subset A_n \otimes A_n$. Indeed, we have $\Delta(x_i) = x_i \otimes 1 + 1 \otimes x_i + \sum y_r \otimes y_s$, where $\dim y_r, \dim y_s < \dim x_i$. Therefore, $y_r, y_s \in A_{i-1} \subset A_i$. We assume that $x_n \notin A_{n-1}$ (otherwise, the generator x_n can be eliminated).

According to Theorem 6.38, the algebra generated by x_n is the polynomial algebra $F[x_n]$ (if the dimension of x_n is even) or the exterior algebra $\Lambda(x_n)$ (if the dimension of x_n is odd). Therefore, since the algebra A is graded commutative and associative, there is a natural epimorphism $A_{n-1} \otimes F[x_n] \to A_n$ (if the dimension of x_n is even) or $A_{n-1} \otimes \Lambda(x_n) \to A_n$ (if the dimension of x_n is odd). It is sufficient to prove that this epimorphism is a monomorphism (and then apply induction on n).

Consider the ideal I generated by x_n^2 and the elements of A_{n-1} of positive dimension. It consists of all expressions of the form $\sum a_k x_n^k$, where $a_k \in A_{n-1}$ and the zero-dimensional components of a_0 and a_1 vanish. Clearly, $x_n \notin I$ because any element of I of dimension $\dim x_n$ must belong to A_{n-1} (the ideal I has no elements of the form λx_n, where $\lambda \in F$).

Consider the composition

$$A_n \xrightarrow{\Delta} A_n \otimes A_n \xrightarrow{p} A_n \otimes (A_n/I),$$

where p is the natural projection. If $a \in A_{n-1}$ is an element of positive dimension, then $\Delta(a) = a \otimes 1 + 1 \otimes a + \sum a_i \otimes a_j$, where $a_j \in A_{n-1}$ has positive dimension. Therefore, $p\Delta(a) = a \otimes 1$ (this equality holds for any $a \in F$). Moreover, $\Delta(x_n) = x_n \otimes 1 + 1 \otimes x_n + \sum y_r \otimes y_s$, where $y_s \in A_{n-1}$ is an element of positive dimension. Hence $p\Delta(x_n) = x_n \otimes 1 + 1 \otimes \bar{x}_n$, where $\bar{x}_n$ is the image of x_n in A_n/I.

First, consider the case where the element x_n has even dimension. Suppose that there is a nontrivial relation $\sum a_k x_n^k = 0$. Let us apply the map $p\Delta$ to this relation. Taking into account the equality $\bar{x}_n^2 = 0$, we obtain

$$\begin{aligned} 0 &= \sum (a_k \otimes 1)(x_n \otimes 1 + 1 \otimes \bar{x}_n)^k \\ &= \sum a_k x_n^k \otimes 1 + \sum k a_k x_n^{k-1} \otimes \bar{x}_n. \end{aligned}$$

By assumption, we have $\sum a_k x_n^k = 0$; hence $\sum k a_k x_n^{k-1} \otimes \bar{x}_n = 0$. It follows that $\sum k a_k x_n^{k-1} = 0$. Indeed, as mentioned above, we have $x_n \in I$, whence $\bar{x}_n \neq 0$. The degree of the relation $\sum k a_k x_n^{k-1} = 0$ is lower than that of the initial relation $\sum a_k x_n^k = 0$. Clearly, this relation is nontrivial because the field has characteristic zero. Choosing an initial relation of minimum degree, we immediately obtain a contradiction.

Now consider the case where the dimension of x_n is odd. Suppose that there is a nontrivial relation $a_0 + a_1 x_n = 0$. Applying the map $p\Delta$ to it, we obtain

$$0 = a_0 \otimes 1 + a_1 \otimes 1(x_n \otimes 1 + 1 \otimes \bar{x}_n) = (a_0 + a_1 x_n) \otimes 1 + a_1 \otimes \bar{x}_n.$$

By assumption, $a_0 + a_1 x_n = 0$; therefore, $a_1 \otimes \bar{x}_n = 0$. As in the preceding case, we have $\bar{x}_n \neq 0$. Hence $a_1 = 0$ and $a_0 = 0$. □

Corollary 1. *Let G be a Lie group. Then the cohomology algebra $H^*(G; \mathbb{R})$ is an exterior algebra on generators of odd dimensions.*

Proof. The polynomial algebra is infinite-dimensional; hence a finite-dimensional algebra cannot contain a polynomial subalgebra. □

Corollary 2. *If G is a simply connected Lie group, then $H^2(G; \mathbb{R}) = 0$.*

Proof. Since G is simply connected, $H^1(G; \mathbb{R}) = 0$. Therefore, all generators in the cohomology algebra $H^*(G; \mathbb{R})$ are of dimension at least 3. □

Hints and Solutions

1. Obviously, the required isomorphism holds even at the level of relative chains.

2. This commutative diagram can be regarded as a short exact sequence of chain complexes $0 \to C'_* \to C_* \to C''_* \to 0$, where C'_* is the chain complex $\cdots \to 0 \to 0 \to A \xrightarrow{\alpha} A' \to 0$; the complexes C_* and C''_* are defined similarly. The nontrivial homology groups of the complex C'_* are precisely $\operatorname{Ker}\alpha$ and $\operatorname{Coker}\alpha$.

3. The proof of Theorem 1.13 applies to the case under consideration without any substantial changes, except that *isomorphism* should be replaced by *monomorphism* or *epimorphism*.

4. First, let us prove that φ_2 induces a homomorphism of the required form. Take $x \in \operatorname{Ker}(\varphi_3\alpha_2)$. Clearly, $\varphi_2 x \in \operatorname{Im}\varphi_2$. It remains to show that $\varphi_2 x \in \operatorname{Im}\beta_1 = \operatorname{Ker}\beta_2$. But $0 = \varphi_3\alpha_2 x = \beta_2\varphi_2 x$. If $x \in \operatorname{Ker}\alpha_2$, then $x \in \operatorname{Im}\alpha_1$; therefore, $\varphi_2 x \in \operatorname{Im}\varphi_2\alpha_1$. If $x \in \operatorname{Ker}\varphi_2$, then $\varphi_2 x = 0$. Thus, the map Φ is well defined. It is also clear that Φ is a homomorphism.

Now let us prove that Φ is an epimorphism. Take $y \in \operatorname{Im}\varphi_2 \cap \operatorname{Im}\beta_1$. There exists an $x \in A_2$ for which $\varphi_2 x = y$. Moreover, $\varphi_3\alpha_2 x = \beta_2\varphi_2 x = \beta_2 y = 0$ because $y \in \operatorname{Im}\beta_1 = \operatorname{Ker}\beta_2$. Hence $x \in \operatorname{Ker}(\varphi_3\alpha_2)$.

Finally, let us prove that Φ is a monomorphism. Suppose that $x \in \operatorname{Ker}(\varphi_3\alpha_2)$ and $\varphi_2 x \in \operatorname{Im}(\varphi_2\alpha_1)$, i.e., $\varphi_2 x = \varphi_2\alpha_1 z$ for some $z \in A_1$. Then $x = \alpha_1 z + w$, where $w \in \operatorname{Ker}\varphi_2$. Hence $x \in \operatorname{Ker}\alpha_2 + \operatorname{Ker}\varphi_2$.

5. (a) Consider the isomorphism $f = \eta_1^{-1}\eta_2$. The required equality is equivalent to $f + f^{-1} = 2\,\mathrm{id}$, i.e., $f^2 + \mathrm{id} = 2f$, and the latter equality can be written in the form $(f - \mathrm{id})(f - \mathrm{id}) = 0$.

Let us show that $(f-\mathrm{id})\alpha_2 = 0$ and $\alpha_3(f-\mathrm{id}) = 0$. The former equality is derived as follows: $\eta_1^{-1}\eta_2\alpha_2 = \eta_1^{-1}\varphi_2\beta_2 = \eta_1^{-1}\eta_1\alpha_2 = \alpha_2$. To prove the latter, note that $\varphi_4\alpha_3 = \beta_3\eta_1$, i.e., $\alpha_3 = \varphi_4^{-1}\beta_3\eta_1$, whence $\alpha_3\eta_1^{-1}\eta_2 = \varphi_4^{-1}\beta_3\eta_1\eta_1^{-1}\eta_2 = \varphi_4^{-1}\beta_3\eta_2 = \varphi_4^{-1}\varphi_4\alpha_3 = \alpha_3$.

Thus, $\mathrm{Im}(f-\mathrm{id}) \subset \mathrm{Ker}\,\alpha_3 = \mathrm{Im}\,\alpha_2$ and $\mathrm{Im}\,\alpha_2 \subset \mathrm{Ker}(f-\mathrm{id})$; therefore, $(f-\mathrm{id})^2 = 0$.

(b) First, suppose that such a homomorphism δ exists. We set $\eta_2 = \eta_1 + \eta_1\delta$. The first two conditions ensure the commutativity of the diagram, and the last condition implies that η_2 is a homomorphism of rings. Let us show that η_2 is a monomorphism. If $\eta_2(x) = 0$, then $\eta_1(x) = \eta_1(-\delta(x))$; therefore, $x = -\delta(x)$, which means that $\delta(x) = -\delta^2(x) = 0$ because $\mathrm{Im}\,\delta \subset \mathrm{Ker}\,\alpha_3 = \mathrm{Im}\,\alpha_2 \subset \mathrm{Ker}\,\delta$. Let us show that η_2 is an epimorphism. Suppose that $y \in B_3$ and $y = \eta_1(x)$. Then $\eta_2(x-\delta(x)) = \eta_1(x-\delta(x)) - \eta_1(\delta(x) - \delta^2(x)) = \eta_1(x) = y$.

Now suppose that there exists a diagram with two different isomorphisms η_1 and η_2. Consider $\delta = \eta_1^{-1}\eta_2 - \mathrm{id}$. It is easy to show that δ has all the required properties.

(c) The commutative diagrams

$$\begin{array}{ccccccccc}
0 & \longrightarrow & \mathbb{Z}_2 & \xrightarrow{\varphi} & \mathbb{Z}_2 \times \mathbb{Z}_2 & \xrightarrow{\psi} & \mathbb{Z}_2 & \longrightarrow & 0 \\
 & & \downarrow{\scriptstyle \mathrm{id}} & & \downarrow{\scriptstyle \eta_i} & & \downarrow{\scriptstyle \mathrm{id}} & & \\
0 & \longrightarrow & \mathbb{Z}_2 & \xrightarrow{\varphi} & \mathbb{Z}_2 \times \mathbb{Z}_2 & \xrightarrow{\psi} & \mathbb{Z}_2 & \longrightarrow & 0,
\end{array}$$

where $i = 1, 2$, $\eta_1 = \mathrm{id}$, $\eta_2(x,y) = (y,x)$, $\varphi(x) = (x,x)$, and $\psi(x,y) = x+y$, give the required example.

6. Let us represent the torus T^2 as the union of simplicial complexes K_0 and K_1 and choose generators a and b in $H_1(K_0 \cap K_1)$ (see Figure H.1). The exact sequence

$$H_1(K_0 \cap K_1) \xrightarrow{(j_0,-j_1)} H_1(K_0) \oplus H_1(K_1) \longrightarrow H_1(T^2) \longrightarrow \tilde{H}_0(K_0 \cap K_1) \longrightarrow 0$$

gives an exact sequence

$$0 \longrightarrow \mathrm{Im}\,\varphi \longrightarrow H_1(T^2) \longrightarrow \mathbb{Z} \longrightarrow 0,$$

where $\varphi = (j_0, -j_1)$. The group $H_1(K_0 \cap K_1)$ consists of elements of the form $na + mb$. The homomorphism φ takes any such element to the pair $(nj_0(a) + mj_0(b), -nj_1(a) - mj_1(b))$. But $j_0(a) = j_0(b)$ and $j_1(a) = j_1(b)$. Therefore, φ has the form $(m,n) \mapsto (n+m, -n-m)$. Hence $\mathrm{Ker}\,\varphi = \mathbb{Z}$ and $\mathrm{Im}\,\varphi = \mathbb{Z}$. Since the exact sequence

$$0 \longrightarrow \mathbb{Z} \longrightarrow H_1(T^2) \longrightarrow \mathbb{Z} \longrightarrow 0$$

splits, we have $H_1(T^2) \cong \mathbb{Z} \oplus \mathbb{Z}$.

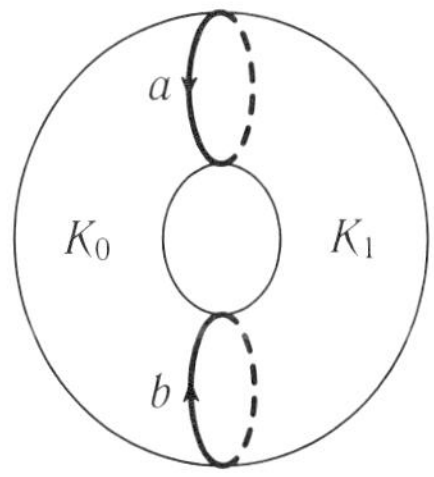

Figure H.1. The torus T^2

From the exact sequence

$$0 \longrightarrow H_2(T^2) \longrightarrow H_1(K_0 \cap K_1) \xrightarrow{(j_0,-j_1)} H_1(K_0) \oplus H_1(K_1)$$

we obtain an exact sequence

$$0 \longrightarrow H_2(T^2) \longrightarrow \operatorname{Ker} \varphi \longrightarrow 0.$$

Therefore, $H_2(T^2) \cong \operatorname{Ker} \varphi \cong \mathbb{Z}$.

7. According to Problem 37 in Part I, we have $S^p \times S^q / S^p \vee S^q \approx S^{p+q}$. Therefore, $H_k(S^p \times S^q, S^p \vee S^q) \cong H_k(S^p \times S^q \cup C(S^p \vee S^q)) \cong H_k(S^{p+q})$ for $k \geq 1$.

8. Writing the exact sequence for the pair $(S^p \times S^q, S^p \vee S^q)$ and applying Problem 7, we see that $H_k(S^p \times S^q) \cong \mathbb{Z}$ for $k = 0, p, q, p+q$ (if $p = q$, then $H_p(S^p \times S^q) \cong \mathbb{Z} \oplus \mathbb{Z}$); the remaining groups are trivial.

9. (a) Suppose that K_0 is the closed ε-neighborhood of the given knot K and K_1 is the closure of $S^3 \setminus K_0$. Then $K_0 \cup K_1 = S^3$, and $K_0 \cap K_1 = T^2$ is the 2-torus. We must calculate the homology of $S^3 \setminus K \sim K_1$. We assume that the sphere S^3 is triangulated so that its triangulation induces triangulations of K_0 and K_1. First, note that the space K_1 is homotopy equivalent to a simplicial complex containing no simplices of dimension ≥ 3. Indeed, we can successively kill all of the 3-simplices in K_1, starting with the boundary. Thus, $H_i(K_1) = 0$ for $i \geq 3$. Clearly, $H_0(K_1) = \mathbb{Z}$.

Let us write the Mayer–Vietoris sequence:

$$H_{i+1}(K_0) \oplus H_{i+1}(K_1) \xrightarrow{i_*} H_{i+1}(S^3)$$
$$\xrightarrow{\partial_*} H_i(T^2) \xrightarrow{j_*} H_i(K_0) \oplus H_i(K_1) \xrightarrow{i_*} H_i(S^3).$$

For $i = 2$, we obtain

$$0 \longrightarrow \mathbb{Z} \xrightarrow{\partial_*} \mathbb{Z} \longrightarrow H_2(K_1) \longrightarrow 0.$$

The homomorphism ∂_* is as follows. Any generator α of the group $H_3(S^3)$ can be represented as a sum of simplices with compatible orientations into which the sphere S^3 is decomposed. These simplices are divided into two parts according to whether they belong to K_0 or K_1. We define $\partial_*\alpha$ to be the common boundary of these parts. Clearly, ∂_* is an isomorphism. Therefore, $H_2(K_1) = 0$.

For $i = 1$, we obtain

$$0 \longrightarrow \mathbb{Z} \oplus \mathbb{Z} \xrightarrow{j_*} \mathbb{Z} \oplus H_1(K_1) \longrightarrow 0.$$

The map j_* takes a parallel on the torus to a generator of the group $H_1(K_0) = \mathbb{Z}$; therefore, a meridian is mapped to a generator of $H_1(K_1)$. Hence $H_1(K_1) = \mathbb{Z}$, and a generator of $H_1(K_1)$ is represented by a small circle in a plane transversally intersecting the knot (see Figure H.2).

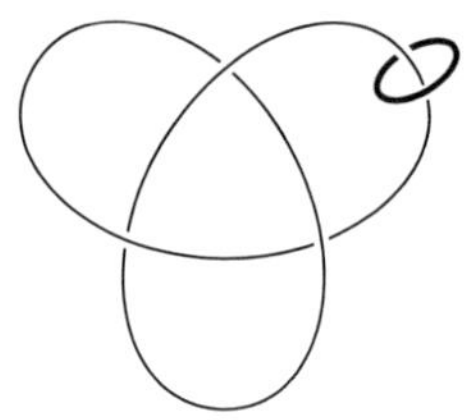

Figure H.2. A generator of the homology group of the complement of a knot

(b) As in (a), we obtain exact sequences

$$0 \longrightarrow \mathbb{Z} \xrightarrow{\partial_*} \mathbb{Z}^n \longrightarrow H_2(K_1) \longrightarrow 0$$

and

$$0 \longrightarrow \mathbb{Z}^{2n} \xrightarrow{j_*} \mathbb{Z}^n \oplus H_1(K_1) \longrightarrow 0.$$

We have $\partial_*(1) = (1, \ldots, 1)$; hence $H_2(K_1) = \mathbb{Z}^{n-1}$. The map j_* takes parallels of n tori to generators of the group $H_1(K_0) = \mathbb{Z}^n$; therefore, meridians of these n tori are mapped to generators of $H_1(K_1)$. Thus, $H_1(K_1) = \mathbb{Z}^n$, and generators of this group are represented by small circles put on the link components.

10. Let us write the Mayer–Vietoris sequence for $K_0 = M^n \setminus \operatorname{int} D^n$ and $K_1 = D^n$:

$$H_k(S^{n-1}) \longrightarrow H_k(M^n \setminus \operatorname{int} D^n) \oplus H_k(D^n) \longrightarrow H_k(M^n) \longrightarrow H_{k-1}(S^{n-1}).$$

If $k \geq 1$, then $H_k(D^n) = 0$, and if $2 \leq k \leq n-2$, then $H_k(S^{n-1}) = 0$ and $H_{k-1}(S^{n-1}) = 0$; for $k = 1$, we consider reduced homology groups.

11. Let K' be the barycentric subdivision of the simplicial complex K. Consider the simplicial map $f\colon K' \to N(\mathcal{L})$ defined as follows. Each simplex Δ in the complex K is covered by one of the subcomplexes L_i. To the barycenter v of Δ we assign the minimum index i. We must show that if K' contains a simplex with vertices $v_0, v_1, \ldots, v_k$ which correspond to subcomplexes $L_{i_0}, L_{i_1}, \dots, L_{i_k}$, then $N(\mathcal{L})$ contains a simplex with vertices $L_{i_0}, \dots, L_{i_k}$, i.e., $L_{i_0} \cap \dots \cap L_{i_k} \neq \varnothing$. We can assume that in the initial simplicial complex K, v_0 is a vertex, v_1 is the midpoint of an edge, v_2 is the center of a 2-face, etc. Then $v_m \in L_{i_m}$ implies $v_0, \dots, v_{m-1} \in L_{i_m}$. Therefore, $v_{i_0} \in L_{i_0} \cap \dots \cap L_{i_k}$.

Let us prove that f induces an isomorphism of homology groups. We use induction on n. For $n = 1$, the complex $K = L_1$ is acyclic, and $N(\mathcal{L}) = *$ consists of one point. Thus, the basis of induction is obvious. The induction step is as follows. Suppose that $\mathcal{L}_1 = \{L_1, \dots, L_n\}$, $K_1 = L_1 \cup \dots \cup L_n$, and $K_2 = L_{n+1}$. Consider the subcomplex $N_1 = N(\mathcal{L}_1)$ of $N(\mathcal{L})$ and let N_2 be the closed star of the vertex $(n+1)$. By construction, f takes K_1' to N_1, and by the induction hypothesis, the restriction of f to K_1' induces an isomorphism in homology. The restriction of f to K_2' also induces an isomorphism in homology because the complex $K_2 = L_{n+1}$ is acyclic and N_2 is a cone with vertex $(n+1)$.

Let $M_i = L_i \cup L_{n+1}$. Then $\mathcal{M} = \{M_1, \dots, M_n\}$ is a cover of the set $K_1 \cap K_2$, to which the induction hypothesis applies. Therefore, f induces an isomorphism between the homology groups of the complexes $K_1 \cap K_2$ and $N(\mathcal{M}) = N_1 \cap N_2$. Let us write the Mayer–Vietoris exact sequence:

$$
\begin{array}{ccccc}
H_k(K_1 \cap K_2) & \longrightarrow & H_k(K_1) \oplus H_k(K_2) & \longrightarrow & H_k(K) \\
\downarrow \cong & & \downarrow \cong & & \downarrow \\
H_k(N_1 \cap N_2) & \longrightarrow & H_k(N_1) \oplus H_k(N_2) & \longrightarrow & H_k(N)
\end{array}
$$

$$
\begin{array}{ccccc}
\longrightarrow & H_{k-1}(K_1 \cap K_2) & \longrightarrow & H_{k-1}(K_1) \oplus H_{k-1}(K_2) \\
& \downarrow \cong & & \downarrow \cong \\
\longrightarrow & H_{k-1}(N_1 \cap N_2) & \longrightarrow & H_{k-1}(N_1) \oplus H_{k-1}(N_2) .
\end{array}
$$

The pairs of the left and right maps are isomorphisms; hence, by the five lemma, the middle map is an isomorphism as well.

12. In case (b), consider the diagram

$$
\begin{array}{ccccccccc}
0 & \longrightarrow & A & \xrightarrow{\varphi} & B & \xrightarrow{\psi} & C & \longrightarrow & 0 \\
& & \downarrow \mathrm{id}_A & & \downarrow \Phi\oplus\psi & & \downarrow \mathrm{id}_C & & \\
0 & \longrightarrow & A & \longrightarrow & A \oplus C & \longrightarrow & C & \longrightarrow & 0 ,
\end{array}
$$

where $(\Phi \oplus \psi)(b) = (\Phi(b), \psi(b))$. It is easy to show that this diagram is commutative; it suffices to apply the relations $\psi\varphi = 0$ and $\Phi\varphi = \mathrm{id}_A$. Therefore, according to the five lemma, $\Phi \oplus \psi$ is an isomorphism.

In case (c), consider the diagram

$$\begin{array}{ccccccccc} 0 & \longrightarrow & A & \xrightarrow{\ \varphi\ } & B & \xrightarrow{\ \psi\ } & C & \longrightarrow & 0 \\ & & \Big\uparrow{\scriptstyle \mathrm{id}_A} & & \Big\uparrow{\scriptstyle \varphi+\Psi} & & \Big\uparrow{\scriptstyle \mathrm{id}_C} & & \\ 0 & \longrightarrow & A & \longrightarrow & A\oplus C & \longrightarrow & C & \longrightarrow & 0, \end{array}$$

where $(\varphi+\Psi)(a,c)=\varphi(a)+\Psi(c)$. This diagram is commutative because $\psi\varphi=0$ and $\psi\Psi=\mathrm{id}_C$.

13. (a) The map $\pi_k(B)\xrightarrow{s_*}\pi_k(E)\xrightarrow{p_*}\pi_k(B)$ is the identity; therefore, p_* is an epimorphism. For $k=n+1$, the map $\partial_*\colon\pi_{n+1}(B)\to\pi_n(F)$ is zero. Thus, we obtain an exact sequence $0\to\pi_n(F)\xrightarrow{i_*}\pi_n(E)\xrightarrow{p_*}\pi_n(B)\to 0$, and s_* is a right inverse of the homomorphism p_*.

(b) The map $\pi_k(F)\xrightarrow{i_*}\pi_k(E)\xrightarrow{r_*}\pi_k(F)$ is the identity; therefore, i_* is a monomorphism. For $k=n-1$, the map $\partial_*\colon\pi_n(B)\to\pi_{n-1}(F)$ is zero. Thus, we obtain an exact sequence $0\to\pi_n(F)\xrightarrow{i_*}\pi_n(E)\xrightarrow{p_*}\pi_n(B)\to 0$, and r_* is a left inverse of the homomorphism i_*.

(c) Since the space F is contractible in E, any spheroid in F is contractible in E. Therefore, the map $\pi_n(F)\to\pi_n(E)$ is zero. Thus, we obtain an exact sequence $0\to\pi_n(E)\xrightarrow{p_*}\pi_n(B)\xrightarrow{\partial_*}\pi_{n-1}(F)\to 0$. Let us show that the homomorphism ∂_* has a right inverse.

Take a homotopy $f_t\colon E\to E$ for which $f_0|_F=\mathrm{id}_F$ and $f_1(E)=e_0$ is a singleton; we can assume that $e_0\in F$. Let $\varphi\colon S^{n-1}\to F$ be a spheroid to F. Then $f_t\varphi$ can be considered as an n-spheroid in B. Homotopic spheroids in F correspond to homotopic spheroids in B; thus, we obtain a homomorphism $\pi_{n-1}(F)\to\pi_n(B)$. Clearly, this homomorphism is a right inverse for ∂_*.

14. Each homomorphism $\varphi\in\mathrm{Hom}(A,\mathbb{Z})$ determines the homomorphism $\tilde{\varphi}\in\mathrm{Hom}(mA,\mathbb{Z})$ defined by $\tilde{\varphi}(ma)=\varphi(a)$; the same formula defines the inverse correspondence. We only need to verify that if $ma=0$, then $\varphi(a)=0$. Suppose that $\varphi(a)=k$. Then $0=\varphi(ma)=mk$, and $k=0$.

15. Let us construct a sequence $a_1,a_2,a_3,\ldots\in\mathbb{Q}/\mathbb{Z}$ as follows. The element a_1 is arbitrary; the element a_2 is such that $2a_2=a_1$; the element a_3 is such that $3a_3=a_2$, etc. After the element a_{n-1} is chosen, there are n possible choices of a_n. Therefore, the set of such sequences is uncountable. Indeed, even if there were only two choices at each step, we would obtain an uncountable set of sequences (the set of all dyadic fractions is uncountable).

For each sequence $a_1,a_2,a_3,\ldots\in\mathbb{Q}/\mathbb{Z}$, we define a homomorphism $f\colon\mathbb{Q}\to\mathbb{Q}/\mathbb{Z}$ by setting $f\big(\frac{1}{n!}\big)=a_n$. The equality $n!\,f\big(\frac{1}{n!}\big)=f(1)$ holds because $n!\,a_n=a_1$. Clearly, if $f\big(\frac{1}{n!}\big)$ is given and f is a homomorphism, then $f\left(\frac{1}{n}\right) = (n-1)!\,f\left(\frac{1}{n!}\right)$ is defined; therefore, $f\left(\frac{m}{n}\right)$ is defined for any integer m.

16. For the group T, there is a free resolution of the form

$$0 \to \mathbb{Z}^r \xrightarrow{i} \mathbb{Z}^r \to T \to 0,$$

where $i(e_k) = m_k e_k$ for the standard generators of $\mathbb{Z}^r$. The group $\operatorname{Ext}(A, \mathbb{Z})$ is the quotient of $\operatorname{Hom}(\mathbb{Z}^r, \mathbb{Z}) \cong \mathbb{Z}^r$ modulo the image of $\operatorname{Hom}(\mathbb{Z}^k \oplus \mathbb{Z}^r, \mathbb{Z})$ under the map induced by $(0, i)$, i.e., modulo the image of $\operatorname{Hom}(\mathbb{Z}^r, \mathbb{Z})$ under the map i_* induced by i. Let $\varepsilon_k \in \operatorname{Hom}(\mathbb{Z}^r, \mathbb{Z})$ be the element dual to e_k, i.e., such that $\varepsilon_k(e_j) = \delta_{kj}$. Then $i_*(\varepsilon_k) = m_k \varepsilon_k$. Therefore, the quotient of $\operatorname{Hom}(\mathbb{Z}^r, \mathbb{Z})$ by the image of i_* is isomorphic to T.

17. Using free resolutions

$$0 \to R \to F \to A \to 0$$

and

$$0 \to R' \to F' \to B \to 0$$

and the canonical isomorphism

$$A \otimes G \cong G \otimes A,$$

we obtain the commutative diagram

$$\begin{array}{ccccccccccc}
 & & & & & & & & 0 & & \\
 & & & & & & & & \downarrow & & \\
 & & & & 0 & & 0 & & \operatorname{Tor}(B, A) & & \\
 & & & & \downarrow & & \downarrow & & \downarrow & & \\
 & & 0 & \longrightarrow & R \otimes R' & \longrightarrow & F \otimes R' & \longrightarrow & A \otimes R' & \longrightarrow & 0 \\
 & & & & \downarrow & & \downarrow & & \downarrow & & \\
 & & 0 & \longrightarrow & R \otimes F' & \longrightarrow & F \otimes F' & \longrightarrow & A \otimes F' & \longrightarrow & 0 \\
 & & & & \downarrow & & \downarrow & & \downarrow & & \\
0 & \longrightarrow & \operatorname{Tor}(A, B) & \longrightarrow & R \otimes B & \longrightarrow & F \otimes B & \longrightarrow & A \otimes B & \longrightarrow & 0 \\
 & & & & \downarrow & & \downarrow & & \downarrow & & \\
 & & & & 0 & & 0 & & 0. & &
\end{array}$$

We construct an isomorphism $\operatorname{Tor}(A, B) \to \operatorname{Tor}(B, A)$ in the same way as the connecting homomorphism in the exact sequence for a pair. For convenience, we introduce a uniform notation for groups and homomorphisms:

$$\begin{array}{ccccccccc}
 & & 0 & & 0 & & & & \\
 & & \downarrow & & \downarrow & & & & \\
0 & \longrightarrow & X_{11} & \xrightarrow{\beta_{11}} & X_{12} & \xrightarrow{\beta_{12}} & X_{13} & \longrightarrow & 0 \\
 & & \downarrow{\scriptstyle \alpha_{11}} & & \downarrow{\scriptstyle \alpha_{12}} & & \downarrow{\scriptstyle \alpha_{13}} & & \\
0 & \longrightarrow & X_{21} & \xrightarrow{\beta_{21}} & X_{22} & \xrightarrow{\beta_{22}} & X_{23} & \longrightarrow & 0 \\
 & & \downarrow{\scriptstyle \alpha_{21}} & & \downarrow{\scriptstyle \alpha_{22}} & & \downarrow{\scriptstyle \alpha_{23}} & & \\
 & & X_{31} & \xrightarrow{\beta_{31}} & X_{32} & \xrightarrow{\beta_{32}} & X_{33} & \longrightarrow & 0 \\
 & & \downarrow & & \downarrow & & \downarrow & & \\
 & & 0 & & 0 & & 0\,. & &
\end{array}$$

We must construct an isomorphism $\operatorname{Ker}\alpha_{13} \to \operatorname{Ker}\beta_{31}$.

Take $x_{13} \in \operatorname{Ker}\alpha_{13}$. Using the surjectivity of β_{12}, we choose an element $x_{12} \in X_{12}$ for which $\beta_{12}(x_{12}) = x_{13}$. We have $\beta_{22}\alpha_{12}(x_{12}) = \alpha_{13}\beta_{12}(x_{12}) = \alpha_{13}(x_{13}) = 0$; therefore, for the chosen element x_{12}, there exists a unique x_{21} such that $\beta_{21}(x_{21}) = \alpha_{12}(x_{12})$. The element x_{12} itself is not uniquely determined. But if x'_{12} is another element for which $\beta_{12}(x_{12}) = x_{13}$, then $\beta_{12}(x_{12} - x'_{12}) = 0$; therefore, $x_{12} - x'_{12} = \beta_{11}(x_{11})$ for some $x_{11} \in X_{11}$.

We pass from x_{21} to $x_{31} = \alpha_{21}(x_{21}) \in X_{31}$. This element is determined uniquely. Indeed, $\beta_{21}(x_{21} - x'_{21}) = \alpha_{12}(x_{12} - x'_{12}) = \alpha_{12}\beta_{11}(x_{11}) = \beta_{21}\alpha_{11}(x_{11})$. But β_{21} is a monomorphism; therefore, $x_{21} - x'_{21} = \alpha_{11}(x_{11})$, whence $\alpha_{21}(x_{21} - x'_{21}) = \alpha_{21}\alpha_{11}(x_{11}) = 0$.

Let us show that $x_{31} \in \operatorname{Ker}\beta_{31}$. By construction, $\beta_{31}(x_{31}) = \beta_{31}\alpha_{21}(x_{21}) = \alpha_{22}\beta_{21}(x_{21}) = \alpha_{22}\alpha_{12}(x_{12}) = 0$.

The map $\operatorname{Ker}\beta_{31} \to \operatorname{Ker}\alpha_{13}$ is constructed is a similar way. Clearly, the element $x_{31} = \alpha_{21}(x_{21})$ is mapped to x_{13}; thus, the constructed maps are mutually inverse.

18. For the periodic group T, there exists a resolution of the form $0 \to F \xrightarrow{i} F \xrightarrow{p} T \to 0$, where F is a free Abelian group. Indeed, suppose that the epimorphism $p\colon F \to T$ takes each generator f_α to an element of order n_α. Then we set $i(f_\alpha) = n_\alpha f_\alpha$. If $\operatorname{Ext}(T, \mathbb{Z}) = 0$, then the map $\operatorname{Hom}(F, \mathbb{Z}) \xrightarrow{\tilde{i}} \operatorname{Hom}(F, \mathbb{Z})$ is an isomorphism. Suppose that $\varphi \in \operatorname{Hom}(F, \mathbb{Z})$ and $\varphi(f_\alpha) = k_\alpha$. Then $\tilde{i}\varphi(f_\alpha) = n_\alpha k_\alpha$. Since $\tilde{i}$ is an isomorphism, we have $n_\alpha = \pm 1$. Thus, i is an isomorphism, and hence $T = 0$.

19. Consider homomorphisms $\alpha\colon G \to A$ and $\beta\colon G \to B$. If $\varphi\alpha = 0$, then $\alpha = 0$ because φ is a monomorphism. Therefore, $\tilde{\varphi}$ is a monomorphism. Let us show that $\operatorname{Im}\tilde{\varphi} = \operatorname{Ker}\tilde{\psi}$, i.e., $\beta = \varphi\alpha$ for some α if and only if

$\psi\beta = 0$. If $\beta = \varphi\alpha$, then $\psi\beta = \psi\varphi\alpha = 0$ because $\psi\varphi = 0$. If $\psi\beta(g) = 0$, then $\beta(g) \in \operatorname{Ker}\psi = \operatorname{Im}\varphi$, i.e., $\beta(g) = \varphi(a)$, and the element $a \in A$ is determined uniquely because φ is a monomorphism. Thus, we can set $\alpha(g) = \varphi^{-1}\beta(g)$.

20. (a) According to Problem 19, it suffices to prove that $\tilde{\psi}$ is an epimorphism. In other words, for any homomorphism $\gamma\colon F \to C$, there exists a homomorphism $\beta\colon F \to B$ such that $\gamma = \psi\beta$. Let $\{f_\alpha\}$ be a basis in the group F. We set $\beta(f_\alpha) = b_\alpha$, where b_α is any element of B for which $\psi(b_\alpha) = \gamma(f_\alpha)$; such an element exists because ψ is an epimorphism. Any homomorphism of a free group is determined by its values on generators.

(b) According to Theorem 1.19, it suffices to prove that $\tilde{\varphi}$ is an epimorphism. In other words, if $A \subset B$ is a subgroup, then any homomorphism $\alpha\colon A \to G$ can be extended to a homomorphism $\beta\colon B \to G$. Take $x \in B \setminus A$. Let us extend the homomorphism α to the group generated by x and A as follows:

(1) if $mx \notin A$ for all $m \in \mathbb{N}$, then we set $\beta(x) = 0$;

(2) if $mx \in A$ for some $m \in \mathbb{N}$, then we take the minimum number $n \in \mathbb{N}$ with this property and set $\beta(x) = g$, where g is the element of G for which $ng = \alpha(nx)$.

Now, the required homomorphism can be constructed by induction (if the group B/A is not finitely generated, then the induction is transfinite).

21. We shall repeatedly use Problem 4; so, for convenience, we introduce the following notation. Let Σ be a commutative diagram

$$\begin{array}{ccc} A_1 & \xrightarrow{\alpha} & A_2 \\ \downarrow{\scriptstyle\varphi_1} & & \downarrow{\scriptstyle\varphi_2} \\ B_1 & \xrightarrow{\beta} & B_2\,. \end{array}$$

We set $\operatorname{Im}\Sigma = (\operatorname{Im}\varphi_2 \cap \operatorname{Im}\beta)/\operatorname{Im}(\varphi_2\alpha)$ and $\operatorname{Ker}\Sigma = \operatorname{Ker}(\varphi_2\alpha)/(\operatorname{Ker}\alpha + \operatorname{Ker}\varphi_1)$. In this notation, the assertion of Problem 4 is that $\operatorname{Ker}\Sigma_1 \cong \operatorname{Im}\Sigma_2$, where Σ_1 is the left square and Σ_2 is the right square. Note that writing the same commutative diagram as

$$\begin{array}{ccc} A_1 & \xrightarrow{\varphi_1} & B_1 \\ \downarrow{\scriptstyle\alpha} & & \downarrow{\scriptstyle\beta} \\ A_2 & \xrightarrow{\varphi_2} & B_2\,, \end{array}$$

we obtain the same $\operatorname{Im}\Sigma$ and $\operatorname{Ker}\Sigma$ because $\beta\varphi_1 = \varphi_2\alpha$.

Consider now the following commutative diagram with exact rows and columns:

$$\begin{array}{ccccccc}
\mathrm{Hom}(A,B) & \longrightarrow & \mathrm{Hom}(A,G) & \longrightarrow & \mathrm{Hom}(A,H) & \longrightarrow & \widetilde{\mathrm{Ext}}(A,B) \\
\downarrow & & \downarrow & \Sigma_2 & \downarrow & \Sigma_1 & \downarrow \\
\mathrm{Hom}(F,B) & \longrightarrow & \mathrm{Hom}(F,G) & \longrightarrow & \mathrm{Hom}(F,H) & \longrightarrow & 0 \\
\downarrow & \Sigma_4 & \downarrow & \Sigma_3 & \downarrow & & \\
\mathrm{Hom}(R,B) & \longrightarrow & \mathrm{Hom}(R,G) & \longrightarrow & \mathrm{Hom}(R,H) & & \\
\downarrow & \Sigma_5 & \downarrow & & & & \\
\mathrm{Ext}(A,B) & \longrightarrow & 0\,. & & & &
\end{array}$$

The two zeros in this diagram appear because the group F is free and G is divisible.

It is easy to show that $\mathrm{Ker}\,\Sigma_1 = \widetilde{\mathrm{Ext}}(A,B)$ and $\mathrm{Ker}\,\Sigma_5 = \mathrm{Ext}(A,B)$. Indeed, in both cases we deal with diagrams of the form

$$\begin{array}{ccc}
A_1 & \xrightarrow{\alpha} & A_2 \\
{\scriptstyle\varphi_1}\downarrow & & \downarrow{\scriptstyle\varphi_2} \\
B_1 & \xrightarrow{\beta} & 0\,,
\end{array}$$

where φ_1 is a monomorphism (here A_2 is $\widetilde{\mathrm{Ext}}(A,B)$ or $\mathrm{Ext}(A,B)$). But $\mathrm{Ker}\,\Sigma = A_1/\,\mathrm{Ker}\,\alpha$ for such a diagram, because $\varphi_2\alpha = 0$ and $\mathrm{Ker}\,\varphi_1 = 0$.

Using Problem 4, we obtain $\widetilde{\mathrm{Ext}}(A,B) \cong \mathrm{Ker}\,\Sigma_1 \cong \mathrm{Im}\,\Sigma_2 \cong \mathrm{Ker}\,\Sigma_3 \cong \mathrm{Im}\,\Sigma_4 \cong \mathrm{Ker}\,\Sigma_5 \cong \mathrm{Ext}(A,B)$.

Note that as a byproduct, we have obtained yet another interpretation of the group $\mathrm{Ext}(A,B)$, namely, $\mathrm{Ext}(A,B) = \mathrm{Ker}\,\Sigma_3$.

22. Take a projective resolution $0 \to R \to F \to X \to 0$ and an injective resolution $0 \to X \to G \to H \to 0$ for the group X. Using these resolutions, we construct the commutative diagrams

$$\begin{array}{ccccccccc} 0 & \longrightarrow & \mathrm{Hom}(F,A) & \longrightarrow & \mathrm{Hom}(F,B) & \longrightarrow & \mathrm{Hom}(F,C) & \longrightarrow & 0 \\ & & \downarrow & & \downarrow & & \downarrow & & \\ 0 & \longrightarrow & \mathrm{Hom}(R,A) & \longrightarrow & \mathrm{Hom}(R,B) & \longrightarrow & \mathrm{Hom}(R,C) & \longrightarrow & 0 \end{array}$$

and

$$\begin{array}{ccccccccc} 0 & \longleftarrow & \mathrm{Hom}(A,G) & \longleftarrow & \mathrm{Hom}(B,G) & \longleftarrow & \mathrm{Hom}(C,G) & \longleftarrow & 0 \\ & & \downarrow & & \downarrow & & \downarrow & & \\ 0 & \longleftarrow & \mathrm{Hom}(A,H) & \longleftarrow & \mathrm{Hom}(B,H) & \longleftarrow & \mathrm{Hom}(C,H) & \longleftarrow & 0 \end{array}$$

with exact rows.

The exact sequences

$$0 \longrightarrow \mathrm{Hom}(X,A) \longrightarrow \mathrm{Hom}(F,A) \longrightarrow \mathrm{Hom}(R,A) \longrightarrow \mathrm{Ext}(X,A) \longrightarrow 0$$

and

$$0 \longrightarrow \mathrm{Hom}(A,X) \longrightarrow \mathrm{Hom}(A,G) \longrightarrow \mathrm{Hom}(A,H) \longrightarrow \mathrm{Ext}(A,X) \longrightarrow 0$$

imply that the kernel and cokernel of the map $\mathrm{Hom}(F,A) \to \mathrm{Hom}(R,A)$ are $\mathrm{Hom}(X,A)$ and $\mathrm{Ext}(X,A)$, respectively, and the kernel and cokernel of the map $\mathrm{Hom}(A,G) \to \mathrm{Hom}(A,H)$ are $\mathrm{Hom}(A,X)$ and $\mathrm{Ext}(A,X)$. Applying Problem 2, we obtain the required result.

23. According to the universal coefficient theorem, we have

$$H^1(K;\mathbb{Z}) \cong \mathrm{Hom}(H_1(K),\mathbb{Z}) \oplus \mathrm{Ext}(H_0(K),\mathbb{Z}).$$

But the group $H_0(K)$ is torsion-free; therefore $\mathrm{Ext}(H_0(K),\mathbb{Z}) = 0$. Clearly, if T_1 is a finite group, then $\mathrm{Hom}(T_1,\mathbb{Z}) = 0$ and $\mathrm{Hom}(T_1 \oplus \mathbb{Z}^r,\mathbb{Z}) \cong \mathbb{Z}^r$.

24. According to the universal coefficient theorem, we have

$$H^i(K;\mathbb{Z}) \cong \mathrm{Hom}(\mathbb{Z}^{n_i} \oplus T_i,\mathbb{Z}) \oplus \mathrm{Ext}(\mathbb{Z}^{n_{i-1}} \oplus T_{i-1},\mathbb{Z}) \cong \mathbb{Z}^{n_i} \oplus T_{i-1}$$

because $\mathrm{Ext}(\mathbb{Z}^{n_{i-1}} \oplus T_{i-1},\mathbb{Z}) \cong T_{i-1}$ by Problem 16.

25. First, let us prove the assertion about torsions. Consider the two sequences of homomorphisms

$$C_{i+1} \xrightarrow{\partial_{i+1}} C_i \xrightarrow{\partial_i} C_{i-1}$$

and

$$C^{i+1} \xleftarrow{\delta^i = \partial^*_{i+1}} C^i \xleftarrow{\delta^{i-1} = \partial^*_i} C^{i-1},$$

where C_k is a free group and $C^k = \mathrm{Hom}(C_k,\mathbb{Z}) \cong C_k$. By a theorem of Smith (see, e.g., [**104**]), multiplying the matrix of the map ∂_{i+1} by integer matrices with determinants ± 1 on the left and on the right, we can reduce this matrix to a matrix in which the diagonal contains $1, \dots, 1, d_1, \dots, d_l$ ($d_j \geq 2$) and all of the remaining elements (both diagonal and off-diagonal) are zero. It is easy to see that $T_i \cong \mathbb{Z}_{d_1} \oplus \cdots \oplus \mathbb{Z}_{d_l}$. To calculate T^i, we must reduce the matrix of the map $\delta^{i-1} = \partial_i^*$ to the Smith normal form. It is obtained from the matrix of ∂_i by transposition; therefore, the group T^i, which is calculated by using the operator ∂_i^*, is isomorphic to the group T_{i-1}, which is calculated by using the operator ∂_i.

Now let us prove the assertion about the ranks of groups. Note that the torsion subgroup T_i is completely determined by the operator ∂_{i+1}; however, to calculate the rank of the ith homology group, we need both operators ∂_{i+1} and ∂_i. Suppose that the diagonal of the Smith normal form of the matrix of ∂_i contains a_i nonzero elements. Then

$$\begin{aligned} \operatorname{rk} \operatorname{Ker} \partial_i &= \operatorname{rk} C_i - a_i, & \operatorname{rk} \operatorname{Im} \partial_{i+1} &= a_{i+1}, \\ \operatorname{rk} \operatorname{Ker} \delta^i &= \operatorname{rk} C_i - a_{i+1}, & \operatorname{rk} \operatorname{Im} \delta^{i-1} &= a_i. \end{aligned}$$

Therefore,

$$\operatorname{rk}(\operatorname{Ker} \partial_i / \operatorname{Im} \partial_{i+1}) = \operatorname{rk} C_i - a_i - a_{i+1} = \operatorname{rk}(\operatorname{Ker} \delta^i / \operatorname{Im} \delta^{i-1}).$$

Indeed, in the quotient by the image, every 1 on the diagonal kills one direct summand $\mathbb{Z}$, and every $d > 1$ replaces one summand $\mathbb{Z}$ by $\mathbb{Z}_d$, which makes no contribution to the rank.

26. Since the groups $H_0(X)$ and $H_0(Y)$ are torsion-free, $\operatorname{Ext}(H_0(X), \mathbb{Z}) = \operatorname{Ext}(H_0(Y), \mathbb{Z}) = 0$. Thus, the universal coefficient theorem gives natural isomorphisms $H^1(X) \to \operatorname{Hom}(H_1(X), \mathbb{Z})$ and $H^1(Y) \to \operatorname{Hom}(H_1(Y), \mathbb{Z})$. The zero homomorphism $f_*\colon H_1(X) \to H_1(Y)$ induces the zero homomorphism $\operatorname{Hom}(H_1(Y), \mathbb{Z}) \to \operatorname{Hom}(H_1(X), \mathbb{Z})$; therefore, after natural identifications we obtain the zero homomorphism $H^1(Y) \to H^1(X)$.

27. Consider the same Mayer–Vietoris sequence as in the solution of Problem 10. If M^n is a closed orientable manifold, then the connecting homomorphism $H_n(M^n) \to H_{n-1}(S^{n-1})$ takes the fundamental class $[M^n]$ to the fundamental class $[S^{n-1}]$; therefore, it is an isomorphism. It follows that the map $H_{n-1}(M^n \setminus \operatorname{int} D^n) \to H_{n-1}(M^n)$ is an isomorphism.

The closedness of the manifold M^n is essential. For example, suppose $M^n = D_1^n$ is the n-disk containing a smaller disk D^n. Then $M^n \setminus \operatorname{int} D^n \sim S^{n-1}$, and therefore $H_{n-1}(M^n \setminus \operatorname{int} D^n) = \mathbb{Z} \neq 0 = H_{n-1}(M^n)$.

The orientability of M^n is essential as well. For example, if $M^n = \mathbb{R}P^2$, then $\mathbb{R}P^2 \setminus \operatorname{int} D^2 \sim S^1$, and $H_1(\mathbb{R}P^2 \setminus \operatorname{int} D^2) = H_1(S^1) = \mathbb{Z} \neq \mathbb{Z}_2 = H_1(\mathbb{R}P^2)$.

28. We have the commutative diagram

$$\begin{array}{ccc} H_{n+1}(\Sigma S^n) & \xrightarrow{\cong} & H_n(S^n) \\ \big\downarrow{\scriptstyle (\Sigma f)_*} & & \big\downarrow{\scriptstyle f_*} \\ H_{n+1}(\Sigma S^n) & \xrightarrow{\cong} & H_n(S^n), \end{array}$$

where the horizontal arrows are isomorphisms of suspensions. Therefore, if f_* is multiplication by d, then so is $(\Sigma f)_*$.

29. Clearly, $H_0 = \mathbb{Z}_p$ and $H_2 = \mathbb{Z}_p$ (in the orientable case) or $H_2 = 0$ (in the nonorientable case). We have

$$H_1(K;\mathbb{Z}_p) \cong (H_1(K) \otimes \mathbb{Z}_p) \oplus \operatorname{Tor}(H_0(K), \mathbb{Z}_p) \cong H_1(K) \otimes \mathbb{Z}_p$$

because $H_0(K) = \mathbb{Z}$. Therefore, $H_1(nT^2;\mathbb{Z}_p) = \mathbb{Z}_p^{2n}$ and $H_1(mP^2;\mathbb{Z}_p) = \mathbb{Z}_p^{m-1}$.

30. Clearly, $H^0 = \mathbb{Z}$. We have

$$H^1(K;\mathbb{Z}) \cong \operatorname{Hom}(H_1(K), \mathbb{Z}) \oplus \operatorname{Ext}(H_0(K), \mathbb{Z}) \cong \operatorname{Hom}(H_1(K), \mathbb{Z})$$

because $H_0(K) = \mathbb{Z}$. Thus, $H^1(nT^2;\mathbb{Z}) = \mathbb{Z}^{2n}$ and $H^1(mP^2;\mathbb{Z}) = \mathbb{Z}^{m-1}$. Finally,

$$H^2(K;\mathbb{Z}) \cong \operatorname{Hom}(H_2(K), \mathbb{Z}) \oplus \operatorname{Ext}(H_1(K), \mathbb{Z}).$$

Thus, $H^2(nT^2;\mathbb{Z}) \cong \operatorname{Hom}(\mathbb{Z},\mathbb{Z}) \cong \mathbb{Z}$ and $H^2(mP^2;\mathbb{Z}) \cong \operatorname{Ext}(H_1(mP^2), \mathbb{Z}) \cong \operatorname{Ext}(\mathbb{Z}_2, \mathbb{Z}) \cong \mathbb{Z}_2$.

31. Obviously, $H \cup X = S^3$ and $H \cap X = \partial H$ is a sphere with g handles. Therefore, the Mayer–Vietoris sequence for X and H has the form

$$\cdots \longrightarrow H_2(S^3) \longrightarrow H_1(\partial H) \longrightarrow H_1(X) \oplus H_1(H) \longrightarrow H_1(S^3) \longrightarrow \cdots.$$

Here $H_2(S^3) = H_1(S^3) = 0$; hence $H_1(\partial H) \cong H_1(X) \oplus H_1(H)$, i.e., $\mathbb{Z}^{2g} \cong H_1(X) \oplus \mathbb{Z}^g$. Thus, $H_1(X) \cong \mathbb{Z}^g$.

The groups $H_1(H)$ and $H_1(X)$ are isomorphic to the kernel and the image of the homomorphism $i_*\colon H_1(\partial H) \to H_1(H)$, respectively. The kernel of i_* is generated by the meridians of the boundary of the handlebody, and the image is generates by its parallels. Therefore, the group $H_1(H)$ is generated by circles put on the handles of H.

32. The torus T^n can be represented as the n-cube with identified respective points of opposite faces. In the n-torus, each k-face is determined by $n - k$ equations of the form $x_i = \pm 1$. Therefore, the number of k-faces is equal to $2^{n-k}\binom{n}{k}$. After identifications in the cube, we obtain $\binom{n}{k}$ k-cells in the torus. Simlarly to the case of the 2-torus, all boundary homomorphisms are zero. Therefore, the homology of the n-torus coincides with that of the chain complex

$$\mathbb{Z} \longrightarrow \mathbb{Z}^n \longrightarrow \mathbb{Z}^{\binom{n}{2}} \longrightarrow \cdots \longrightarrow \mathbb{Z}^{\binom{n}{2}} \longrightarrow \mathbb{Z}^n \to \mathbb{Z}$$

with zero homomorphisms. Hence $H_k(T^n) = \mathbb{Z}^{\binom{n}{k}}$.

33. The integral cohomology groups are easy to calculate if the integral homology groups are known (see Problem 24). Example 10 shows that

$$H^k(\mathbb{R}P^n;\mathbb{Z}) = \begin{cases} \mathbb{Z} & \text{for } k=0 \text{ and } k=n, \text{ where } n \text{ odd;} \\ \mathbb{Z}_2 & \text{for even } k,\ 0<k\le n; \\ 0 & \text{otherwise.} \end{cases}$$

34. Taking 1-cells a and b and attaching them to two 2-cells along the words a^5b^{-3} and $b^3(ab)^{-2}$, we obtain a CW-complex X. The chain complex for calculating the cell homology of X has the form $\mathbb{Z}^2 \xrightarrow{\partial} \mathbb{Z}^2 \to 0$, where the homomorphism ∂ is determined by the matrix $\left(\begin{smallmatrix} 5 & -2 \\ -3 & 1 \end{smallmatrix}\right)$. The determinant of this matrix equals -1; therefore, ∂ is an isomorphism. Thus, the CW-complex X is acyclic.

To prove that X is noncontractible, it suffices to show that the group $\pi_1(X)$ is nontrivial. The group $\pi_1(X)$ is defined by the generators a and b and the relations $a^5 = b^3 = (ab)^2$. It admits a nontrivial homomorphism to the group of proper motions of the dodecahedron. Namely, let A be a rotation of the dodecahedron through $2\pi/5$ about an axis passing through the center of a face, and let B be a rotation through $2\pi/3$ about an axis passing through a vertex (the particular axes and directions of rotation are specified in Figure H.3). Then AB is a rotation through π about an axis

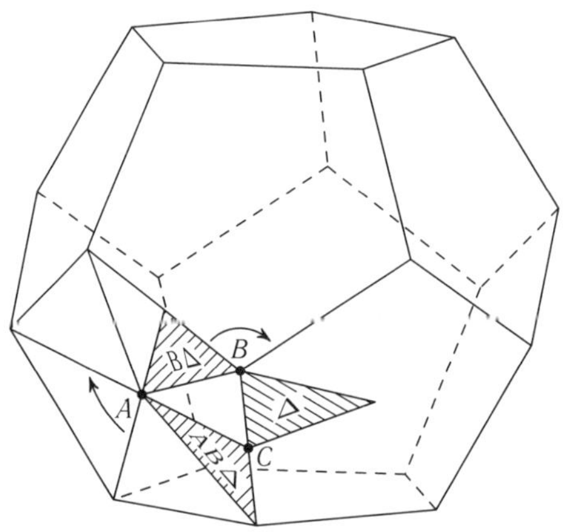

Figure H.3. The rotations of the dodecahedron.

passing through C (the midpoint of an edge). Thus, $A^5 = B^3 = (AB)^2$ is the identity transformation. Therefore, the map which takes a to A and b to B can be extended to a homomorphism of the groups.

35. The universal coefficient theorem implies $\mathbb{Z}^{b_k} \oplus T^k \cong \mathbb{Z}^{a_k} \oplus T_{k-1}$, i.e., $b_k = a_k$ and $T^k \cong T_{k-1}$. In particular, $T^1 \cong T_0 = 0$. By the Poincaré duality theorem, we have $a_k = b_{n-k}$ and $T_k \cong T^{n-k}$. Therefore, $T_k \cong T^{n-k} \cong T_{n-k-1}$ and $T^k \cong T_{n-k} \cong T^{n-k+1}$.

36. Let $H_k(M^n) = H_k \oplus T_k$, where H_k is a free Abelian group and T_k is the torsion subgroup. Applying Problem 35 to the manifold M^n, we obtain $H_k = H_{n-k}$ for $k = 1, 2, \ldots, n-1$ and $T_k = T_{n-k-1}$ for $k = 1, 2, \ldots, n-2$. Applying Problem 35 and the suspension isomorphism to the manifold ΣM^n, we obtain $H_{n-1} = 0$, $H_{k-1} = H_{n-k}$ for $k = 2, \ldots, n-1$, $T_{n-2} = 0$, and $T_{k-1} = T_{n-k-1}$ for $k = 2, \ldots, n-2$. Therefore, $H_1 = \cdots = H_{n-1} = 0$ and $T_1 = \cdots = T_{n-2} = 0$. Moreover, $T_{n-1} = 0$ for any closed orientable manifold M^n.

37. Consider the unit vector perpendicular to the plane of the diagram and directed upward; we treat it as a point of the sphere S^2. The points of the torus mapped to this point are in one-to-one correspondence with the crosses at which the curve J passes under K. It is seen from Figure H.4 that different types of crosses correspond to different signs of the Jacobian of f.

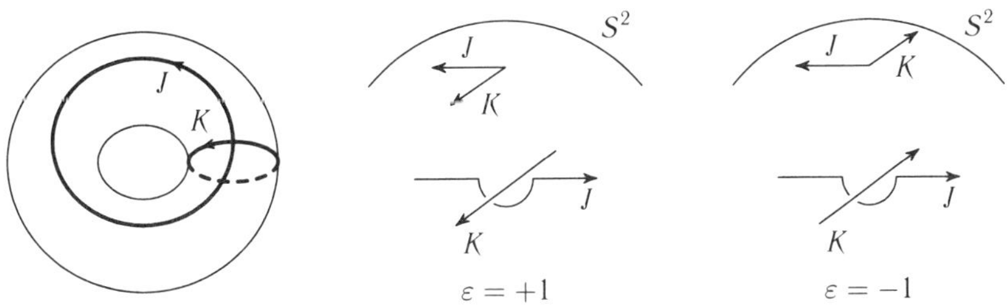

Figure H.4. The map f

To ensure that $\deg f = \operatorname{lk}(J, K)$, we choose the orientations as follows. The basis e_1, e_2, where e_1 and e_2 are the tangent vectors to J and K, determines the positive orientation of the torus. A basis $\varepsilon_1, \varepsilon_2$ determines the positive orientation of the sphere if the basis $\varepsilon_1, \varepsilon_2, \varepsilon_3$, where ε_3 is an outward normal, has positive orientation.

38. Refining the triangulation of L if necessary, we can assume that each simplex from L is completely covered by the neighborhood U involved in the definition of the covering. For the triangulation of K we can take the preimages of the simplices from L. Let us denote the number of i-simplices in these triangulations of K and in L by $a_i(K)$ and $a_i(L)$, respectively. Then $a_i(K) = na_i(L)$; therefore, $\chi(K) = n\chi(L)$.

39. Let b'_k and b''_k be the number of k-simplices in K and L, respectively, and let b_k be the number of k-cells in $K \times L$. Then $(-1)^k b_k = \sum_{p+q=k} (-1)^p b'_p (-1)^q b''_q$ because $b_k = \sum_{p+q=k} b'_p b''_q$. Therefore,

$$\chi(K \times L) = \sum (-1)^k b_k = \left(\sum (-1)^p b'_p\right)\left(\sum (-1)^q b''_q\right) = \chi(K)\chi(L).$$

40. If $f^n = \mathrm{id}_K$ and the map f has no fixed points, then we can construct an n-fold covering $p\colon K \to K/{\sim}$, where $x \sim f(x) \sim f^2(x) \sim \cdots \sim f^n(x) = x$. If the triangulation of the complex K is sufficiently fine, then the simplices Δ and $f(\Delta)$ do not intersect. Such a triangulation induces a triangulation of $K/{\sim}$. According to Problem 38, we have $\chi(K) = n\chi(K/{\sim}) \equiv 0 \pmod n$.

41. Clearly, $\operatorname{Im} f_i \cong V_i / \operatorname{Ker} f_i$; therefore,

$$\dim V_i = \dim \operatorname{Im} f_i + \dim \operatorname{Ker} f_i.$$

Thus,

$$\begin{aligned}\sum (-1)^i \dim V_i &= \sum (-1)^i \dim \operatorname{Im} f_i + \sum (-1)^i \dim \operatorname{Ker} f_i \\ &= \sum (-1)^i \dim \operatorname{Im} f_i + \sum (-1)^i \dim \operatorname{Im} f_{i+1} = 0.\end{aligned}$$

42. Let K_0 be the ε-neighborhood of the given link, and let K_1 be the closure of $S^2 \setminus K_0$. Then $\chi(S^3) = \chi(K_0) + \chi(K_1) - \chi(K_0 \cap K_1)$. Here $\chi(S^3) = 0$, $\chi(K_0) = 0$, and $\chi(K_0 \cap K_1) = 0$ because $K_0 \cap K_1$ consists of one or several tori. Hence $\chi(K_1) = 0$, as required.

43. Let us write the exact sequence of a pair for homology with coefficients in some field:

$$\cdots \longrightarrow H_i(B) \longrightarrow H_i(A) \longrightarrow H_i(A, B) \longrightarrow H_{i-1}(B) \longrightarrow \cdots.$$

This sequence contains only finitely many nonzero terms, so we can use Problem 41. As a result, we obtain

$$\sum (-1)^i \dim H_i(B) + \sum (-1)^{i+1} \dim H_i(A) + \sum (-1)^i \dim H_i(A, B) = 0.$$

44.* Let us enumerate the i-faces of the complex K by $1, 2, \dots, f_i$, where f_i is the number of i-faces, and consider the number $p_{ij}(k)$ of j-faces that intersect the kth i-face F^i_k. The number of j-faces disjoint from this face is $f_j - p_{ij}(k)$; therefore,

$$\alpha_{ij} = \sum_{k=1}^{f_i} (f_j - p_{ij}(k)) = f_i f_j - \sum_{k=1}^{f_i} p_{ij}(k). \tag{1}$$

Lemma. *For each face F^i_k, $\sum_{j=0}^{n} (-1)^j p_{ij}(k) = (-1)^n$.*

Proof. Let us endow the manifold M^n with a Riemannian metric and consider the set Q_ε of all points of M^n at distance at most ε from F^i_k. If ε is sufficiently small, then Q_ε is homeomorphic to the n-disk and Q_ε contains no vertices of K different from those of the face F^i_k.

The intersections of the faces of K with $\partial Q_\varepsilon \approx S^{n-1}$ determine a CW-complex structure on S^{n-1}. The number of $(j-1)$-cells in the CW-complex thus obtained is equal to the number of j-faces of K intersecting the face F^i_k and not contained in F^i_k. Therefore,

$$p_{ij}(k) = q_{j-1} + r_j, \tag{2}$$

where q_{j-1} is the number of $(j-1)$-cells in the CW-complex ∂Q_ε (it is assumed that $q_{-1} = 0$) and r_j is the number of j-faces in the convex polyhedron F^i_k (it is assumed that $r_i = 1$ and $r_j = 0$ for $j > i$).

Clearly, $\sum_{j=0}^n (-1)^j r_j = \chi(D^k) = 1$ and $\sum_{j=0}^{n-1} (-1)^j q_j = \chi(S^{n-1}) = 1 + (-1)^{n-1}$. Using (2), we obtain

$$\sum_{j=0}^n (-1)^j p_{ij}(k) = \sum_{j=0}^n (-1)^j q_{j-1} + \sum_{j=0}^n (-1)^j r_j = -1 - (-1)^{n-1} + 1 = (-1)^n,$$

as required. □

Formula (1) implies

$$\begin{aligned}\Psi(K) &= \sum_{i=0}^n \sum_{j=0}^n (-1)^{i+j} \alpha_{ij} \\ &= \sum_{i=0}^n \sum_{j=0}^n (-1)^{i+j} f_i f_j - \sum_{i=0}^n \sum_{j=0}^n \sum_{k=1}^{f_i} (-1)^{i+j} p_{ij}(k).\end{aligned}$$

The first sum can be calculated directly:

$$\sum_{i=0}^n \sum_{j=0}^n (-1)^{i+j} f_i f_j = \left(\sum_{i=0}^n (-1)^i f_i \right)^2 = \chi^2(M^n).$$

To calculate the second sum, we use the lemma:

$$\begin{aligned}\sum_{i=0}^n \sum_{j=0}^n \sum_{k=1}^{f_i} (-1)^{i+j} p_{ij}(k) &= \sum_{i=0}^n (-1)^i \sum_{k=1}^{f_i} \sum_{j=0}^n (-1)^j p_{ij}(k) \\ &= \sum_{i=0}^n (-1)^i \sum_{k=1}^{f_i} (-1)^n = (-1)^n \sum_{i=0}^n (-1)^i f_i = (-1)^n \chi(M^n).\end{aligned}$$

As a result, we obtain $\Psi(K) = \chi^2(M^n) - (-1)^n \chi(M^n)$. But if n is odd, then $\chi(M^n) = 0$. Therefore, $\Psi(K) = \chi^2(M^n) - \chi(M^n)$.

45. Consider the following equivalence relation on M^k: $x \sim y$ if the normal subspaces at x and y coincide (i.e., x and y belong to the same normal subspace). The map $M^k \to N^k/{\sim}$ is a covering because each normal plane intersects M^k transversally.

46. It is sufficient to prove that for coefficients in $\mathbb{Z}_2$, the intersection number of M^k and every normal subspace is 0. Any $(n-k)$-dimensional subspace in $\mathbb{R}^n$ can be transformed into any other $(n-k)$-dimensional subspace by a motion; therefore, the intersection number of M^k and an $(n-k)$-dimensional subspace does not depend on the subspace. But for any compact manifold, there exists a subspace disjoint from it.

47. Choose a point $a \in M^k$ such that the function $f(x) = \|x-a\|^2$, $x \in M^k$, has no degenerate critical points. The critical points of this function are the intersection points of M^k with the normal subspace at a. Let r_i be the number of critical points with intersection number i. Then $\sum_{i=0}^k r_i = r$ and $\sum_{i=0}^k (-1)^i r_i = \chi(M^k)$. Clearly, $r_0 > 0$. Therefore, $-r < \chi(M^k) \leq r$. But $\chi(M^k) = r\chi(N^k)$, where N^k is the manifold introduced in the solution of Problem 45. Hence $-1 < \chi(N^k) \leq 1$. Thus, $\chi(N^k)$ is equal to 0 or 1, i.e., $\chi(M^k)$ is equal to 0 or r.

48. Solving Problem 47, we proved that if M^k is a closed manifold admitting a transnormal embedding and $\chi(M^k) \neq 0$, then $r_i = 0$ for all odd i. Therefore, M^k is homotopy equivalent to a *CW*-complex containing no cells of odd dimension.

49. It is sufficient to prove the required assertion in the case where f is a simplicial map. In this case, the chain map $f_k\colon C_k(K;\mathbb{R}) \to C_k(K;\mathbb{R})$ is determined by a matrix A with integer elements, and the chain map f_k^p is determined by the matrix A^p.

Lemma. *Suppose that A is a square matrix with integer elements and p is a prime. Then* $\operatorname{tr}(A^p) \equiv \operatorname{tr} A \pmod p$.

Proof. Let $f(x) = \det(xI - A)$, where I is the identity matrix. Any polynomial f with integer coefficients determines a polynomial $\bar{f}$ over the field $\mathbb{F}_p$ (that is, the finite field with p elements). Adding the roots $x_1, \dots, x_n$ of $\bar{f}$ to $\mathbb{F}_p$, we obtain the field $\mathbb{F}_{p^k}$ with some k.

For any elements $x, y \in \mathbb{F}_{p^k}$, we have $(x+y)^p = x^p + y^p$ because $\binom{p}{m} \equiv 0 \pmod p$ for $m = 1, 2, \dots, p-1$. Therefore,

$$x_1^p + \dots + x_n^p = (x_1 + \dots + x_n)^p = x_1 + \dots + x_n$$

(the second equality follows from $x_1 + \dots + x_n \in \mathbb{F}_p$). But $x_1 + \dots + x_n = \operatorname{tr} A \pmod p$ and $x_1^p + \dots + x_n^p = \operatorname{tr}(A^p) \pmod p$. The former equality is obvious, and the latter holds because if $f(x) = (x-\lambda_1)\cdots(x-\lambda_n)$, then $\lambda_1^p + \dots + \lambda_n^p$ can be expressed as a polynomial with integer coefficients in the coefficients of the polynomial f, and $x_1^p + \dots + x_n^p$ can be expressed similarly in terms of the coefficients of $\bar{f}$. □

This lemma immediately implies the required assertion.

50. This is so because the definition of cup product for relative cohomology coincides with that for absolute cohomology. The only point to be noted is that $L \cup \varnothing = L \cup L$.

51. (a) The exact cohomology sequence

$$H^{p_i}(K, L_i) \longrightarrow H^{p_i}(K) \longrightarrow H^{p_i}(L_i) = 0$$

shows that the map $H^{p_i}(K, L_i) \to H^{p_i}(K)$ is an epimorphism, that is, $H^{p_i}(K, L_i)$ has an element β_i mapped to α_i. Moreover, $\beta_1 \smile \cdots \smile \beta_n$ is mapped to $\alpha_1 \smile \cdots \smile \alpha_n$. But $\beta_1 \smile \cdots \smile \beta_n \in H^*\big(K, \bigcup_{i=1}^n L_i\big) = 0$.

(b) The required assertion follows from (a) because the suspension can be represented as the union of two contractible spaces (cones).

52. Let $\alpha|_{S^n}$ be the image of α under the natural homomorphism $H^*(S^n, M^k) \to H^*(S^n)$. Then $\alpha \smile \beta - (\alpha|_{S^n}) \smile \beta = \alpha \smile (\beta|_{S^n})$. If $0 < \dim \alpha < n$, then $\alpha|_{S^n} = 0$, and if $0 < \dim \beta < n$, then $\beta|_{S^n} = 0$. It remains to consider the case of $\dim \alpha = \dim \beta - n$. In this case, we have $\alpha \smile \beta \in H^{2n}(S^n, M^k) = 0$.

53. (a) Suppose that G and H are Abelian groups and $\varphi \colon G \to H$ and $\psi \colon H \to G$ are homomorphisms such that $\psi\varphi = \mathrm{id}_G$. Then $H = \operatorname{Im}\varphi \oplus \operatorname{Ker}\psi$. Indeed, any element $h \in H$ can be represented as $h = \varphi\psi h + (h - \varphi\psi h)$, where $\varphi\psi h \in \operatorname{Im}\varphi$ and $h - \varphi\psi h \in \operatorname{Ker}\psi$. Moreover, if $x = \varphi g$ and $\psi x = 0$, then $g = \psi\varphi g = \psi x = 0$; therefore, $g = 0$, and hence $x = 0$.

Applying this assertion to the homomorphisms $i_* \colon H_*(A) \to H_*(X)$, $r_* \colon H_*(X) \to H_*(A)$, $r^* \colon H^*(A) \to H^*(X)$, and $i^* \colon H^*(X) \to H^*(A)$, we obtain the required result.

(b) The equality $r^*\alpha \smile r^*\beta = r^*(\alpha \smile \beta)$ implies that $\operatorname{Im} r^*$ is a subring, and the equality $i^*(\alpha \smile \beta) = i^*\alpha \smile i^*\beta$ implies that if $i^*\alpha = 0$ or $i^*\beta = 0$, then $i^*(\alpha \smile \beta) = 0$.

54. First, note that $H^p(X \vee Y) = H^p(X) \oplus H^p(Y)$ for $p \geq 1$. The existence of natural retractions $X \vee Y \to X$ and $X \vee Y \to Y$ implies that $H^*(X)$ and $H^*(Y)$ are ideals in $H^*(X \vee Y)$ (see Problem 53). Therefore, $(x + y) \smile (x' + y') = x \smile x' + y \smile y'$ for any cohomology classes $x, x' \in H^*(X)$ and $y, y' \in H^*(Y)$ of positive dimension. Indeed, $x \smile y' \in H^*(X) \cap H^*(Y)$, and therefore $x \smile y' = 0$.

55. Suppose that there exists a retraction $r \colon S^p \times S^q \to S^p \vee S^q$. Then, according to Problem 53, we have $H^*(S^p \times S^q) = \operatorname{Ker} i^* \oplus \operatorname{Im} r^*$. Let 1, α^p, β^q, and $\alpha^p \smile \beta^q$ be generators of the nontrivial cohomology groups of $S^p \times S^q$. It is easy to see that $\operatorname{Ker} i^* = H^{p+q}(S^p \times S^q)$. Therefore, $\operatorname{Im} r^*$

consists of all linear combinations of 1, α^p, and β^q. Thus, $\alpha^p \smile \beta^q \notin \operatorname{Im} r^*$, and $\operatorname{Im} r^*$ is not a subring. We have obtained a contradiction.

56. (a) Let $\alpha_{(n)}$ and $\alpha_{(m)}$ be the generators of the groups $H^1(\mathbb{R}P^n;\mathbb{Z}_2)$ and $H^1(\mathbb{R}P^m;\mathbb{Z}_2)$. Suppose that there exists a retraction $r\colon \mathbb{R}P^n \to \mathbb{R}P^m$. Then, according to Problem 53, the group $H^*(\mathbb{R}P^n;\mathbb{Z}_2)$ is the direct sum of $\operatorname{Ker} i^*$ and $\operatorname{Im} r^*$. Clearly, $i^*(\alpha_{(n)}) = \alpha_{(m)}$; therefore, $\operatorname{Ker} i^*$ consists of $0, \alpha_{(n)}^{m+1}, \dots, \alpha_{(n)}^n$, and $\operatorname{Im} r^*$ consists of $0, 1, \alpha_{(n)}, \alpha_{(n)}^2, \dots, \alpha_{(n)}^m$. This is not a subring, while $\operatorname{Im} r^*$ must be a subring by Problem 53.

(b) The solution is similar to that of (a).

57. (a) Take the same bases $\alpha_1, \dots, \alpha_n, \beta_1, \dots, \beta_n$ and $\alpha_1', \dots, \alpha_m', \beta_1', \dots, \beta_m'$ of $H^1(M_2^2)$ and $H^1(M_1^2)$ as in Theorem 2.3. Let $\Lambda = \alpha_1 \smile \beta_1$ and $\Lambda' = \alpha_1' \smile \beta_1'$ be generators of the groups $H^2(M_2^2)$ and $H^2(M_1^2)$. If

$$f_1^*(\alpha_1) = a_1\alpha_1' + \dots + a_m\alpha_m' + b_1\beta_1' + \dots + b_m\beta_m'$$

and

$$f_1^*(\beta_1) = c_1\alpha_1' + \dots + c_m\alpha_m' + d_1\beta_1' + \dots + d_m\beta_m',$$

then

$$f^*(\Lambda) = f^*(\alpha_1 \smile \beta_1) = (a_1d_1 + \dots + a_md_m - b_1c_1 - \dots - b_mc_m)\Lambda'.$$

(b) The homomorphism h that takes α_1 to α_1' and β_2 to β_1' is not induced by any continuous map $f\colon M_1^2 \to M_2^2$ because $f^*(\alpha_1) \smile f^*(\beta_2) = f^*(\alpha_1 \smile \beta_2) = 0$.

58. (a) Clearly, $H^0(S^k \times S^k) \cong H^{2k}(S^k \times S^k) \cong \mathbb{Z}$ and $H^k(S^k \times S^k) \cong \mathbb{Z} \oplus \mathbb{Z}$, and all of the remaining cohomology groups are trivial. Let $\alpha, \beta \in H^k(S^k \times S^k)$ be the cohomology classes dual (in the sense of linear algebra) to the homology classes of the cycles $S^k \times \{x_0\}$ and $\{y_0\} \times S^k$. Then $\alpha \smile \beta = \Lambda$, where Λ is a generator of $H^{2k}(S^k \times S^k)$ because $S^k \times \{x_0\}$ and $\{y_0\} \times S^k$ intersect transversally in only one point, namely, (y_0, x_0). Let us show that $\alpha \smile \alpha = 0$ and $\beta \smile \beta = 0$. The cycle $S^k \times \{x_0\}$ is homologous to $S^k \times \{x_0'\}$; to see this, it suffices to consider the chain $S^k \times I$, where I is a path joining x_0 and x_0'. The cycles $S^k \times \{x_0\}$ and $S^k \times \{x_0'\}$ are disjoint; therefore, $\alpha \smile \alpha = 0$. Similarly, $\beta \smile \beta = 0$.

(b) First, note that the cocycle $\alpha + \beta$ is dual to the diagonal and that $(\alpha+\beta) \smile (\alpha+\beta) = \alpha \smile \beta + \beta \smile \alpha$. For k odd, $\alpha \smile \beta = -\beta \smile \alpha$; in this case, the self-intersection number of the diagonal vanishes. If k is even, then $\alpha \smile \beta + \beta \smile \alpha = 2\alpha \smile \beta = 2\Lambda$, and therefore the diagonal has self-intersection number ± 2. The sign depends on the orientations of the diagonal and $S^k \times S^k$. The orientation of the diagonal is canonically determined by that of S^k. For even k, the orientation of the sphere $S^k \times S^k$ is canonically determined as well because the bases $e_1, \dots, e_k$, $\varepsilon_1, \dots, \varepsilon_k$

and $\varepsilon_1, \dots, \varepsilon_k$, $e_1, \dots, e_k$ have the same orientation. Under this choice of orientations, the sign is plus.

59. Let α and β be the cohomology classes dual (in the sense of linear algebra) to the homology classes of the cycles $S^m \times \{x_0\}$ and $\{y_0\} \times S^n$. Then $\alpha \smile \beta$ is a generator of $H^{n+m}(S^n \times S^m)$. Thus, it is sufficient to prove that $f^*(\alpha) = 0$ and $f^*(\beta) = 0$. But $f^*(\alpha) \in H^m(S^{n+m}) = 0$ and $f^*(\beta) \in H^n(S^{n+m}) = 0$.

69. Any contractible manifold is orientable; therefore, we have the Lefschetz isomorphism $H_k(M^n, \partial M^n) \cong H^{n-k}(M^n)$. Hence $H_n(M^n, \partial M^n) = \mathbb{Z}$ and $H_k(M^n, \partial M^n) = 0$ for $k \neq n$. It follows from the exact sequence for the pair $(M^n, \partial M^n)$ that $H_k(\partial M^n) \cong H_k(M^n)$ for $k < n-1$ and $H_{n-1}(\partial M^n) \cong H_n(M^n, \partial M^n) = \mathbb{Z}$.

70. Consider the following commutative (up to sign) diagram for homology and cohomology groups with coefficients in the field $\mathbb{Q}$:

$$\begin{array}{ccccc} H_2(M^3, \partial M^3) & \xrightarrow{\partial} & H_1(\partial M^3) & \xrightarrow{i_*} & H_1(M^3) \\ \cong \uparrow \frown [M^3] & & \cong \uparrow \frown [\partial M^3] & & \cong \uparrow \frown [M^3] \\ H^1(M^3) & \xrightarrow{i^*} & H^1(\partial M^3) & \xrightarrow{\delta} & H^2(M^3, \partial M^3). \end{array}$$

Since we consider homology and cohomology with coefficients in a field, the maps i_* and i^* are dual to each other in the sense of linear algebra.

Any linear map $A\colon U \to V$ determines the exact sequence

$$0 \longrightarrow \operatorname{Ker} A \longrightarrow U \xrightarrow{A} V \longrightarrow \operatorname{Coker} A \longrightarrow 0.$$

Passing to dual spaces, we obtain the exact sequence

$$0 \longleftarrow (\operatorname{Ker} A)^* \longleftarrow U^* \xleftarrow{A^*} V^* \longleftarrow (\operatorname{Coker} A)^* \longleftarrow 0.$$

Thus, $\operatorname{Coker}(A^*) = (\operatorname{Ker} A)^*$. Therefore, $\dim \operatorname{Coker} i^* = \dim \operatorname{Ker} i_*$, which means that $\dim \operatorname{Im} \partial = \dim \operatorname{Ker} i_* = \dim \operatorname{Coker} i^* = \dim \operatorname{Coker} \partial$. As a result, we obtain $\dim \operatorname{Im} \partial = \dim(H_1(\partial M^3)/\operatorname{Im} \partial) = \dim H_1(\partial M^3) - \dim \operatorname{Im} \partial$, i.e., $\dim \operatorname{Im} \partial = \frac{1}{2} \dim H_1(\partial M^3)$, as required.

71. According to the Alexander duality theorem, we have

$$\tilde{H}_{n-k-1}(X) \cong \tilde{H}^k(\underbrace{S^{n-2} \sqcup \dots \sqcup S^{n-2}}_{m}) = \begin{cases} \mathbb{Z}^m & \text{if } k = n-2, \\ \mathbb{Z}^{m-1} & \text{if } k = 0, \\ 0 & \text{otherwise.} \end{cases}$$

Therefore,

$$H_i(X) = \begin{cases} \mathbb{Z} & \text{if } i = 0, \\ \mathbb{Z}^m & \text{if } i = 1, \\ \mathbb{Z}^{m-1} & \text{if } i = n-1, \\ 0 & \text{otherwise.} \end{cases}$$

72. According to the Alexander duality theorem, we have $\tilde{H}_i(X) \cong \tilde{H}^{n-i}(S^p)$ for $0 \le i \le n$. Moreover, $H_{n+1}(X) = 0$ because the space X is homotopy equivalent to an $(n+1)$-manifold with boundary. Thus, $\tilde{H}_i(X) = \mathbb{Z}$ for $i = n - p$, and all the remaining reduced homology groups of X are trivial.

73. According to the Alexander duality theorem, we have $\tilde{H}_i(X) \cong \tilde{H}^{n-i}(S^p \sqcup S^q)$ for $0 \le i \le n$. Moreover, $H_i(X) = 0$ for $i \ge n+1$.

60. (a) Let $\alpha_{(n)}$ and $\alpha_{(m)}$ be the generators of the groups $H^1(\mathbb{R}P^n; \mathbb{Z}_2)$ and $H^1(\mathbb{R}P^m; \mathbb{Z}_2)$. If $f^*(\alpha_{(n)}) = 0$, then $f^*(\alpha^k_{(n)}) = 0$, where $\alpha^k_{(n)} = \alpha_{(n)} \smile \cdots \smile \alpha_{(n)}$ is the generator of $H^k(\mathbb{R}P^n; \mathbb{Z}_2)$. Suppose that $f^*(\alpha_{(n)}) \ne 0$, i.e., $f^*(\alpha_{(n)}) = \alpha_{(m)}$. Then $f^*(\alpha^{n+1}_{(n)}) = \alpha^{n+1}_{(m)} \ne 0$ because $n+1 \le m$. On the other hand, $\alpha^{n+1}_{(n)} = 0$. We have obtained a contradiction.

(b) The map f_* is dual to f^*.

61. Suppose that such a map g exists. Then the diagram

$$\begin{array}{ccc} S^m & \xrightarrow{g} & S^n \\ \downarrow{\scriptstyle p_m} & & \downarrow{\scriptstyle p_n} \\ \mathbb{R}P^m & \xrightarrow{f} & \mathbb{R}P^n, \end{array}$$

where p_m and p_n are double coverings, is commutative. According to Problem 60(a), the map $f_*^{(H)}\colon H_1(\mathbb{R}P^m; \mathbb{Z}_2) \to H_1(\mathbb{R}P^n; \mathbb{Z}_2)$ is zero; therefore, so is the map $f_*^{(\pi)}\colon \pi_1(\mathbb{R}P^m; \mathbb{Z}_2) \to \pi_1(\mathbb{R}P^n; \mathbb{Z}_2)$. Indeed, if $n = 1$, then the homomorphism $f_*\colon \mathbb{Z}_2 \to \mathbb{Z}$ is zero for obvious algebraic reasons, and if $n > 1$, then we can identify the generators of the groups $H_1(\mathbb{R}P^n; \mathbb{Z}_2)$ and $\pi_1(\mathbb{R}P^n)$ and the generators of the groups $H_1(\mathbb{R}P^m; \mathbb{Z}_2)$ and $\pi_1(\mathbb{R}P^m)$. After such an identification, we obtain $f_*^{(\pi)} = f_*^{(H)}$.

The map f induces the zero homomorphism of fundamental groups; hence it has a lifting

$$\begin{array}{ccc} & & S^n \\ & \overset{\tilde{f}}{\nearrow} & \downarrow{\scriptstyle p_n} \\ \mathbb{R}P^m & \xrightarrow{f} & \mathbb{R}P^n. \end{array}$$

We have $p_n\tilde{f}p_m = fp_m = p_ng$. For a point $x \in S^m$, either $g(x) = \tilde{f}p_m(x)$ or $g(-x) = \tilde{f}p_m(x) = \tilde{f}p_m(-x)$. Since the maps g and $\tilde{f}p_m$ coincide at one point, they coincide everywhere because the map of covering spaces that covers a given map of the bases of coverings is completely determined by the image of one point.

This is impossible because g takes x and $-x$ to different points, whereas p_m takes them to the same point.

62. (a) This problem is solved by using Theorems 2.11 and 2.12, which describe multiplication in the cohomology rings of the spaces $\mathbb{R}P^n$ and $\mathbb{C}P^n$. In addition, Problem 51 should be applied.

(b) The sets $\{(x_0 : x_1 : \cdots : x_n),\ x_i \neq 0\}$ are contractible (they are homeomorphic to open disks) and cover $\mathbb{R}P^n$ ($\mathbb{C}P^n$)

63. Let α be a generator of the group $H^1(\mathbb{C}P^n)$. Then $f^*(\alpha) = \lambda\alpha$, where $\lambda \in \mathbb{Z}$. Therefore, $f^*(\alpha^k) = (f^*\alpha)^k = \lambda^k\alpha^k$. In calculating $\Lambda(f)$, we can use maps of both homology and cohomology groups. Hence $\Lambda(f) = 1 + \lambda + \lambda^2 + \cdots + \lambda^n$.

64. According to Problem 63, we have $\Lambda(f) = 1 + \lambda + \lambda^2 + \cdots + \lambda^n = \frac{\lambda^{n+1}-1}{\lambda-1}$. Therefore, if $\Lambda(f) = 0$, then $\lambda^{n+1} = 1$ and $\lambda \neq 1$. Thus, $\lambda = -1$, and the number n is odd.

65. As in the solution of Problem 63, we have $f^*\alpha^n = \lambda^n\alpha^n$. Hence $f_*[\mathbb{C}P^n] = \lambda^n[\mathbb{C}P^n]$.

66. (a) If n is even, then $\deg f = \lambda^n \geq 0$ by Problem 65.

(b) Let us show that the diffeomorphism $\mathbb{C}P^m \to \mathbb{C}P^m$ specified by $(z_0 : \cdots : z_m) \mapsto (\bar{z}_0 : \cdots : \bar{z}_m)$ in homogeneous coordinates is orientation-reversing for odd m. Indeed, in a neighborhood of the fixed point $(1 : 0 : \cdots : 0)$, this diffeomorphism coincides with the map $\mathbb{C}^m \to \mathbb{C}^m$ defined by $(z_1, \ldots, z_m) \mapsto (\bar{z}_1, \ldots, \bar{z}_m)$. This map considered as a map $\mathbb{R}^{2m} \to \mathbb{R}^{2m}$ has determinant $(-1)^m$.

67. The cohomology groups (with coefficients in $\mathbb{Z}$) of the given spaces are isomorphic, but their cohomology rings are different: multiplication in the ring $H^*(\mathbb{C}P^2)$ is nontrivial, whereas in the ring $H^*(S^2 \vee S^4)$ it is trivial.

The space $\mathbb{C}P^2$ is obtained by attaching D^4 to S^2 via the map $\partial D^4 \to S^2$, which is a Hopf fibration. If this map were null-homotopic, then the space thus obtained would be homotopy equivalent to $S^2 \vee S^4$.

68. The cohomology groups of the given spaces are $\mathbb{Z}$ in dimensions 0, m, n, and $m+n$ (if $m = n$, then the cohomology group in dimension $m = n$ is $\mathbb{Z}\oplus\mathbb{Z}$); the remaining cohomology groups are trivial. For the space $S^m \times S^n$, the product of generators of the m- and n-dimensional cohomology groups is a generator of the $(m+n)$-dimensional cohomology group. For the space $S^m \vee S^n \vee S^{n+m}$, multiplication in cohomology is trivial.

74. Let us write the Mayer–Vietoris sequence for $K_1 = M_1^{4n} \setminus \operatorname{int} D^{4n}$ and $K_2 = M_2^{4n} \setminus \operatorname{int} D^{4n}$:

$$H_{2n}(S^{4n-1}) \to H_{2n}(K_1) \oplus H_{2n}(K_1) \to H_{2n}(M_1^{4n} \# M_2^{4n}) \to H_{2n-1}(S^{4n-1}).$$

Here $H_{2n}(S^{4n-1}) = 0$ and $H_{2n-1}(S^{4n-1}) = 0$; therefore, $H_{2n}(K_1) \oplus H_{2n}(K_1) \cong H_{2n}(M_1^{4n} \# M_2^{4n})$. Moreover, according to Problem 10, we have $H_{2n}(K_i) \cong H_{2n}(M_i^{4n})$. Clearly, there are $2n$-cycles that generate $H_{2n}(M_i^{4n})$ and are disjoint from D^{4n} (the disk D^{4n} can be chosen to be arbitrarily small). Therefore, the intersection form of the manifold $M_1^{4n} \# M_2^{4n}$ is the direct sum of the intersection forms of the manifolds M_1^{4n} and M_2^{4n}.

75. Let z^p and w^q be cochains with values in G which represent the cohomology classes a^p and b^q. Then $\delta(z^p \smile w^q)$, $\delta(z^p) \smile w^q$, and $z^p \smile \delta(w^q)$ is a cochain representing the classes $\beta^*(a^p \smile b^q)$, $\beta^*(a^p) \smile b^q$, and $a^p \smile \beta^*(b^q)$. Clearly, $\delta(z^p \smile w^q) = \delta(z^p) \smile w^q + (-1)^p z^p \smile \delta(w^q)$.

76. Obviously, $\operatorname{Tor}(H_p(M^p), H_{q-1}(N^q)) = 0$ because $H_p(M^p) = 0$ or $\mathbb{Z}$. Therefore, according to the Künneth theorem, we have $H_{p+q}(M^p \times N^q) \cong H_p(M^p) \otimes H_q(N^q)$. This group is isomorphic to $\mathbb{Z}$ if and only if $H_p(M^p) = \mathbb{Z}$ and $H_q(N^q) = \mathbb{Z}$.

77. Suppose that $S^n = M^p \times N^q$, where M^p and N^q are manifolds of dimensions $p, q \geq 1$. Clearly, these manifolds must be closed. Moreover, according to Problem 76, they must be orientable. Therefore, by the Künneth theorem, the group $H_p(S^n)$ contains the subgroup $\mathbb{Z} = H_p(M^p) \otimes H_0(N^q)$, and $0 < p < n$. This is impossible.

78. The fundamental groups of both spaces coincide with $\mathbb{Z}_2$, and all of the remaining homotopy groups for the sphere and projective space of the same dimension are isomorphic. For homology, it suffices to prove that the homology groups with coefficients in $\mathbb{Z}_2$ are different. By the Künneth theorem, the dimension of the space $\bigoplus_{i \geq 0} H_i(X \times Y; \mathbb{Z}_2)$ is equal to the product of the dimensions of $\bigoplus_{i \geq 0} H_i(X; \mathbb{Z}_2)$ and $\bigoplus_{i \geq 0} H_i(Y; \mathbb{Z}_2)$. Hence the dimensions of $\bigoplus_{i \geq 0} H_i(S^n \times \mathbb{R}P^m; \mathbb{Z}_2)$ and $\bigoplus_{i \geq 0} H_i(S^m \times \mathbb{R}P^n; \mathbb{Z}_2)$ are equal to $2m+2$ and $2n+2$, respectively.

79. Clearly, $\pi_1(S^2 \times \mathbb{R}P^\infty) \cong \mathbb{Z}_2 \cong \pi_1(\mathbb{R}P^2)$ and $\pi_k(S^2 \times \mathbb{R}P^\infty) \cong \pi_k(\mathbb{R}P^2)$ for $k \geq 2$ because $\pi_k(P^\infty) = 0$ and $\pi_k(S^2) \cong \pi_k(\mathbb{R}P^2)$ for $k \geq 2$.

According to the Künneth theorem, the group $H_i(S^2 \times \mathbb{R}P^\infty; \mathbb{Z}_2)$ contains a subgroup isomorphic to $H_0(S^2; \mathbb{Z}_2) \otimes H_i(\mathbb{R}P^\infty; \mathbb{Z}_2) \cong \mathbb{Z}_2$; in particular, this group is nontrivial for any i. On the other hand, $H_i(\mathbb{R}P^2; \mathbb{Z}_2) = 0$ for $i \geq 3$.

80. *Solution* 1. The space $A * B$ is the quotient of $A \times [0,1] \times B$ by the following equivalence relation: $(a_1, t_1, b_1) \sim (a_2, t_2, b_2)$ if either $t_1 = t_2 = 0$ and $a_1 = a_2$ or $t_1 = t_2 = 1$ and $b_1 = b_2$. Consider the set $\bar{A} \subset A * B$ of all points for which $t \leq 1/2$ and the set $\bar{B} \subset A * B$ of all points for which $t \geq 1/2$. The spaces A and B can be identified with the subsets of $A * B$ consisting of the points for which $t = 0$ and $t = 1$. Under this identification, A and B become deformation retracts of $\bar{A}$ and $\bar{B}$. Clearly, $\bar{A} \cap \bar{B} = A \times B$. Thus, we obtain the Mayer–Vietoris exact sequence

$$\cdots \xrightarrow{\varphi} H_{r+1}(A*B) \to H_r(A\times B) \xrightarrow{\psi} H_r(\bar{A})\oplus H_r(\bar{B}) \xrightarrow{\varphi} H_r(A*B) \to \cdots .$$

The map φ is induced by the inclusions $A \subset A * B$ and $B \subset A * B$. It is easy to see that both inclusions are null-homotopic. For example, for the inclusion $A \subset A * B$, the homotopy is defined by $f_t(a) = (a, t, b_0)$, where b_0 is a point of B. Thus, the above exact sequence has the form

$$0 \longrightarrow H_{r+1}(A * B) \longrightarrow H_r(A \times B) \xrightarrow{\psi} H_r(\bar{A}) \oplus H_r(\bar{B}) \longrightarrow 0.$$

The group $H_r(A \times B)$ is found using the Künneth theorem. It remains to calculate the kernel of ψ. This map is the direct sum of the maps ψ_A and ψ_B, and the composition

$$H_r(A \times B) \xrightarrow{\psi_A} H_r(\bar{A}) \xrightarrow{\cong} H_r(A)$$

is induced by the projection onto the first factor. For $q > 0$, the map $H_{r-q}(A) \otimes H_q(B) \to H_r(A)$ is zero, and $H_r(A) \otimes H_0(B) \to H_r(A) \cong H_r(A) \otimes H_0(*)$ is induced by the map $B \to *$, which has kernel $H_r(A) \otimes \tilde{H}_0(B)$. For the map ψ_B, the argument is similar.

Solution 2. Recall that $A * B \sim \Sigma(A \wedge B)$, where $A \wedge B = A \times B / A \vee B$ (see Part I, Problem 49). Therefore,

$$H_{r+1}(A * B) \cong \tilde{H}_r(A \wedge B) \cong H_r(A \times B, A \vee B).$$

Let us write the exact sequence for the pair $(A \times B, A \vee B)$:

$$\cdots \longrightarrow H_r(A \vee B) \longrightarrow H_r(A \times B) \longrightarrow H_r(A \times B, A \vee B) \longrightarrow H_{r-1}(A \vee B) \longrightarrow \cdots .$$

Here

$$H_r(A \times B) = \bigoplus_{i+j=r} (H_i(A) \otimes H_j(B)) \oplus \bigoplus_{i+j=r-1} \operatorname{Tor}(H_i(A), H_j(B)),$$

and $H_r(A \vee B) \to H_r(A \times B)$ is an isomorphism onto the direct term

$$(H_r(A) \otimes H_0(B)) \oplus (H_0(A) \otimes H_r(B)).$$

In particular, this map (and a similar map of the $(r-1)$-dimensional homology groups) is a monomorphism. Therefore, the group $H_r(A \times B, A \vee B)$ is isomorphic to the quotient group $H_r(A \times B)/H_r(A \vee B)$. Taking the quotient modulo the subgroup $H_r(A \vee B)$ is equivalent to passing to reduced homology.

81. The Künneth theorem implies that the graded cohomology groups of the given spaces are isomorphic to the quotients of the additive groups of the polynomial rings $\mathbb{Z}_2[x_1, \dots, x_k]$ and $\mathbb{Z}[x_1, \dots, x_k]$ modulo the relations $x_1^{n_1+1} = \dots = x_k^{n_k+1} = 0$. These groups are generated by elements of the form $\alpha_1^{m_1} \times \dots \times \alpha_k^{m_k}$, $0 \le m_i \le n_i$. Here α_i is a generator of the one-dimensional cohomology group in the real case and of the two-dimensional cohomology group in the complex case. By Theorem 2.27 not only the groups but also the rings are isomorphic. In the complex case, the sign $(-1)^{q_1 p_2}$ can be ignored because all nonzero elements have even dimension.

82. Theorem 2.27 expresses multiplication in the image of the natural monomorphism $H^*(K) \otimes H^*(L) \to H^*(K \times L)$ in terms of those in the rings $H^*(K)$ and $H^*(L)$. But if one of the groups $H^*(K)$ and $H^*(L)$ is torsion-free, then this monomorphism is an isomorphism.

84. If there exists such a nondegenerate bilinear map, then the number $\binom{n}{k}$ is even for all k satisfying the inequalities $0 < k < n$. Therefore, $(a+b)^n \equiv a^n + b^n \pmod 2$ for all integers a and b.

Clearly, $(a+b)^2 \equiv a^2 + b^2 \pmod 2$. Therefore, $(a+b)^{2^k} \equiv a^{2^k} + b^{2^k} \pmod 2$. Suppose that $n = 2^k m$, where m is an odd number larger than 1. Then

$$\begin{aligned}(a+b)^n &\equiv (a^{2^k} + b^{2^k})^m \\ &\equiv a^n + m a^{(m-1)2^k} b^{2^k} + \frac{m(m-1)}{2} a^{(m-2)2^k} b^{2 \cdot 2^k} + \cdots \pmod 2.\end{aligned}$$

Here $m \not\equiv 0 \pmod 2$.

85. We set $f(x, y) = (z_1, \dots, z_{r+s-1})$, where

$$\sum_{k=1}^{r+s-1} z_k t^{k-1} = \left(\sum_{i=1}^{r} x_i t^{i-1} \right) \left(\sum_{j=1}^{s} y_j t^{j-1} \right).$$

If the product of two polynomials is the zero polynomial, then one of these polynomials must be zero. Therefore, the bilinear map f is nondegenerate.

We have constructed a nondegenerate bilinear map $\mathbb{R}^r \times \mathbb{R}^s \to \mathbb{R}^{r+s-1}$. Similarly, we can construct a nondegenerate bilinear map $\mathbb{C}^r \times \mathbb{C}^s \to \mathbb{C}^{r+s-1}$, which is a nondegenerate bilinear map $\mathbb{R}^{2r} \times \mathbb{R}^{2s} \to \mathbb{R}^{2r+2s-2}$. Moreover, we can use not only the complex numbers $\mathbb{C}$, but also the quaternions $\mathbb{H}$ and the octonions (Cayley numbers) $\mathbb{O}$.

86. (a) Let X be an acyclic CW-complex. Then the space X is path-connected. Therefore, according to Problem 50 in Part I, the space ΣX is simply connected, and hence $H_1(\Sigma X) = 0$. We have $H_{i+1}(\Sigma X) = H_i(X) = 0$ for $i \geq 1$. By the Hurewicz theorem, all homotopy groups of ΣX are trivial. Thus, the space ΣX is contractible by the Whitehead theorem.

(b) An example of a noncontractible acyclic CW-complex was given in the solution of Problem 34.

87. The space $S^p \times S^q$ can be represented as a CW-complex with cells of dimensions 0, p, q, and $p+q$. Therefore, $S^p \vee S^q$ is the k-skeleton of $S^p \times S^q$ if $\max\{p, q\} \leq k \leq p+q-1$. The map f corresponds to attaching the cell D^{p+q} to the $(p+q-1)$-skeleton. The homotopy class of the map $S^{p+q-1} \xrightarrow{f} S^p \vee S^q$ is the obstruction to extending the identity map $S^p \vee S^q \to S^p \vee S^q$ from the $(p+q-1)$-skeleton of $S^p \times S^q$ to the $(p+q)$-skeleton. Suppose that such an extension $g\colon S^p \times S^q \to S^p \vee S^q$ exists. Let α^p, β^q and $\bar{\alpha}^p$, $\bar{\beta}^q$ be the cohomology classes in $H^*(S^p \vee S^q)$ and $H^*(S^p \times S^q)$ corresponding to S^p and S^q. Then $\bar{\alpha}^p = g^*\alpha^p$ and $\bar{\beta}^q = g^*\beta^q$. Therefore, $\bar{\alpha}^p \smile \bar{\beta}^q = g^*(\alpha^p \smile \beta^q)$. But $\bar{\alpha}^p \smile \bar{\beta}^q \neq 0$, whence $\alpha^p \smile \beta^q = 0$.

88. Consider the following cell decomposition of the sphere S^∞, which contains m cells of each dimension:

$$
\begin{aligned}
&(e^{2\pi ik/m}, 0, 0, \dots);\\
&(e^{2\pi i\theta}, 0, 0, \dots), \quad \frac{k}{m} < \theta < \frac{k+1}{m};\\
&(w_1, \rho e^{2\pi ik/m}, 0, 0, \dots), \quad 0 < \rho \leq 1, \quad |w_1|^2 = 1 - \rho^2;\\
&(w_1, \rho e^{2\pi i\theta}, 0, 0, \dots), \quad 0 < \rho \leq 1, \quad |w_1|^2 = 1 - \rho^2, \quad \frac{k}{m} < \theta < \frac{k+1}{m};\\
&(w_1, w_2, \rho e^{2\pi ik/m}, 0, 0, \dots), \quad 0 < \rho \leq 1, \quad |w_1|^2 + |w_2|^2 = 1 - \rho^2;\\
&(w_1, w_2, \rho e^{2\pi i\theta}, 0, 0, \dots), \quad 0 < \rho \leq 1, \quad |w_1|^2 + |w_2|^2 = 1 - \rho^2,\\
&\qquad \frac{k}{m} < \theta < \frac{k+1}{m};
\end{aligned}
$$

....

This cell decomposition is compatible with the equivalence relation on S^∞; therefore, it gives a cell decomposition of the lens space L_m^∞ containing one cell of each dimension. The boundary homomorphisms are calculated in precisely the same way as for the three-dimensional lens spaces. As a result, we obtain the chain complex

$$\dots \xrightarrow{0} \mathbb{Z} \xrightarrow{\times m} \mathbb{Z} \xrightarrow{0} \mathbb{Z} \xrightarrow{\times m} \mathbb{Z} \xrightarrow{0} \mathbb{Z}.$$

The homology of this complex is as required.

89. Let $\tilde{M}^3$ be the universal covering of the manifold M^3. Then $\pi_1(\tilde{M}^3) = 0$ and $\pi_2(\tilde{M}^3) = \pi_2(M) = 0$. The group $\pi_1(M^3)$ is infinite; therefore, the manifold $\tilde{M}^3$ is noncompact. Hence $H_3(\tilde{M}^3) = 0$. Clearly,

$H_k(\tilde{M}^3) = 0$ for $k \geq 3$. Applying the Hurewicz theorem, we obtain $\pi_i(\tilde{M}^3) = 0$ for all i. According to the Whitehead theorem, the manifold $\tilde{M}^3$ is contractible, which means that $\pi_i(M^3) = \pi_i(\tilde{M}^3) = 0$ for $i \geq 2$.

90. Suppose that the group π contains an element of finite order. Then it contains the cyclic subgroup $\mathbb{Z}_m$ generated by this element. Let $\tilde{X} \to X$ be the covering that corresponds to the subgroup $\mathbb{Z}_m \subset \pi$. Then $\tilde{X}$ is a $K(\mathbb{Z}_m, 1)$ space; therefore, it is homotopy equivalent to L^∞_m. This is a contradiction because, on one hand, $\tilde{X}$ is a finite-dimensional simplicial complex, and on the other hand, according to Problem 88, $\tilde{X}$ has nontrivial homology of arbitrarily high dimensions.

91. Obviously, the required map π exists for the trivial bundle. Now, suppose that such a map π exists for the bundle E. Consider the map $f\colon E \to B \times \mathbb{R}^n$ defined by $f(e) = (p(e), \pi(e))$. According to Theorem 3.13, this map determines an isomorphism from E to the trivial bundle $B \times \mathbb{R}^n$.

92. Introducing a Riemannian metric, we can assume the trivializations to be orthonormal. Take some trivialization of the bundle TS^3. Any trivialization of TS^3 is uniquely determined by a map $S^3 \to \mathrm{SO}(3)$. A homotopy of trivializations is a homotopy of such maps. It remains to note that $\mathrm{SO}(3) \approx \mathbb{R}P^3$ (see the solution of Problem 91 in Part I) and $\pi_3(\mathbb{R}P^3) = \pi_3(S^3) = \mathbb{Z}$.

93. Take a trivialization of the bundle TS^3. Any vector field on S^3 without singular points is uniquely determined by a map $S^3 \to \mathbb{R}^3 \setminus \{0\} \sim S^2$. A homotopy of vector fields without singular points is a homotopy of such maps. Clearly, $\pi_3(S^2) = \mathbb{Z}$.

94. We prove the required assertion by induction on k. For $k = 1$, this assertion is obvious. Suppose that it is true for the products $S^{n_1} \times \cdots \times S^{n_{k-1}}$. We assume the sphere S^{n_k} to be embedded in $\mathbb{R}^{n_k+1}$. The normal bundle to S^{n_k} in $\mathbb{R}^{n_k+1}$ is trivial; hence so is the normal bundle to S^{n_k} in $\mathbb{R}^{n_1+\cdots+n_k+1}$. The fiber of this trivial bundle is $\mathbb{R}^{n_1+\cdots+n_{k-1}+1}$. By the induction hypothesis, $S^{n_1} \times \cdots \times S^{n_{k-1}}$ can be embedded in this space. As a result, we obtain the required embedding.

95 [**124**]. We use Examples 38 and 39.

If all numbers n_i are even, then

$$\chi(S^{n_1} \times \cdots \times S^{n_k}) = \chi(S^{n_1}) \cdots \chi(S^{n_k}) = 2^k \neq 0.$$

Therefore, any vector field on the manifold $S^{n_1} \times \cdots \times S^{n_k}$ has a singular point, and hence this manifold is not parallelizable.

Now let us prove that the manifold $S^{n_1} \times \cdots \times S^{n_k}$ with $k \geq 2$ is parallelizable provided that one of the numbers n_i is odd. For $k = 2$, this assertion is an obvious corollary of the following lemma.

Lemma. *Let M_1 and M_2 be closed manifolds such that the Whitney sum of the tangent bundle τ_{M_i} and the one-dimensional trivial bundle is trivial and $\chi(M_1) = 0$. Then the manifold $M_1 \times M_2$ is parallelizable.*

Proof. Let η^n_B be the trivial n-dimensional bundle over B. By assumption, $\chi(M_1) = 0$; hence there exists a vector field without singular points on M_1. It determines two bundles over M_1, the trivial one-dimensional bundle $\eta^1_{M_1}$ and its orthogonal complement, which we denote by α. We have $\dim\alpha = \dim M_1 - 1 = m_1 - 1$ and $\tau_{M_1} = \alpha \oplus \eta^1_{M_1}$.

For any vector bundles ξ_1 and ξ_2 over B and any continuous map $f\colon B_1 \to B$, the bundle $f^*(\xi_1 \oplus \xi_2)$ is isomorphic to $(f^*\xi_1) \oplus (f^*\xi_2)$. Moreover, $f^*\eta^n_B = \eta^n_{B_1}$. Therefore,

$$\begin{aligned}\tau_{M_1\times M_2} &\cong \pi_1^*(\tau_{M_1}) \oplus \pi_2^*(\tau_{M_2}) \cong \pi_1^*(\alpha \oplus \eta^1_{M_1}) \oplus \pi_2^*(\tau_{M_2})\\ &\cong \pi_1^*(\alpha) \oplus \eta^1_{M_1\times M_2} \oplus \pi_2^*(\tau_{M_2}) \cong \pi_1^*(\alpha) \oplus \pi_2^*(\eta^1_{M_1} \oplus \tau_{M_2})\\ &\cong \pi_1^*(\alpha) \oplus \eta^{m_2+1}_{M_1\times M_2} \cong \pi_1^*(\alpha \oplus \eta^1_{M_1})\eta^{m_2}_{M_1\times M_2} \cong \pi_1^*(\tau_{M_1}) \oplus \eta^{m_2}_{M_1\times M_2}\\ &\cong \pi_1^*(\tau_{M_1} \oplus \eta^1_{M_1}) \oplus \eta^{m_2-1}_{M_1\times M_2} \cong \eta^{m_1+1}_{M_1\times M_2} \oplus \eta^{m_2-1}_{M_1\times M_2} \cong \eta^{m_1+m_2}_{M_1\times M_2}. \qquad \square\end{aligned}$$

For $k > 2$, we use induction on k. Suppose that n_1 is odd. Then we can consider $M_1 = S^{n_1} \times S^{n_{k-1}}$ and $M_2 = S^{n_k}$. By the induction hypothesis, M_1 is parallelizable; therefore, $\tau_{M_1} \oplus \eta^1_{M_1}$ is the trivial bundle.

96. We shall not check that $V(n,k)$ is a manifold; we shall only calculate its dimension. The map $V(n,k) \to V(n,k-1)$ deleting the last vector from each k-tuple $v_1, \dots, v_k$ is a locally trivial bundle with fiber S^{n-k}. Indeed, the unit vector v_k belongs to the $(n-k+1)$-space that is the orthogonal complement of the subspace generated by $v_1, \dots, v_{k-1}$. Clearly, $V(n,1) \approx S^{n-1}$. We obtain the sequence of bundles

$$V(n,k) \xrightarrow{S^{n-k}} V(n,k-1) \xrightarrow{S^{n-k-1}} \cdots \xrightarrow{S^{n-2}} V(n,1) = S^{n-1}.$$

Thus, the dimension of the manifold $V(n,k)$ is equal to

$$(n-1) + (n-2) + \cdots + (n-k) = nk - \frac{k(k+1)}{2}.$$

97. The manifold $V(n+2,2)$ has dimension $2n+1$. According to Theorem 3.16, it is $(n-1)$-connected and

$$\pi_n(V(n+2,2)) = \begin{cases}\mathbb{Z} & \text{if } n \text{ is even,}\\ \mathbb{Z}_2 & \text{if } n \text{ is odd.}\end{cases}$$

By the Hurewicz theorem, we have $H_n(V(n+2,2)) \cong \pi_n(V(n+2,2))$ and $H_k(V(n+2,2)) = 0$ for $0 < k < n$. We have calculated half of the homology groups. The remaining groups can be found using Poincaré duality. As a result, we obtain

$$H_k(V(n+2,2)) = \begin{cases} \mathbb{Z} & \text{if } k = 0,\ n,\ n+1,\ 2n+1, \\ 0 & \text{otherwise,} \end{cases}$$

and if n is odd, then

$$H_k(V(n+2,2)) = \begin{cases} \mathbb{Z} & \text{if } k = 0,\ 2n+1, \\ \mathbb{Z}_2 & \text{if } k = n, \\ 0 & \text{otherwise.} \end{cases}$$

98. We represent M^4 as a CW-complex and construct three independent sections of the tangent bundle by induction on the skeletons. The first nontrivial obstruction occurs when we extend these sections over the 2-skeleton. This obstruction is the class w_2. Therefore, an extension over the 2-skeleton exists if and only if $w_2 = 0$. The obstruction to extending sections over the 3-skeleton belongs to the cohomology with coefficients in $\pi_2(V(4,3))$. But $V(4,3) \approx \mathrm{SO}(4)$ and $\pi_2(\mathrm{SO}(4)) = 0$ (this it is seen from the exact sequence of the bundle $\mathrm{SO}(4) \to S^3$ with fiber $\mathrm{SO}(3)$ because $\pi_2(\mathrm{SO}(3)) = \pi_2(\mathbb{R}P^3) = 0$). Therefore, this obstruction vanishes.

Obviously, three independent sections can be extended from the boundary of the 4-disk to this disk with the center removed.

If independent sections over M^n with several punctures $x_1, \dots, x_n$ are given, then we construct independent sections over M^n with one puncture x_0. To this end, we connect the point x_0 ith $x_1, \dots, x_n$ by disjoint paths and contract these paths to x_0.

101. Let $\xi_1 = (u_1, v_1)$ and $\xi_2 = (u_2, v_2)$ be vectors in the space $T_xM \times T_xM \cong T_{(x,x)}(M \times M)$. We set $(\xi_1, \xi_2) = (u_1, u_2) + (v_1, v_2)$.

Take $u, v \in T_xM$. The vector $\xi = (u, v) \in T_{(x,x)}(M \times M)$ is tangent to $d(M)$ if and only if $u = v$, i.e., $\xi = (u, u)$. The vector $\eta = (v, w)$ is orthogonal to all vectors (u, u) if and only if $v = -w$, i.e., $\eta = (v, -v)$. The map $\tau_M \to \nu_{d(M)}$ taking each vector $v \in T_xM$ to $(v, -v) \in T_{(x,x)}(M \times M)$ is an isomorphism of vector bundles.

102. The sphere with g handles is orientable; therefore, $w_1 = 0$. Its Euler characteristic is even; therefore, $w_2 = 0$.

The equalities $w_1 = 0$ and $w_2 = 0$ can also be proved by using the triviality of the normal bundle for the standard embedding of the sphere with g handles in $\mathbb{R}^3$.

103. For the sphere to which m Möbius bands are attached, the parity of the Euler characteristic coincides with that of the number m. Therefore,

$w_2 = m\Lambda$, where Λ is the generator of the two-dimensional cohomology group, i.e., $w_2 = 0$ for even m and $w_2 = \Lambda$ for odd m.

For the coefficient group $\mathbb{Z}_2$, the one-dimensional cohomology group of the sphere with m Möbius bands attached has a basis $\alpha_1, \dots, \alpha_m$, where the cohomology classes $\alpha_1, \dots, \alpha_m$ are dual to cycles such that traverses along them reverse orientation; therefore, $w_1 = \alpha_1 + \dots + \alpha_m$.

Finally, $\alpha_i \alpha_j = \delta_{ij}\Lambda$, whence $w_1^2 = m\Lambda = w_2$.

104. Let ν_{M^n} be the normal bundle, and let $\alpha = w_1(\nu_{M^n})$. Then $w(M^n)(1+\alpha) = 1$; therefore, $w_1+\alpha = 1, w_2+w_1\alpha = 0, \dots, w_n+w_{n-1}\alpha = 0$. Hence $w_1 = \alpha, w_2 = \alpha^2, \dots, w_n = \alpha^n$.

105. Let $w_k = w_k(\mathbb{R}P^n) = \binom{n+1}{k}\alpha^k$ for $k = 1, \dots, n$. According to Problem 104, we have $w_k = w_1^k$. Therefore, the numbers $\binom{n+1}{k}$, where $k = 1, \dots, n$, are either all even or all odd.

Over the field $\mathbb{Z}_2$, we have $(1+x)^{2^m} = 1+x^{2^m}$. Let $a_0+a_1\cdot 2+\dots+a_m\cdot 2^m$ be the binary decomposition of $n+1$. Then

$$(1+x)^m = (1+x)^{a_0}(1+x^2)^{a_1}\cdots(1+x^{2^m})^{a_m}.$$

If all of the numbers $\binom{n+1}{k}$ are even, then $a_0 = \dots = a_{m-1} = 0$ and $a_m = 1$; this means that $n+1 = 2^m$. If all of these numbers are odd, then $a_0 = \dots = a_m = 1$; this means that $n+1 = 2^{m+1}-1$.

106. If n is even, then the cohomology class $w_n(\mathbb{R}P^n) = (n+1)\alpha^n$, where α is the generator of $H^1(\mathbb{R}P^n; \mathbb{Z}_2)$, is nonzero. Therefore, the Stiefel–Whitney number $w_n[\mathbb{R}P^n]$ is nonzero. The cohomology class $w_1(\mathbb{R}P^n) = (n+1)\alpha = \alpha$ is nonzero as well and $\alpha^n \neq 0$. Therefore, the Stiefel–Whitney number $w_1^n[\mathbb{R}P^n]$ is nonzero.

107. If $n = 2k-1$, then $w(\mathbb{R}P^n) = (1+a)^{2k} = (1+a^2)^k$; hence $w_i(\mathbb{R}P^n) = 0$ for any odd i. But any monomial $w_1^{r_1}\cdots w_n^{r_n}$, where $r_1+2r_2+\dots+nr_n = 2k-1$, contains a factor w_i with odd i.

108. The *only if* part is obvious; let us prove that if the bundle $\xi^k \oplus \varepsilon^1$ is trivial, then so is ξ^k. Since both bundles $\xi^k \oplus \varepsilon^1$ and ε^1 are orientable, so is ξ^k. We assume that all fibers of $\xi^k \oplus \varepsilon^1$ and ξ^k are endowed with orientations. Each fiber of $\xi^k \oplus \varepsilon^1$ is identified with $\mathbb{R}^{k+1}$, and each fiber of ξ^k is an oriented k-dimensional subspace in $\mathbb{R}^{k+1}$. The space of oriented k-dimensional subspaces in $\mathbb{R}^{k+1}$ is homeomorphic to the sphere S^k. Consider the bundle γ_+^k over S^k whose fiber over each point $x \in S^k$ is the oriented k-dimensional subspace of $\mathbb{R}^{k+1}$ corresponding to this point. Let $f\colon B \to S^k$ be the map taking every point $b \in B$ to the fiber of ξ^k over this point. Clearly, $\xi^k = f^*\gamma_+^k$. Since the dimension of the complex B is less than k, it follows that the map f is null-homotopic. Therefore, the bundle ξ^k is trivial.

109. (a) This is an obvious corollary of Problem 108.

(b) Let ν^k be the normal bundle for an embedding $M^n \subset \mathbb{R}^{n+k}$. Then $\tau_{M^n} \oplus \nu^k \cong \theta^{n+k}$; therefore, $\varepsilon^1 \oplus \tau_{M^n} \oplus \nu^k \cong \varepsilon^{n+k+1}$. We consider the situation in which $k \geq n+1$. Suppose that the manifold M^n is stably parallelizable. Then the bundle $\varepsilon^1 \oplus \tau_{M^n}$ is trivial. Hence, according to Problem 108, the bundle ν^k is trivial as well. Conversely, if ν^k is trivial, then so is $\varepsilon^1 \oplus \tau_{M^n}$.

110. The space $E(P\gamma_n^k)$ consists of all pairs (l, Π^k), where $l \subset \Pi^k$. The space $E(P\bar{\gamma}_n^{k-1})$ consists of all pairs (l, Π^{k-1}), where $l \perp \Pi^{k-1}$. The map $E(P\gamma_n^k) \to E(P\bar{\gamma}_n^{k-1})$ takes each pair (l, Π^k) to the pair (l, Π^{k-1}), where Π^{k-1} is the orthogonal complement to l in Π^k. The inverse map takes each pair (l, Π^{k-1}) to the pair (l, Π^k), where Π^k is the space spanned by l and Π^{k-1}.

111. Suppose that some finite group acts freely on $\mathbb{R}^n$. Any finite group contains the subgroup $G = \mathbb{Z}_p$ with some prime p. The group G acts freely on $\mathbb{R}^n$. This action induces an action of G on S^n such that it has the unique fixed point ∞, i.e., $(S^n)^G = \{\infty\}$. On the other hand, Theorem 3.56 implies that the space $(S^n)^G$ is a homology sphere, while the topological space consisting of one point is not a homology sphere.

112. According to Theorem 3.62, we have $\operatorname{Sq} \alpha^{2^m} = (\alpha + \alpha^2)^{2^m} = \alpha^{2^m} + \alpha^{2^{m+1}}$ because $\binom{2^m}{i} \equiv 0 \pmod 2$ for $1 \leq i \leq 2^m - 1$. Therefore, $\operatorname{Sq}(\alpha + \alpha^2 + \alpha^4 + \cdots) = (\alpha + \alpha^2) + (\alpha^2 + \alpha^4) + (\alpha^4 + \alpha^8) + \cdots = \alpha$.

113. *Solution* 1. Theorem 3.32 implies the existence of a monomorphism

$$f^* \colon H^*(G(\infty, n); \mathbb{Z}_2) \to H^*(\mathbb{R}P^\infty \times \cdots \times \mathbb{R}P^\infty; \mathbb{Z}_2) \cong \mathbb{Z}_2[\alpha_1, \ldots, \alpha_n],$$

where $f^* w_m = \sigma_m(\alpha_1, \ldots, \alpha_n)$ is the mth elementary symmetric function. Moreover, $\operatorname{Sq}^0 \alpha_i = \alpha_i$, $\operatorname{Sq}^1 \alpha_i = \alpha_i^2$, and $\operatorname{Sq}^r \alpha_i = 0$ for $r > 1$. Therefore, by the Cartan formula,

$$\begin{aligned} \operatorname{Sq}^k(\alpha_{i_1} \cdots \alpha_{i_m}) &= \alpha_{i_1} \operatorname{Sq}^k(\alpha_{i_2} \cdots \alpha_{i_m}) + \alpha_{i_1}^2 \operatorname{Sq}^{k-1}(\alpha_{i_2} \cdots \alpha_{i_m}) \\ &= \cdots = \sum \alpha_{j_1}^2 \cdots \alpha_{j_k}^2 \alpha_{j_{k+1}} \cdots \alpha_{j_m}, \end{aligned}$$

where the summation is over all permutations of $i_1, \ldots, i_m$ for which $j_1 < \cdots < j_k$ and $j_{k+1} < \cdots < j_m$. Therefore,

$$\operatorname{Sq}^k(f^* w_m) = \sum \alpha_{i_1}^2 \cdots \alpha_{i_k}^2 \alpha_{i_{k+1}} \cdots \alpha_{i_m}.$$

Since f^* is a monomorphism, it suffices to prove that

(3)
$$\sum \alpha_{i_1}^2 \cdots \alpha_{i_k}^2 \alpha_{i_{k+1}} \cdots \alpha_{i_m} \equiv \sum_{i=0}^{k} \binom{m-k+i-1}{i} \sigma_{k-i} \sigma_{m+i} \pmod 2.$$

Let us apply induction on the number n of variables. For $n = 1$, we have the obvious identities $\alpha_1 = \sigma_0\sigma_1$ and $\alpha_1^2 = \sigma_1^2$. The induction step is as follows. If $k = m$, then congruence (3) takes the form $\sum(\alpha_{i_1}\cdots\alpha_{i_k})^2 \equiv (\sigma_k)^2 \pmod 2$; obviously, this is true. Suppose that $k < m$. Consider a similar sum $\sum^* \alpha_{i_1}^2\cdots\alpha_{i_k}^2\alpha_{i_{k+1}}\cdots\alpha_{i_m}$ for the $n-1$ variables $\alpha_1,\dots,\alpha_{n-1}$. Then the left-hand side of (3) takes the form

$$\alpha_n^2\sum\nolimits^* \alpha_{i_1}^2\cdots\alpha_{i_{k-1}}^2\alpha_{i_{k+1}}\cdots\alpha_{i_{m-1}} + \alpha_n\sum\nolimits^* \alpha_{i_1}^2\cdots\alpha_{i_k}^2\alpha_{i_{k+1}}\cdots\alpha_{i_{m-1}} + \sum\nolimits^* \alpha_{i_1}^2\cdots\alpha_{i_k}^2\alpha_{i_{k+1}}\cdots\alpha_{i_m}.$$

Let σ_m^* be the mth elementary function of the variables $\alpha_1,\dots,\alpha_{n-1}$. Then $\sigma_m = \alpha_n\sigma_{m-1}^* + \sigma_m^*$; therefore, the right-hand side of (3) takes the form

$$\begin{aligned}
&\binom{m-1}{k}(\alpha_n\sigma_{m+k-1}^* + \sigma_{m+k}^*)\\
&\qquad+\sum_{i=0}^{k-1}\binom{m-k+i-1}{k}(\alpha_n\sigma_{k-i-1}^* + \sigma_{k-i}^*)(\alpha_n\sigma_{m+i-1}^* + \sigma_{m+i}^*)\\
&\quad= \alpha_n^2\sum_{i=0}^{k-1}\binom{m-k+i-1}{i}\sigma_{k-i-1}^*\sigma_{m+i-1}^*\\
&\qquad+\alpha_n\sum_{i=0}^{k}\left(\binom{m-k+i-1}{i}+\binom{m-k+i-2}{i-1}\right)\sigma_{k-i}^*\sigma_{m+i-1}^*\\
&\qquad+\sum_{i=0}^{k}\binom{m-k+i-1}{i}\sigma_{k-i}^*\sigma_{m+i}^*;
\end{aligned}$$

here we set $\binom{m-k-2}{-1} = 0$. To complete the proof, it remains to use the induction hypothesis. For the coefficients of α_n^2 and α_n^0, the required relation follows at once. For the coefficient of α_n, note that

$$\binom{m-k+i}{i} + \binom{m-k+i-1}{i-1} \equiv \binom{m-k+i-1}{i} \pmod 2$$

because

$$\binom{m-k+i-1}{i-1} + \binom{m-k+i-1}{i} = \binom{m-k+i}{i}.$$

Solution 2. It is easy to obtain the required relation for the bundle γ^1 over $\mathbb{R}P^\infty$. Indeed, $w_1(\gamma^1) = \alpha$ is the generator of $H^1(\mathbb{R}P^\infty;\mathbb{Z}_2)$. Hence the proof reduces to the obvious equality $\operatorname{Sq}^1\alpha = \alpha^2$. Suppose that the required relation is valid for the bundles ξ and $w_i = w_i(\xi)$. According to the Whitney formula, we have $w_m(\xi\times\gamma^1) = w_m\times 1 + w_{m-1}\times\alpha$. The Cartan formula implies $\mathrm{Sq}(w_{m-1}\times\alpha)=(\mathrm{Sq}^k w_{m-1})\times\alpha+(\mathrm{Sq}^{k-1}w_{m-1})\times\alpha^2$. Therefore,

$$\begin{aligned}\mathrm{Sq}^k(w_m(\xi\times\gamma^1)) &= w_kw_m\times 1+\binom{m-k}{1}w_{k-1}w_{m+1}\times 1\\ &\quad+\dots+\binom{m-1}{k}w_0w_{m+k}\times 1\\ &\quad+w_kw_{m-1}\times\alpha+\binom{m-k-1}{1}w_{k-1}w_m\times\alpha\\ &\quad+\dots+\binom{m-2}{k}w_0w_{m+k-1}\times\alpha\\ &\quad+w_{k-1}w_{m-1}\times\alpha^2+\binom{m-k}{1}w_{k-2}w_m\times\alpha^2\\ &\quad+\dots+\binom{m-2}{k-1}w_0w_{m+k-2}\times\alpha^2.\end{aligned}$$

On the other hand, $w_i(\xi\times\gamma^1)w_j(\xi\times\gamma^1)=(w_i\times 1+w_{i-1}\times\alpha)(w_j\times 1+w_{j-1}\times\alpha)=w_iw_j\times 1+(w_{i-1}w_j+w_iw_{j-1})\times\alpha+w_{i-1}w_{j-1}\times\alpha^2$. Now it is easy to check the required relation for the bundle $\xi\times\gamma^1$. For the terms of the form $\dots\times 1$ and $\dots\times\alpha^2$, it is seen directly. For the terms of the form $\dots\times\alpha$, we must show that, modulo 2, $1+\binom{m-k}{1}=\binom{m-k-1}{1}$, $\binom{m-k}{1}+\binom{m-k+1}{2}=\binom{m-k}{2}$, etc. These equalities follow from the equalities $-1+\binom{m-k}{1}=\binom{m-k-1}{1}$, $-\binom{m-k}{1}+\binom{m-k+1}{2}=\binom{m-k}{2}$, etc., because the sign is inessential when residues modulo 2 are considered.

Thus, the required formula is valid for the bundle $\gamma^1\times\dots\times\gamma^1$. Therefore, it is valid for the bundle γ^n because the natural map $H^*(G(n,k);\mathbb{Z}_2)\to H^*(\mathbb{R}P^\infty\times\dots\times\mathbb{R}P^\infty;\mathbb{Z}_2)$ is a monomorphism. Any n-dimensional bundle ξ is a pullback of γ^n; therefore, the required relation holds for ξ as well.

114. From the cone CY, we remove the cone whose lateral side is one half of that of CY. Then we apply the excision theorem, after which we can use the homotopy invariance of the singular homology of a pair (see Theorem 1.14 on p. 16; for singular homology, the proof is precisely the same).

115. Let us write the exact sequence for the pair $(X\cup CY, CY)$:

$$H_i(CY)\longrightarrow H_i(X\cup CY)\longrightarrow H_i(X\cup CY, CY)\longrightarrow \tilde{H}_{i-1}(CY).$$

Here $H_i(CY)=\tilde{H}_{i-1}(CY)=0$ because the cone CY is contractible. Therefore, $H_i(X\cup CY)\cong H_i(X\cup CY, CY)$ for $i\ge 1$. It remains to apply Problem 114.

116. For $i=0$, the required assertion is obvious. Suppose that $i\ge 1$. Then $H_i(X,Y)\cong H_i(X\cup CY)$ by Problem 115. Since the complex CY is contractible, we have $X\cup CY\sim(X\cup CY)/CY=X/Y$.

117. No, this is not true. The set $U \subset \mathbb{R}^n$ determined by the equation $x_n = 0$ is closed. It is homeomorphic to the set $V \subset \mathbb{R}^n$ determined by the equation $x_n = 0$ and the inequality $x_1^2 + \cdots + x_{n-1} < 1$. This set is not closed.

119. Let us represent S^n as $\partial\Delta^{n+1}$, where $\Delta^{n+1} = [v_0, v_1, \dots, v_{n+1}]$, and consider the sets $\operatorname{st} v_i$ for $i = 0, 1, \dots, n+1$. Any $r \le n+1$ vertices $v_{i_1}, \dots, v_{i_r}$ span a simplex $[v_{i_1}, \dots, v_{i_r}]$. Moreover, $\bigcup_{k=1}^{r} \operatorname{st} v_{i_k}$ is the star of this simplex; this is a contractible set.

120. We calculated the cohomology groups of the manifold $V(2n+2, 2)$ in the solution of Problem 97. Let us write the Gysin sequence for the bundle $V(2n+2,2) \xrightarrow{S^1} G_+(2n+2,2)$. Starting from the end, we consequtively obtain $H^{4n} = \mathbb{Z}$, $H^{4n-1} = 0$, $H^{4n-2} = \mathbb{Z}$, $H^{4n-3} = 0, \dots, H^{2n+1} = 0, H^{2n} = \mathbb{Z} \oplus \mathbb{Z}$. After this, we can apply Poincaré duality. As a result for $0 \le k \le 4n$, we obtain

$$H^k(G_+(2n+2,2)) = \begin{cases} \mathbb{Z} & \text{if } k \text{ is even, } k \ne 2n, \\ \mathbb{Z} \oplus \mathbb{Z} & \text{if } k = 2n, \\ 0 & \text{if } k \text{ odd.} \end{cases}$$

121. (a) According to the Wu theorem, we have $w_1 = v_1, w_2 = v_2 + \operatorname{Sq} v_1, \dots, w_{k+1} = v_{k+1} + \operatorname{Sq}^1 v_k + \operatorname{Sq}^2 v_{k-1} + \cdots$. Therefore, if $w_1 = \cdots = w_k = 0$, then $v_1 = \cdots = v_k = 0$ and $v_{k+1} = w_{k+1}$.

(b) If $w_1 = \cdots = w_k = 0$, then $v_1 = \cdots = v_k = 0$, and if $n = 2k$ or $2k+1$, then $v_i = 0$ for $i > k$. Therefore, $w_i = 0$ for $i > k$.

122. In addition to the dimensions 0 and $4k$, the manifold M^{4k} has nontrivial homology and cohomology groups only in dimension $2k$, and it has no torsion.

The Wu class has only one component of positive dimension, namely, v_{2k}. According to Problem 121, we have $w_{2k} = v_{2k}$. Clearly, $\langle \operatorname{Sq} \alpha^{2k}, [M^{4k}] \rangle = \langle \alpha^{2k} \smile \alpha^{2k}, [M^{4k}] \rangle$ and $\langle \alpha^{2k} \smile v, [M^{4k}] \rangle = \langle \alpha^{2k} \smile v_{2k}, [M^{4k}] \rangle$. Therefore, the class v_{2k} is completely determined by the equality $\alpha^{2k} \smile \alpha^{2k} = \alpha^{2k} \smile v_{2k}$ (for all α^{2k}). In particular, $v_{2k} = 0$ if and only if $\alpha^{2k} \smile \alpha^{2k} = 0$ for all $\alpha^{2k} \in H^{2k}(M^{4k}; \mathbb{Z}_2)$.

Now we consider a cohomology class a^{2k} over $\mathbb{Z}$. It can be naturally associated with the cohomology class α^{2k} over $\mathbb{Z}_2$ (by reducing the value assigned to each chain modulo 2). The number $\langle a^{2k} \smile a^{2k}, [M^{4k}] \rangle$ is even if and only if $\alpha^{2k} \smile \alpha^{2k} = 0$.

123. It is seen from the proof of Theorem 3.26 that if $n = 2^k$, then $\bar{w}(\mathbb{R}P^n) = 1 + \alpha + \alpha^2 + \cdots + \alpha^{n-1}$; therefore, $\bar{w}_{n-1}(\mathbb{R}P^n) \ne 0$. On the other hand, if a manifold M^n is embedded in $\mathbb{R}^{n+k}$, then $\bar{w}_k(M^n) = 0$. Thus, if $n = 2^k$, then $\mathbb{R}P^n$ cannot be embedded in $\mathbb{R}^{2n-1}$.

124. Let $G = \mathbb{Z}_{n_1} \oplus \cdots \oplus \mathbb{Z}_{n_k}$. For a presentation matrix of this module we can take the diagonal matrix $A = \operatorname{diag}(n_1, \dots, n_k)$. The ideal $\mathcal{E}_1$ is generated by its determinant, which equals $n_1 \cdots n_k = |G|$.

125. For the knot under consideration, we can choose f_1 and f_2 as shown in Figure 4 (with the only difference that the lowest cross in this figure should be replaced by three crosses). Only the linking number $\operatorname{lk}(f_2, f_2^+)$ changes and becomes equal to 2. Therefore, $A = \begin{pmatrix} 1 & -1 \\ 0 & 2 \end{pmatrix}$, and the Alexander polynomial is $2(t^2 - 2t + 1) + t = 2t^2 - 3t + 2$.

126. (a) We choose f_1 and f_2 as shown in Figure H.5. It is easy to verify that, for these f_1 and f_2, the Seifert matrix is $\frac{1}{2}\begin{pmatrix} p+q & q+1 \\ q-1 & q+r \end{pmatrix}$. Hence the Alexander polynomial is $\frac{1}{4}((pq + qr + rp)(t^2 - 2t + 1) + t^2 + 2t + 1)$.

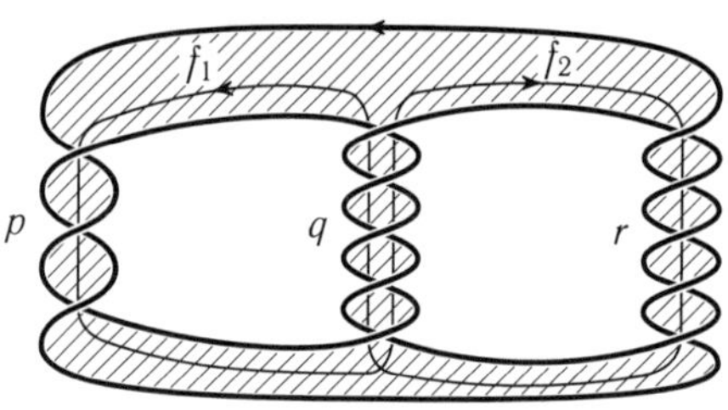

Figure H.5. The choice of f_1 and f_2

(b) If $p = -3$, $q = 5$, and $r = 7$, then $pq + qr + rp = -1$. Therefore, in this case, the Alexander polynomial equals t.

127. The diagram of the link $L_0 = L$ can be represented as shown in Figure H.6. The links L_+ and L_- are isotopic; hence $\nabla_{L_+} = \nabla_{L_-}$, and therefore $z\nabla_L = \nabla_{L_+} - \nabla_{L_-} = 0$.

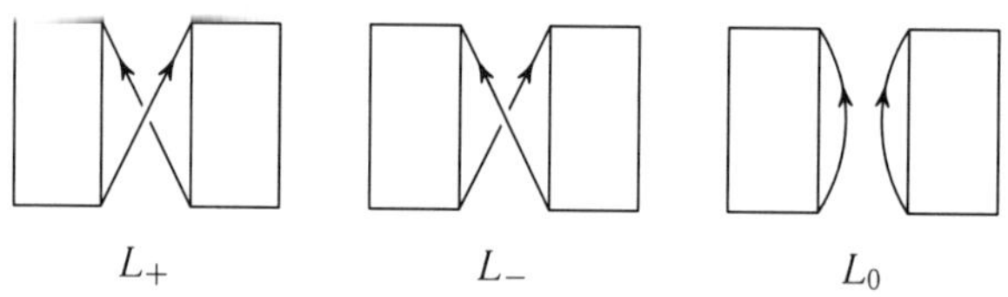

Figure H.6. The diagrams of the three links

128. It is easy to show that $n(L_+) = n(L_-) = n(L_0) \pm 1$ (see Figure H.7). Thus, if the required assertion is true for two of the three links L_+, L_-, and L_0, then it is also true for the third link. The assertion is true for the trivial m-component link (if $m > 1$, then the Conway polynomial is zero); hence it is true for any link.

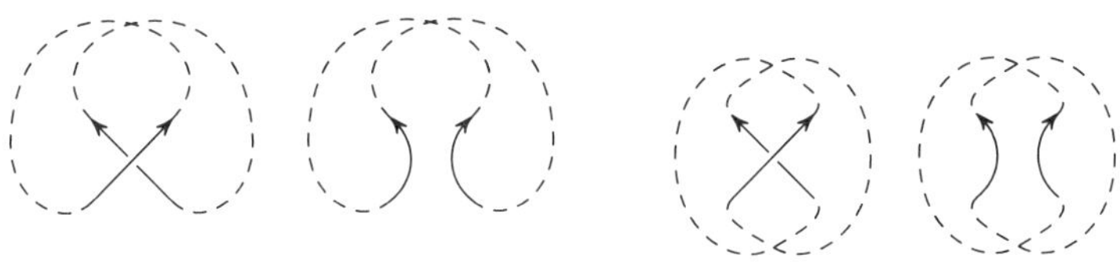

Figure H.7. The proof of the equality $n(L_+) = n(L_-) = n(L_0) \pm 1$

129. Clearly, $a_0(L) = \nabla_L(0)$, therefore, the relation $\nabla_{L_+}(z) - \nabla_{L_-}(z) = z\nabla_{L_0}(z)$ implies $a_0(L_+) = a_0(L_-)$. Changing types of crosses, we can transform any link into the trivial link (with the same number of components). Hence $a_0(L) = a_0(L')$, where L' is the trivial link with the same number of components as in L. For the unknot, we have $a_0 = 1$, whereas for the trivial m-component link with $m > 1$, we have $a_0 = 0$.

130. We prove the required assertion by induction on $n(L)$. For $n(L) = 1$, there is nothing to prove. Clearly, $n(L_0) \geq n(L_\pm) - 1$; therefore, the induction hypothesis implies $a_i(L_0) = 0$ for $i < n(L_\pm) - 2$. Thus, $a_i(L_+) = a_i(L_-)$ for $i < n(L_\pm) - 1$. Changing types of crosses, we can transform any link into the trivial link with the same number of components. Such changes do not affect $a_i(L)$ with $i < n(L) - 1$; therefore, $a_i(L) = 0$ for $i < n(L) - 1$.

131. Suppose that L_+ and L_- are two-component links and both components are present at the distinguished cross. Then L_0 is a knot; therefore, $a_0(L_0) = 1$ according to Problem 129(a). Thus, $a_1(L_+) - a_1(L_-) = a_0(L_0) = 1$. Clearly, $\operatorname{lk}(L_{1+}, L_{2+}) - \operatorname{lk}(L_{1-}, L_{2-}) = 1$. Changing types of crosses, we can transform any two-component link into a two-component link with unlinked components. For such a link, we have $a_1(L) = \operatorname{lk}(L_1, L_2)$ because $a_1(L) = 0$ and $\operatorname{lk}(L_1, L_2) = 0$. Therefore, this equality is true for any two-component link.

132. The relation $\nabla_{L_+}(z) - \nabla_{L_-}(z) = z\nabla_{L_0}(z)$ implies $a_2(L_+) - a_2(L_-) = a_1(L_0)$. If L_0 is a two-component link, then $a_1(L_0) = \operatorname{lk}(L_1, L_2)$ by Problem 131.

133. In the proof of Theorem 6.17, it was shown that $TM^n = N_*(TS^n)$. If the bundle TS^n is trivial, then so is any of its pullbacks.

134. (a) The action of each w_i on U_a is defined by

$$(u_0t_0, \dots, u_nt_n) \mapsto (u_0t_0, \dots, w_iu_it_i, \dots, u_nt_n).$$

Under such an action, the ith face $\partial_i e$, which is determined by the equation $t_i = 0$, is fixed. Therefore, $(1 - w_i)\partial_i e = 0$, which means that

$$\partial\varepsilon = \prod(1 - w_i)\sum(-1)^i\partial_i e = 0.$$

(b) First, suppose that $n = 0$. Let $w = n_0 + n_1 w_0 + \cdots + n_{a_0} w^{a_0 - 1}$, where $n_0, \ldots, n_{a_0}$ are integers. Then

$$w(1 - w_0) = (n_0 - n_{a_0}) + (n_1 - n_0)w_0 + \cdots + (n_{a_0} - n_{a_0 - 1})w^{a_0 - 1};$$

therefore, an element w belongs to the kernel of the given map if and only if $n_0 = n_1 = \cdots = n_{a_0 - 1} = n_{a_0}$.

Now, suppose that $n = 1$. Again, let $w = n_0 + n_1 w_0 + \cdots + n_{a_0} w^{a_0 - 1}$; however, this time, the n_i are linear combinations of $1, w_1, \ldots, w_1^{a_1 - 1}$ with integer coefficients rather than simply integers. Consider $w' = w(1 - w_1) = n_0' + n_1' w_0 + \cdots + n_{a_0}' w^{a_0 - 1}$, where $n_i' = n_i - n_i w$. The equality $w'(1 - w_0) = 0$ is again equivalent to $n_0' = n_1' = \cdots = n_{a_0}'$, i.e.,

$$w(1 - w_1) = n'(1 + w_0 + w_0^2 + \cdots + w_0^{a_0 - 1}).$$

It follows that $w = a(1 + w_0 + w_0^2 + \cdots + w_0^{a_0 - 1}) + b$, where $b(1 - w_1) = 0$, i.e., $b = c(1 + w_1 + w_1^2 + \cdots + w_1^{a_1 - 1})$. The further argument is similar.

(c) The simplicial complex U_a has no $(n+1)$-cells; hence its n-dimensional homology group U_a coincides with the n-cycle group. Therefore, $H_n(U_a)$ contains the subgroup $\mathbb{Z}[\mathbb{Z}_a]\varepsilon \cong \mathbb{Z}[\mathbb{Z}_a]/I_a$. The elements specified in the statement of the problem form a basis in this subgroup. Clearly, the rank of the subgroup equals $(a_0 - 1) \cdots (a_n - 1)$, i.e., coincides with the rank of the entire group $H_n(U_a)$. Hence $H_n(U_a) \cong \mathbb{Z}[\mathbb{Z}_a]\varepsilon$.

136. Let x be an arbitrary element of G. According to Corollary 2 of Theorem 6.29, we have $x = \exp A$ for some matrix A. We set $g = \exp(\frac{1}{n}A)$. Then $g^n = x$.

Bibliography

[1] M. Adachi, *Embeddings and Immersions* (Providence, RI, Amer. Math. Soc., 1993).

[2] J. F. Adams, *Lectures on Lie Groups* (New York–Amsterdam, Benjamin, 1969).

[3] P. M. Akhmetiev, D. Repovš, and A. B. Skopenkov, "Embedding products of low-dimensional manifolds into $\mathbb{R}^n$," Topology Appl. **113**, 7–12 (2001).

[4] J. W. Alexander, "A proof of the invariance of certain constants of analysis situs," Trans. Amer. Math. Soc. **16**, 148–154 (1915).

[5] J. W. Alexander, "A proof and extension of the Jordan–Brouwer separation theorem," Trans. Amer. Math. Soc. **23**, 333–349 (1922).

[6] J. W. Alexander, "On the connectivity ring of an abstract space," Ann. Math. **37**, 419–421 (1936).

[7] F. Apéry, "La surface de Boy," Advan. Math. **61**, 185–266 (1986).

[8] C. Arf, "Untersuchungen über quadratische Formen in Körpern der Characteristik 2," J. Reine Angew. Math. **183**, 148–167 (1941).

[9] V. N. Aznar, "On the Chern classes and the Euler characteristic for nonsingular complete intersections," Proc. Amer. Math. Soc. **78**, 143–148 (1980).

[10] M. G. Barratt and J. Milnor, "An example of anomalous singular homology," Proc. Amer. Math. Soc. **13**, 293–297 (1962).

[11] R. Bott and L. W. Tu, *Differential Forms in Algebraic Topology* (New York, Springer, 1989).

[12] G. E. Bredon, *Sheaf Theory* (New York, McGraw-Hill, 1967).

[13] G. E. Bredon, *Introduction to Compact Transformation Groups* (New York, Academic Press, 1972).

[14] L. E. J. Brouwer, "Sur le théorème de M. Jordan dans l'espace à n dimensions," C. R. Acad. Sci. Paris. **153**, 542–544 (1911).

[15] L. E. J. Brouwer, "On looping coefficients," Proc. Akad. Wetensch. Amsterdam **15**, 113–122 (1912).

[16] L. E. J. Brouwer, "Über Abbildungen von Mannigfaltigkeiten," Math. Ann. **71**, 97–115 (1912).

[17] L. E. J. Brouwer, "Beweis des Invarianz des n-dimensionalen Gebietes," Math. Ann. **71**, 305–313 (1912); **72**, 55–56 (1912).

[18] L. E. J. Brouwer, "Beweis des Jordanschen Satzes für den n-dimensionalen Raum," Math. Ann. **71**, 314–319 (1912).

[19] L. E. J. Brouwer, "Über Jordansche Mannigfaltigkeiten," Math. Ann. **71**, 320–327 (1912).

[20] R. Brown, "Locally flat embeddings of topological manifolds," Ann. Math. **75**, 331–341 (1962).

[21] E. Čech, "Multiplication on a complex," Ann. Math. **37**, 681–697 (1936).

[22] B.-Y. Chen and K. Ogiue, "Some implications of the Euler–Poincaré characteristic for complete intersection manifolds," Proc. Amer. Math. Soc. **44**, 1–8 (1974).

[23] S. S. Chern, *Complex Manifolds* (Recife, Instituto de Fisica e Matematica, Univ. Recife, 1959).

[24] R. Connelly, "A new proof of Brown's collaring theorem," Proc. Amer. Math. Soc. **27**, 180–182 (1971).

[25] J. H. Conway, "An enumeration of knots and links, and some of their properties," in *Computational Problems in Abstract Algebra* (NY, Pergamon, 1970), pp. 229–244.

[26] J. H. Conway and C. McA. Gordon, "Knots and links in spatial graphs," J. Graph Theory **7**, 445–453 (1983).

[27] R. H. Crowell and R. H. Fox, *Introduction to Knot Theory* (Boston, Mass., Ginn, 1963).

[28] M. L. Curtis and J. Dugundji, "A proof of de Rham's theorem," Fund. Math. **68**, 265–268 (1970).

[29] H. E. Debrunner, "Helly type theorems derived from basic singular homology," Amer. Math. Monthly **77**, 375–380 (1970).

[30] A. Dold, "A simple proof of the Jordan–Alexander complement theorem," Amer. Math. Monthly **100**, 856–857 (1993).

[31] S. Eilenberg, "Cohomology and continuous mappings," Ann. Math. **41**, 231–251 (1940).

[32] S. Eilenberg and S. MacLane, "Relations between homology and homotopy groups of spaces, I, II," Ann. Math. **46**, 480–509 (1945); **51**, 514–533 (1950).

[33] S. Eilenberg and S. MacLane, "Acyclic models," Amer. J. Math. **75**, 189–199 (1953).

[34] S. Eilenberg and S. MacLane, "On the groups $H(\pi, n)$, I, II, III," Ann. Math. **58**, 55–106 (1953); **60**, 49–139, 513–557 (1954)

[35] S. Eilenberg and J. A. Zilber, "On products of complexes," Amer. J. Math. **75**, 200–204 (1953).

[36] E. E. Floyd, "On periodic maps and the Euler characteristics of associated spaces," Trans. Amer. Math. Soc. **72**, 138–147 (1952).

[37] A. T. Fomenko and D. B. Fuks, *A Course in Homotopic Topology* (Moscow, Nauka, 1989) [in Russian].

[38] R. H. Fox, "Free differential calculus. I. Derivation in the free group ring," Ann. Math. **57**, 547–560 (1953).

[39] R. H. Fox, "Free differential calculus, II," Ann. Math. **59**, 196–210 (1954).

[40] R. Godement, *Topologie algébrique et théorie des faisceaux* (Paris, Hermann, 1958).

[41] M. Greenberg, *Lectures on Algebraic Topology* (New York, Benjamin, 1967).

[42] W. Greub, S. Halperin, and R. Vanstone, *Connections, Curvature, and Cohomology*, Vol. 1: *De Rham Cohomology of Manifolds and Vector Bundles* (New York, Academic Press, 1972).

[43] W. Greub, S. Halperin, and R. Vanstone, *Connections, Curvature, and Cohomology,* Vol. 2: *Lie Groups, Principal Bundles, and Characteristic Classes* (New York, Academic Press, 1973).

[44] P. A. Griffiths and J. W. Morgan, *Rational Homotopy Theory and Differential Forms* (Boston, Mass., Birkhäuser, 1981).

[45] M. Gromov, "Volume and bounded cohomology," Publ. Math. IHES. **56**, 5–100 (1982).

[46] W. Gysin, "Zur Homologie Theorie des Abbildungen und Faserungen von Mannigfaltigkeiten," Comm. Math. Helv. **14**, 61–121 (1941).

[47] A. Hatcher, *Algebraic Topology* (Cambridge, Cambridge Univ. Press, 2002).

[48] E. Helly, "Über Mengen konvexer Körper mit gemeinschaftlichen Punkten," Jber. Deutsch. Math. Verein. **32**, 175–176 (1923).

[49] E. Helly, "Über Systeme von abgeschlossenen Mengen mit gemeinschaftlichen Punkten," Monatsh. Math. Phys. **37**, 281–302 (1930).

[50] P. J. Hilton and U. Tammbach, *A Course in Homological Algebra* (New York, Springer, 1971).

[51] F. Hirzebruch, *Topological Methods in Algebraic Geometry* (New York, Springer, 1966).

[52] F. Hirzebruch, "Division algebras and topology," in H.-D. Ebbinghaus *et al. Numbers* (New York, Springer, 1991), pp. 281–302.

[53] F. Hirzebruch and K. H. Mayer, $O(n)$*-Mannigfaltigkeiten, exotische Sphäre und Singularitäten* (Berlin–New York, Springer, 1968).

[54] *Homotopy theory of differential forms,* Mathematics: Recent Publications in Foreign Science, **25** (Mir, Moscow, 1981) [in Russian].

[55] H. Hopf, "Über die Curvatura integra geschlossener Hyperflächen," Math. Ann. **95**, 340–367 (1926).

[56] H. Hopf, "Vektorfelder in n-dimensionalen Mannigfaltigkeit," Math. Ann. **96**, 225–250 (1927).

[57] H. Hopf, "A new proof of Lefschetz formula on invariant points," Proc. Nat. Acad. Sci. U.S.A. **14**, 149–153 (1928).

[58] H. Hopf, "Über die Abbildungen der dreidimensionalen Sphäre auf die Kugelfläche," Math. Ann. **104**, 639–665 (1931).

[59] H. Hopf, "Die Klassen der Abbildungen der n-dimensionalen Polyeder auf die n-dimensionalen Sphäre," Comment. Math. Helv. **5**, 39–54 (1933).

[60] H. Hopf, "Über die Abbildungen von Sphären auf Sphären niedriger Dimensionen," Fund. Math. **25**, 427–440 (1935).

[61] H. Hopf, "Über die Topologie der Gruppen-Mannigfaltigleiten und ihre Verallgemeinerungen," Ann. Math. **13**, 22–52 (1940).

[62] H. Hopf, "Ein topologischer Beitrag zur reelen Algebra," Comment. Math. Helv. **13**, 219–239 (1941).

[63] Sze-tsen Hu, *Homotopy Theory* (New York–London, Academic Press, 1959).

[64] W. Hurewicz, "Homotopie- und Homologiegruppen," Proc. Akad. Wetensch. Amsterdam **38**, 521–528 (1935).

[65] D. Husemoller, *Fibre Bundles* (New York, McGraw-Hill, 1966).

[66] B. Iversen, *Cohomology of Sheaves* (Berlin, Springer, 1986).

[67] D. M. Kan and G. W. Whitehead, "On the realizability of singular cohomology groups," Proc. Amer. Math. Soc. **12**, 24–25 (1961).

[68] R. M. Kane, *The Homology of Hopf Spaces* (Amsterdam, North-Holland, 1988).

[69] L. H. Kauffman, "The Conway polynomial," Topology **20**, 101–108 (1981).

[70] M. A. Kervaire and J. W. Milnor, "On 2-spheres in 4-manifolds," Proc. Nat. Acad. Sci. U.S.A. **47**, 1651–1657 (1961).

[71] M. A. Kervaire and J. W. Milnor, "Groups of homotopy spheres, I," Ann. Math. **77**, 504–537 (1963).

[72] A. Kolmogoroff, "Über die Dualität im Aufbau der kombinatorischen Topologie," Matem. Sb. **1**, 97–102 (1936).

[73] A. Kolmogoroff, "Homologiering des Komplexes und des local-bikompakten Räumes," Matem. Sb. **1**, 701–705 (1936).

[74] H. Künneth, "Über die Bettische Zahlen einer Produktmannigfaltigkeit," Math. Ann. **90**, 65–85 (1923).

[75] H. Künneth, "Über die Torsionzahlen von Produktmannigfaltigkeiten," Math. Ann. **91**, 125–134 (1924).

[76] H.-F. Lai, "On the topology of the even-dimensional complex quadrics," Proc. Amer. Math. Soc. **46**, 419–425 (1974).

[77] S. Lefschetz, "Intersections and transformations of complexes and manifolds," Trans. Amer. Math. Soc. **28**, 1–49 (1926).

[78] S. Lefschetz, *Topology*, New York, 1930.

[79] J. Levine, "An algebraic classification of some knots of codimension two," Comment. Math. Helv. **45**, 185–198 (1970).

[80] A. Libgober, "Alexander polynomial of plane algebraic curves and cyclic multiple planes," Duke Math. J. **49**, 833–851 (1982).

[81] W. B. Lickorish, *An Introduction to Knot Theory* (New York, Springer, 1997).

[82] S. MacLane, *Homology* (Berlin–New York, Springer, 1967).

[83] D. G. Malm, "A note on exact sequences," Proc. Amer. Math. Soc. **14**, 637–639 (1963).

[84] W. Mayer, "Über abstrakte Topologie," Monatsh. Math. und Physik. **36**, 1–42, 219–258 (1929).

[85] M. D. Meyerson, "Representing homology classes of closed orientable surfaces," Proc. Amer. Math. Soc. **61**, 181–182 (1976).

[86] J. W. Milnor, "Construction of universal bundles, II," Ann. Math. **63**, 430–436 (1956).

[87] J. W. Milnor, "On the immersions of n-manifolds in $(n+1)$-space," Comment. Math. Helv. **30**, 275–284 (1956).

[88] J. W. Milnor, "On simply connected 4-manifolds," in *Symposium internacional de topologia algebraica* (México, UNAM, 1958), pp. 122–128.

[89] J. W. Milnor, "A procedure for killing the homotopy groups of differentiable manifolds," in *Proc. Sympos. Pure Math.,* Vol. 3 (Providence, RI, Amer. Math. Soc., 1961), pp. 39–55.

[90] J. Milnor, *Morse Theory* (Princeton, NJ, Princeton Univ. Press, 1963).

[91] J. W. Milnor, "Microbundles. I," *Topology,* **3**, suppl. 1, 53–81 (1964).

[92] J. Milnor and J. Stasheff, *Characteristic Classes* (Princeton, Princeton Univ. Press, 1979).

[93] M. Mimura and H. Toda, *Topology of Lie groups* (Providence, RI, Amer. Math. Soc., 1991).

[94] R. E. Mosher and M. C. Tangora, *Cohomology Operations and Applications in Homotopy Theory* (New York–London, Harper & Row, 1968).

[95] H. J. Munkholm, "Simplices of maximal volume in hyperbolic space, Gromov's norm, and Gromov's proof of Mostow's rigidity theorem (following Thurston)," in *Topology Symposium (Siegen, 1979),* Lecture Notes in Math. **788** (Berlin, Springer, 1980), pp. 109–124.

[96] J. R. Munkres, *Elements of algebraic topology* (Reading, Mass., Addison-Wesley, 1984).

[97] E. Noether, "Ableitung der Elementarteilertheorie aus der Gruppentheorie," Jahresbericht Deutschen Math. Verein. **34**, 104 (1926).

[98] P. Olum, "Non-abelian cohomology and van Kampen's theorem," Ann. Math. **68**, 658–668 (1958).

[99] G. F. Paechter, "On the groups $\pi_r(V_{mn})$. I, II, III, IV, V," Quart. J. Math. Oxford, Ser. 2 **7** (28), 249–265 (1956); **9** (33), 8–27 (1958); **10** (37), 17–37 (1959); **10** (40), 241–260 (1959); **11** (41), 1–16 (1960).

[100] H. O. Peitgen, "On the Lefschetz number for iterates of continuous mappings," Proc. Amer. Math. Soc. **54**, 441–444 (1976).

[101] F. Pham, "Formules de Picard–Lefschetz généralisées et ramification des intégrales," Bull. Soc. Math. France. **93**, 333–367 (1965).

[102] E. Pitcher, "Inequalities of critical point theory," Bull. Amer. Math. Soc. **64**, 1–30 (1958).

[103] A. Polombo, "Classes de Chern," Astérisque **58**, 51–75 (1978).

[104] V. V. Prasolov, *Problems and Theorems in Linear Algebra* (Providence, RI, Amer. Math. Soc., 1994).

[105] V. V. Prasolov, *Intuitive Topology* (Providence, RI, Amer. Math. Soc., 1995).

[106] V. V. Prasolov, "The Seifert surface," in *Mathematical Education,* Ser. 3, No. 3 (Moscow, MTsNMO, 1999), pp. 116–126 [in Russian].

[107] V. V. Prasolov and A. B. Skopenkov, "The Ramsey theory of knots and links," in *Mathematical Education,* Ser. 3, No. 9 (Moscow, MTsNMO, 2005), pp. 108–115 [in Russian].

[108] V. V. Prasolov and A. B. Sossinsky, *Knots, Links, Braids and 3-Manifolds* (Providence, RI, Amer. Math. Soc., 1997).

[109] G. de Rham, "Sur l'analysis situs des variétés à n dimensions," J. Math. Pures et Appl., Ser. 9 **10**, 115–120 (1931).

[110] S. A. Robertson, "On transnormal manifolds," Topology **6**, 117–123 (1967).

[111] V. A. Rokhlin, "Homotopy groups," Uspekhi Mat. Nauk **1** (5–6), 175–223 (1946).

[112] J. J. Rotman, *An Introduction to Algebraic Topology* (New York–Berlin, Springer, 1988).

[113] H. Samelson, "On Poincaré duality," J. Anal. Math. **14**, 323–336 (1965).

[114] H. Samelson, "On de Rham's theorem," Topology **6**, 427–432 (1967).

[115] R. Schön, "Acyclic models and excision," Proc. Amer. Math. Soc. **59**, 167–168 (1976).

[116] J.-P. Serre, "Homologie singulière des espaces fibrés. Applications," Ann. of Math. **54**, 425–505 (1951).

[117] J.-P. Serre, "Cohomologie modulo 2 des complexes d'Eulenberg–MacLane," Comment. Math. Helv. **27**, 198–232 (1953).

[118] J.-P. Serre, "Groupes d'homotopie et classes de groupes abéliens," Ann. of Math. **28**, 258–294 (1953).

[119] D. B. Shapiro, *Composition of Quadratic Forms* (Berlin–New York, Walter de Gruyter, 2000).

[120] M.-H. Shih, "A combinatorial Lefschetz fixed-point formula," J. Combin. Theory Ser. A **61**, 123–129 (1992).

[121] I. Singer and J. Thorpe, *Lecture Notes on Elementary Topology and Geometry* (Glenview, Ill., Scott, Foresman, and Co., 1967).

[122] P. A. Smith, "Transformations of finite period, I," Ann. Math. **39**, 127–164 (1938).

[123] E. H. Spanier, *Algebraic Topology* (New York, McGraw-Hill, 1966).

[124] E. B. Staples, "A short and elementary proof that a product of spheres is parallelizable if one of them is odd," Proc. Amer. Math. Soc. **18**, 570–571 (1967).

[125] N. E. Steenrod, "Products of cocycles and extension of mappings," Ann. Math. **48**, 290–320 (1947).

[126] N. E. Steenrod, "Cohomology invariants of mappings," Ann. Math. **50**, 954–988 (1949).

[127] N. E. Steenrod, *The Topology of Fibre Bundles* (Princeton, NJ, Princeton Univ. Press, 1951).

[128] N. E. Steenrod, "Homology groups of symmetric groups and reduced power operations," Proc. Nat. Acad. Sci. U.S.A. **39**, 213–223 (1953).

[129] N. E. Steenrod, "Cohomology operation derived from the symmetric group," Comm. Math. Helv. **31**, 195–218 (1957).

[130] N. E. Steenrod, *Cohomology Operations.* Lectures by N. E. Steenrod written and revised by D. B. A. Epstein (Princeton, NJ, Princeton Univ. Press, 1962).

[131] N. Steenrod and S. Eilenberg, *Foundations of Algebraic Topology* (Princeton, NJ, Princeton Univ. Press, 1952).

[132] E. Stiefel, "Richtungsfelder und Fernparallelismus in Mannigfaltigkeit," Comm. Math. Helv. **8**, 3–51 (1936).

[133] E. Stiefel, "Über Richtungsfelder in den projektiven Räumen," Comm. Math. Helv. **13**, 201–218 (1941).

[134] R. E. Stong, *Notes on Cobordism Theory* (Princeton, NJ, Princeton Univ. Press, 1968).

[135] D. Sullivan, "Infinitesimal computations in topology," Publ. Math. IHES. **47**, 269–332 (1977).

[136] R. M. Switzer, *Algebraic Topology—Homotopy and Homology* (New York, Springer, 1975).

[137] R. Thom, "Espaces fibrés en spheres et carrés de Steenrod," Ann. Sci. École Norm. Super. **69**, 109–181 (1952).

[138] R. Thom, "Les classes charactéristiques de Pontryagin des variétés triangulées," in *Symposium internacional de topologia algebraica* (México, UNAM, 1958), pp. 54–67.

[139] W. Thurston, *The Geometry and Topology of 3-Manifolds* (Princeton, NJ, Princeton Univ. Press, 1978).

[140] V. A. Vassiliev, *Introduction to Topology* (Providence, RI, Amer. Math. Soc., 2001).

[141] J. W. Vick, *Homology Theory* (New York, Academic Press, 1973).

[142] L. Vietoris, "Über die Homologiegruppen der Vereinigung zweier Komplexe," Monatsh. Math. Phys. **37**, 159–162 (1930).

[143] Ph. L. Wadler, "On pairs of nonintersecting faces of cell complexes," Proc. Amer. Math. Soc. **51**, 438–440 (1975).

[144] H. C. Wang, "The homology groups of the fibre-bundles over a sphere," Duke Math. J. **16**, 33–38 (1949).

[145] Ch. A. Weibel, "History of homological algebra," in *History of Topology,* Ed. by I. M. James (Amsterdam, Elsevier, 1999), pp. 797–836.

[146] A. Weil, "Sur les theorems de de Rham," Comment. Math. Helv. **26**, 119–145 (1952).

[147] O. Wells R., Jr., *Differential Analysis on Complex Manifolds* (Englewood Cliffs, NJ, Prentice-Hall, 1973).

[148] Sh. Weinberger, "Oliver's formula and Minkowski's theorem," in *Algebraic and Geometric Topology (New Brunswick, NJ, 1983).* Lecture Notes in Math., **1126** (Berlin, Springer, 1985), pp. 420–421.

[149] H. Whitney, "Sphere spaces," Proc. Nat. Acad. Sci. U.S.A. **21**, 462–468 (1935).

[150] H. Whitney, "The maps of an n-complex into an n-sphere," Duke Math. J. **3**, 51–55 (1937).

[151] H. Whitney, "On products in a complex," Ann. Math. **39**, 397–432 (1938).

[152] H. Whitney, "The self-intersections of a smooth n-manifold in $2n$-space," Ann. Math. **45**, 220–246 (1944).

[153] H. Whitney, "The self-intersections of a smooth n-manifold in $(2n+1)$-space," Ann. Math. **45**, 247–293 (1944).

[154] H. Whitney, *Geometric Integration Theory* (Princeton, NJ, Princeton Univ. Press, 1957).

[155] Wen-Tsün Wu, "Classes caractéristiques et i-carrés d'une variété," C. R. Acad. Sci. Paris. **230**, 508–511 (1950).

Index

G-complex
  regular, 174
  simplicial, 173
G-space, 173
H-space, 358
$K(\pi, n)$ space, 122
S-equivalent matrices, 314
r-transnormal embedding, 56

action
  effective, 177
  free, 183
  simplicial, 173
acyclic
  functor, 103
  model, 104
    theorem, 104
  simplicial complex, 6
  support theorem, 7
admissible set, 231
Alexander
  duality, 81
  ideal, 308
  polynomial, 308
    in Conway's normalization, 315
  theorem, 205
Alexander–Pontryagin duality, 271
Alexander–Whitney diagonal approximation, 105, 214
algebra
  Hopf, 361
    connected, 361
  Lie, 344
    cohomology, 349
algebraically trivial map, 120
almost parallelizable manifold, 141
anticommutativity of cup product, 63
Arf
  invariant
    of a knot, 319
    of a link, 320
    of a quadratic form, 317
  theorem, 318
associated sheaf, 265
attaching a handle, 312
augmentation, 6, 17
axiom
  dimension, 203, 204
  exactness, 203, 204
  excision, 200, 203, 204
    noncommutative, 224
  homotopy, 203, 204
axiomatic approach to Stiefel–Whitney classes, 159

basis
  of a free Abelian group, 4
  of a module, 305
  symplectic, 317
Betti numbers, 3
bi-invariant form, 348
bilinear map
  nondegenerate, 109
  of Abelian groups, 27
Bockstein homomorphism, 14, 92, 142
Borromean rings, 85
  multidimensional, 87
Borsuk–Ulam theorem, 76
boundary, 2
  homomorphism, 2
  of a simplex, 1

braid
  colored, 122
  group, 123
bundle
  associated with a divisor, 339
  conjugate, 170
  dual, 171
  induced, 133
  nonorientable, 143
  orientable, 143
  vector, 131
    equivalent, 132
    isomorphic, 132
    smooth, 131
    stably equivalent, 148
  with structure group, 272

canonical
  map, 9
  vector bundle, 153
cap product, 70
Cartan formula, 189
Cartesian product of Abelian groups, 21
category, 103
  with models, 103
Čech cohomology, 267, 268
cellular homology, 37, 210
centralizer, 351
chain, 2
  complex, 4
    free, 4
    nonnegative, 4
    ordered, 60
    total, 60
  homotopy, 5, 196
  map, 4
  relative, 12
  with closed supports, 48
characteristic
  class
    Chern, 163
    Chern of a complex manifold, 171
    Euler, 144
    Pontryagin, 172
    Stiefel–Whitney, 141
  Euler, 51, 90
    of a pair, 55, 182
Chern characteristic class, 163
  of a complex manifold, 171
class
  Chern, 163
    of a complex manifold, 171
  Euler, 144
  fundamental, 36
    cohomology, 125
  Pontryagin, 172
  primitive homology, 49
  Stiefel–Whitney, 141
    of a manifold, 150
  Thom, 237, 255
  Wu, 259
closed form, 277
coboundary, 22
  formula, 186
  homomorphism, 267
cochain, 21, 266
  difference, 117
  relative, 23
  with compact supports, 48
cocycle, 22
  noncommutative, 222
    cohomologous, 222
coefficient group, 203, 204
cofinal set, 266
cohomology
  Čech, 267, 268
  cross product, 106
  de Rham, 277
  fundamental class, 125
  group, 22
  Lie algebra, 349
  noncommutative, 222, 274
  operation, 127
  reduced, 23
  relative, 23
  singular, 202
  with compact supports, 48
  with local coefficients, 136
cokernel, 15
collar theorem, 78, 247
colored braid group, 122
commutator, 346
  of vector fields, 276
  subgroup, 112
complete intersection, 341
complex
  chain, 4
  Stiefel manifold, 162
  vector bundle, 162
complexification, 172
conjugate bundle, 170
connected
  Hopf algebra, 361
  sum of manifolds, 336
connecting homomorphism, 12, 14
consistent family, 264
constant presheaf, 264
contravariant functor, 103
Conway polynomial, 316
coproduct, 360
covariant functor, 103
cross product, 106
cup product, 59, 62

cycle, 2
  homologous, 3

degree of a map, 36
Dehn twist, 50
de Rham
  cohomology, 277
  theorem, 289
diagonal approximation, 105, 184, 214
  Alexander–Whitney, 105, 214
difference cochain, 117
differential form
  closed, 277
  exact, 277
  polynomial
    on a complex, 296
    on a simplex, 296
  smooth, 296
dimension axiom, 203, 204
direct
  limit, 265
  product of vector bundles, 134
  sum
    of Abelian groups, 21
    of bundles, 134
directed set, 264
  of Abelian groups, 265
divisible group, 32
domain invariance theorem, 205
double point, 325
dual
  bundle, 171
  Stiefel–Whitney class, 149
duality
  Alexander, 81
  Alexander–Pontryagin, 271
  Poincaré, 44
dunce hat, 115

effective action, 177
Eilenberg theorem, 117
Eilenberg–MacLane space, 122
Eilenberg–Zilber theorem, 213
element
  regular, 353
  singular, 353
elementary ideal, 307
embedding
  r-transnormal, 56
  transnormal, 56
equivalent
  microbundles, 254
  vector bundles, 132
equivariant map, 173
Euler
  characteristic, 51, 90
    of a pair, 55, 182
  class, 144
exact
  form, 277
  sequence
    Mayer–Vietoris, 18
    of a pair, 12
    Smith, 181
    split, 24
exactness axiom, 203, 204
excision
  axiom, 200, 203, 204
    noncommutative, 224
  isomorphism, 12
  theorem, 198
extraordinary (co)homology theory, 204

five lemma, 15
form
  bi-invariant, 348
  closed, 277
  exact, 277
  intersection, 88, 260
  left-invariant, 348
  polynomial
    on a complex, 296
    on a simplex, 296
  quadratic over $\mathbb{Z}_2$, 317
  right-invariant, 348
  Seifert, 304
  smooth, 296
formula
  Cartan, 189
  coboundary, 186
  of universal coefficients, 33
  Thom, 257
  Whitney, 147
  Wu, 193
free
  action, 183
  chain complex, 4
  functor, 103
  module, 305
  resolution of an Abelian group, 28
functor
  acyclic, 103
  contravariant, 103
  covariant, 103
  free, 103
fundamental class, 36
  cohomology, 125
  of a topological manifold, 232
    with boundary, 249

generalized (co)homology theory, 204
germ, 265
Gromov norm, 221

group
  cohomology, 22
    reduced, 23
    relative, 23
    singular, 202
  colored braid, 122
  divisible, 32
  free Abelian, 4
  homology
    of a chain complex, 4
    singular, 196
  Lie, 344
  of braids, 123
  of coefficients, 203, 204
  periodic, 31
  ring, 180
  simplicial homology, 3
  Smith homology, 181
Gysin sequence, 256

Helly's theorem, 208, 209
homologous cycles, 3
homology
  cellular, 37, 210
  disk, 183
  group
    of a chain complex, 4
    simplicial, 3
    Smith, 181
  primitive class, 49
  reduced, 17
  relative, 12
  sequence
    for a triple, 15
    of a pair, 12
  singular, 196
  sphere, 45, 81, 183
  with closed supports, 48
homomorphism
  Bockstein, 14, 92, 142
  boundary, 2
  connecting, 12, 14
  Hurewicz, 112
  of presheaves, 263
  restriction, 263
  transfer, 179
homotopic
  trivializations, 133
  vector fields, 133
homotopy
  axiom, 203, 204
  chain, 5, 196
Hopf
  algebra, 361
  fibration, 219
  invariant, 219
  theorem, 337, 362
Hopf–Whitney theorem, 119
Hurewicz
  homomorphism, 112
  theorem, 113
hyperbolic manifold, 222

ideal
  Alexander, 308
  elementary, 307
induced bundle, 133
infinite
  cyclic covering, 304
  Grassmann manifold, 153
infinite-dimensional lens space, 123
injective resolution, 32
integral form, 280
intersection
  complete, 341
  form, 88, 260
  number, 42
invariance
  of a boundary, 206
  of a domain, 205
invariant
  Arf
    of a knot, 319
    of a link, 320
    of a quadratic form, 317
  Hopf, 219
isomorphic vector bundles, 132
isomorphism
  excision, 12
  Lefschetz, 81
    for topological manifolds, 250
  of bundles with structure group, 273
  Poincaré, 44
    for de Rham cohomology, 285
    for topological manifolds, 243
    with local coefficients, 138
  suspension, 20, 201
  Thom, 238

Künneth theorem, 99
  for relative homology, 215
  for singular homology, 213
  relative, 215
Kan–Whitehead theorem, 129
Kolmogorov–Alexander multiplication, 59

Lefschetz
  fixed point theorem, 56
  isomorphism, 81
    for topological manifolds, 250
  number, 56
left-invariant
  form, 348
  vector field, 346

lemma
  on extension, 297
  Poincaré, 283
lens space, 93
  infinite-dimensional, 123
Leray–Hirsh theorem, 168
Lie
  algebra, 344
  group, 344
line bundle associated with a divisor, 339
linking number, 46, 83
local system of groups, 136

manifold
  almost parallelizable, 141
  Grassmann infinite, 153
  hyperbolic, 222
  parallelizable, 132, 260, 338
  Schubert, 157
  stably parallelizable, 156
  Stiefel, 139
    complex, 162
  topological with boundary orientable, 248
map
  algebraically trivial, 120
  chain, 4
  equivariant, 173
  splitting, 167
Massey triple product, 84
matrix
  presentation, 305
  Seifert, 304
maximal torus, 351
Mayer–Vietoris sequence, 18, 200
  for de Rham cohomology, 278
  for de Rham cohomology with compact supports, 279
  noncommutative, 224
  relative, 20, 202
microbundle, 254
  equivalent, 254
  tangent, 254
Milnor theorem, 338
Minkowski theorem, 177
model, 103
  acyclic, 104
module
  finitely generated, 305
  free, 305
Moore space, 128
morphism, 103
Morse inequality, 210
multidimensional Borromean rings, 87
multiplication, 358
  Kolmogorov–Alexander, 59

natural transformation, 103
naturality
  of cap product, 71
  of Stiefel–Whitney classes, 146
noncommutative
  cocycle, 222
  cohomology, 222, 274
  excision axiom, 224
  Mayer–Vietoris sequence, 224
nondegenerate
  bilinear map, 109
  quadratic form over $\mathbb{Z}_2$, 317
nonnegative chain complex, 4
nonorientable bundle, 143
normal degree of an immersion, 335
normalizer, 351
number
  Betti, 3
  intersection, 42
  Lefschetz, 56
  linking, 46, 83
  self-intersection, 326, 327
  Stiefel–Whitney, 152

object of a category, 103
obstruction, 116
  to extending sections, 138
ordered chain complex, 60
orientable
  bundle, 143
  topological manifold, 232
    with boundary, 248
orientation
  of a topological manifold, 232
    with boundary, 248
  system of groups, 136
oriented topological manifold, 232

parallelizable manifold, 132, 260, 338
partition of an integer, 158
periodic group, 31
Poincaré
  duality, 44
  isomorphism, 44
    for de Rham cohomology, 285
    for topological manifolds, 243
    with local coefficients, 138
  lemma, 283
  theorem, 112
point
  double, 325
  self-intersection, 325
polynomial
  Alexander, 308
    in Conway's normalization, 315
  Conway, 316

differential form
on a complex, 296
on a simplex, 296
Pontryagin
characteristic class, 172
theorem, 152
presentation matrix, 305
presheaf, 263
constant, 264
primitive homology class, 49
product
Massey triple, 84
of Abelian groups
Cartesian, 21
tensor, 27
tensor of chain complexes, 97
vector bundle, 132
projective
resolution, 32
projectivization of a vector bundle, 167
pullback, 133

quadratic form over $\mathbb{Z}_2$, 317
nondegenerate, 317

rank of a Lie group, 353
reduced
cohomology, 23
homology, 17
regular
G-complex, 174
element, 353
immersion, 325
relative
chain, 12
cochain, 23
cohomology, 23
homology, 12
Künneth theorem, 215
Mayer–Vietoris sequence, 20, 202
resolution
injective, 32
projective, 32
restriction homomorphism, 263
right-invariant form, 348
ring
Borromean, 85
group, 180
roots, 355

Schubert manifold, 157
section, 265
of a bundle, 131
zero, 132
Seifert
form, 304
knot, 310
matrix, 304
surface, 305
self-intersection
number, 326, 327
point, 325
sequence
exact of a pair, 12
Gysin, 256
Mayer–Vietoris, 18, 200
for de Rham cohomology, 278
for de Rham cohomology with compact supports, 279
noncommutative, 224
relative, 20, 202
Smith exact, 181
set
admissible, 231
cofinal, 266
directed, 264
of Abelian groups, 265
sheaf, 264
associated with a presheaf, 265
generated by a presheaf, 265
signature
of a manifold, 90
of a product, 108
Thom theorem, 91
simplex
boundary, 1
singular, 195
simplicial
G-complex, 173
action, 173
complex acyclic, 6
homology group, 3
volume, 221
singular
cohomology, 202
element, 353
homology, 106
simplex, 195
skein relation, 316
skew-commutativity of cup product, 63
Smith
exact sequence, 181
homology group, 181
theorem, 183
smooth
differential form
on a complex, 296
on a simplex, 296
triangulation, 289
vector bundle, 131
space
$K(\pi, n)$, 122
Eilenberg–MacLane, 122
lens, 93

Moore, 128
split exact sequence, 24
splitting map, 167
stably
equivalent vector bundles, 148
parallelizable manifold, 156
Steenrod square, 188
Steenrod's five lemma, 15
Steenrod–Eilenberg axioms, 203, 204
Stiefel
manifold, 139
complex, 162
theorem, 260
Stiefel–Hopf theorem, 109
Stiefel–Whitney
class
characteristic, 141
dual, 149
of a manifold, 150
total, 149
number, 152
Stokes theorem, 281
strong Whitney embedding theorem, 325
sum
connected of manifolds, 336
direct
of Abelian groups, 21
of bundles, 134
Whitney, 134
support of a chain, 6
suspension isomorphism, 20, 201
symplectic basis, 317

tangent microbundle, 254
tensor product
of Abelian groups, 27
of chain complexes, 97
theorem
acyclic model, 104
acyclic support, 7
Alexander, 205
Arf, 318
Borsuk–Ulam, 76
chain homotopy, 5
collar for smooth manifolds, 78
de Rham, 289
simplicial, 299
domain invariance, 205
Eilenberg, 117
Eilenberg–Zilber, 213
excision, 198
Helly's, 208, 209
Hopf, 337, 362
Hopf–Whitney, 119
Hurewicz, 113
Künneth, 99
for relative homology, 215
for singular homology, 213
relative, 215
Kan–Whitehead, 129
Lefschetz
fixed point, 56
isomorphism, 81
Leray–Hirsh, 168
Milnor, 338
Minkowski, 177
on a collar, 247
on acyclic
models, 104
supports, 7
on domain invariance, 205
Poincaré, 112
Pontryagin, 152
Smith, 183
Stiefel, 260
Stiefel–Hopf, 109
Stokes, 281
Thom, 257
signature, 91
Whitney
duality, 149
strong embedding, 325
Wu, 259
Thom
class, 237, 255
formula, 257
isomorphism, 238
theorem, 257
signature, 91
topological
generator, 350
manifold
orientable, 232
oriented, 232
with boundary orientable, 248
torsion subgroup, 44
torus, 345
maximal, 351
total
chain complex, 60
Stiefel–Whitney class, 149
transfer homomorphism, 179
transition function, 272
transnormal embedding, 56
transversality, 73
triple
homology sequence, 15
Massey product, 84
trivial vector bundle, 132

universal coefficient
formulas, 33
theorem, 33

vector
  bundle, 131
    canonical, 153
    complex, 162
    equivalent, 132
    isomorphic, 132
    product, 132
    smooth, 131
    stably equivalent, 148
    trivial, 132
  field
    homotopic, 133
    left-invariant, 346
Whitney
  formula, 147
  sum, 134
  theorem
    duality, 149
    strong embedding, 325
  trick, 329
Wu
  class, 259
  formula, 193
  theorem, 259

zero section, 132

Titles in This Series

84 **Charalambos D. Aliprantis and Rabee Tourky,** Cones and order, 2007
83 **Wolfgang Ebeling,** Functions of several complex variables and their singularities (translated by Philip G. Spain), 2007
82 **Serge Alinhac and Patrick Gérard,** Pseudo-differential operators and the Nash–Moser theorem (translated by Stephen S. Wilson), 2007
81 **V. V. Prasolov,** Elements of homology theory, 2007
80 **Davar Khoshnevisan,** Probability, 2007
79 **William Stein,** Modular forms, a computational approach (with an appendix by Paul E. Gunnells), 2007
78 **Harry Dym,** Linear algebra in action, 2007
77 **Bennett Chow, Peng Lu, and Lei Ni,** Hamilton's Ricci flow, 2006
76 **Michael E. Taylor,** Measure theory and integration, 2006
75 **Peter D. Miller,** Applied asymptotic analysis, 2006
74 **V. V. Prasolov,** Elements of combinatorial and differential topology, 2006
73 **Louis Halle Rowen,** Graduate algebra: Commutative view, 2006
72 **R. J. Williams,** Introduction the the mathematics of finance, 2006
71 **S. P. Novikov and I. A. Taimanov,** Modern geometric structures and fields, 2006
70 **Seán Dineen,** Probability theory in finance, 2005
69 **Sebastián Montiel and Antonio Ros,** Curves and surfaces, 2005
68 **Luis Caffarelli and Sandro Salsa,** A geometric approach to free boundary problems, 2005
67 **T.Y. Lam,** Introduction to quadratic forms over fields, 2004
66 **Yuli Eidelman, Vitali Milman, and Antonis Tsolomitis,** Functional analysis, An introduction, 2004
65 **S. Ramanan,** Global calculus, 2004
64 **A. A. Kirillov,** Lectures on the orbit method, 2004
63 **Steven Dale Cutkosky,** Resolution of singularities, 2004
62 **T. W. Körner,** A companion to analysis: A second first and first second course in analysis, 2004
61 **Thomas A. Ivey and J. M. Landsberg,** Cartan for beginners: Differential geometry via moving frames and exterior differential systems, 2003
60 **Alberto Candel and Lawrence Conlon,** Foliations II, 2003
59 **Steven H. Weintraub,** Representation theory of finite groups: algebra and arithmetic, 2003
58 **Cédric Villani,** Topics in optimal transportation, 2003
57 **Robert Plato,** Concise numerical mathematics, 2003
56 **E. B. Vinberg,** A course in algebra, 2003
55 **C. Herbert Clemens,** A scrapbook of complex curve theory, second edition, 2003
54 **Alexander Barvinok,** A course in convexity, 2002
53 **Henryk Iwaniec,** Spectral methods of automorphic forms, 2002
52 **Ilka Agricola and Thomas Friedrich,** Global analysis: Differential forms in analysis, geometry and physics, 2002
51 **Y. A. Abramovich and C. D. Aliprantis,** Problems in operator theory, 2002
50 **Y. A. Abramovich and C. D. Aliprantis,** An invitation to operator theory, 2002
49 **John R. Harper,** Secondary cohomology operations, 2002
48 **Y. Eliashberg and N. Mishachev,** Introduction to the h-principle, 2002
47 **A. Yu. Kitaev, A. H. Shen, and M. N. Vyalyi,** Classical and quantum computation, 2002

46 **Joseph L. Taylor,** Several complex variables with connections to algebraic geometry and Lie groups, 2002

45 **Inder K. Rana,** An introduction to measure and integration, second edition, 2002

44 **Jim Agler and John E. McCarthy,** Pick interpolation and Hilbert function spaces, 2002

43 **N. V. Krylov,** Introduction to the theory of random processes, 2002

42 **Jin Hong and Seok-Jin Kang,** Introduction to quantum groups and crystal bases, 2002

41 **Georgi V. Smirnov,** Introduction to the theory of differential inclusions, 2002

40 **Robert E. Greene and Steven G. Krantz,** Function theory of one complex variable, third edition, 2006

39 **Larry C. Grove,** Classical groups and geometric algebra, 2002

38 **Elton P. Hsu,** Stochastic analysis on manifolds, 2002

37 **Hershel M. Farkas and Irwin Kra,** Theta constants, Riemann surfaces and the modular group, 2001

36 **Martin Schechter,** Principles of functional analysis, second edition, 2002

35 **James F. Davis and Paul Kirk,** Lecture notes in algebraic topology, 2001

34 **Sigurdur Helgason,** Differential geometry, Lie groups, and symmetric spaces, 2001

33 **Dmitri Burago, Yuri Burago, and Sergei Ivanov,** A course in metric geometry, 2001

32 **Robert G. Bartle,** A modern theory of integration, 2001

31 **Ralf Korn and Elke Korn,** Option pricing and portfolio optimization: Modern methods of financial mathematics, 2001

30 **J. C. McConnell and J. C. Robson,** Noncommutative Noetherian rings, 2001

29 **Javier Duoandikoetxea,** Fourier analysis, 2001

28 **Liviu I. Nicolaescu,** Notes on Seiberg-Witten theory, 2000

27 **Thierry Aubin,** A course in differential geometry, 2001

26 **Rolf Berndt,** An introduction to symplectic geometry, 2001

25 **Thomas Friedrich,** Dirac operators in Riemannian geometry, 2000

24 **Helmut Koch,** Number theory: Algebraic numbers and functions, 2000

23 **Alberto Candel and Lawrence Conlon,** Foliations I, 2000

22 **Günter R. Krause and Thomas H. Lenagan,** Growth of algebras and Gelfand-Kirillov dimension, 2000

21 **John B. Conway,** A course in operator theory, 2000

20 **Robert E. Gompf and András I. Stipsicz,** 4-manifolds and Kirby calculus, 1999

19 **Lawrence C. Evans,** Partial differential equations, 1998

18 **Winfried Just and Martin Weese,** Discovering modern set theory. II: Set-theoretic tools for every mathematician, 1997

17 **Henryk Iwaniec,** Topics in classical automorphic forms, 1997

16 **Richard V. Kadison and John R. Ringrose,** Fundamentals of the theory of operator algebras. Volume II: Advanced theory, 1997

15 **Richard V. Kadison and John R. Ringrose,** Fundamentals of the theory of operator algebras. Volume I: Elementary theory, 1997

14 **Elliott H. Lieb and Michael Loss,** Analysis, 1997

13 **Paul C. Shields,** The ergodic theory of discrete sample paths, 1996

12 **N. V. Krylov,** Lectures on elliptic and parabolic equations in Hölder spaces, 1996

11 **Jacques Dixmier,** Enveloping algebras, 1996 Printing

For a complete list of titles in this series, visit the AMS Bookstore at **www.ams.org/bookstore/**.